PRACTICAL PETROLEUM GEOCHEMISTRY FOR EXPLORATION AND PRODUCTION

PRACTICAL PETROLEUM GEOCHEMISTRY FOR EXPLORATION AND PRODUCTION

HARRY DEMBICKI, Jr.

ELSEVIER

AMSTERDAM • BOSTON • HEIDELBERG • LONDON
NEW YORK • OXFORD • PARIS • SAN DIEGO
SAN FRANCISCO • SINGAPORE • SYDNEY • TOKYO

Elsevier
Radarweg 29, PO Box 211, 1000 AE Amsterdam, Netherlands
The Boulevard, Langford Lane, Kidlington, Oxford OX5 1GB, United Kingdom
50 Hampshire Street, 5th Floor, Cambridge, MA 02139, United States

Library of Congress Cataloging-in-Publication Data
A catalog record for this book is available from the Library of Congress

British Library Cataloguing-in-Publication Data
A catalogue record for this book is available from the British Library

ISBN: 978-0-12-803350-0

Publisher: Candice Janco
Acquisition Editor: Amy Shapiro
Editorial Project Manager: Tasha Frank
Production Project Manager: Maria Bernard
Cover Designer: Matthew Limbert

Typeset by TNQ Books and Journals

Dedication

I dedicate this book to my father for instilling in me a need to know why and passing along to me a good work ethic, to my son Mischa for being a great sounding board for my ideas and for turning into a fine man who I am proud of, and to my wife Judie, my rock and the love of my life, I could not have done this without you.

Contents

Acknowledgments ix

1. Introduction

Introduction 1
A Brief History of Petroleum Geochemistry 2
Definitions 4
Organic Chemistry Review 6
Stable Isotope Review 15
References 16

2. The Formation of Petroleum Accumulations

Introduction 19
Incorporating Organic Matter Into Sediments 20
Kerogen Formation 25
Source Rock Deposition 29
Maturation and Hydrocarbon Generation 33
Petroleum Migration 45
Origin of Nonhydrocarbon Gases 50
Coals as Oil-Prone Source Rocks 53
Summary 55
References 56

3. Source Rock Evaluation

Definitions and Fundamental Concepts 62
Sample Collection 63
Total Organic Carbon 66
Rock-Eval Pyrolysis 67
Solvent Extraction, S-A-R-A Analysis, and Extract Data 80
Gas Chromatography 82
Headspace Gas Analysis 88
Pyrolysis–Gas Chromatography 91
Kerogen Isolation 95
Elemental Analysis 96
Vitrinite Reflectance 98
Alternative Reflectance Method 109
Visual Kerogen Typing 111
Thermal Alteration Index 113
Kerogen Fluorescence 115
Conodont Alteration Index 117
Wireline Log Interpretations 118
Using Outcrop Samples 125
Strategies in Source Rock Evaluation 126
References 129

4. Interpreting Crude Oil and Natural Gas Data

Introduction 136
Bulk Properties of Crude Oil and Natural Gas 136
Phase Behavior 139
Crude Oil and Natural Gas Alteration 141
Oil-to-Oil and Oil-to-Source Rock Correlations 150
Crude Oil Inversion 160
Strategies and Obstacles in Oil Correlation and Oil Inversion Studies 170
Natural Gas Data 173
The Source of Natural Gas: Biogenic Versus Thermogenic 173
The Maturity of Thermogenic Natural Gas 176
Gas-to-Gas and Gas-to-Source Rock Correlations 178
Strategies and Obstacles in Interpreting Gas Data 183
References 184

5. Reservoir Geochemistry

Introduction 189
Pay Zone Detection 190
High-Molecular Weight Waxes 199
Asphaltenes 200
Reservoir Continuity 202
Production Allocation 207
Production Problems and Periodic Sampling 208
Monitoring Enhanced Oil Recovery 210
Reservoir Souring 212
Strategies in Reservoir Geochemistry 212
References 213

6. Surface Geochemistry

Introduction 217
Microseepage 219
Direct Indicators of Hydrocarbon Microseepage 220
Indirect Indicators of Hydrocarbon Microseepage 222
Microseepage Survey Design and Interpretation 224
Onshore Macroseepage 229
Offshore Macroseepage 230
Locating Potential Seafloor Seep Sites 231
Sampling Potential Seafloor Seep Sites 240
Analyzing Seafloor Sediments for Thermogenic Hydrocarbons 241
Sea Surface Slicks 244
References 248

7. Unconventional Resources

Introduction 253
Coalbed Methane 254
Shale Gas 259
Shale Oil 265
Hybrid Systems 267
References 269

8. Basin Modeling

Introduction 273
Burial History 275
Thermal History 281
Modeling Maturation, Hydrocarbon Generation, and Expulsion 286
Modeling Migration 292
Predicting Preservation 294
1-D Model Results 295
Model Validation 299
Sensitivity Analysis 300
Volumetric Estimations 302
The Role of Basin Modeling in Unconventional Plays 305
References 305

9. Petroleum System Concepts and Tools

Introduction 309
Elements and Processes 310
Temporal Aspects 312
Spatial Aspects 315
Plays and Prospects 317
A Working Petroleum System 318
Risking 320
References 322

Index 325

Acknowledgments

I thank Martin S. Rutstein and Warren G. Meinschein, my undergraduate and graduate school advisors, respectively, for guiding me along those early steps toward becoming a scientist and geochemist. I also thank all of my organic geochemistry and petroleum industry colleagues for sharing their ideas and learnings that feed my quest for knowledge. And I am grateful to Conoco Inc., Marathon Oil Company, and Anadarko Petroleum Corporation for the opportunities provided to me to develop and practice my craft while in their employ.

CHAPTER

1

Introduction

OUTLINE

Introduction 1

A Brief History of Petroleum Geochemistry 2

Definitions 4
- *Petroleum* 4
- *Geochemistry* 4
- *Petroleum System* 5
- *Sedimentary Organic Matter* 5
- *Other Sedimentary Organic Deposits* 6

Organic Chemistry Review 6
- *Covalent Bonds* 6
- *Hydrocarbons* 7
- *Aromatic Hydrocarbons* 10
- *N-S-O Compounds* 12
- *Asphaltenes* 12
- *Reactions* 14

Stable Isotope Review 15

References 16

INTRODUCTION

In the realm of petroleum exploration and production, the geosciences have long been referred to as G & G, geology and geophysics. However, petroleum geochemistry has long been a major contributor to finding oil and gas and deserves to be recognized as the third "G" along with geology and geophysics. It is the intent of this volume to demonstrate the importance of petroleum geochemistry by explaining how it can be applied to a variety of exploration and production problems, in both conventional and unconventional plays, and the role of petroleum geochemistry in basin modeling and petroleum systems analysis. By the end, it is hoped that the reader will think about petroleum geosciences as G, G, & G.

But before delving into the theoretical underpinnings and applications of petroleum geochemistry in subsequent chapters, this chapter will begin with a brief history of the science to provide a perspective for how it came to be what it is today. This will be followed by some fundamental definitions so the discussion can begin with some common language. The chapter will then conclude with a review of some important aspects of organic chemistry and relevant concepts in stable isotopes.

Before starting, a few sentences are needed to manage expectations. This is not a petroleum geochemistry book intended for petroleum geochemists. It is also not an exhaustive review

Practical Petroleum Geochemistry for Exploration and Production
http://dx.doi.org/10.1016/B978-0-12-803350-0.00001-5

of all the concepts and techniques of the subjects covered or all the subtle nuisances of data interpretations. Nor is it a "cookbook" with "recipes" that geologists and geophysists can use to do their own interpretations. Although many readers will be capable of doing some simple interpretations for themselves, the opportunity to make serious errors will still exist. It is instead a reference book for non-specialist geoscientists to gain a better understanding of the value and potential applications of petroleum geochemistry to their exploration and production projects. After reading this book, geologists and geophysists will be better equipped to read and understand geochemistry reports, ask probing questions of their geochemists, and apply the finding from the geochemistry reports to their exploration and/or production projects. With that in mind, let us begin.

A BRIEF HISTORY OF PETROLEUM GEOCHEMISTRY

Petroleum geochemistry is a relatively young science, tracing its roots back to the 1934 discovery of chlorophyll-like structures in crude oil by Albert Treibs (1934). While many petroleum geologists as early as the late 19th century believed oil was derived from organic matter in sediments, Treibs' findings were an undeniable proof of the organic origin of crude oil (Durand, 2003). By the 1950s, major oil companies had begun research programs to learn more about oil and gas, especially how it forms and moves about in the subsurface. An indication of the importance given to petroleum geochemistry by the petroleum industry is the 1958 patent issued for a method for prospecting for petroleum using source rocks granted to Hunt and Meinert (1958).

By the early 1960s, professional societies and research conferences on organic geochemistry were established, and the first book was published that was devoted solely to this science (Breger, 1963). It was also in the 1960s that advances in analytical tools, such as the development of gas chromatography and improvements in mass spectrometry, began providing more detailed data on the distribution and structure of the organic compounds in sediments and crude oils. These new data lead to the development of the concept of biological maker compounds, or biomarker (Eglinton and Calvin, 1967), chemical fossils which would become important tools for oil–source rock correlations and oil–oil correlations.

In the late 1960s through early 1970s, major advances in understanding of the hydrocarbon generation process were made leading to the oil window concept. The need to understand the thermal maturity of sediments and the composition of the kerogen were also recognized. And migration theories were given serious attention. By the mid-1970s, the Rock-Eval instrument was developed and available (Espitalie et al., 1977). This standardized pyrolysis method for source-rock characterization and evaluation would become an industry standard in petroleum geochemistry that is still used today. The late 1970s also saw the publication of *Petroleum Formation and Occurrence* by Tissot and Welte (1978) and *Petroleum Geochemistry and Geology* by Hunt (1979), the first textbooks on petroleum geochemistry.

The 1980s brought the expansion of pyrolysis techniques, the proliferation of biomarker applications, and a better understanding of petroleum migration. It was also when basin modeling became a mainstream application. Prior work on basin modeling used over simplified time–temperature relationships, such as Connan (1974), or relied on complex calculations that required large mainframe computers. Then, Waples (1980) provided a simple method of estimating maturity based on the work of Lopatin (1971). It allowed geologists to make

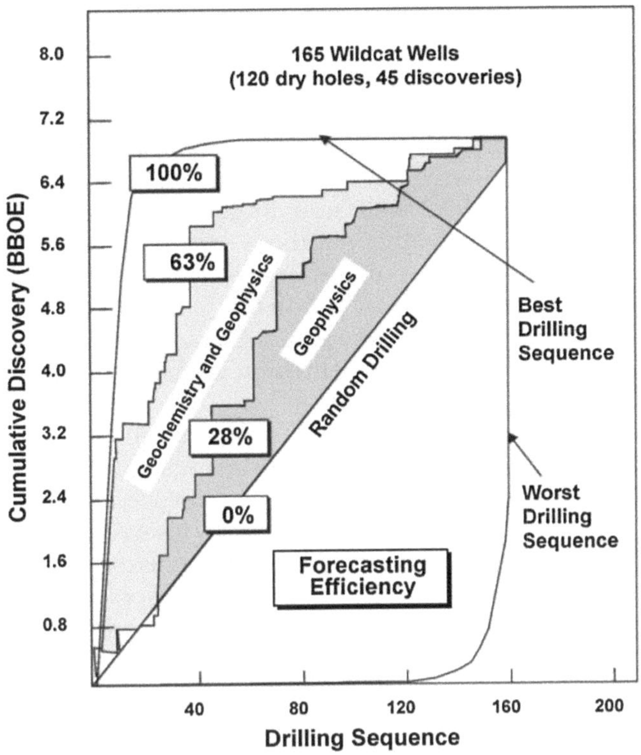

FIGURE 1.1 An assessment of approaches to exploration, by Sluijk and Parker (1986), comparing random drilling, exploration based on trap size from geophysics, and the use of geophysics in conjunction with petroleum geochemistry.

models using graph paper and a handheld calculator. With the concurrent introduction of personal computers, basin modeling was adapted to this new tool and quickly became a standard method for petroleum geochemistry and exploration.

Another significant development in the 1980s was the publication of a paper by Sluijk and Parker (1986) that addressed the value of petroleum geochemistry in exploration. As shown in Fig. 1.1, they considered three cases: random drilling, exploration based on trap size only from geophysical data, and exploration using geophysics in conjunction with petroleum geochemistry for a set of 165 exploration prospects. While using only trap size, an exploration efficiency of 28% for making a discovery could be achieved. But using the combination of geophysics and geochemistry, exploration efficiency was increased to 63%. This clearly demonstrated that petroleum geochemistry is an effective tool for reducing exploration risk.

Along with the advancements in the 1980s, there were negative factors for petroleum geochemistry. Mergers starting in the early to mid-1980s and a downturn in the petroleum industry that began in 1986 lead to reduced research budgets and signaled the beginning of the end of most industry labs. While petroleum geochemists in industry still did research, it was with very limited funds and an increasing reliance on contract labs for analytical services. As research at the corporate level was dwindling, the industry philosophy was that academia, government institutes, and contract labs would make up the shortfall. To this end, there was an emphasis on joint industry projects, investigation carried out by academia and contract labs funded by a group of industry partners. But the lack of large volumes of data, diverse sample material, and experience found in companies limited the success of these projects. Much of these shortcomings are still evident today.

With the industry downturn extending into the 1990s, more emphasis in reservoir applications of petroleum geochemistry was being seen. Exploration budget were still tight and companies looked to "getting another barrel out" of already discovered reserves. Investigating reservoir continuity, making production allocations from commingled production, and diagnosing production problems (Kaufman et al., 1990) were now important parts of the petroleum geochemist's tool kit. This emphasis on reservoir applications continues today.

Another significant development in the 1990s was the publication by Magoon and Dow (1994) on the concept of petroleum systems. While the idea of petroleum systems had been around since the 1970s (e.g., Dow, 1974), formalizing the concept brought additional focus on the role petroleum geochemistry has in understanding the petroleum system's workings (more on petroleum system later in this chapter and in Chapter 9).

From 2000 to the present, petroleum geochemistry continues to advance on all fronts. However, the largest efforts are being directed toward the so-called unconventional plays. In shale gas plays, the source rock is also the reservoir requiring a different approach to understand the source rock and how it functions (Passey et al., 2010). In tight oil reservoir plays, more attention needs to be given to fluid properties and phase behavior for successful exploitation (Dembicki, 2014). Whatever problems the future holds in petroleum exploration and production, innovative applications of the concept and methods of petroleum geochemistry will continue to contribute to solutions.

For more details of the history of the petroleum geochemistry, the reader is referred to Kvenvolden (2002), Hunt et al. (2002), Durand (2003), Kvenvolden (2006), and Dow (2014).

DEFINITIONS

Petroleum

Petroleum is a naturally occurring material in the earth composed predominantly of chemical compounds of carbon and hydrogen with and without other nonmetallic elements such as sulfur, nitrogen, and oxygen. It is formed by the diagenesis of sedimentary organic matter transforming the biological input into sediments first into kerogen, then the constituents of petroleum, with the final product being an inert carbon residue. Petroleum may exist as gas, liquid, or solid depending on the nature of its chemical constituents and temperature and pressure conditions where it exists. The main forms of petroleum are: *natural gas*, which does not condense into a liquid at surface conditions; *condensate*, which is gaseous at reservoir temperature and pressure and condenses into a liquid at the surface; and *crude oil*, the liquid part of petroleum. Often within the oil and gas industry, the term *hydrocarbon* is substitute for petroleum, crude oil, and/or natural gas.

Geochemistry

Geochemistry is defined as the study of the processes that control the abundance, composition, and distribution of chemical compounds and isotopes in geologic environments. *Organic*

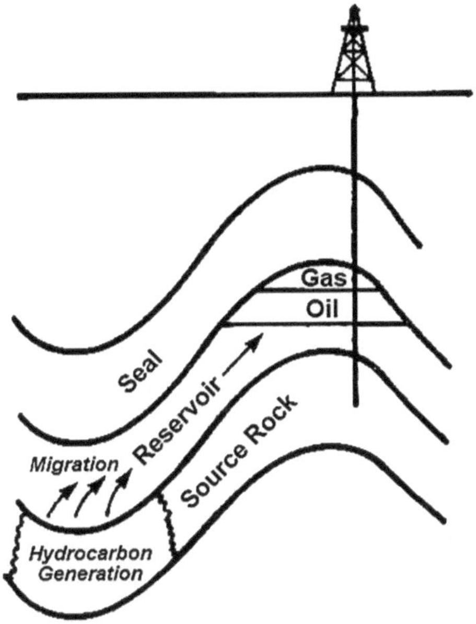

FIGURE 1.2 Schematic of a petroleum system.

geochemistry is simply the subdiscipline of geochemistry that focuses on organic (carbon bearing) compound found in geologic environments. *Petroleum geochemistry* is the practical application of organic geochemistry to the exploration for and production of petroleum.

Petroleum System

A *petroleum system*, shown schematically in Fig. 1.2, is defined by a genetic relationship linking a source rock to all oil and gas it has generated, and it consists of all the geologic elements and processes that are essential for the formation of a petroleum accumulation (Magoon and Dow, 1994). The geologic elements needed are source rock, reservoir, seal, and overburden, while the geologic processes involved are trap formation, generation, migration, accumulation, and preservation. There are also stratigraphic, temporal, and spatial components to a *petroleum system*, such that the elements and process must occur at the right place at the right time to produce a petroleum accumulation. The key element from a petroleum geochemistry point of view is the *source rocks*, rocks from which petroleum has been generated or is capable of being generated. More about *source rocks* can be found in the next chapter, while more about *petroleum systems* can be found in Chapter 9.

Sedimentary Organic Matter

Organic matter found in sediments and sedimentary rocks is typically subdivided into two categories: *bitumen*, that part of the sedimentary organic matter that is soluble in common organic solvents; and *kerogen*, that part of the sedimentary organic matter that is insoluble in common organic solvents. The term *bitumen* has different connotations depending on what type of sedimentary rock it is found in. In fine-grained sediments, *bitumen* is indigenous to the rock and may be preserved organic matter from the depositional environment or the product of generation. In coarse-grained sediments (reservoir rocks), *bitumen* usually refers to residual crude oil dispersed in the sediment. *Bitumen* is also used to refer to solid vein-filling material, pitch, tar, and asphalt.

Kerogen, the insoluble organic matter preserved in sedimentary rocks, is a complex material derived from the breakdown and diagenesis of the components of plants, animals, and bacteria deposited in the sediment. The chemical composition of *kerogen* is variable and depends on the organic material incorporated into the sediment and the chemical processes involved in its diagenetic alteration and polymerization. For additional information on the formation of kerogen, see the discussion of Kerogen Formation in Chapter 2.

Another form of insoluble organic matter is *pyrobitumen*. It is solidified bitumen that is the insoluble residue remaining after residual bitumen in source rocks or oil in reservoir rocks is cracked in situ to gas.

Other Sedimentary Organic Deposits

In addition to oil and gas deposits, there are several forms of sedimentary rock very rich in organic matter. *Coal* is a combustible sedimentary rock containing at least ~75% by weight organic matter (~60% by weight total organic carbon). Most, but not all, coals are from the accumulation and preservation of higher plant materials, usually in a swamp environment. An *oil shale* is an immature organic-rich, oil-prone source rock which can be heated to yield oil. *Tar sands* are sandstone reservoirs containing viscous heavy oil usually at or near the earth's surface.

ORGANIC CHEMISTRY REVIEW

Covalent Bonds

As stated earlier, petroleum is composed of organic (carbon bearing) chemical compounds made of predominantly carbon and hydrogen with and without other nonmetallic elements such as sulfur, nitrogen, and oxygen. Organic compounds range in size and complexity from the simple one carbon gas methane to the complex geopolymer kerogen, with a molecular weight of 50,000 or more. All of these compounds are based on building molecular structures with covalent bonds. Covalent bonds involve the sharing of a pair of electrons by two atoms. By sharing electrons, both atoms can fill their outer electron shells leading to more stability for both the atoms and the resulting molecule.

Looking at the examples of carbon–hydrogen and carbon–carbon covalent bonding in Fig. 1.3, the sharing of electrons can be demonstrated. Carbon has two electron shells, an inner shell with two electrons and an outer shell with four electrons. While the inner electron shell is stable, the outer electron shell would like to have four more electrons to reach a stable state with eight. Hydrogen has only one electron shell containing a single electron. A stable configuration for hydrogen would be to have two electrons, like the inner shell of the carbon atom. Methane is formed by bringing four hydrogen atoms into close proximity with a carbon atom.

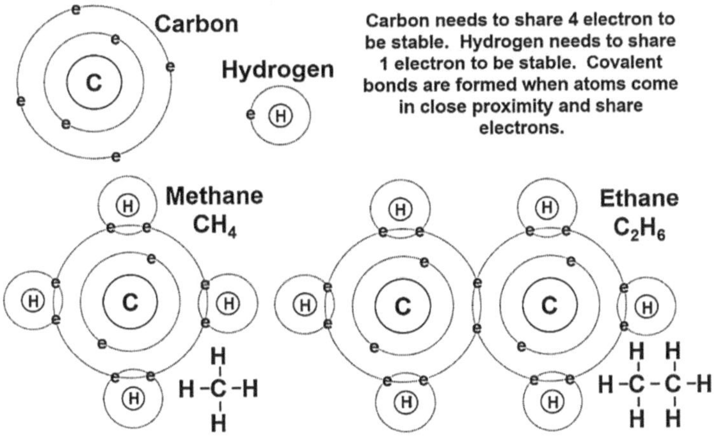

FIGURE 1.3 Carbon–hydrogen and carbon–carbon covalent bonds in methane and ethane.

The sharing of a pair of electrons consists of one electron from the carbon pairing with the one electron from one of the hydrogens. The final result from the electron sharing is eight electrons in the outer electron shell of the carbon and two electrons in the electron shell of each hydrogen, with each pair of shared electrons constituting a covalent bond.

Sharing of electrons can also occur between two carbon atoms as shown in the ethane molecule in Fig. 1.3. In this case, two carbon atoms come into close proximity to share a pair of electrons along with six hydrogen atoms. This is essentially taking two methane molecules, removing one hydrogen from each and forming the carbon–carbon bond between the remains of the two molecules. This process could be continued to form larger molecules.

Because it is not efficient to put in all the electrons of the atoms while drawing a compound, organic chemists devised a shorthand for depicting a covalent bond. The sharing of two electrons is usually shown as a bar connecting the atoms. In Fig. 1.3, methane and ethane are shown in this style just below and to the right of the full electron representations.

Hydrocarbons

The first class of organic compounds to discuss is the hydrocarbons. They are by far the most abundant class of compounds in petroleum. Hydrocarbons are made exclusively of carbon and hydrogen. Saturated hydrocarbons, also called alkanes or paraffins, are hydrocarbons that contain only carbon-to-carbon single bonds. Unsaturated hydrocarbons are hydrocarbons that contain at least one carbon-to-carbon double or triple bond. While carbon and hydrogen can only share one pair of electrons, carbon and carbon can share up to three in one bond. The sharing of two pairs of electrons between carbon atoms results in a double bond, shown by two bars between the carbons, and the sharing of three pairs of electrons makes a triple bond, shown by three bars between the carbons, as shown in Fig. 1.4. As the number of pairs of shared electrons increases, the nuclei of the two carbon atoms come closer together. As a result, the nuclei begin to repel each other making double and triple bonds less stable than the carbon–carbon single bond. This is reflected in that very few carbon–carbon triple bonds are found in naturally occurring substances. Carbon–carbon double are common

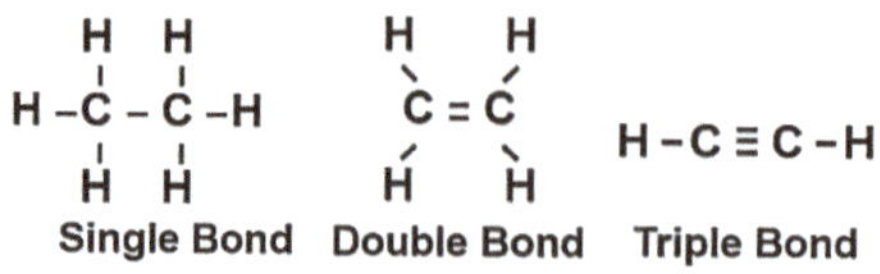

FIGURE 1.4 Carbon–carbon single, double, and triple bonds.

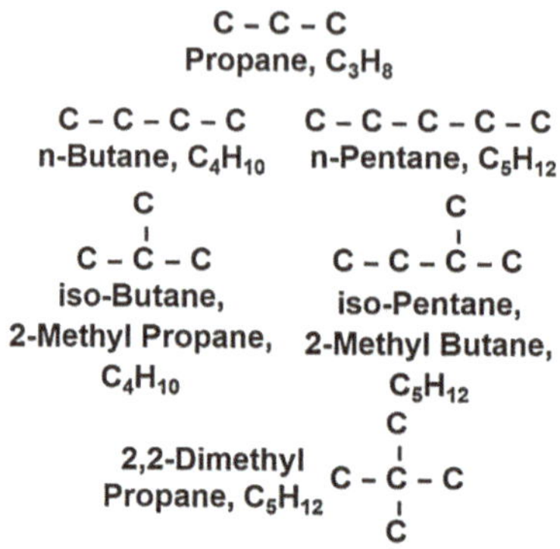

FIGURE 1.5 Some examples of small hydrocarbon molecules to illustrate the concept of isomers.

in some biological materials but not very common in geological materials, the exception being aromatic hydrocarbons which will be discussed shortly.

Hydrocarbons can exist as linear chains of carbons atoms, branched chains of carbon atoms, one or more rings of carbon atoms, or as a combination of ringed and chain structures. Many of the common hydrocarbons have so-called trivial names. This is typical for many of the smaller less complex hydrocarbons. But as the structural complexity of hydrocarbons increases, trivial names are often replaced by more formal scientific names that are arrived at by a set of strict rules based on the number and type of atoms involved, the structure of the compound, and the type of bonds present. The purpose of the scientific names is that as compound size and complexity increases, the name should provide needed information about the compound's structure so that it can be drawn based solely on the formal name.

Some simple examples of hydrocarbons are shown in Fig. 1.5. Propane is a simple three-carbon saturated hydrocarbon. The structure cannot be any more complex for this three-

carbon chain. However, if one carbon is added to form butane, the number of possible structures increases. It is now possible to have the butane exist as a straight-chained (normal), n-butane, and as isobutane (the trivial name) or 2-methyl propane, which is the three-carbon propane chain with a methyl group (methane less one hydrogen atom), located at the middle or number 2 carbon. These two compounds are called isomers, which means they have the same number of carbon and hydrogen atoms but are arranged in different structures. Starting with pentane on the right side of Fig. 1.5, a similar pattern can be observed. It starts with n-pentane and the isomer isopentane (the trivial name), or 2-methyl butane, a four-carbon chain, butane, with a methyl group located at the number 2 carbon. In addition, another isomer 2, 2 dimethyl propane is also a possible structural configuration. With increasing number of carbons, the number of possible isomer also increases. In addition to methyl groups as

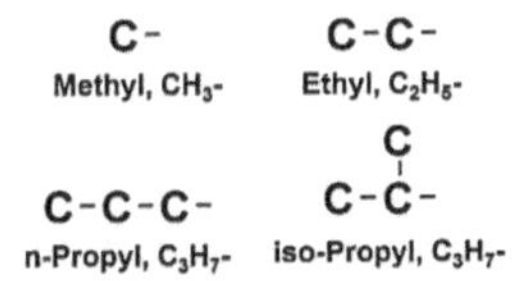

FIGURE 1.6 Common small side chains for hydrocarbon molecules.

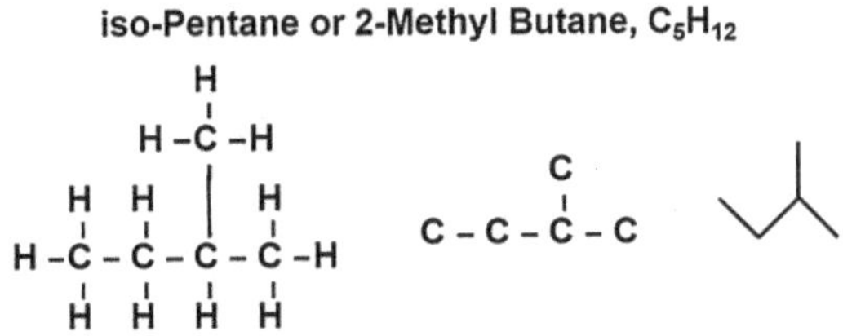

FIGURE 1.7 Different forms of structural notations for hydrocarbon molecules.

side chains to the main straight-chained structure, there can be ethyl groups (ethane less one hydrogen atom), n-propyl groups (propane less one hydrogen atom), and greater, as well as branched side chains such as an isopropyl group, as shown in Fig. 1.6.

For the compounds shown in Figs. 1.5 and 1.6, a different notation is used to depict the compounds' structures. As compounds become larger and their structures become more complex, it is not efficient to put in all the hydrogens while drawing a compound. In this notation, organic chemists simply do not show them, the assumption being, there are sufficient hydrogen atoms in place to match up with any unpaired carbon electrons not already involved in carbon–carbon bonds. The sharing of a pair of electrons is still shown as a single bar connecting the atoms. But as organic compounds become larger, even showing the carbon becomes cumbersome and inefficient, giving rise to the "stick" notation. The example in Fig. 1.7 for isopentane (2-methyl butane, C_5H_{12}) shows three single sticks connected in a zigzag pattern with a fourth stick connected to one of the intersections. This notation indicates that there is a carbon at the end of each "stick" and that these single "sticks" represent carbon–carbon bonds. The three single "sticks" connected in a zigzag pattern represent four carbons in the butane chain, and the fourth "stick" connected to one of the intersections represents the methyl group side chain. As in the previous notation scheme, there is an assumption that there are sufficient hydrogen atoms in place to match up with any unpaired carbon electrons not already involved in carbon–carbon bonds. Two parallel "sticks" would indicate a double bond between the carbons. This notation system is used extensively with the large straight-chained, branched-chained, and cyclic hydrocarbons.

The straight and branched-chained saturated hydrocarbon can get quite large, commonly up to 60 carbons or greater. However, most of the compounds used in petroleum geochemistry are in the C_1–C_{35} range, primarily due to analytical considerations. Branching can be very simple or very complex, as shown in the examples in Fig. 1.8. Starting with the straight-chained n-heptadecane (n-C_{17}), commonly encountered hydrocarbons, such as 2-methyl heptadecane and 3-methyl heptadecane, are formed simply by adding a methyl group to the number 2 or number 3 carbon in the base chain. However, more complex branched-chained compounds, such as pristane (2, 6, 10, 14-tetramethyl pentadecane) and phytane (2, 6, 10,

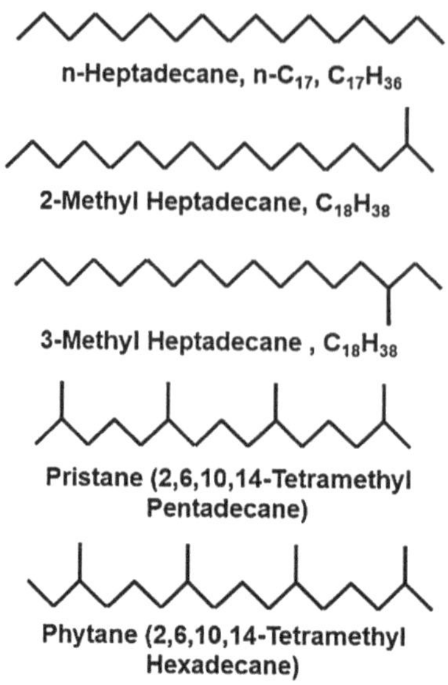

FIGURE 1.8 Some examples of straight-chained and branched saturated hydrocarbons.

14-tetramethyl hexadecane), are also abundant. The structural simplicity or complexity of molecules such as these contribute to the geochemical information they carry, as will be discussed in subsequent chapters.

Saturated hydrocarbons containing one or more rings of carbon atoms in their structures are called cycloalkanes or naphthenes. These rings usually consist of five or six carbons, with six-carbon rings being the most stable and, therefore, the most common. Geochemically significant naphthenes can consist of one to six rings. Some examples of one to four rings structures are shown in Fig. 1.9. In addition to the rings, one or more side chains can be added to structure at any of the carbons. Two examples of this are shown by the sterane and hopane structures in Fig. 1.10. These geochemically significant naphthene groups also demonstrate how a series of related compounds are formed by simply varying the side chain at one location, R, on the base molecule.

Aromatic Hydrocarbons

Aromatic hydrocarbons are a special class of unsaturated hydrocarbon based on a six carbon ring moiety called benzene. The saturated hydrocarbon cyclohexane is transformed into the aromatic hydrocarbon benzene by adding three alternating carbon–carbon double bonds, as shown in Fig. 1.11. The benzene structure can have two arrangements of these double bonds, shown by the pair of benzene molecules in the middle of Fig. 1.11. In nature, these arrangements of bonds rapidly alternate in the benzene structure. Because of this rapid changing or resonating of the position of the three carbon–carbon double bonds, benzene is usually represented as a hexagon with a circle in the center, as shown on the right side of Fig. 1.11. As discussed earlier, carbon–carbon double bonds are usually considered less stable than carbon–carbon single bonds. However, the resonating alternating double bounds in benzene distribute the electron sharing over all six carbons and impart more stability to the molecule.

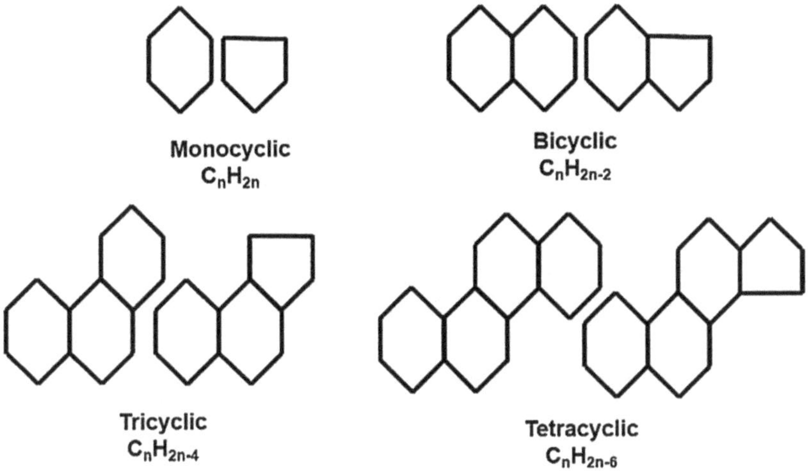

FIGURE 1.9 Some examples of five- and six-membered ring cyclic saturated hydrocarbons, also called naphthenes or cycloalkanes.

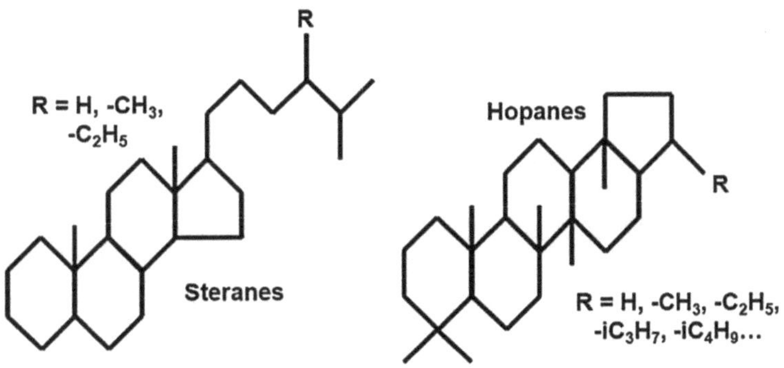

FIGURE 1.10 Structures of steranes and hopanes as examples of some of the variations in related molecules that can be achieved from different side chains.

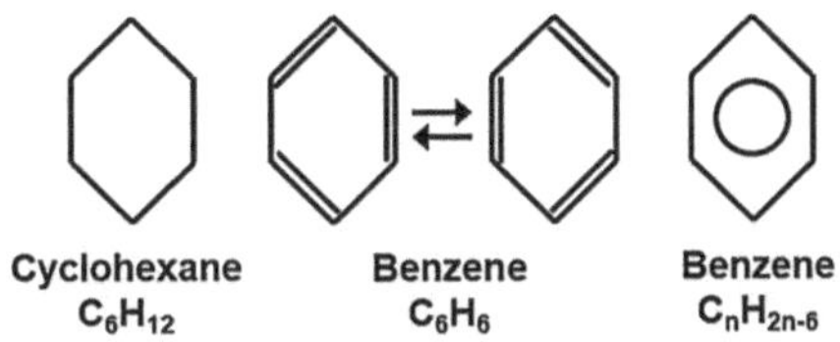

FIGURE 1.11 The benzene structure.

The basic benzene ring structure can be used to build much large molecules by attaching saturated hydrocarbon chains (either straight or branched) or by building multiple ring structures, similar to the naphthenes. These ring compounds, as illustrated in Fig. 1.12, may consistent of all aromatic rings, pure aromatic compounds, or a mix of aromatic and saturated rings, naphthenoaromatic compounds. In the pure aromatic structures, the resonating stabilization of the benzene unit is extended to the entire structure. And just like in

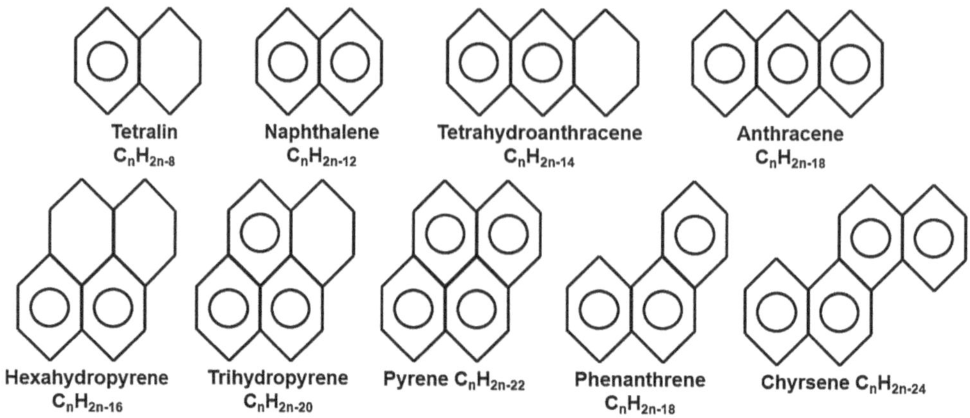

FIGURE 1.12 Some examples of basic pure aromatic and naphthenoaromatic structures.

the naphthenes, one or more side chains can be added to any of the carbons in any of the aromatic ring structures.

N-S-O Compounds

N-S-O compounds, or resins, are organic compounds that contain nitrogen, sulfur, or oxygen in addition to the carbon and hydrogen. In many crude oils the N-S-Os represent only a small portion of the oil as compared to the saturated and aromatic hydrocarbons; however, there are instances where the N-S-Os are in higher concentration. Some examples of N-S-O compounds commonly found in bitumen and crude oil are given in Fig. 1.13. The smaller less complex compounds are more typically found during sediments diagenesis and early generation. These include the alcohols, acids, mercaptans, sulfides, and disulfides. The remainder of the compounds, mainly ringed structures, are more abundant later in the generation history of the sediment. Because of analytical difficulties, most of the N-S-O compounds are not frequently used in making geochemical interpretations. The exception to this is the thiophene group, which is used in deciphering crude oil origins, as discussed in Chapter 4.

Asphaltenes

Asphaltenes are a high-molecular weight component of crude oils and bitumen which is insoluble in n-heptane. Chemically, they are polyaromatic nuclei linked by aliphatic chains or rings and functional groups with molecular weights in the range of 1000–10,000 (Pelet et al., 1986). A representation of a typical asphaltene structure is shown in Fig. 1.14. It should be remembered that while the exact composition will vary from one asphaltene molecule to another, the molecular composition of the asphaltenes of a particular crude oil or bitumen exhibit a high degree of similarity (Behar and Pelet, 1985). Asphaltenes were originally thought to be small fragments of kerogen after some thermal degradation (Louis and Tissot, 1967). However, Pelet et al. (1986) showed that asphaltenes could also be derived from the condensation processes similar to those that form kerogen (see discussion

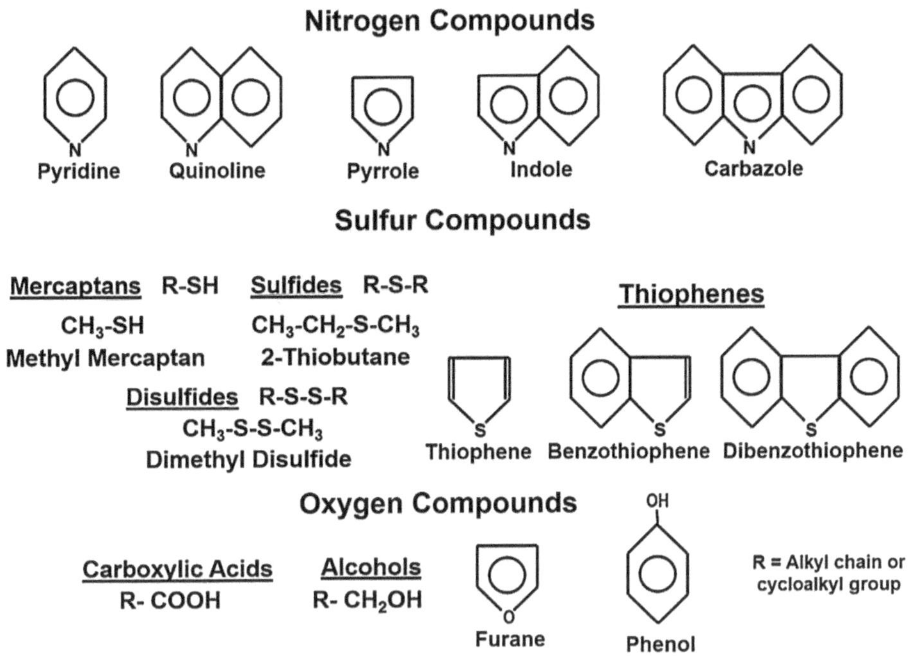

FIGURE 1.13 Some examples of nitrogen-, oxygen-, and sulfur-containing organic compounds commonly found in crude oils and bitumen.

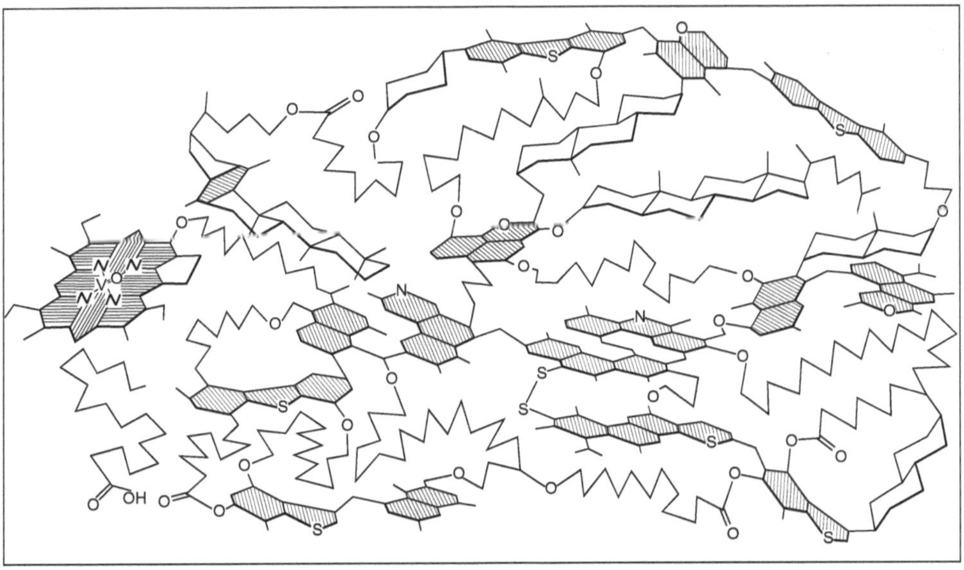

FIGURE 1.14 A representation of the structure of a typical asphaltene molecule. Chain aliphatic structures are shown as "saw-toothed" lines, naphthenic structures are shown as unshaded polygons, and aromatic structures are the shaded polygons. Oxygen is shown as O, hydrogen as H, and nitrogen as N. *From Pelet, R., Behar, F., Monin, J.C., 1986. Resins and asphaltenes in the generation and migration of petroleum. Organic Geochemistry 10, 481–498.*

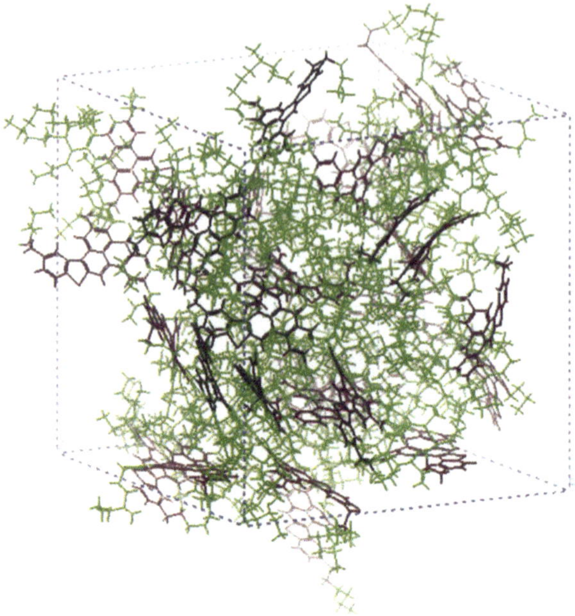

FIGURE 1.15 A 3-dimensional representation of an asphaltene aggregate. The different colors and shades of gray represent individual asphaltene molecules that have come together to form the aggregate. *From Mullins, O.C., 2003. Asphaltenes and polycyclic aromatic hydrocarbons. In: Oral Presentation at the Stanford Synchrotron Radiation Laboratory (SSRL) 30th Annual Users' Meeting, October 8–10, 2004, Menlo Park, CA. http://www-ssrl.slac.stanford.edu/conferences/ssrl30/mullins.pdf.*

of Kerogen Formation in Chapter 2). While asphaltenes exist as free molecules in colloidal solution within crude oils and bitumens, these solutions are often unstable and can be perturbed causing asphaltene molecules to come out of solution and cluster together in aggregates. A model of an asphaltene aggregate is shown in Fig. 1.15. These aggregates can have a molecular weight of a few tens of thousands to nearly a million. Clusters of asphaltene aggregates can reduce porosity and clog permeability in reservoir rocks (more on this in Chapter 4).

Reactions

Organic compounds are subject to a variety of reactions that can alter their composition and structure. Some of the reactions that occur in geological environments are shown in Fig. 1.16. Hydrolysis is the adding of water at a carbonyl group, such as an ester linkage, to create a carboxylic acid and a hydroxyl group. Decarboxylation is the loss of a carboxylic acid group creating carbon dioxide and an alkyl group. The carbon dioxide may then be dissolved in the sediment pore waters or go to the formation of carbonate cements. Dehydroxylation is the addition of hydrogen to a hydroxyl (alcohol) group creating water and an alkyl group. Reduction is the conversion of a carbon–carbon double bond to a carbon–carbon single bond by the addition of hydrogen. And cracking is the breaking of a carbon–carbon bond creating

Hydrolysis: $R\text{-}COO\text{-}R + H_20 \rightarrow R\text{-}COOH + R\text{-}OH$

Decarboxylation: $R\text{-}CH_2\text{-}COOH \rightarrow R\text{-}CH_3 + CO_2$

Dehydroxylation: $R\text{-}CH_2\text{-}OH + H_2 \rightarrow R\text{-}CH_3 + H_2O$

Reduction: $R\text{-}CH{=}CH_2 + H_2 \rightarrow R\text{-}CH_2\text{-}CH_3$

Cracking: $R\text{-}CH_2\text{-}CH_2\text{-}CH_3 + H_2 \rightarrow R\text{-}CH_3 + CH_3\text{-}CH_3$

FIGURE 1.16 Some of the reactions that sedimentary organic matter may undergo during diagenesis and hydrocarbon generation.

two smaller compounds. If sufficient extra hydrogen is present, the two smaller compounds will be saturated, as shown. If hydrogen is not present, an unsaturated or aromatic compound may form. While hydrolysis, decarboxylation, dehydroxylation, and reduction are reactions observed mainly in the early diagenesis of organic matter in sediments, cracking is an important reaction in the hydrocarbon generation process.

STABLE ISOTOPE REVIEW

Elements are the primary constituents of all matter and as such cannot be chemically broken down into simpler constituents. They are composed of the protons, electrons, and neutrons with each element being distinguished from the others by the number of protons in its nucleus. Protons are positively charged. To balance this charge, the number of electrons in an element equals the number of protons. The neutrons have no charge and can vary in number. Isotopes are simply forms of an element that have different numbers of neutrons in the nuclei of their atoms.

A relevant example is carbon, which has three natural isotopes: carbon 12, ^{12}C, six protons and six neutrons; carbon 13, ^{13}C, six protons and seven neutrons; and carbon 14, ^{14}C, six protons and eight neutrons. Some isotopes, like ^{14}C, are radioactive and decay to form a different element or isotope plus a high-energy particle. Stable isotopes, like ^{12}C and ^{13}C, do not decay but are subject to changes in their relative concentration due to chemical, physical, and biological processes. For example, biological processes, such as photosynthesis, tend to favor utilization of ^{12}C over ^{13}C, and different types of plants have a stronger preference for ^{12}C than others. This can result in carbon isotope ratios distinctive for specific plants groups. Nonbiological processes also show an isotopic preference. There is a kinetic isotope effect observed when cleaving methane (methyl groups) from larger organic molecules during cracking. It is energetically more favorable to remove a ^{12}C methyl group over ^{13}C methyl group. As a result, methane formed early in the process of cleaving of methyl groups will have higher amount of the ^{12}C isotope than methane formed later.

The stable isotopes of hydrogen, nitrogen, sulfur, and oxygen behave similarly to carbon. These isotopes along with carbon are shown in Fig. 1.17 with their natural abundance and the base ratios of the isotopes used in geochemical studies. While all these stable isotopes can contribute to petroleum geochemistry interpretations, the carbon and hydrogen isotope ratios are used most often.

While the actual ratios of stable isotopes are useful, they are small numbers, usually ranging from three to five decimal places. To make it easier to see changes in the isotope ratios,

Carbon	**Sulfur**
$^{12}C = 98.99\%$	$^{32}S = 95.02\%$
$^{13}C = 1.11\%$	$^{33}S = 0.75\%$
$^{13}C/^{12}C = 1/89$	$^{34}S = 4.21\%$
Hydrogen	$^{36}S = 0.02\%$
$^{1}H = 99.9844\%$	$^{34}S/^{32}S = 1/22.5$
$^{2}H = 0.0156\%$	**Oxygen**
^{2}H = Deuterium, D	$^{16}O = 99.763\%$
D/H = 1/6400	$^{17}O = 0.0375\%$
Nitrogen	$^{18}O = 0.1995\%$
$^{14}N = 99.63$	$^{18}O/^{16}O = 1/500$
$^{15}N = 0.37$	
$^{15}N/^{14}N = 1/269$	

FIGURE 1.17 The stable isotopes of carbon, hydrogen, nitrogen, sulfur, and oxygen with their natural abundance and the base ratios of the stable isotopes used in geochemical studies.

$$\delta^{13}C = \left[\frac{^{13}C/^{12}C \text{ sample}}{^{13}C/^{12}C \text{ standard}} - 1 \right] \times 1000$$

FIGURE 1.18 An example of δ notation used to report stable isotope ratios. This example is for carbon, and similar δ notations are used for reporting hydrogen, nitrogen, sulfur, and oxygen stable isotope ratios.

δ notation has been adopted. An example of δ notion for stable carbon isotopes is shown in Fig. 1.18. The isotopic ratio is normalized to a standard and is expressed in parts per 1000 (‰). A negative $\delta^{13}C$ value indicates the sample is depleted in ^{13}C relative to the standard, and a positive value indicates an enrichment in ^{13}C. The standard for carbon, PDB, is the carbonate mineral in a belemnite fossil from the Upper Cretaceous Pee Dee Formation from South Carolina. For hydrogen and oxygen stable isotope measurements, the standard is Standard Mean Ocean Water, SMOW. For sulfur the standard is the Cañon Diablo Troilite, CDT, and for nitrogen the standard is atmospheric air, AIR.

Carbon and hydrogen isotopes are used extensively in petroleum geochemistry, especially for interpreting the origin of natural gases and comparison of crude oils, as discussed in Chapter 4. Oxygen, sulfur, and nitrogen stable isotopes are not routinely used for interpretations.

References

Behar, F., Pelet, R., 1985. Pyrolysis-gas chromatography applied to organic geochemistry: structural similarities between kerogens and asphaltenes from related rock extracts and oils. Journal of Analytical and Applied Pyrolysis 8, 173–187.

Breger, I.A. (Ed.), 1963. Organic Geochemistry. Pergamon Press, Oxford. 658 p.

Connan, J., 1974. Time-temperature relation in oil genesis. American Association of Petroleum Geologists Bulletin 58, 2516–2521.

Dembicki, H., 2014. Challenges to black oil production from shales. In: Oral Presentation at the American Association of Petroleum Geologists Geoscience Technology Workshop, Hydrocarbon Charge Considerations in Liquid-Rich Unconventional Petroleum Systems, November 5, 2013. American Association of Petroleum Geologists Search and Discovery, Vancouver, BC, Canada. Article #80355(2014) http://www.searchanddiscovery.com/documents/2014/80355dembicki/ndx_dembicki.pdf.

Dow, W.G., 1974. Application of oil correlation and source rock data to exploration in Williston basin. AAPG Bulletin 58 (7), 1253–1262.

Dow, W.G., 2014. Musings on the history of petroleum geochemistry – from My Perch. In: Oral Presentation at the American Association of Petroleum Geologists Annual Convention and Exhibition, Houston, Texas, April 6–9, 2014. American Association of Petroleum Geologists Search and Discovery. Article #80375(2014) http://www.searchanddiscovery.com/documents/2014/80375dow/ndx_dow.pdf.

Durand, B., 2003. A history of organic geochemistry. Oil & Gas Science and Technology – Revue IFP 58, 203–231.

Eglinton, G., Calvin, M., 1967. Chemical fossils. Scientific American 216, 32–43.

Espitalie, J., Madec, M., Tissot, B., Mennig, J.J., Leplat, P., 1977. Source rock characterization method for petroleum exploration. In: Proceedings of the 9th Annual Offshore Technology Conference, vol. 3, pp. 439–448.

Hunt, J.M., Meinert, R., 1958. Petroleum prospecting. U.S. Patent 2,854,396.

Hunt, J.M., 1979. Petroleum Geochemistry and Geology. W. H. Freeman, New York. 617 pp.

Hunt, J.M., Philp, R.P., Kvenvolden, K.A., 2002. Early developments in petroleum geochemistry. Organic Geochemistry 33, 1025–1052.

Kaufman, R.L., Ahmed, A.S., Elsinger, R.J., 1990. Gas chromatography as a development and production tools for fingerprinting oils from individual reservoirs: applications in the Gulf of Mexico. In: GCSSEPM Foundation Ninth Annual Research Conference Proceedings, October 1, 1990, pp. 263–282.

Kvenvolden, K.A., 2002. History of the recognition of organic geochemistry in geoscience. Organic Geochemistry 33, 517–521.

Kvenvolden, K.A., 2006. Organic geochemistry—A retrospective of its first 70 years. Organic Geochemistry 37, 1–11.

Lopatin, N.V., 1971. Temperature and geologic time as factors in coalification. Izvestiya Akademii Nauk. Seriya Khimicheskaya 3, 95–106.

Louis, M., Tissot, B., 1967. Influence de la température et de la pression sur la formation des hydrocarbures dans les argiles à kérogène. In: 7th World Petroleum Congress, Mexico, vol. 2, pp. 47–60.

Magoon, L.B., Dow, W.G., 1994. The petroleum system. In: Magoon, L.B., Dow, W.G. (Eds.), The Petroleum System: From Source to Trap, American Association of Petroleum Geologists Memoir 60, pp. 3–24.

Mullins, O.C., 2003. Asphaltenes and polycyclic aromatic hydrocarbons. In: Oral Presentation at the Stanford Synchrotron Radiation Laboratory (SSRL) 30th Annual Users' Meeting, October 8–10, 2004, Menlo Park, CA. http://www-ssrl.slac.stanford.edu/conferences/ssrl30/mullins.pdf.

Passey, Q., Bohacs, K., Esch, W., Klimentidis, R., Sinha, S., 2010. From oil-prone source rock to gas producing shale reservoir - geological and petrophysical characterization of unconventional shale-gas reservoirs. In: CPS/SPE International Oil & Gas Conference and Exhibition in China, Paper SPE 131350- MS29 p.

Pelet, R., Behar, F., Monin, J.C., 1986. Resins and asphaltenes in the generation and migration of petroleum. Organic Geochemistry 10, 481–498.

Sluijk, D., Parker, J.R., 1986. Comparison of predrilling predictions with postdrilling outcomes, using Shell's prospect appraisal system. In: Association of Petroleum Geologists Studies in Geology 21: Oil and Gas Assessment: Methods and Applications American, pp. 55–58.

Tissot, B.P., Welte, D.H., 1978. Petroleum Formation and Occurrence: A New Approach to Oil and Gas Exploration. Springer-Verlag, New York. 538 p.

Treibs, A., 1934. The occurrence of chlorophyll derivatives in an oil shale of the upper Triassic. Annalen 517, 103–114.

Waples, D.W., 1980. Time and temperature in petroleum formation: application of Lopatin's method to petroleum exploration. American Association of Petroleum Geologists Bulletin 64 (6), 916–926.

CHAPTER

2

The Formation of Petroleum Accumulations

OUTLINE

Introduction	19	**Origin of Nonhydrocarbon Gases**	50
Incorporating Organic Matter Into Sediments	20	*Carbon Dioxide*	50
Kerogen Formation	25	*Nitrogen*	51
Source Rock Deposition	29	*Hydrogen Sulfide*	51
Maturation and Hydrocarbon Generation	33	*Helium*	52
Petroleum Migration	45	*Hydrogen*	53
		Coals as Oil-Prone Source Rocks	53
		Summary	55
		References	56

INTRODUCTION

In order to find petroleum accumulations, it is necessary to understand how they form. From a geochemical perspective, the first issue to address is where the petroleum comes from. Some workers in the field (e.g., Kudryavtsev, 1951; Porfir'ev, 1974) have suggested that petroleum could come from purely abiogenic processes deep in the lower crust or mantle. Processes such as Fischer–Tropsch reactions, as shown in Fig. 2.1, where carbon monoxide could combine with hydrogen to form hydrocarbons, were suggested as possible mechanisms (Szatmari, 1989). After much study, most geochemists feel that although these reactions could occur in the very deep subsurface and produce small volumes of low-molecular weight hydrocarbons, they cannot account for the volumes of petroleum observed in sedimentary basins or the diversity in the molecular structure and molecular weight range observed in petroleum (Glasby, 2006).

The prevailing theory, embraced by the majority of geochemists, is that petroleum is derived from the transformation of preexisting organic matter of biological origin that has

http://dx.doi.org/10.1016/B978-0-12-803350-0.00002-7

$$CO + H_2 \xrightarrow{\text{Catalyst}} CH_3(CH_2)_nCH_3 + \text{other HC's} + H_2O$$

FIGURE 2.1 The Fischer–Tropsch reaction for producing hydrocarbons from carbon monoxide and hydrogen. Note the catalyst required for the reaction.

been incorporated into sediments (Erdman, 1965). This biogenic origin of petroleum is supported by the complexity of molecular structures and the molecular weight range observed in petroleum, as well as petroleum's stable carbon isotope composition, observed optical activity, and the presence of preserved biological structures in some of its compounds (Eglinton, 1969; Speers and Whitehead, 1969). This large body of evidence overwhelmingly supports this concept.

While all sediments may contain some organic matter, there are certain sediments that contain high concentrations of organic matter. It is these organic matter-rich sediments that are responsible for the generation of hydrocarbons that form accumulations of oil and gas. These sediments are called source rocks. And while the biogenic origin of petroleum identifies the organic matter in source rocks as the "feedstock" for oil and gas, it also raises additional questions. How is this biological material incorporated into and preserved in sediments? What kind of depositional conditions and settings are more conducive for organic matter preservation and formation of source rocks? How is sedimentary organic matter then converted into the components of petroleum? And, once it is formed in source rocks, how does oil and gas move about in the subsurface and eventual gather to form an accumulation in a reservoir/trap system? This chapter will answer these questions and provide an overall framework for applying petroleum geochemistry to exploration and production problems.

INCORPORATING ORGANIC MATTER INTO SEDIMENTS

Source rocks are fine-grained sedimentary rocks containing relatively high concentrations of organic matter deposited in aqueous depositional settings. In the schematic of a marine depositional environment in Fig. 2.2, organic matter in sediments can come from three primary sources. The autochthonous organic matter is the product of biological activity in the water column above the depositional site. It is primarily the product of photosynthetic organisms such as algae and phytoplankton in the photic zone in the upper part of the water column (Gagosian, 1983). These primary producers can be consumed by zooplankton and other organisms in the water column that may also contribute to the sediment. Allochthonous organic matter has been transported some lateral distance from where it formed before being incorporated into the sediments. It is often the product of terrestrial higher plants contributing biomass to fluvial system that carry this organic matter to the depositional environment (Gagosian, 1983). A small portion may also be delivered to aqueous depositional systems by eolian processes (Gagosian and Peltzer, 1986). The third source is recycled or reworked organic matter. This is preexisting sedimentary organic matter derived from the erosion and redeposition of older sedimentary rocks. Of these three types, the autochthonous and allochthonous contributions are important in the development of source rocks. The reworked organic matter is nearly always degraded to the point where it has little or no capacity for being converted into oil and gas.

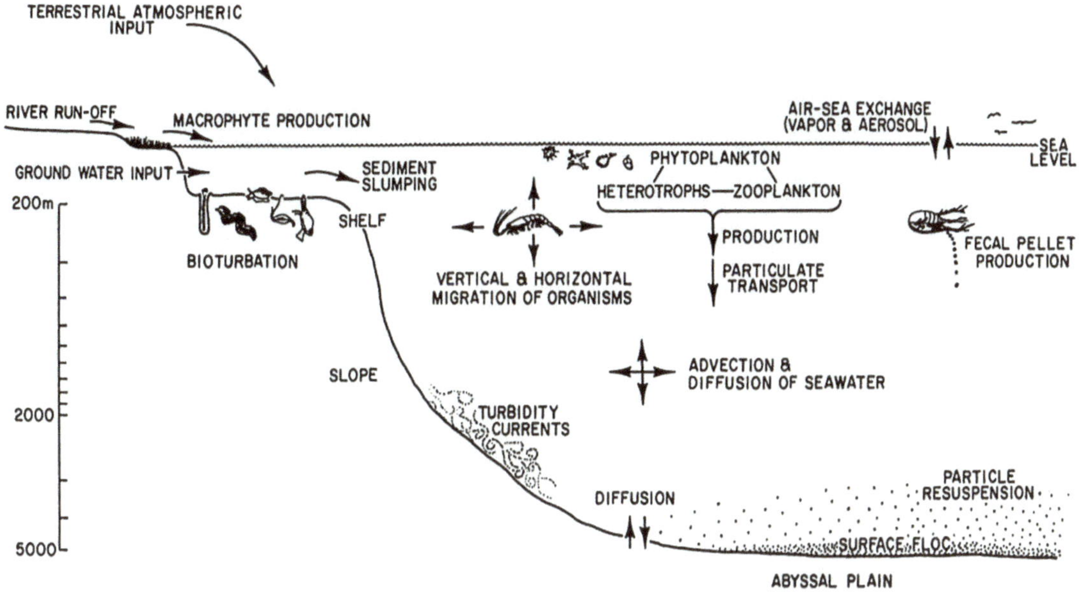

FIGURE 2.2 A schematic of transport mechanisms introducing organic matter to a marine deposition environment (Gagosian, 1983).

To get sufficient amounts of organic matter contributed to sediments to form a source rock, there must be high primary biological productivity in or near the depositional environment (Calvert, 1987). Primary biological productivity is controlled by a series of factors including the amount of solar radiance received (related to latitude), the nutrient supply (e.g., nitrogen, phosphorous, and iron), and the water supply for terrestrial environments. Looking at a map of global distribution of marine and terrestrial primary production in Fig. 2.3, high biological productivity is concentrated in equatorial regions for terrestrial environments and the continental margins and polar regions for marine environments. Terrestrial productivity in the equatorial regions is driven by the solar energy and water supply there. More moderate terrestrial productivity in the middle latitudes reflects the seasonal nature of primary biological production in these areas. Primary biological productivity on continental margins in marine environments is driven by nutrient influx by fluvial systems. The outflow from river systems continually replenishes these nutrients. In contrast, open ocean areas are depleted in nutrient due to lack of recharge resulting in little primary productivity. The polar regions benefit from the upwelling of nutrient-rich ocean bottom waters which supports high seasonal biological productivity. Upwelling of nutrient-rich ocean bottom waters can also impact the productivity of middle and lower latitude oceans. This is especially prevalent along the west coasts of continents due to prevailing winds.

While this picture of primary biological productivity limits to the present, it does give clues as to where high primary productivity might occur in the geologic past. Source rocks will be more likely deposited on continental margins and not in deep ocean environments. Highest productivity will likely occur in the equatorial regions. However, middle and high latitude regions should not be discounted because more than adequate primary productivity

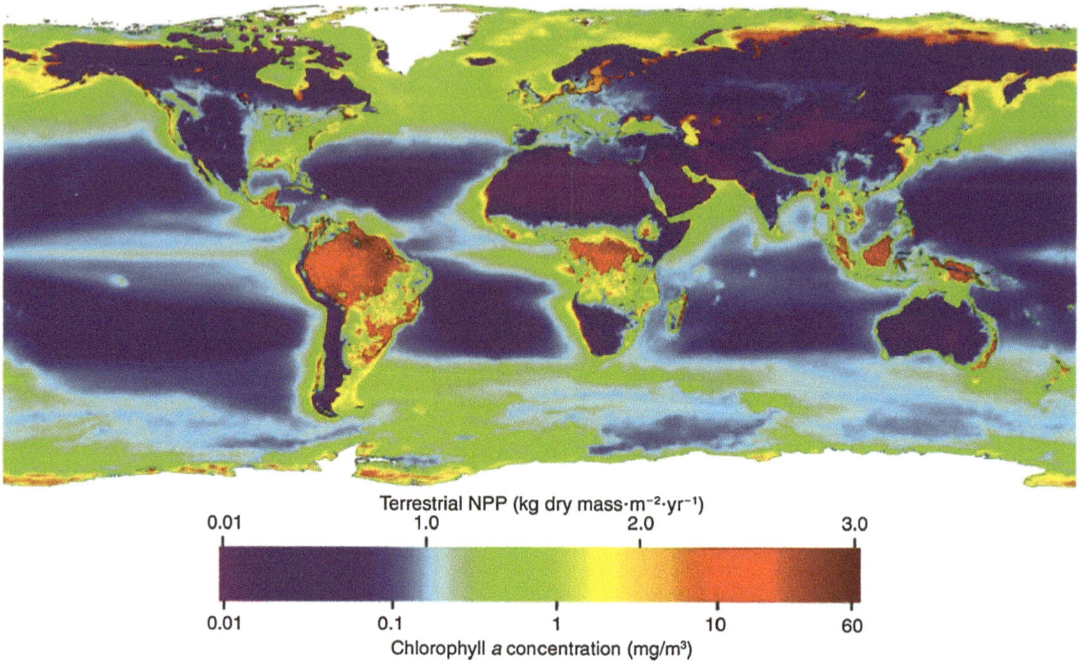

FIGURE 2.3 Global distribution of marine and terrestrial primary production *(From Huston, M.A., Wolverton, S., 2009. The global distribution of net primary production: resolving the paradox. Ecological Monographs 79, 343–377.)*. Terrestrial primary production is based on the Carnegie Ames Stanford Approach carbon model (Field et al., 1998). Ocean primary production (to 30-m depth) is inferred from chlorophyll a concentrations from SeaWIFS satellite data. Ocean chlorophyll a concentration is highly correlated with net primary production.

can occur there. And, areas of upwelling can increase productivity and the potential for source rock development.

Primary biological productivity is not the only factor to be considered in the development of a source rock. The organic matter needs to get into the sediment and be preserved in order to have a chance at forming a source rock. Once the autochthonous and allochthonous organic matter are in the water column above the depositional environment, they need to make their way to the sediment–water interface, become buried in sediment, and converted into a stable form in order to become part of a source rock. Transport through the water column needs to be quick. The longer the organic matter takes to reach the sediment–water interface, the more opportunity it has to be degraded by inorganic chemical processes such as oxidation, or consumed by organisms. One mechanism for rapidly transporting autochthonous organic matter to the seafloor is by fecal pellets. Phytoplankton in the photic zone can be consumed by zooplankton and excreted in fecal pellets. The zooplankton is typically not a very efficient digester and much of the organic matter from the phytoplankton is conserved. The fecal pellets are denser, more ballistic than the original phytoplankton and will rapidly settle to the seafloor as a component of marine snow.

Once at the sediment–water interface, preservation of organic matter is a function of the amount of oxidant available, the consumer organism population, and the burial (sedimentation) rate. With respect to the amount of oxidant available, preservation of the organic matter

	Organic Matter (Wt %)
Siltstone	1.79
Clay > 2μ	2.08
Clay < 2μ	6.50

FIGURE 2.4 An example of the variation in organic matter content with grain size in the Viking Shale, Cretaceous, western Canada. Tissot and Welte (1984) after Hunt (1963).

is dependent on the location of the oxic/anoxic boundary with respect to the sediment–water interface (Demaion and Moore, 1980). If the oxic/anoxic boundary is substantially below the sediment–water interface, oxidative processes, as well as aerobic biological activity, can consume any organic matter at the sediment surface. This condition will also encourage bioturbation that will consume more of the organic matter and further oxygenating the sediments. If the organic matter is left at or near the sediment–water interface long enough (slow sedimentation rates), it can be highly degraded or totally destroyed. If the oxic/anoxic boundary is at or just below the sediment–water interface, oxidation and biologic activity may be limited if the sedimentation rate is high enough to minimize the residence time of the organic matter at the interface (Didyk et al., 1978; Demaison and Moore, 1980). This provides a better opportunity for preservation. However, when the oxic/anoxic boundary is somewhere in the water column above the sediment–water interface (anoxic bottom waters), conditions for organic matter preservation are optimal (Didyk et al., 1978; Demaison and Moore, 1980). Not only are oxidative processes halted, but the anoxic bottom waters limit the biological activity to less efficient anaerobic organisms (Demaison and Moore, 1980). Therefore, the best depositional environments for the preservation of organic matter and the formation of source rocks have an anoxic water column.

Grain size is important for organic matter preservation. Looking at the organic matter content of different grain size fractions in the Viking shale of western Canada (Fig. 2.4), it is clear that as grain size decreases, the organic matter content increases. This is due to the influence of grain size on the development of anoxia just below the sediment–water interface (Hunt, 1963). As shown in Fig. 2.5, coarser grained sediment, such as sands and silt, can allow the circulation of bottom waters into the sediments (Tissot and Welte, 1984). This circulation from the water column can replenish oxygen in the interstitial waters allowing oxidative processes to continue below the sediment–water interface and encourage aerobic biological activity. In finer grained sediments, such as fine clays and carbonate muds, circulation of bottom waters into the sediments is highly restricted. This encourages the development of localized anaerobic environments in the sediment's interstitial spaces, promoting preservation and limiting biological activity that may consume the organic matter (Tissot and Welte, 1984).

How quickly the organic matter becomes buried in the sediment is also a significant factor in preservation. As shown in Fig. 2.6, if sedimentation rates are too low, the residence time for the organic matter at the sediment–water interface will be longer and allow degradation processes to continue and provide more opportunity for consumption by bottom grazing organisms (Ibach, 1982). If the sedimentation rates are too high, the organic matter may become diluted by the sediment and its concentration may not be high enough for a source rock to develop (Ibach, 1982). From studies of source rock occurrences, a sedimentation rate of approximately 1 mm/year appears to be most conducive for source rock development (Ibach, 1982; Calvert, 1987; Kelts, 1988; Bohacs et al., 2005).

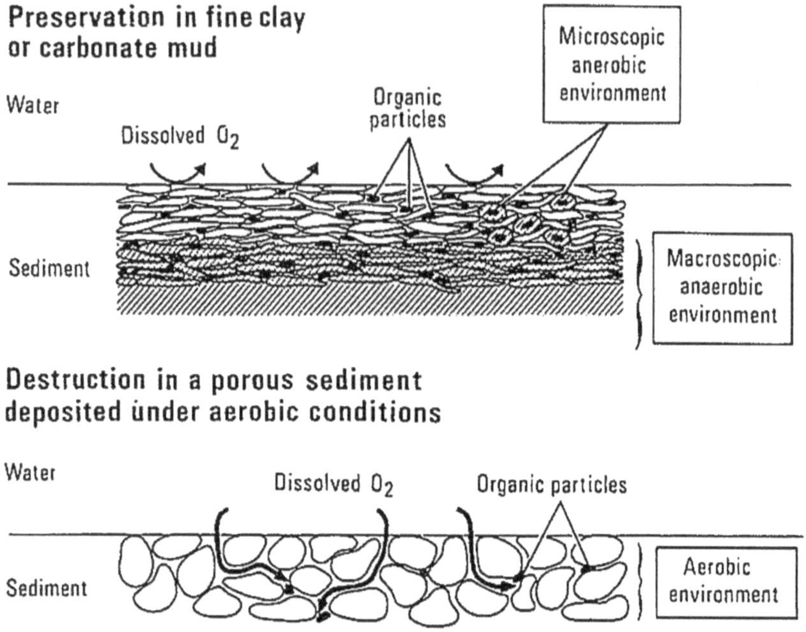

FIGURE 2.5 Comparison of preservation in fine-grained sediments versus coarse-grained sediments. *From Tissot, B.P., Welte, D.H., 1984. Petroleum Formation and Occurrence. New York, Springer-Verlag, 699 p.*

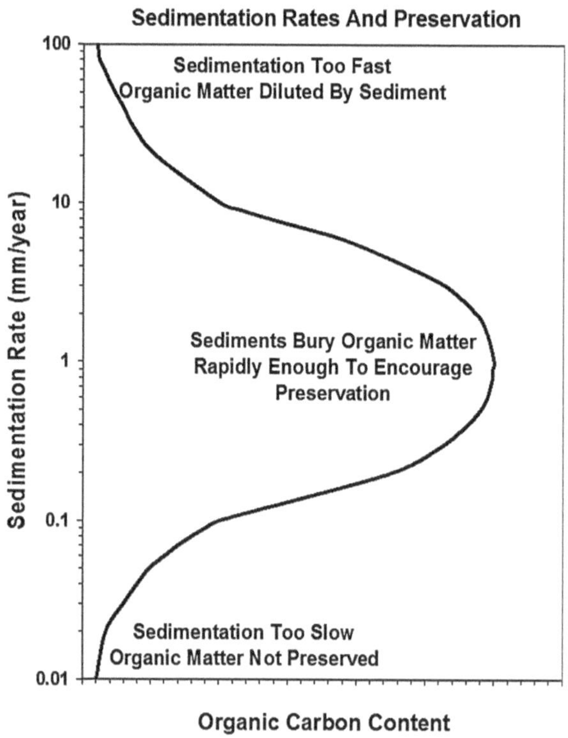

FIGURE 2.6 The influence of sedimentation rate on the preservation of organic matter. *Modified from Environments of deposition of lacustrine petroleum source rocks - an introduction. In: Fleet, A.J., Kelts, K., Talbot, M.R. (Eds.), Lacustrine Petroleum Source Rocks. Geological Society of London Special Publication 40, p. 3–26.*

KEROGEN FORMATION

The organic matter that initially goes into the sediment may have experienced some amount of alteration depending on the conditions in the water column and at the sediment–water interface. But once this organic matter is incorporated into the sediment, it begins a major transformation from biological organic matter into geological organic matter. Chemical processes such as hydrolysis, reduction, and oxidation as well as microbial activity begin to break down the large molecules and biopolymers into smaller organic compounds.

These compounds can then follow one of two main pathways, forming either solvent-soluble or solvent-insoluble sedimentary organic materials, as shown in Fig. 2.7. A small portion of the biological organic matter may go through additional alteration by reduction, dehydration, and decarboxylation to form an initial preserved bitumen, while the bulk of this organic material undergoes diagenetic condensation and polymerization to form kerogen (Welte, 1974; Tissot and Welte, 1984; Tegelaar et al., 1989). The bitumen is the solvent-soluble portion of the organic matter preserved in the sediments. Unlike petroleum, this bitumen is indigenous to the rock in which it is found, is formed early in the diagenetic process, and often contains some compounds that are indicative of the organisms that contributed organic matter to the sediment. In contrast, the kerogen is the solvent-insoluble portion of the organic matter preserved in sediments. It is a complex material with a variable composition and structure that may be converted into oil and gas under the proper subsurface conditions. The variability in the composition and structure of the kerogen is controlled by the type of organisms that contributed to the sediments and by how well the organic matter was preserved (Vandenbroucke and Largeau, 2007). As a result, not all kerogens are created equal. Some kerogens will be capable of generating oil while other kerogens may only be able to generate gas or, in some cases, nothing at all (inert kerogen). What controls whether a kerogen will be oil-prone or gas-prone is its hydrogen content and the type of structures (chemical moieties) it contains (Durand and Espitalie, 1973; Tissot et al., 1974). In order to be oil-prone, kerogen must be rich in hydrogen and contain the structures within the kerogen that can give rise to the large, complex molecules observed in crude oil. In contrast, gas-prone kerogen is less rich in hydrogen and only needs to contain small, simple structure to make the compounds found in natural gas.

As stated above, the nature of the kerogen is controlled by both the type of organisms that contributed organic matter to the sediments and by how well the organic matter was

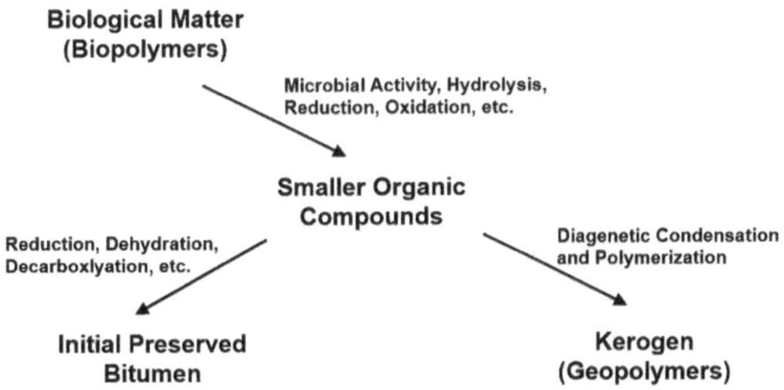

FIGURE 2.7 Schematic of the diagenetic conversion of biological organic matter to geological organic matter after the biological organic matter is incorporated into sediments.

FIGURE 2.8 The relationship between the type of organic matter incorporated into the sediment and preservation conditions in determining the type of hydrocarbon generating potential the resulting kerogen will have.

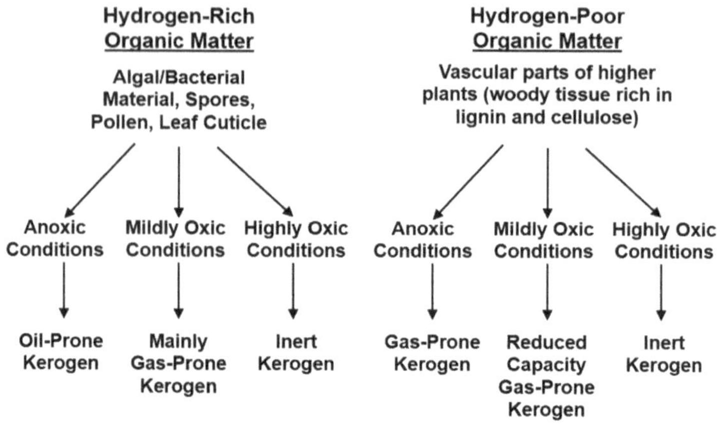

preserved. Biological organic matter rich in hydrogen includes hydrocarbons, waxes, fats, and lipids. This hydrogen-rich organic matter is generally thought to be the results of contributions from algae, bacteria, leaf cuticle, spores, and pollen to the sediments (Hunt, 1996). Hydrogen-poor organic matter comes from materials like cellulose and lignin that makes up the structural or woody tissue in the vascular parts of higher plants (Hunt, 1996). If the organic matter is deposited under anoxic conditions, the hydrogen content of the organic matter will be preserved as introduced into the sediments. But hydrogen content of the organic matter can also be reduced by the preservation conditions. If hydrogen-rich organic matter is poorly preserved under oxic conditions, the hydrogen content will be reduced (Demaison and Moore, 1980). This can result in organic matter that might have been oil-prone actually becoming gas-prone or even inert kerogen. Similarly, hydrogen-poor organic matter that could have become gas-prone kerogen can be degraded by poor preservation under oxic conditions resulting in a reduced gas generating capacity or an inert kerogen. This concept is summarized in Fig. 2.8.

From these observations, an initial classification of kerogen would seem to result in three types: oil-prone, gas-prone, and inert. However, it was recognized that oil-prone kerogens could be subdivided into kerogens that generate waxy oil and kerogens that generate naphthenic oils. This gave rise to the chemical classification system for kerogens used today. These basic hydrocarbon-generating kerogens are Type I, II, and III as described by Tissot et al. (1974). Type I kerogen has a high initial H/C and low initial O/C atomic ratios. It is derived primarily from algal material deposited mainly in lacustrine environments; however, it has also been recognized to form in brackish water and lagoonal environments as well (Revill et al., 1994). Type I kerogen produces mainly waxy oil. Type II kerogen has a moderately high H/C and moderate O/C atomic ratios in its initial state. It is derived from autochthonous organic matter deposited under reducing conditions in marine environments, but can also contain an allochthonous higher plant component (leaf cuticle, spores, pollen). In contrast to Type I kerogen, Type II produces mainly naphthenic oil. Type III kerogen has a low initial H/C and high initial O/C atomic ratios. It is derived from terrestrial higher plant debris and/or aquatic organic matter deposited in an oxidizing environment that produces mainly gas.

Additional insight into the nature of these kerogens came from the investigations of Behar and Vandenbroucke (1987) and resulted in the structural models of Types I, II, and III found in Fig. 2.9. The structure of Type I kerogen shows an abundance of long-chained aliphatic structures with few naphthenic ring (cyclic aliphatic structures) or aromatic structures. This would account for the generation of waxy oil from Type I kerogen. The structure also showed the cross-linking between chemical moieties within the kerogen to be dominated by —C—C— bonds. The classic example of a Type I kerogen in this classification is the Eocene Green River Shale. The Type II kerogen structure, in Fig. 2.9, shows more abundant naphthenic rings (cyclic aliphatic structures) than Type I kerogen. This would account for the generation of naphthenic oil from Type II kerogen. Cross-linking between chemical moieties contains more —C—O— bonds and fewer —C—C— bonds than found in Type I. The classic example of

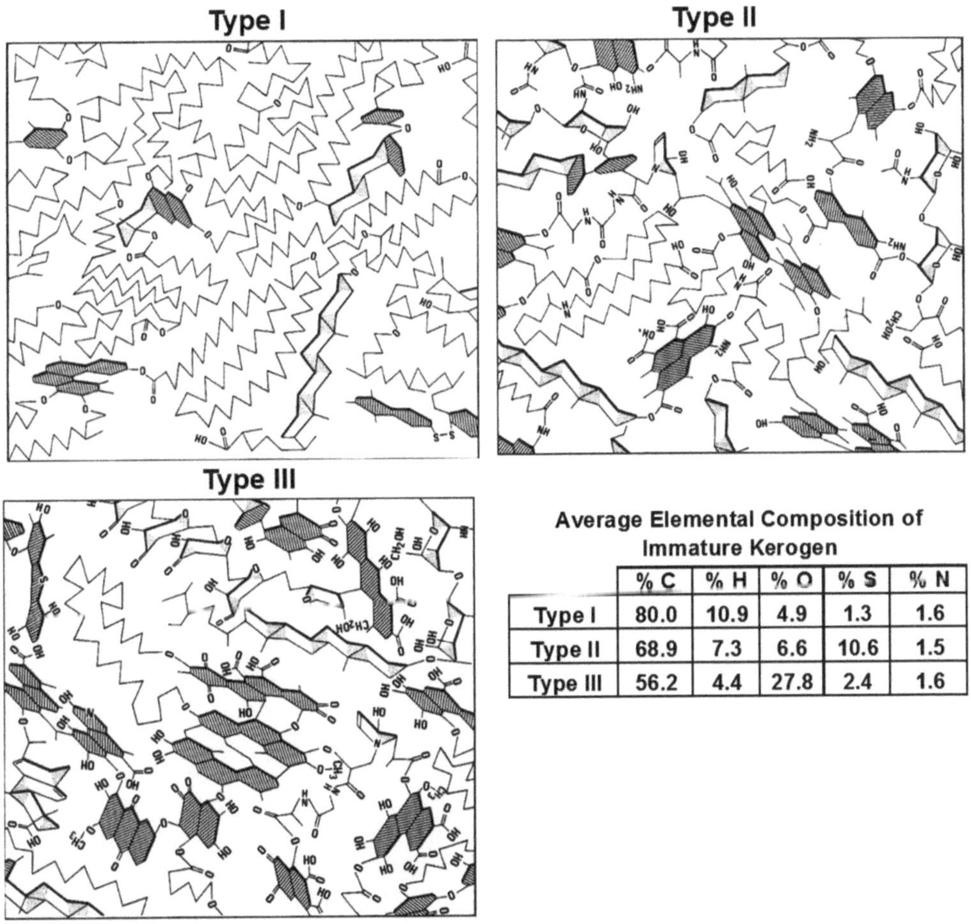

Average Elemental Composition of Immature Kerogen

	% C	% H	% O	% S	% N
Type I	80.0	10.9	4.9	1.3	1.6
Type II	68.9	7.3	6.6	10.6	1.5
Type III	56.2	4.4	27.8	2.4	1.6

FIGURE 2.9 Structural modeling for the three main kerogen types as elucidated by Behar and Vandenbroucke (1987). Chain aliphatic structures are shown as "saw-toothed" lines, naphthenic structures are shown as unshaded polygons, and aromatic structures are the shaded polygons. *O*, oxygen; *H*, hydrogen; *N*, nitrogen. Average elemental composition of the immature kerogens is also provided.

a Type II kerogen in this classification is the Toarcian Shale of the Paris Basin. The Type III structure shows more abundant aromatic and short-chained aliphatic structures than Type I or Type II. The lack of abundant long-chained aliphatic and large naphthenic ring structures limits the ability of Type III kerogen to generate oil. The cross-linking within Type III consists of both —C—O— and —C—C— bonds, but they are often associated with the aromatic rings, increasing their stability and making the cross-links more difficult to break. The classic example of a Type III kerogen in this classification is the Mahakam Delta humic kerogen. The structures and cross-linking present in the kerogen in Fig. 2.9 are important considerations in the hydrocarbon generation process as discussed later in this chapter.

While the three types of hydrocarbon-generating kerogen were adequate in most situations to describe the reactive organic matter in sediments, it was also necessary to include the nongenerating organic matter. This inert kerogen, when present, could dilute the reactive kerogen and therefore needed to be accounted for. As a result, Type IV kerogen was added (Tissot and Welte, 1984). Type IV kerogen has a very low initial H/C ratio and a variable initial O/C atomic ratio. It is a product of severe alteration and/or oxidation of organic matter in the depositional environment and is essentially inert with no hydrocarbon-generating potential.

As the study of kerogens continued, a variation of Type II kerogen was recognized. This form of Type II kerogen was initially recognized in the Miocene Monterey Shale of California and termed Type II-S. Type II-S kerogen has a high initial H/C and low initial O/C atomic ratios. It is derived from autochthonous organic matter deposited under highly reducing conditions in marine environments most often associated with upwelling conditions (Orr, 1986). The kerogen's sulfur content is high (8–14 wt%). Sulfur has substituted for oxygen in the cross-linking in the kerogen structure as a result of anaerobic microbial contributions (e.g., sulfur bacterial) to the organic matter (Williams, 1984). These —C—S— bonds are relatively weak as compared to —C—C— and —C—O— bonds resulting in earlier generation of high sulfur naphthenic oil (Orr, 1986).

Similar to Type II, organic sulfur-rich variants have been recognized for Type I and Type III kerogens. Type I-S has sulfur substituting for carbon in the cross-linking in the kerogen structure (Sinninghe Damste et al., 1993). The formation of Type I-S is linked to hypersaline conditions in the depositional environment (Sinninghe Damste et al., 1993; Carroll and Bohacs, 2001) and appears to be the result of diagenetic sulfurization using abundant sulfate (Sheng et al., 1987). Due to the abundance of —C—S— bonds, it is likely to generate earlier than conventional Type I (Carroll and Bohacs, 2001). Type III-S has been described from Tertiary age brown coals (Sinninghe Damste et al., 1992). The Type III-S kerogen has a high atomic S/C (0.04) and yields abundant sulfur compounds upon heating. While interesting, Type I-S and Type III-S kerogens are rare, occurring in limited quantity in only a few basins worldwide, and are not an important overall contributor to petroleum accumulations.

Conditions during the deposition of source rock are not homogeneous. Seasonal variation and climate cycles can have profound impacts on the type and the amount of organic matter that is introduced into sediment. As a result, very few source rocks contain only one type of kerogen. Most contain mixtures of two or more types. Typically, they are mixtures of primarily Type I and Type III or Type II and Type III. It is likely that there is also a small amount of Type IV kerogen present in most sediments. The presence of Type IV kerogen is usually discounted unless its concentration is high. What is important to remember is that mixed kerogens are common and their presence needs to be accounted for when describing source rocks and the hydrocarbons they might generate.

SOURCE ROCK DEPOSITION

From the discussion above, the optimum conditions for source rock deposition begins with high primary biological productivity in and around the depositional environment. This organic matter should be rich in hydrogen with major contributions from algal/bacterial material, spores, pollen, and leaf cuticle. Oxic/anoxic boundary in the depositional environment should be near or above the sediment–water interface to promote good organic matter preservation. And the sediments being deposited should be fine-grained sediments, such as very fine silt to clay or carbonate muds (<~4 μm), at intermediate sedimentation rates (~1 mm/year) that will bury and protect the organic matter without diluting it. So where can these conditions exist that might lead to source rock deposition?

In the marine environment, Demaison and Moore (1980) recognized a series of depositional settings where source rock might be deposited linked to organic matter preservation. Those ideas were further refined by Demaison et al. (1983) into two basic depositional settings: the ventilated open ocean on the shelf and slope; and silled basins including small basins, epicontinental basins; and lagoons. Observations made on the silled basin setting are also applicable to large lakes. In these depositional settings as shown in Fig. 2.10, it is essential that anoxic conditions occur in the water column above the sediment–water interface to insure organic matter preservation by limiting benthic scavenging and restricting microbial activity to anaerobic organisms. If oxic waters persist, the amount of organic matter preserved will be limited and the organic matter will have either a gas-prone nature (Type III kerogen) or no hydrocarbon generating potential (Type IV kerogen).

In the open ocean settings depicted in Fig. 2.10A, anoxia can be the result of the formation of oxygen-minimum layers. This can occur when high organic productivity in surface waters introduces large volumes of organic matter into the water column. Oxygen demand by organisms trying to consume this organic matter depletes the dissolved oxygen content of the water which can lead to anoxic conditions mainly in intermediate water depths (Dow, 1978). This depletion of oxygen in bottom waters may be exasperated if the bottom waters are not being replenished by oxygen-rich colder denser water derived from polar regions (Demaison and Moore, 1980). The development of oxygen-minimum layers are often seasonal events suggesting that development of source rocks in oxygen-minimum settings requires persistent annual reoccurrence of these conditions.

Upwelling is another mechanism for the development of oxygen-minimum zones in open ocean settings (Demaison and Moore, 1980; Parrish, 1982). Upwelling currents are often formed where the direction of prevailing winds is near perpendicular to coasts (Ziegler et al., 1979), such as observed today along the coastline of Peru. The winds impinging on the coast result in a net transport of surface water offshore. As the surface water moves away from the shore, it is replaced from below resulting in the upwelling. Upwelling can also occur in the open ocean at the convergence of two water masses and in response to seasonal monsoonal winds. Upwelling waters can be nutrient-rich and result in high organic productivity

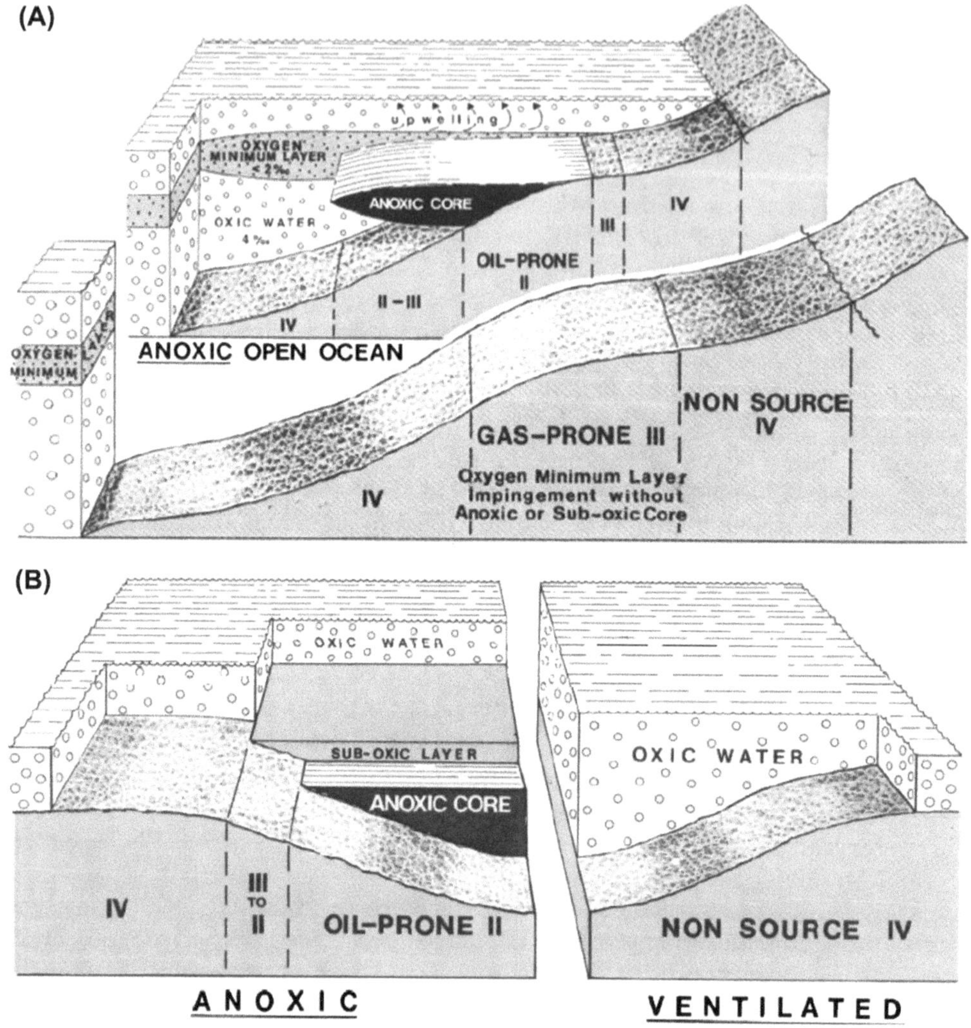

FIGURE 2.10 Potential depositional settings for source rock formation in the marine environment. The numbers II, III, and IV refer to the chemical kerogen type expected. (A) Ventilated open ocean. (B) Silled basin. *After Demaison, G.J., Hoick, A.J.J., Jones, R.W., Moore, G.T., 1983. Predictive source bed stratigraphy; a guide to regional petroleum occurrence. Proceedings of the 11th World Petroleum Congress, vol. 2. John Wiley & Sons, Ltd., London, p. 17.*

in surface waters. Large volumes of organic matter may be introduced into the water column which can deplete the dissolved oxygen content of the water and lead to anoxic conditions, as discussed above.

In silled basins settings (Fig. 2.10B), water column stratification can lead to the development of anoxic bottom water (Demaison et al., 1983) when there are near-horizontal density boundaries within the water masses due to stable temperature or salinity. Water balance is also important (Demaison and Moore, 1980). As shown in Fig. 2.11, a negative water balance leads to an oxic water column, usually in arid regions, when evaporation exceeds fluvial water input, such as the

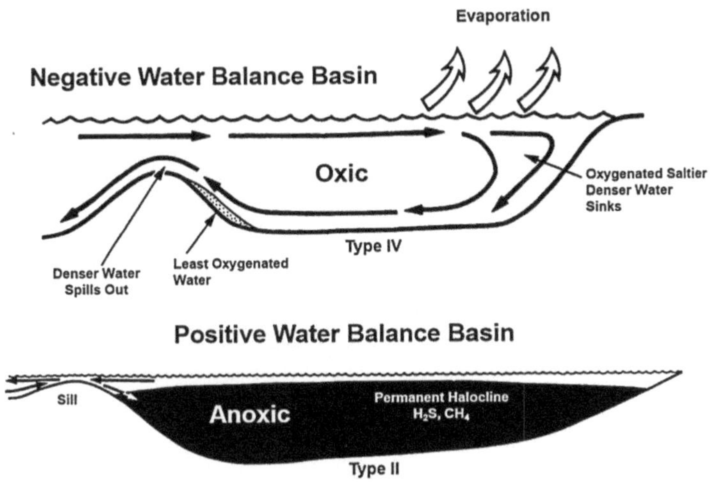

FIGURE 2.11 Schematic of negative and positive water balance in silled basins and their effect on developing oxic and anoxic conditions. *After Demaison, G.J., Moore, G.T., 1980. Anoxic environments and oil source bed genesis. American Association of Petroleum Geologists Bulletin 64, 1179–1209.*

Mediterranean Sea. The evaporation causes the surface waters to have higher salinity and higher density. This denser water eventually sinks to mix and oxygenate the water column. An anoxic silled basin, like the Black Seas, has a positive water balance (Fig. 2.11), where fluvial water input exceeds evaporation. These fresher, less dense fluvial waters remain in the surface layers and contributing to the formation of a stratified water column. Lack of mixing eventually leads to the development of anoxic bottom waters conducive to organic matter preservation. These same conditions can lead to anoxic bottoms waters in epicontinental seas and lagoonal settings.

Silled basins can also form on continental shelves and slopes, as observed in north central Gulf of Mexico. Salt movement there has resulted in the formation of numerous intraslope "mini-basin." If these basins are large and deep enough with bathymetric conditions around the margins restricting or preventing water circulation into the basin, anoxic conditions can develop (Williams and Lerche, 1987).

Large lakes are analogous to silled basins (Demaison et al., 1983). In low latitudes, large lakes are prone to developing density stratification leading to potential anoxic conditions (Demaison and Moore, 1980), such as those postulated for the Eocene age Lake Gosiute and Lake Uinta that gave rise to the Green River Formation. In contrast to marine settings, the kerogen type expected to develop in anoxic lakes is Type I or a mixture of Type I and III. In mid-latitudes, seasonal changes in climatic conditions leading to the overturning of water columns in lakes (Swain, 1970). Cold oxygenated waters from fluvial input also enhance the oxic conditions in these lakes.

In deltaic settings as shown in Fig. 2.12, the proximal depositional facies are likely to receive organic input mainly from fluvial transported organic matter and coastal swamps that are dominated by terrestrial higher plant material (Barker, 1982). In the onshore portion of the delta, this organic matter input may be substantial and result in coal deposits. In the near shore (proximal) facies, this terrestrial higher plant material will likely to be subjected to oxidizing conditions and produce Type III/IV kerogens (Kosters et al., 2000). In the more distal delta depositional setting, fluvial discharge from the delta brings nutrients to the marine waters encouraging phytoplankton growth so that the organic input may then be dominated by marine phytoplankton and algal material. The higher organic matter productivity in the

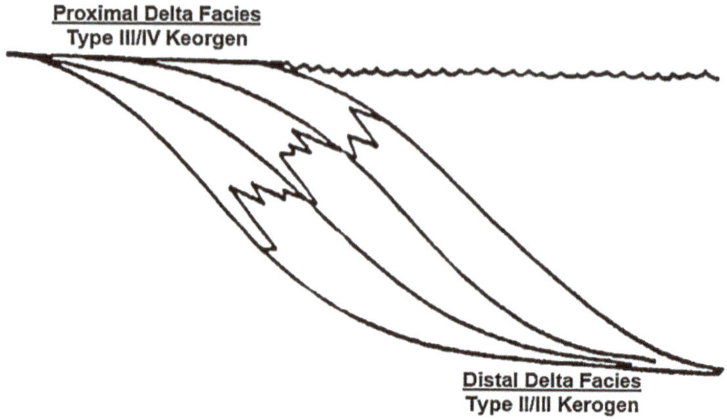

FIGURE 2.12 Schematic view of the distribution of kerogen types in a prograding delta depositional setting. *Modified after Barker, C., 1982. Oil and gas on passive continental margins. In: Watkins, J.S., Drake, C.L. (Eds.), Studies in Continental Margin Geology. American Association of Petroleum Geologists Memoir, vol. 34, p. 549–565.*

water column may help promote more reducing conditions (anoxia), similar to upwelling, and Type II/III kerogens will be produced (Barker, 1982; Kosters et al., 2000).

Lacustrine deposition in rift basins may also be conducive to source rock deposition. In rift basin, a series of narrow half-grabens develop along the rift axis separated by accommodation zones (Younes and McClay, 2002). The half-graben subbasins have opposite dip directions occur in alternating half-grabens. Lake development in the half-grabens is variable responding to differences in geometry of the subbasins. In addition, multiple sediment inputs are possible in each half-graben from both the flexural margin and the rift shoulder. Source rock quality is linked to the depositional setting within the rift basin, as shown in Fig. 2.13. Deep lacustrine sediments have the potential for containing oil-prone Type I kerogen (Katz, 1990; Lin et al., 2001), while shallow lacustrine sediments are less anoxic and usually contain both oil-prone Type I and gas-prone Type III kerogens (Carroll and Bohacs, 2001). Shallow fluvial and alluvial sediments are deposited under mainly oxic conditions and the preserved organic matter in these sediments, including coals, contains gas-prone mostly Type III kerogen (Carroll and Bohacs, 2001; Lin et al., 2001). Source rock quality and distribution can be highly variable from one half-graben to another. The presence of good source rocks in one half-graben does not assure that source rocks will be present in adjacent half-grabens (Morley, 1999).

In addition to source rock development associated with the lacustrine phase in rift basins, early marine incursions into rifts can be favorable for source rock deposition, such as observed during the opening of the South Atlantic (Heilbron et al., 2000). The narrow rift basin geometry can result in restricted water circulation in the early phases of the marine incursion that leads to the development of anoxic conditions. If high enough primary organic productivity exists, source rocks may be deposited.

To summarize, depositional settings where source rocks have been observed to occur include: open marine anoxic settings associated with areas of upwelling or where oxygen-minimum zones form; areas where stratified water columns develop, such as deep areas in epicontinental seas, silled basins, high salinity lagoons, and anoxic lakes; distal parts of deltas; rift-associated deepwater lacustrine settings; and early marine incursion into rift basins. Although these depositional settings may give rise to anoxic conditions and oil-prone source rock deposition, these basin geometries do not automatically imply the presence of oil-prone source beds. For example, in the present oceans, we find not all oxygen minimum and

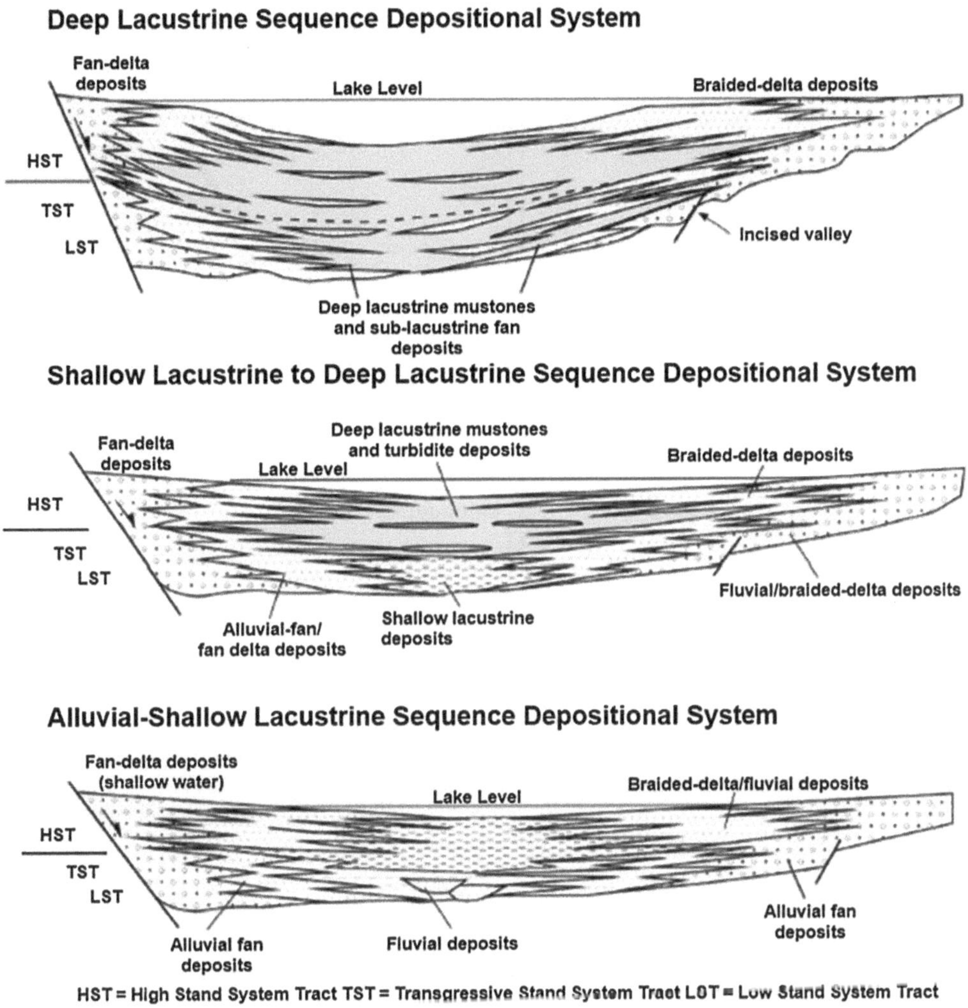

FIGURE 2.13 Source rock quality and distribution in various lacustrine settings. *After Lin, C., Eriksson, K.A., Li, S., Wang, Y., Ren, J., Zhang, Y., 2001. Sequence architecture, depositional systems, and controls on development of lacustrine basin fills in part of the Erlian Basin, northeast China. American Association of Petroleum Geologists Bulletin 85 (11), 2017–2043.*

upwelling zones produce anoxic environments. There is still a need to have adequate primary organic matter production of the appropriate type feeding into these depositional systems, as well as anoxic condition, in order for source rock deposition to occur.

MATURATION AND HYDROCARBON GENERATION

Once organic matter is incorporated into sediment and kerogen is formed, the next step in developing a petroleum accumulation is the conversion of the organic matter into the hydrocarbons and other compounds that make up petroleum. This conversion is a by-product of

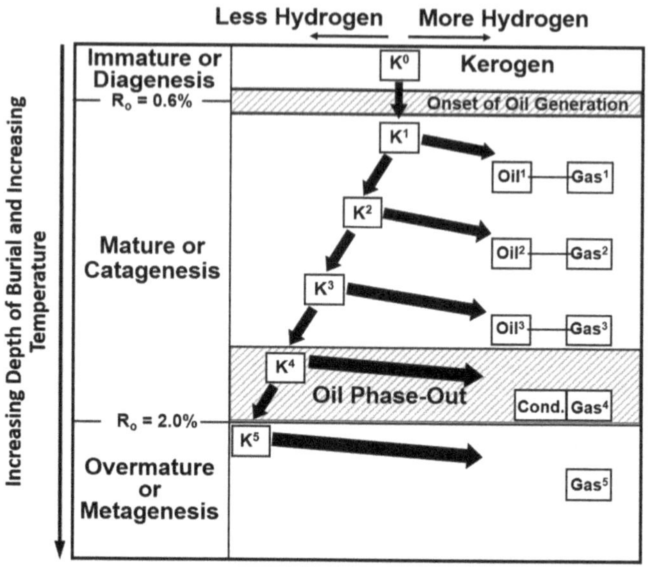

FIGURE 2.14 A simplified representation of the dynamic changes experienced by sedimentary organic matter as a result of progressive thermal maturation from immature to overmature. *Ro*, percent vitrinite reflectance. *Modified after Horsfield B., Rullkotter, J., 1994. Diagenesis, catagenesis, and metagenesis of organic matter. In: Magoon, L.B., Dow, W.G. (Eds.), The Petroleum System–From Source to Trap, vol. 60. American Association of Petroleum Geologists Memoir, p. 189–199.*

the maturation of the sediment's organic matter. Maturation, sometimes called thermal maturation, is the process of chemical changes in sedimentary organic matter under the influence of increasing temperature over geologic time due to burial. Fig. 2.14 (modified after Horsfield and Rullkotter, 1994) provides an overview of these processes showing the coevolution of kerogen, oil, and gas during maturation. Starting in the immature (diagenesis) stage, the kerogen is in its original state. Any gas or bitumen present is preserved from the depositional environment. As the kerogen enters into the mature (catagenesis) stage, the onset of hydrocarbon generation has been surpassed. The kerogen's composition starts to change as it begins to produce oil (bitumen) and gas. As the hydrocarbon generation continues, the kerogen composition continues to change as more oil (bitumen) and gas are generated. At the same time the oil (bitumen) and gas being formed are also changing composition, reflecting the changes in the parent kerogen. Eventually, the oil (bitumen) generating capacity of the kerogen becomes depleted during the oil phase-out stage. Instead, gas and a light hydrocarbons condensate phase are being formed. Finally, as the sediments reach the overmature (metagenesis) stage, only limited gas can be generated until the hydrogen available in the kerogen becomes exhausted.

It is possible to demonstrate the dynamic nature of the maturation process by examining a few examples of the series of changes observed in the kerogen, gas, and oil/bitumen. Looking at the kerogen, changes in the both the structure and chemical composition can be detected. In Fig. 2.15, the structural changes in Type II kerogen from immature to overmature states have been modeled by Behar and Vandenbroucke (1987) based on observations from field examples. The immature state (Fig. 2.15A) has a more random orientation of the chemical moieties and more hydrogen-rich organic matter as evidenced by the abundance of aliphatic structures. As maturity increases (Fig. 2.15B), there is an increase in aromatic structures, reflecting a loss of hydrogen, and there are indications that these aromatic structures are beginning to take on preferred orientations. By the overmature stage (Fig. 2.15C), nearly all the organic material is in aromatic structures and their orientations in preferred alinements has become obvious. At this point, the kerogen is beginning to transition into a graphite-like material. Based on observations of natural maturity series, the other kerogen types follow a similar evolution to the one described here for Type II.

Paralleling these structural changes are chemical changes, for example, the elemental composition of the kerogen. As mentioned above, the hydrogen content of the kerogen will change with increasing thermal maturity. But, in fact, the entire elemental composition of the kerogen changes. This can be readily observed in the van Krevelen diagram, shown in Fig. 2.16. The van Krevelen diagram plots the H/C atomic ratio of kerogen versus the O/C atomic ratio. This diagram was first developed for tracking the chemical evolution of coal macerals with increasing rank (van Krevelen, 1961) and adapted for kerogens (Tissot et al., 1974). In the van Krevelen diagram (Fig. 2.16), the pathways for Type I, II, and III begin separately while the kerogens are in the diagenesis/immature zone. The broad bands for the composition of each kerogen type reflects variation in composition mostly likely due to mixtures of kerogen types present in the samples used, as well as compositional variation in the actual end-member kerogens. As the kerogens evolve with thermal maturity, they enter the principal zone of oil formation. The Type I and II kerogen trends begin to merge as this stage progresses. These basic trends document that hydrogen and oxygen are being depleted in all three kerogens, while the carbon content is being enriched. Finally, as all three kerogen types enter the zone of gas formation, their elemental compositions become indistinguishable from each other.

While the kerogen is evolving during maturation, the gas being generated also goes through changes in composition (Schoell, 1983). As shown schematically in Fig. 2.17, the initial gas present in the sediments is biogenic methane (C_1) without any wet gas (C_2–C_4) components. As the kerogen begins to generate hydrocarbons, wet gas compounds begin to form and reach a maximum concentration at about the same point as peak oil generation. As the thermal maturity continues to increase, the wet gas content begins to decline. This is initially because more methane and less wet gas components are being generated, but eventually the wet gas components are also being destroyed by cracking. At the final stage of the evolution of the gas, only methane will remain.

As the composition of the gas changes with maturation, there are also changes in the isotopic signature of the methane (Schoell, 1983). Initially, the biogenic methane is relatively depleted in the ^{13}C isotope because biological processes prefer the ^{12}C isotope over the ^{13}C isotope. As hydrocarbons begin to be generated, the ^{13}C content of the methane increases slowly at first. This is because during the initial cracking of the kerogen to form methane, it is easier to break the ^{12}C carbon bonds than the ^{13}C bonds. But as the relative amount of ^{13}C in the kerogen increases due to preferential loss of ^{12}C, the ^{13}C content of the methane steadily increases. The ^{13}C content of the methane will continue to increase throughout the maturation/generation history of the kerogen. Details about these compositional and isotopic changes in the generated gas will be discussed in Chapter 4 in the section on interpreting natural gas data.

Compositional changes with increasing thermal maturity are similarly observed in the generated oil/bitumen. An example of this is shown in the gas chromatograms of saturated hydrocarbons, in Fig. 2.18, from three sediments with the same type of organic matter but at different maturities. Chromatogram A in Fig. 2.18 represents the immature stage for this type of organic matter. The hydrocarbon compounds are primarily in the C_{16}–C_{22} range. There is

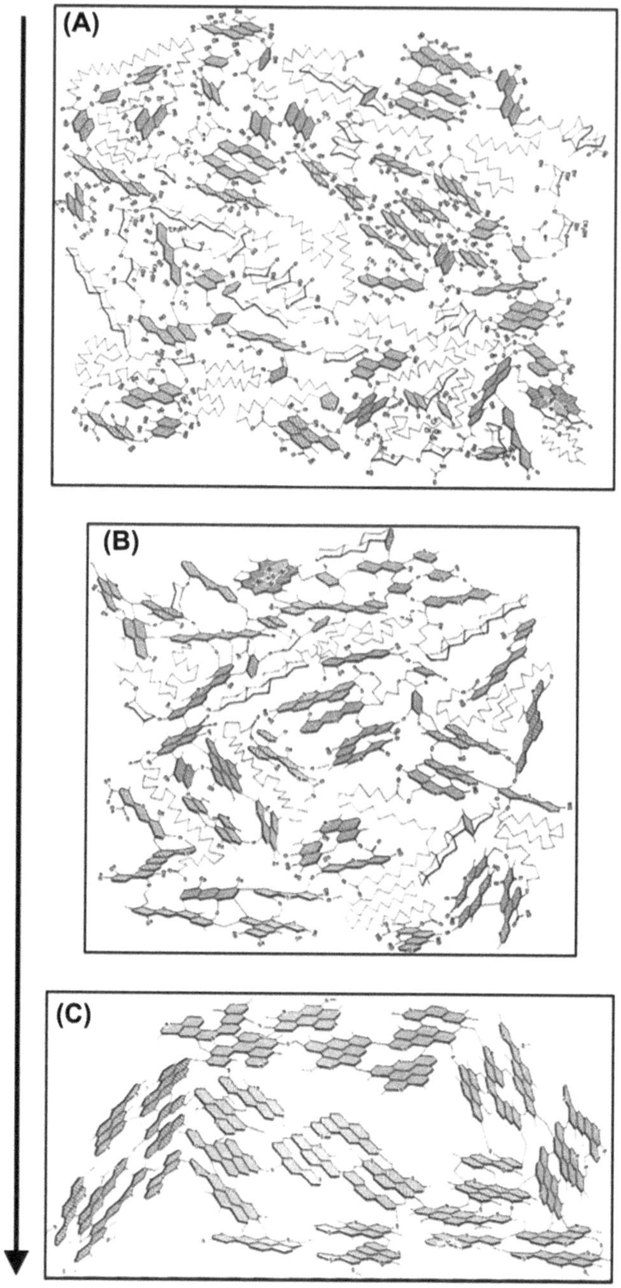

FIGURE 2.15 Model of the evolution of the structure of Type II kerogen with increasing maturity *(Adapted from Behar, F., Vandenbroucke, M., 1987. Chemical modelling of kerogens. Organic Geochemistry 11, 15–24.)*. Chain aliphatic structures are shown as *"saw-toothed" lines*, naphthenic structures are shown as *unshaded polygons*, and aromatic structures are the *shaded polygons*. (A) More random, more hydrogen. (B) Increasing maturity. (C) More ordered, less hydrogen and oxygen.

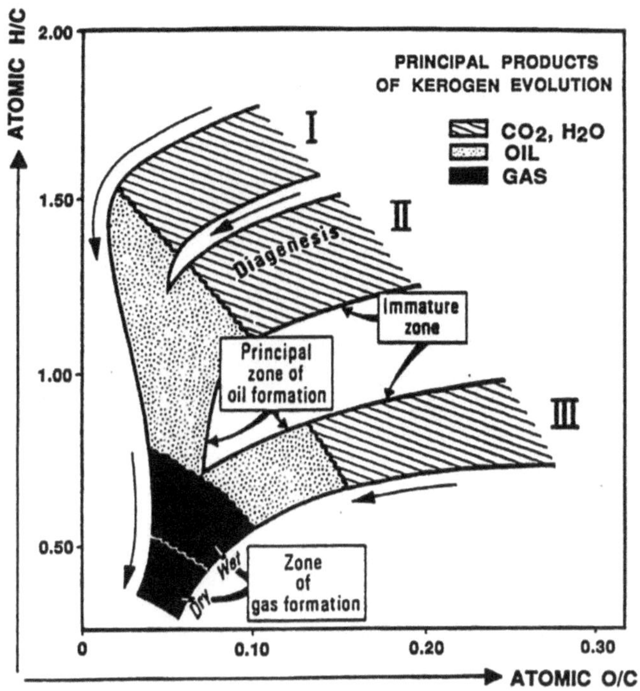

FIGURE 2.16 The van Krevelen Diagram plotting atomic H/C ratios versus atomic O/C ratios adapted for kerogen. *From Tissot, B.P., Welte, D.H., 1984. Petroleum Formation and Occurrence. New York, Springer-Verlag, 699 p.*

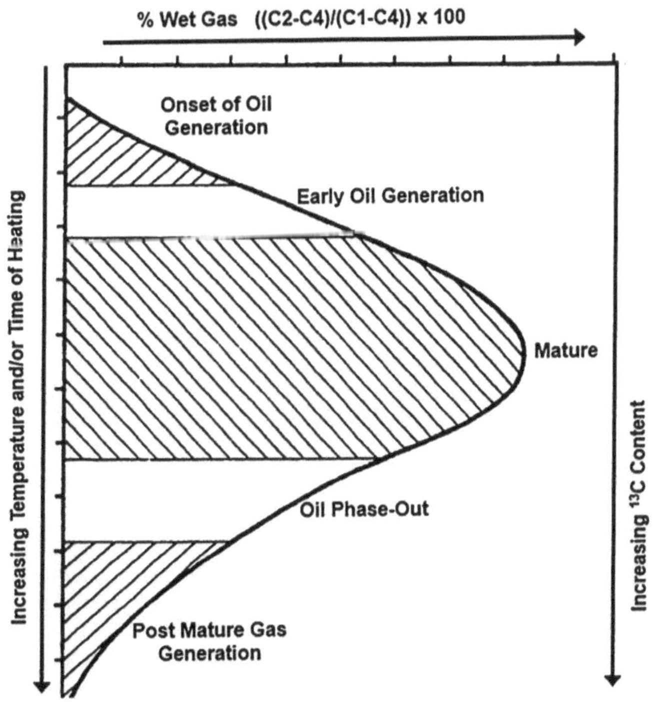

FIGURE 2.17 A schematic representation of changes in wet gas content and methane carbon isotope ratios with increasing maturity.

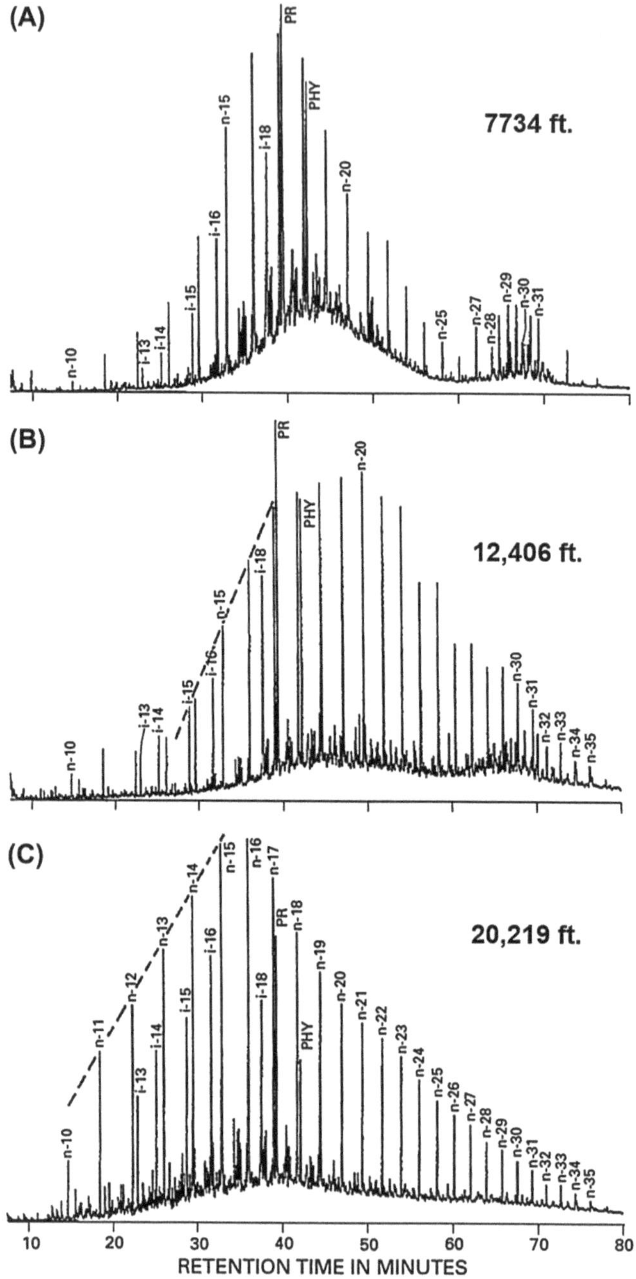

FIGURE 2.18 Saturated hydrocarbon gas chromatograms for three sediments containing the same organic matter type from the Los Angeles–Ventura Basin showing changes in the characteristics of the hydrocarbons with increasing maturity; n-paraffins are labeled with n- and their respective carbon number; isoprenoid hydrocarbons are labeled with i- and their respective carbon number; pristane and phytane are labeled PR and PHY, respectively. (A) Immature. (B) Moderately mature. (C) Mature. *Chromatograms taken from Price, L.C., 2000. Organic metamorphism in the California petroleum basins; Chapter B, insights from extractable bitumen and saturated hydrocarbons. U.S. Geological Survey, Bulletin 2174-B, 33.*

a distinct "hump" of unresolved material, possibly naphthenic hydrocarbons, below these hydrocarbon peaks and a secondary smaller "hump" of unresolved material in the C_{28}–C_{32} range. And there is an odd carbon number predominance in the C_{25}–C_{35} n-paraffins. These features are characteristic of preserved bitumen associated with immature organic matter in sediments. In the moderately mature sample in Fig. 2.18B, there is a noticeable reduction in both the "humps" of unresolved material and the odd carbon number predominance in the C_{25}–C_{35} n-paraffins is nearly gone. There is also an increase in the relative amount C_{15}–C_{18} isoprenoids, as well as n-paraffins in the C_{20}–C_{30} ranges. These changes reflect the addition of hydrocarbons generated from the kerogen to the original bitumen in the sediments indicating the sediments have reached the onset of significant hydrocarbon generation. In the chromatogram of the mature sample (Fig. 2.18C), the "hump" of unresolved material in the C_{28}–C_{32} range and the odd carbon number predominance in the C_{25}–C_{35} n-paraffins have been eliminated. There is a substantial increase in the n-paraffins in the C_{15}–C_{18} range relative to the isoprenoids (compared the dashed line in chromatograms B and C). And, the general peak envelope in the chromatogram has become smoother and more crude oil-like. These characteristics indicate further contributions of generated hydrocarbons to the bitumen with increasing maturity and suggest the sediments are approaching the peak of oil generation. In addition to changes in the hydrocarbons being generated by the kerogen, eventually the oil/bitumen would start to crack and begin to alter its own composition.

The above examples of changes in the kerogen, gas, and bitumen in sediments during thermal maturity are but a few of those that have been documented. It is fortunate that many of these compositional changes in sedimentary organic occur in a systematic fashion during the maturation process. Because of this, they can be used to monitor the progress of the maturation process. These so-called maturation indicators are important for understanding the thermal history of the sediments and how it has influenced changes in the sediment's organic matter and hydrocarbon generation. A more detailed discussion of maturation indicators can be found in Chapter 3.

As mentioned above, some of the chemical changes to the sediment's organic matter during maturation are responsible for the formation of oil and gas. In simple terms, hydrocarbon generation is the alteration of the kerogen in a sediment under the influence of time and temperature to form gas, oil, and a carbon-rich residue (char), as shown schematically in Fig. 2.19. The generated oil can subsequently decompose to form more gas and additional carbon rich residue. Although more complex models for the generation of oil and gas exist, this simple model is sufficient for this discussion (more complex models will be discussed in Chapter 8).

This simple model for the formation of oil and gas is the basis of one of the fundamental concepts in petroleum geochemistry, the oil window. The idea of an oil window started with early observations by a number of researchers (e.g., Louis, 1964; Philippi, 1965) recognizing an exponential increase in hydrocarbon formation relative to the total organic carbon content in shales as a function of depth. Later observations also showed a decrease in liquid hydrocarbon content with a corresponding increase in hydrocarbons gases (e.g., Le Tran, 1972) in deeper settings. These observations formed the basis of the oil window concept. As shown in Fig. 2.20, the oil window is not the product of a single process. Instead, it is the cumulative result of several processes acting concurrently. Kerogen is converted to oil and gas and subsequently oil is converted to gas, as shown in the simple model in Fig. 2.19. Although not yet observed in nature, there is also the thermodynamically theoretical potential for gas to be destroyed (Barker and Takach, 1992). The so-called oil window is therefore the summed total of all these

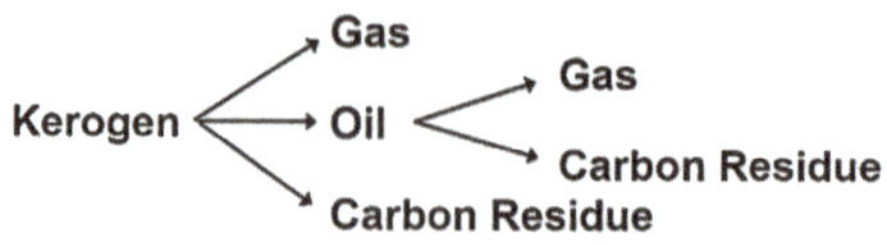

FIGURE 2.19 The simple model for kerogen generation of oil and gas.

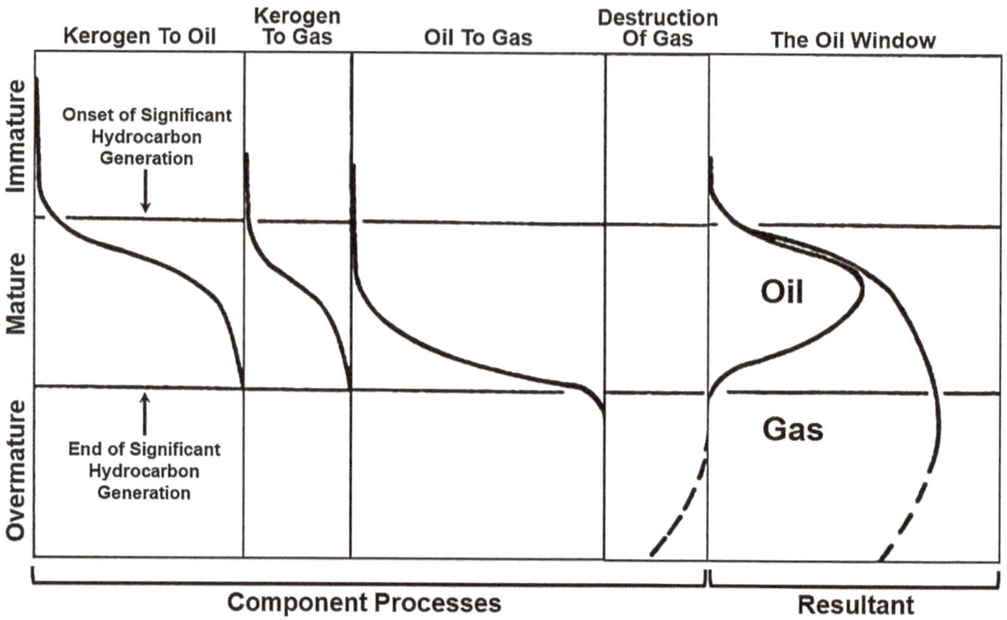

FIGURE 2.20 The oil window concept.

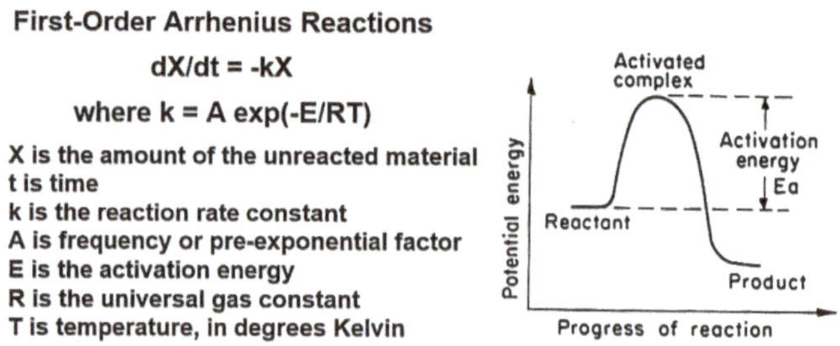

FIGURE 2.21 A summary of first-order Arrhenius reaction kinetics.

processes. Where a source rock is within the oil window is an important consideration when discussing its hydrocarbon generating potential, as will be seen in the following chapter.

The reactions that produce oil and gas have been observed in nature to approximately follow first-order Arrhenius kinetics (Tissot, 1969), as defined in Fig. 2.21. In first-order reactions, it is assumed that the reaction is irreversible. If the amount of the reactive material in Fig. 2.21 is X and time is t, the change in the concentration of the reactant over time, dX/dt, is governed by the reaction rate constant, k. The reaction rate constant is defined as the product of the frequency factor, A, and the exponential function raised to the power of −E/RT. The frequency factor is a statistical estimate of how often the reaction can take place. The activation energy, E, is the amount of energy required to overcome the free energy barrier in order for the reaction to occur, R is the gas constant, and T is the temperature. The progress of chemical reactions governed by first-order Arrhenius kinetics is monitored by tracking the consumption of the reactant, in this case kerogen.

The conversion of kerogen to oil and gas is, however, more complex than the simple generation model shown in Fig. 2.19. While the overall process still approximates first-order Arrhenius kinetics, it is best described as the cumulative result of a series of parallel reactions, as portrayed in Fig. 2.22. Each of the parallel reactions roughly represents the breaking of a different type of chemical bond in the kerogen structure. Initially, the bonds are the cross-links in the kerogen connecting remnant structures, or moieties, preserved from both the altered and unaltered biological input into the sediment. As the processes progress, the bonds that break may be internal or peripheral to these structures and may break the moieties into small fragments.

In a geologic context, the two variables in the equation having the most influence on the reactions are time and temperature. By comparing the progress of the kerogen conversion reaction from a linear increase in time versus a linear increase in temperature, as shown in Fig. 2.23, it is possible to see how the relative impact these two factors have on the process. While the reactions progresses nearly linearly with time, the reactions progresses exponentially with temperature. It is clear that temperature exerts a much larger influence over the reactions than time. This is very similar to the impact of time and temperature on the coalification process observed by Lopatin (1983).

To better understand the hydrocarbon generation of different kerogen types, the kinetic input parameters for each type need to be compared. Fortunately, these parameters can be estimated experimentally. To determine what the main activation energies are for the series of parallel reactions for a given kerogen, methods have been developed to artificially mature kerogens in a series of laboratory heating experiments, as described by Tissot et al. (1987). The kerogens are heated using at least three different heating rates and the amount of hydrocarbons that are generated are recorded with respect to the heating temperature. These data are then used in an iterative statistical procedure to best fit the results to a series of activation energies, corresponding frequency factors, and the amount of the reactive kerogen assigned to each activation energy (e.g., Ungerer and Pelet, 1987; or Braun and Burnham, 1994).

Numerous investigations of the kinetic parameters for hydrocarbon generation for the major chemical kerogen types described above have been carried out over the years. These include the work of Tissot et al. (1987), Braun et al. (1991), Behar et al. (1992), Tegelaar and Noble (1994), Pepper and Corvi (1995), Behar et al. (1997), and many others. To summarize these works, representative examples of the kinetic parameters of the major kerogen types from some of these studies are shown in Fig. 2.24. These kinetic parameters represent ideal end-members of the kerogen types and do not reflect the kerogen mixtures that are encountered in most source rocks. For the parameters shown, Type I kerogen kinetics were adapted from the work done at Lawrence Livermore National Laboratories (e.g., Burnham and Braun, 1985, 1999; Braun et al., 1991), while the Type II and III kerogen kinetics were adapted from studies by the Institut Francais du Petrole (e.g., Tissot et al., 1987; Behar et al., 1992, 1997). The Type II-S kerogen kinetics were adapted from a number of sources (e.g., Tissot et al., 1987; Behar et al., 1997; Jarvie and Lundell, 2001; Lehne and Dieckmann, 2007) as well as the author's own data.

Reaction Kinetics

For heterogeneous materials like kerogen,

X → Y

may actually represent a series of parallel reactions

$X_1 \rightarrow Y_1$
$X_2 \rightarrow Y_2$
$X_3 \rightarrow Y_3$
.
.
$X_n \rightarrow Y_n$

For each parallel reaction,

$X_i \rightarrow Y_i$

A, the frequency factor, is the same, but the activation energy, E_i, is different.

For the reaction of the ith component at a time-dependent temperature T(t)

$dX_i/dt = -X_i A \exp[-E_i/RT(t)]$

and

$dX/dt = \Sigma\, dX_i/dt$

FIGURE 2.22 Modifications to first-order Arrhenius reaction kinetics to accommodate a series of parallel reactions.

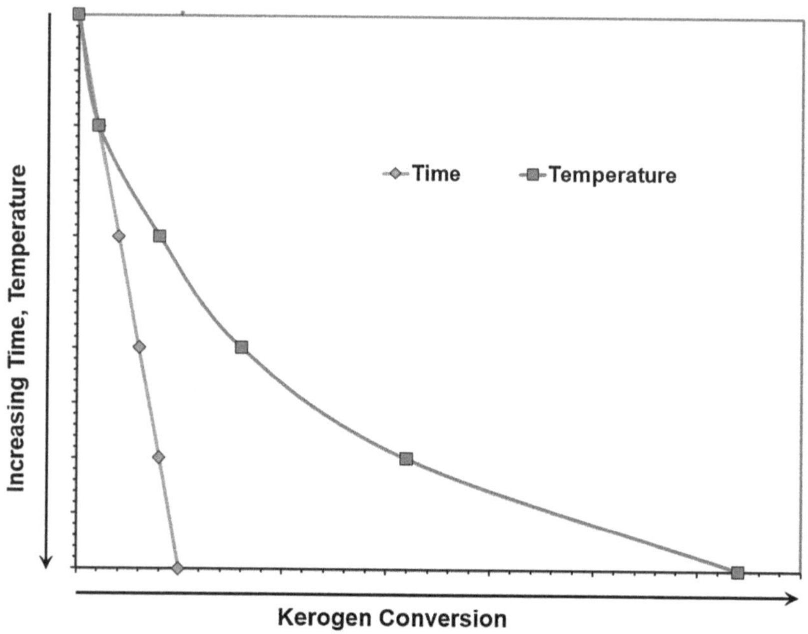

FIGURE 2.23 Comparing the progress of the kerogen conversion reaction from a linear increase in time versus a linear increase in temperature.

A review of the kinetic parameters in Fig. 2.24 is very revealing. The distributions of activation energies indicate that two of the kerogen types, Types I and II-S, are dominated by a single activation energy. The Type II-S kerogen is dominated by a single reaction at about 50 kcal/mol, likely reflecting the abundance of —C—S—C— cross-linking in the kerogen structure. As a result, these kinetic parameters indicate that Type II-S kerogen should begin to generate significant amounts of hydrocarbons at lower temperatures than the other kerogens. Type I kerogen is also dominated by a single activation energy, at 53 kcal/mol. This likely reflects the preponderance of —C—C—C— cross-linking in the kerogen structure and indicates more energy (higher temperature) is needed to break these bonds. In contrast, the Type II and III kerogens have a more "Gaussian-like" distribution of activation energies. This suggests more heterogeneity in the cross-link bond types. Type II kerogen has a mix of —C—S—C—,

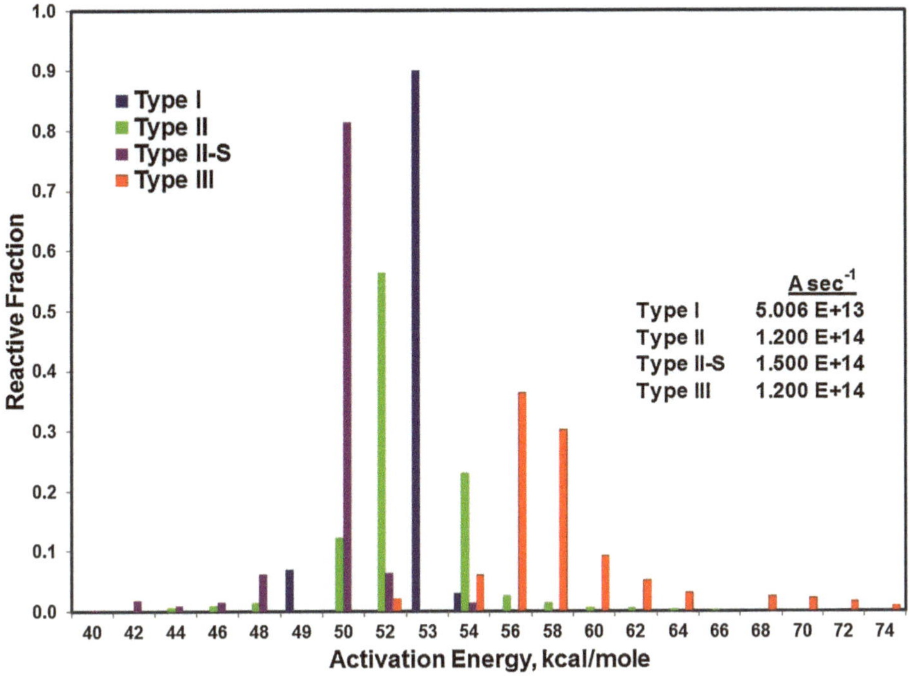

FIGURE 2.24 Representative examples of the kinetic parameters for the four major hydrocarbon generating kerogen types.

—C—O—C—, and —C—C—C— cross-links, with —C—O—C— being more abundant. The distribution of activation energies for Type III kerogen is shifted toward much high energies, which likely reflects the more stable aromatic nature of the kerogen structure for Type III.

Because the frequency factor, A, is a potential compensating factor for differences in activation energy, direct comparison of these distributions should be done with caution. It is more appropriate to compare kinetic parameters based on the results of hydrocarbon generation simulations. To this end, the kinetic parameters for all of the four major chemical kerogen types were used in simulations of hydrocarbon generation using the same constant heating rate (more details of simulating hydrocarbon generation in Chapter 8). The results of these simulations were then used to calculate a transformation ratio for each of the kerogen types. The transformation ratio is defined as the amount of reactive kerogen that has been converted divided by the total amount of reactive kerogen initially available expressed as percentage. When no kerogen has been converted, the transformation ratio is 0, and when all the kerogen has been converted, the transformation ratio is 100. Looking at the calculated transformation ratios for each kerogen type plotted versus temperature, in Fig. 2.25, the relative order of hydrocarbons generation using the kinetic parameters, given in Fig. 2.24, confirms the observations made about bond types. Type II-S generates earlier than the other kerogen types most likely due to dominated by weaker —C—S—C— bonds. Next in the sequence is the Type II kerogen, with its mixed bond types. Third is Type I kerogen, generating later than Type II and rapidly goes to completion most likely due to the predominance of —C—C— bonds. And finally, Type III

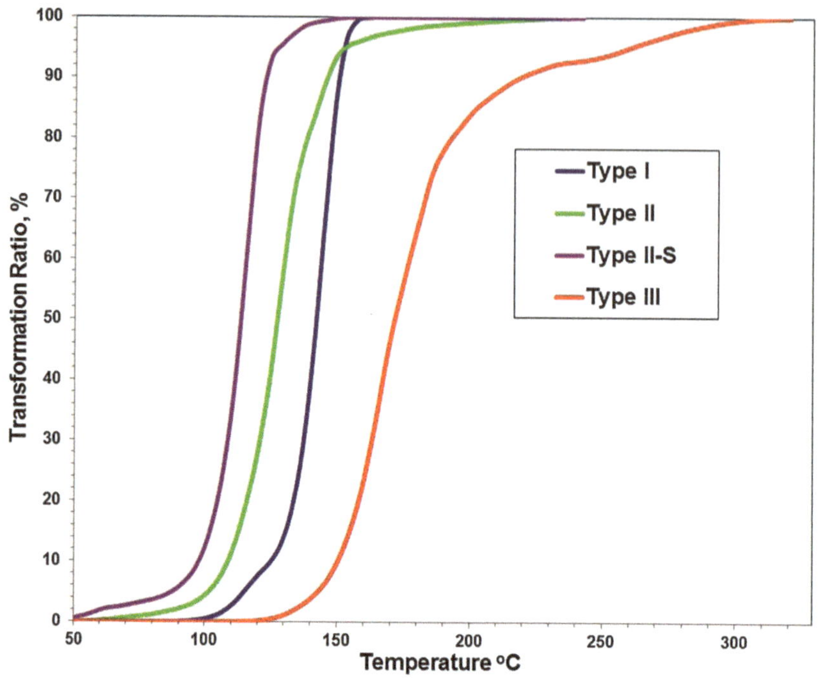

FIGURE 2.25 Results of simulations of hydrocarbon generation for the four major hydrocarbon-generating kerogen types using the kinetic parameters given in Fig. 2.24. All four simulations were done using the same constant heating rate.

kerogen is delayed even more, most likely due to the presence of more stable aromatic structures. Most important, these results mimic what is observed in the subsurface with Type II-S generating early, then Type II followed by Type I, and lastly Type III (Dembicki, 2009).

In addition to delineating the initial stage of the hydrocarbon generation process, kerogen breaking down into oil and gas, first-order Arrhenius reaction kinetics can also be used to describe the second stage of the process, the cracking of oil to gas. As with the formation of oil, there is a hydrogen deficit in the cracking of oil to gas which results in excess carbon left behind as a residue or char. Oil composition, like kerogen composition, should have an influence on the process. And like the breakdown of kerogen to form oil and gas, the cracking of oil to gas is much more complex than our models indicate (Behar et al., 2008). But because the formation of gas through oil cracking is thought to be controlled primarily by the breaking of mainly —C—C— bonds, it is often simulated as a single reaction.

From the discussion of kerogen type in the previous section, some kerogens (Types I, II, and II-S) are oil generators, some kerogens are gas generators (Type III), and some kerogens are essentially inert (Type IV). Does this mean that only oil-prone source rock produces oil accumulations and only gas-prone source rock is responsible for gas accumulations? While oil-prone source rocks are necessary for forming an oil accumulation, gas accumulations can come from either oil-prone or gas-prone source rocks. Although gas-prone source rocks can generate gas once they reach adequate maturity, they may not be the major source of gas in the subsurface. After oil generation is complete, the kerogen in an oil-prone source rock still has a significant capacity for generating hydrocarbon gas. In addition, between 20% and 30% of the oil generated in an oil-prone source rock may be retained in the rock adsorbed on mineral grains and occupying pore spaces. This oil will eventually crack to form a significant amount of gas. Because of the inherent nature of kerogen in oil-prone source rocks and amount of residual oil they may contain, late-stage gas generation from oil-prone source rocks may account for more gas generation than gas-prone source rocks (Dembicki, 2013). This concept is actually the basis for the "unconventional" shale gas plays.

To summarize, kerogen generates oil and gas under the influence of time and temperature. Temperature is the dominant control. Kerogen, oil, and gas compositions change as maturation progresses. Observing the composition of the kerogen, oil, and gas can determine what stage of the hydrocarbon generation process has been achieved. Remember, oil-prone source rocks also generate gas in the gas generation window and are likely to have generated more gas than gas-prone source rocks.

PETROLEUM MIGRATION

The process of migrating the generated petroleum from the source rock to the reservoir/trap begins with part of the generated petroleum from the interstitial spaces (pores) in the source rock moving toward a porous and permeable carrier system. The carrier system may consist of a porous sediment, such as sand or silt, or may be a fault or fracture zone. This process is often referred to as primary migration or expulsion. For many years in the early days of petroleum geochemistry, the mechanism for primary migration was the subject of much speculation. Many ideas were put forth to explain this phenomenon including having the petroleum move by diffusion, in water solution, as a colloidal (micellar) solution, and in gas phase (Tissot, 1987). However, after much study, it is generally believe that oil is expelled from the source rock and moves as a liquid phase (Palciauskas, 1991).

Ungerer et al. (1990) described the basic concept for liquid-phase primary migration, summarized in Fig. 2.26. As oil and gas are generated, they move out into the pore spaces of the source rock displacing the pore water. At some point, a minimum saturation threshold is reached where the areas of oil saturation coalesce and begin to form a contiguous oil-wet migration pathway. As petroleum generation continues, the amount of material above this threshold saturation is available for movement along this pathway, termed expulsion. If this oil-wet migration pathway eventually connects a carrier system, the migrating petroleum may eventually travel to a reservoir/trap and form an accumulation.

There are several factors that can influence if and when expulsion may occur in a source rock. The amount and type of organic matter present are primary controls. The more organic matter that is present, the more oil and gas can be generated. High concentrations of organic matter will allow the minimum saturation threshold to be reached sooner as well as provide more petroleum for expulsion once the oil-wet pathway is formed. In sediments with low concentrations of organic matter, insufficient petroleum may be generated to form an oil-wet pathway and any oil and gas formed would then remain in the source rock.

Kerogen type is also important. As discussed in the previous section, certain kerogen types generate more oil than other. Different kerogen types also generate oil and gas at different points in their time–temperature (thermal) history. In source rocks with a high concentration of

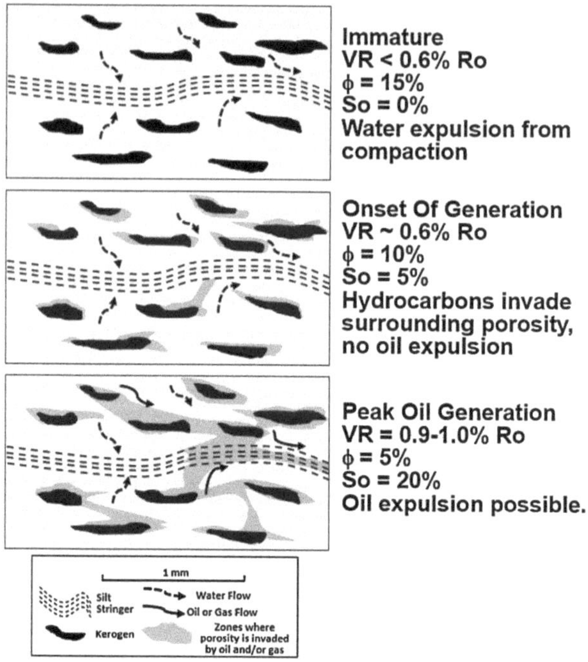

FIGURE 2.26 A conceptual model for the development of contiguous oil-wet migration pathway for hydrocarbon expulsion from source rocks. *VR*, vitrinite reflectance; *ϕ*, porosity; *So*, oil saturation. *After Ungerer, P., Burrus, J., Doligez, B., Chenet, P.Y., Bessis, F., 1990. Basin evaluation by integrated two-dimensional modeling of heat transfer, fluid flow, hydrocarbon generation, and migration. American Association of Petroleum Geologists Bulletin 74, 309–335.*

predominantly oil-prone kerogens (Types I and II), high yields from liquid hydrocarbon generation can easily exceed the threshold saturation to create and maintain contiguous oil-wet pathways. Associated gas generation assists this process by increasing pore pressure and reducing viscosity of the migrating hydrocarbon phase. In source rocks with predominantly gas-prone kerogen (Types III), liquid hydrocarbons may be generated, but in insufficient quantities to reach the threshold saturation and establish contiguous oil-wet pathways. As a result, the liquid hydrocarbons are retained in the pore spaces. Late maturity stage increases in gas generation, as well as the cracking of the retained liquid hydrocarbons to gas, can eventually form a migration pathway to allow gas to be expelled. Thus, gas-prone source rocks are gas-prone as much for their inability to migrate liquid hydrocarbons as for their limited ability to generate them.

Secondary influences on expulsion are the sediment type, sedimentation rate, overpressure, and thickness of the source rock. The porosity and permeability of a given sediment type can have an impact on if and when a contiguous oil-wet migration pathway can form, while sedimentation rate may affect the heating rate of the sediment, which in turn will affect petroleum generation. Slow sedimentation can result in a gradual buildup of generated hydrocarbons in the sediments pore space causing a slow continuous expulsion of hydrocarbons. Rapid sedimentation and the associated rapid increase in temperature may result in accelerated hydrocarbon generation with an associated pulse of hydrocarbon expulsion. High sedimentation rates can also cause under compaction of sediments which may inhibit fluid loss and contribute to overpressure. This excess pore pressure may assist in eventually driving the fluid flow out of the source rock and into the carrier system (Hunt, 1990).

Source bed thickness may be an important influence on expulsion efficiency. Thin source beds may be more efficient at expelling hydrocarbons because of shorter migration distances

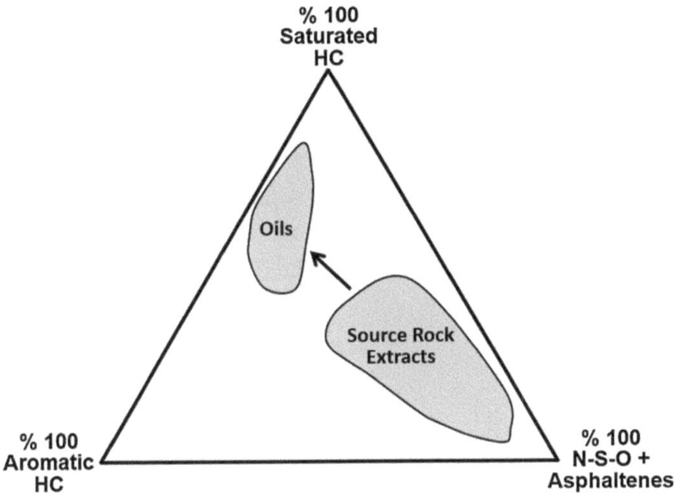

FIGURE 2.27 Change in bulk composition of solvent extraction organic matter (bitumen) from source rocks as compared to crude oils. *After Deroo, G., 1976. Correlations between crude oil and source rocks on the scale of sedimentary basins. Research Centre Pau Societe Nationale des Petrole d'Aquitaine Bulletin 10, 317–335.*

out of the interior of the source rock to the carrier bed. As a source bed thickness, the distance to the carrier bed increases, which may result in some of the hydrocarbons being trapped in the interior of the source rock.

As the generated hydrocarbons begin to be expelled, there is a partitioning of some of the component. Deroo (1976) documented the change in bulk composition of solvent extraction organic matter (bitumen) in source rocks as compared to crude oils. The data plotted in the ternary diagram in Fig. 2.27 show that crude oils are enriched in hydrocarbons, especially the saturated compounds, while the material remaining in the source rock contained more nitrogen, oxygen, and sulfur-bearing (N—S—O) compounds and asphaltenes. This suggests the N—S—Os and asphaltenes are less mobile and perhaps have a higher affinity to the mineral matrix in the source rock than the hydrocarbon fractions. Additional loss of N—S—Os and asphaltenes may also occur along the migration pathway leading to reservoir/trap system.

Once the hydrocarbons have left the source rock and entered into a carrier system, their continued movement in the subsurface is referred to as secondary migration. The main processes governing secondary migration are buoyancy and capillary pressure (Dembicki and Anderson, 1989). The carrier system may consist of a porous sediment (e.g., sandstone or porous carbonate) or intergranular space associated with a fault or fracture zone. Hydrocarbons entering the carrier system begin to accumulate and are held in place by the capillary forces associated with these water-wet intergranular spaces. The buoyancy force is a result of the density contrast between the hydrocarbons and intergranular water. As more hydrocarbons accumulate, the buoyancy force increases and eventually exceeds the capillary pressure in the carrier system allowing the hydrocarbons to move upward. This vertical movement of hydrocarbons forms a continuous oil-wet pathway, similar to the pathways formed in the source rock (Dembicki and Anderson, 1989). The hydrocarbons do not move in mass like water (i.e., Darcy flow). Instead, the migration conduits are restricted in size such that only small amounts of residual hydrocarbons are needed to form and maintain the path, as shown in Fig. 2.28. Because of this, oil can move along these conduits with minimal loss and great efficiency.

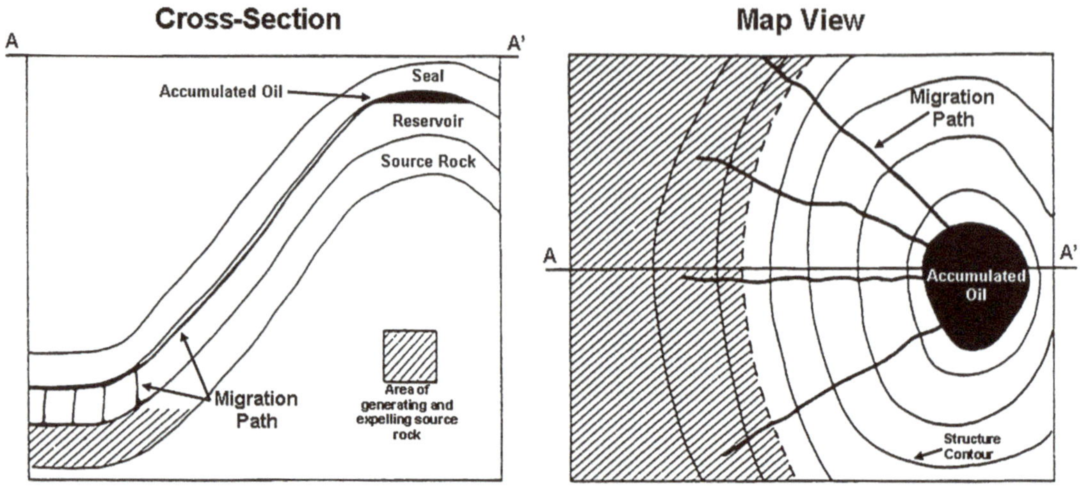

FIGURE 2.28 Cross section and map view of oil migration in a simple anticlinal stricture. *Modified after Dembicki Jr., H., Anderson, M.J., 1989. Secondary migration of oil: experiments supporting efficient movement of separate, buoyant oil phase along limited conduits. American Association of Petroleum Geologists Bulletin 73, 1018–1021.*

When the carrier system enters a reservoir rock, hydrocarbon movement continues along restricted pathways, as shown in Fig. 2.28. Initial movement of the hydrocarbons will be vertical until a permeability barrier is encountered. The permeability barrier may be the top seal, or it may be an intermediary barrier within the reservoir rock itself. Then, updip movement toward the top of the trap will be initiated displacing formation water. When the top of the trap is reached, accumulation will begin.

The process of accumulation, or trap filling, is shown in Fig. 2.29. Hydrocarbons will at first be confined to zones of the highest porosity and permeability (England et al., 1987). Filling may proceed episodically, depending on the hydrocarbon generation history of the source rock, with pulses of petroleum moving along the established migration pathways. As filling progresses, hydrocarbons eventually will occupy areas of lower porosity and permeability in the sediments (England et al., 1987). This progressive filling of high to low porous and permeability may result in isolated water-filled regions within the reservoir. In the final stages of filling, most of the formation water has been displaced and the oil-water contact becomes more uniform.

Once a reservoir is filled if any of the oil and gas leaves the trap and migrates to another trap, this is called remigration. Two scenarios for remigration have been proposed by Schowalter (1979), as illustrated in Fig. 2.30. In structural traps, this may occur when the trap is filled to the spill point. Additional oil and gas entering the trap will displace hydrocarbons already present, spilling them updip to the next trap. If a gas cap is present, the spilled hydrocarbons will be oil. In stratigraphic traps, remigration might take place due a semipermeable seal. Thin or silt-rich zones may be permeable to gas while retaining oil. In these cases, gas may be preferentially leaked updip.

All seals are imperfect and leak to some extent. As a result, some leakage of reservoired hydrocarbons toward the surface should be expected. This leakage is often referred to as tertiary migration, or seepage, and is classified according to the amount of hydrocarbon that

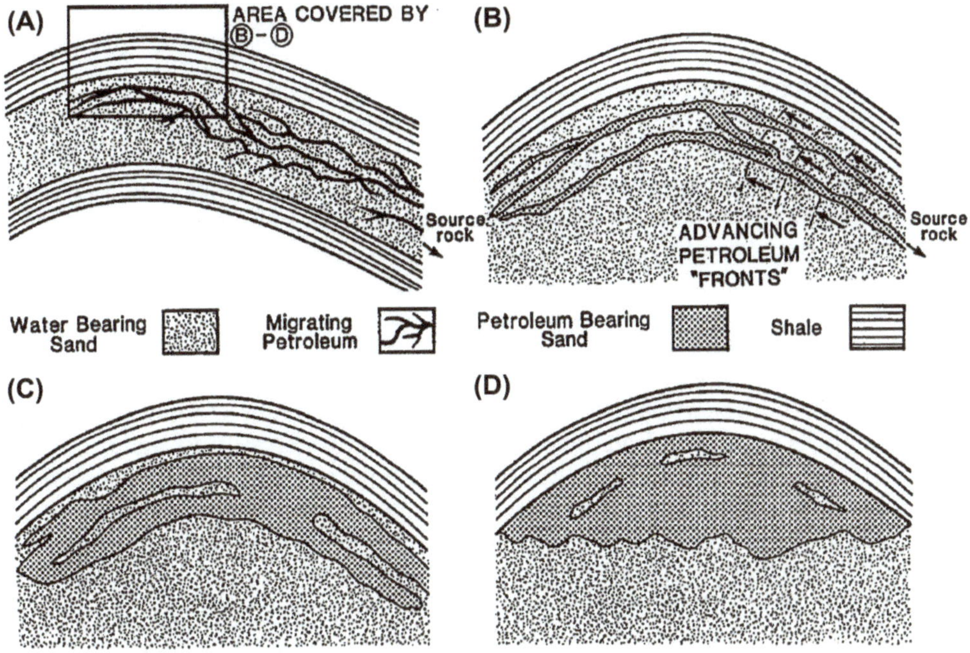

FIGURE 2.29 Schematic of the reservoir/trap filling process. (A) shows the structure at the beginning of filling, the box indicates the focus area shown in (B–D), (B) is a detailed look at early filling along limited migration pathways, (C) illustrates the coalescing of oil migration pathways, and (D) shows the reservoir filling nearing completion. Note the isolated water saturated zones remaining in (D). *From England, W.A., 1994. Secondary migration and accumulation of hydrocarbons. In: Magoon, L.B., Dow, W.G., (Eds.), American Association of Petroleum Geologists Memoir 60: The Petroleum System–From Source to Trap, 211–217.*

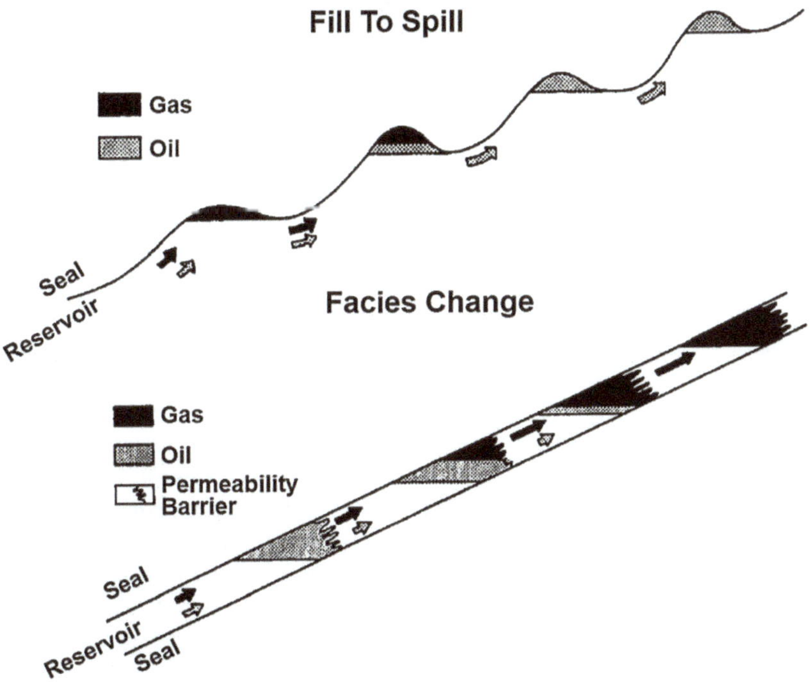

FIGURE 2.30 Two models of remigration and differential entrapment: fill to spill in structural traps and facies change in stratigraphic traps. *Modified from Schowalter, T.T., 1979. Mechanics of secondary hydrocarbon migration and entrapment. American Association of Petroleum Geologists Bulletin 63, 723–760.*

makes it to the surface or near-surface sediments. Low-concentration (just above background) hydrocarbon seepage is usually called microseepage. Microseepage is difficult to detect with any certainty. In contrast, macroseepage is characterized by high concentration of oil and/or gas in near-surface sediments or obvious surface expressions of the leakage, such as visible seepage or mud volcanos. The topic of seepage and its use in petroleum exploration will be covered in greater detail in Chapter 6.

ORIGIN OF NONHYDROCARBON GASES

In addition to the hydrocarbon gases in petroleum accumulations, nonhydrocarbon gases are also found. These include carbon dioxide, nitrogen, hydrogen sulfide, helium, and hydrogen. Some of these gases can be the by-product of the petroleum generation process, while some may be derived from completely independent geochemical pathways. By deciphering the origin of these gases, it is often possible to gain additional insight into hydrocarbon generation and other geochemical processes that may be operating in parallel and improving to the understanding of the petroleum system.

Carbon Dioxide

In the depositional environment, carbon dioxide (CO_2) is produced by aerobic respiration in the sediments. If the sediments are anaerobic, CO_2 could also be formed by microbial oxidation of methane. In either case, the carbon dioxide formed during this early stage of sediment diagenesis is unlikely to significantly contribute to reservoired gases.

There are many sources for the carbon dioxide that is found in reservoired gases. It is a by-product of the oil generation process via the decomposition of oxygen bearing function groups such as carboxyl (—COOH), carbonyl (C=O), and hydroxyl/phenolic (—OH). This occurs mainly in the temperature range from 80 to 120°C (176–248°F). Types III kerogen produces the most CO_2, followed by Type II, while Types I and II-S kerogens produce the least. Coals can be a substantial source of carbon dioxide and are capable of producing up to 75 L of CO_2 per kilogram of coal (Karweil, 1969).

Another source of significant carbon dioxide in a reservoir at low temperatures is related to the biodegradation of crude oil. This will be discussed later in the section on biodegradation in Chapter 4.

At temperatures above 120°C, thermal decomposition of carbonates becomes a more significant source of carbon dioxide gas. In argillaceous sandstone reservoirs with carbonates cements, Smith and Ehrenberg (1989) attributed observed increases in CO_2 with increasing temperature to the interaction of feldspars and clay minerals with the carbonate cements, as shown in Fig. 2.31. This process initiates at temperatures of about 120–140°C (248–284°F) and accelerates as the temperature increases. The potential for these reactions is not confined to the reservoir rock. Similar interaction may also occur in the source rock during late-stage gas generation (Smith and Ehrenberg, 1989).

Direct thermal decomposition of carbonates to produce CO_2 requires temperatures in excess 300°C. This can occur during contact metamorphism when igneous intrusions penetrate into carbonate rocks. Carbonate minerals in contact with magma can produce large quantities of carbon dioxide that can migrate to and accumulate in nearby reservoirs. High

Illite/Kaolinite + Carbonates → Chlorite + CO_2

FIGURE 2.31 Mechanism for thermal decomposition of carbonates in argillaceous sandstone and shale. *As proposed by Smith, J.T., Ehrenberg, S.N., 1989. Correlation of carbon dioxide abundance with temperature in clastic hydrocarbon reservoirs: relationship to inorganic chemical equilibrium. Marine and Petroleum Geology 6, 129–135.*

concentrations of carbon dioxide from magmatic-induced thermal decomposition of carbonates have been documented in many areas worldwide including the Rockies, West Texas, and Indonesia (Thrasher and Fleet, 1995).

Magmatic degassing may also contribute carbon dioxide to reservoirs. This can happen during exsolution of gases from high-volatile magmas. This may occur in tectonically active areas where deep penetrating faults or fractures can access the magma bodies. A large igneous body is required for significant CO_2 contributions. In these instances, the carbon dioxide is associated with radiogenic sourced gases, such as helium and argon.

CO_2 derived from organic matter is usually distinguished from CO_2 derived carbonate decomposition and magmatic outgassing by its carbon isotope ratio. Carbon dioxide from carbonates typically has $\delta^{13}C$ in the range of +4 to −5‰, while magmatic CO_2 is in the −4 to −8‰ range. Carbon dioxide derived from thermal maturation of sedimentary organic matter is usually much more depleted in ^{13}C and falls into the range of −10 to −25‰ $\delta^{13}C$ (Thrasher and Fleet, 1995).

Nitrogen

Nitrogen can come from a variety of sources in geologic environments. Atmospheric nitrogen can be adsorbed on mineral grains in sediments or dissolved in pore waters. Desorption and dissolution of this nitrogen is probably in response to increasing temperature with burial. But it is unlikely that atmospheric nitrogen is significant in hydrocarbon-bearing reservoirs.

An additional source of nitrogen is ammonia. Ammonia is common in interstitial water of sediments and can be easily oxidized to nitrogen by dissolved oxygen reactions with metal oxides, such as ferric oxides. While ammonia in early pore waters is not considered a major source of nitrogen in reservoired gases, the reduction of organic nitrogen compounds by ferric oxide may be a significant process in the subsurface. High nitrogen content gases that have been observed associated with red beds are likely the product of oxidation by ferric oxides (Guseva and Fayngersh, 1973). Nitrogen is fixed in the kerogen structure and can be released concurrent with hydrocarbon generation in the form of nitrogen and nitrogen-bearing organic compounds (Littke et al., 1995). These nitrogen-bearing compounds could be oxidized by ferric oxides to form additional nitrogen. Coals are especially enriched in nitrogen which is freed mainly during the transition from bituminous to anthracite (Hunt, 1996). And finally, nitrogen can come from magmatic and mantle degassing. This is usually associated with volcanic activity and/or deep basement faults and the nitrogen is normally associated with radiogenic sourced gases such as helium and argon.

Hydrogen Sulfide

Hydrogen sulfide (H_2S) is a foul-smelling (rotten egg odor) very reactive gas that is highly toxic. If the depositional environment is anaerobic, bacterial sulfate reduction can produce small quantities of H_2S. Because of the highly reactive nature of this gas, it is unlikely that any of this hydrogen sulfide gas will survive to contribute to a reservoir.

The most common source of hydrogen sulfide gas in reservoirs is sedimentary sulfur. Sulfur is a constituent of most kerogens and elemental sulfur is a common component of sediments that have been dysaerobic or anoxic. During the hydrocarbon generation process, both the organic sulfur bound in kerogen and elemental sulfur associated with the sediment can form hydrogen sulfide and free organic sulfur compounds. Source rocks with kerogens with higher initial sulfur contents, such as Type II and II-S, will produce gases higher in H_2S than Type I and Type III. Because of the highly reactive nature of H_2S, it will readily react with any iron or other transition state metals to form sulfide minerals, such as pyrite. This process of iron scavenging H_2S will occur in both the source rock and reservoirs. There is about 12 times more iron in shales than in fine-grained carbonates and about 3 times more iron in sandstones than in coarse-grained carbonates (Hunt, 1996). As a result, carbonate source rocks will produce more hydrogen sulfide gas and carbonate reservoirs will tend to retain more H_2S, with reservoired accumulations with up to 5% hydrogen sulfide possible from these processes (Le Tran, 1972).

However, reservoired gases with much more than 5% H_2S have been encountered in carbonate reservoirs associated with evaporate sequences, with as much 98% H_2S being reported from the Smackover Trend of the Gulf Coast (Le Tran, 1972). The presence of anhydrite is key to the formation of high H_2S gases by a process called thermochemical sulfate reduction, often referred to as TSR (Orr, 1974), as shown in Fig. 2.32. The sulfate in anhydrite reacts with hydrocarbons to form hydrogen sulfide and carbon dioxide. The CO_2 typically combines with the calcium from the anhydrite to form calcite cement in the reservoir. Excess carbon from the hydrocarbons will go toward the formation of pyrobitumen (Walters et al., 2011). The process is autocatalytic, in that the H_2S that forms also catalyzes the reaction (Orr, 1974). While most of the reservoirs where TSR is observed are at relatively high temperatures (250°F–300°F, 121°C–149°C), Orr (1974) demonstrated these reactions can occur at geologically significant rates at temperatures of only 170°F–250°F (77°C–121°C). One method to confirm the presence of H_2S formed from the TSR reaction is by looking at the sulfur isotopes of the gas. Whereas the $\delta^{34}S$ of H_2S derived from kerogen will reflect the sulfur isotope ratio signature of the organic matter, the $\delta^{34}S$ of H_2S from TSR will be more representative of the anhydrite (Orr, 1977).

Another source of significant hydrogen sulfide in a reservoir is related to reservoir souring during production. This will be discussed later in Chapter 5.

Helium

Most of the helium observed associated with hydrocarbon gases is ^4He. It is the product of alpha decay of heavier radioactive elements, such as uranium, thorium, and radium found in crustal rocks. The other isotope of helium, ^3He, is derived from the mantle and is not very abundant. To determine the origin of helium in a reservoir, it is necessary to measure the $^3He/^4He$ isotope ratio. If the ratio is about 10^{-8}, a sedimentary origin is indicated, while

$$\textbf{Hydrocarbons + Anhydrite} \longrightarrow \textbf{H}_2\textbf{S + Calcite}$$

FIGURE 2.32 Mechanism for thermochemical sulfate reduction. *As proposed by Orr, W.L., 1974. Changes in sulfur content and sulfur isotope ratios during petroleum maturation—study of Big Horn basin Paleozoic oils. American Association of Petroleum Geologists Bulletin 50, 2295–2318.*

$^3He/^4He$ ratios of 10^{-7} to 10^{-5} indicate a mantle origin. While helium is not directly associated with hydrocarbon generation processes, it is a very mobile gas and often follows the same migration pathways used by hydrocarbon gases.

Hydrogen

Hydrogen is not frequently analyzed for or reported in natural gas compositions. In small amounts (<15 mol%), the hydrogen associated with hydrocarbon gases is likely derived from the hydrocarbon generation process. During the cracking of the kerogen, hydrogen is often liberated and may migrate before it has time to react with organic matter. The highest concentrations of hydrogen associated with hydrocarbons are in crude oil fields, while the lowest is in gas fields. This suggests hydrogen production during hydrocarbon generation is early in the kerogen's generation history. Because hydrogen is very mobile and highly reactive, high concentration of hydrogen in the gas associated with a crude oil accumulation suggests the accumulation is actively being charged by recent generation (Hunt, 1996).

Rarely, higher concentrations of hydrogen have also been encountered. Most notable is the Mid-Continent rift basin in Kansas where a series of gas reservoirs were encountered with an average of 35 mol% hydrogen (Coveney et al., 1987). The remainder of this gas was mostly nitrogen, while only trace amounts of hydrocarbons were present. The hydrogen was initially thought to be the result of mantle outgassing. However, isotopic studies of the hydrogen and associated gases suggest that the hydrogen gas resulted from Fe^{+2} oxidation involving olivine and water during serpentinization in the deep crust (Coveney et al., 1987), as shown in Fig. 2.33. The hydrogen has likely migrated up to the sedimentary section along deep vertical faults associated with the rifting.

COALS AS OIL-PRONE SOURCE ROCKS

While the discussion about the formation of petroleum accumulations has focused on the deposition of source rocks and the generation and migration of hydrocarbons from them, little has been said about the role of coals. With the widespread occurrence of coal in sedimentary basins and the high organic matter concentration in coal, the question of whether coal can act as a source rock is very relevant to this discussion of the formation of petroleum accumulations. To begin, a look at the association of hydrocarbon accumulations with coals may provide some insight.

While coals are often found in the same basins as conventional oil and gas accumulations, coal-bearing sequences are most often directly associated with only commercial gas accumulations, sometimes being cited as the source for the gas. In addition, during the mining of coals, it is common to encounter accumulations of methane within the coal seams, but accumulations of liquid hydrocarbons are rarely observed. Only minor liquid hydrocarbon seepage is occasionally reported within liptinite-rich coal seams. And oil staining is not common in sediments adjacent to coal beds. Although a number of papers have indicated coals as the source of some oil accumulations, many of these examples lack conclusive evidence. A few of these oil–coal correlations, such as in the Gippsland Basin (Philp and Gilbert, 1982), have been shown to be the result of the migrating oil extracting geochemical marker from the coal

Olivine + Water → Serpentine + Magnetite + Hydrogen Gas

$$6[(Mg_{1.5}Fe_{0.5})SiO_4] + 7H_2O \rightarrow 3[Mg_3Si_2O_5(OH)_4] + Fe_3O_4 + H_2$$

FIGURE 2.33 Serpentinization process *(As proposed by Coveney Jr, R.M., Goebel, E.D., Zellar, E.J., Dreschhoff, G.A.M., Angino, E.E., 1987. Serpentinization and the origin of hydrogen gas in Kansas. American Association of Petroleum Geologists Bulletin 71, 39–48.)*, for the origin of high concentrations of hydrogen in some of the reservoired gases in the Mid-Continent rift basin.

during migration and not direct sourcing. Overall, these associations of coals and hydrocarbon accumulations suggest that coals may be a source of natural gas, but perhaps not for oil.

However, these observations are far from conclusive and to address this issue more completely, it is necessary to determine if coals can fit the same criteria that have been established for defining source rocks. To accomplish this, there are three issues that must be addressed: can coals contain the right type of organic matter to source oil and gas; is the hydrocarbon generation process in coals the same as observed in conventional source rocks; and is the expulsion/migration process in coals the same as observed in conventional source rocks.

One of the key factors in hydrocarbon generation, as shown above, is the hydrogen content of the organic matter in immature sediments. Looking at coals, they are usually divided into two basic groups: humic coals with low hydrogen contents and sapropelic coals with high hydrogen content. Humic coals are derived from woody plant debris formed under oxic conditions and constitute greater than 80% of all coal (Hunt, 1996). The sapropelic coals are usually subdivided into two subgroups, cannel coals, and boghead coals. Cannel coals are thin and lenticular, often occurring at the bases and tops of coal seams. They contain large amounts of spore and resin material, but make up less than10% of all coals (Hunt, 1996). Boghead coals are lacustrine in origin, contain primarily algal-derived organic matter, and also make up less than 10% of all coals (Hunt, 1996). A comparison of the average hydrogen-to-carbon ratio for immature examples of the major kerogen and coals types, in Fig. 2.34, shows that humic coals are similar to Type III kerogens, cannel coal is similar to Type II kerogen, and boghead coal is similar to Type I kerogen. In addition, studies (e.g., Horsfield et al., 1988) show coals contain chemical moieties similar to those observed in kerogens. These data suggest that coals can contain organic matter that is very similar to source rock kerogens, but the more oil-prone organic matter in cannel and boghead coals are limited in abundance. This raises some doubt about the volumetric significance of coals as oil-prone source rocks.

While hydrocarbon generation in coals has been shown experimentally and by observation to be analogous to generation in source rocks (Brooks and Smith, 1967; Horsfield et al., 1988; Littke et al., 1989; Wilkins and George, 2002, and others), the mechanism for expulsion and migrations of oil from coal is problematic. Porosity and permeability are not an issue with potential expulsion in coals (Littke and Leythaeuser, 1993). Adequate pathways appear to exist and these pathways would be aided by cleat formations as the coal increases in rank. Instead, the high adsorption capacity of coal for both gas and oil appears to be the main obstacle to oil migration out of coals (Hunt, 1991). Pepper (1991) suggests that the mobile petroleum generated in coal may also be entrapped in molecular "cages" preventing expulsion of oil. In both mechanisms, oil is trapped within the coal until increased maturity eventually cracks it to gas which may escape.

Coal Type	H/C Ratio	Kerogen Type	H/C Ratio
Boghead	1.5	Type I	> 1.4
Cannel	1.2-1.3	Type II	1.2-1.4
Humic	0.7-0.8	Type III	0.7-1.0

FIGURE 2.34 A comparison of the hydrogen-to-carbon ratios for immature examples of coals and major kerogen types. *Data compiled from van Krevelen (1961), Tissot et al. (1974), Hunt (1996), Baskin (1997), and Vandenbroucke and Largeau (2007).*

To summarize, most coals do not contain the hydrogen-rich organic matter that would generate sufficient quantities of oil to form a commercial accumulation. Those coals that might have enough hydrogen-rich organic matter will then have to overcome the high adsorption capacity of the coal for liquid hydrocarbons, as well as potential entrapment of oil within the molecular structure of the coal. While coal-sourced oils may be possible, they are likely rare and present higher exploration risk than conventional source rock-generated oils.

SUMMARY

The principles set down in this chapter are the foundation for understanding how geochemical data are used in petroleum exploration and production. As such, some of the more significant points deserve to be reiterated before moving on to discuss how they might be applied to real-world problems.

Petroleum is derived from the transformation of preexisting organic matter of biological origin that has been incorporated into sediments. While all sediments may contain some organic matter, there are certain sediments, called source rocks that contain high concentrations of good quality organic matter. They are responsible for the generation of hydrocarbons that form accumulations of oil and gas.

Optimum conditions for source rock formation begin with high primary biological productivity in and around the depositional environment. This organic matter should be rich in hydrogen from major contributions of algal/bacterial material, spores, pollen, and leaf cuticle. The oxic/anoxic boundary in the depositional environment should be near or above the sediment–water interface to promote good organic matter preservation. And the sediments should be fine-grained, such as very fine silt to clay or carbonate muds (<~4 μm), deposited at intermediate sedimentation rates (~1 mm/year) that will bury and protect the organic matter without diluting it with sediment. Depositional settings where source rocks may develop include: open marine anoxic settings associated with areas of upwelling or oxygen-minimum zones; areas where stratified water columns develop, such as deep areas in epicontinental seas, silled basins, high salinity lagoons, and anoxic lakes; distal parts of deltas; rift-associated deepwater lacustrine settings; and early marine incursion into rifts.

Organic matter in sediments does not remain in its biological form. Instead, it is converted into kerogen, a complex geopolymer whose characteristics are shaped by both the type of organic matter contributed to the sediment and its preservation. There are three basic types of hydrocarbon-generating kerogens: Type I kerogen derived primarily from algal material, producing mainly waxy oil, and usually deposited in lacustrine environments; Type II derived mainly from autochthonous organic matter deposited under reducing conditions in marine environments, producing mainly naphthenic oil; and Type III derived from terrestrial higher plant debris and/or aquatic organic matter deposited in an oxidizing environment, producing mainly gas.

As source rocks become more deeply buried, the organic matter begins to change through a process called maturation. Increasing temperature over geologic time due to burial is the driving mechanism for these changes. Many of the changes in the sedimentary organic occur in a systematic fashion during maturation and can be used to monitor its progress. These maturation indicators provide a context for understanding the thermal history of the sediments and how it has influenced hydrocarbon generation.

Petroleum generation is part of the maturation process. Kerogen generates oil and gas under the influence of time and temperature, with temperature as the dominant control. Oil will eventually break down and generate additional gas. The reactions that generate petroleum approximate first-order Arrhenius kinetics, which provides a means of understanding the process, as well as a method to simulate it. Kinetic properties for the main kerogen types have been determined experimentally and can be related to the kerogen's structure and chemical composition.

After hydrocarbons are generated, they must leave the source rock and move toward a reservoir/trap system to form an accumulation. The initial movement of oil and gas out of the source rock occurs by saturating the pore spaces in the source rock to a critical point where contiguous oil-wet migration pathways form allowing the hydrocarbons to move out of the source rock. Once the hydrocarbons have left the source rock and enter into a carrier system, their continued movement is governed by buoyancy and capillary pressure. The hydrocarbons do not move in mass. Instead, migration occurs along restricted pathways requiring only small amounts of residual hydrocarbons. Initial movement of the hydrocarbons in the carrier system will be vertical until a permeability barrier, such as a top seal, is encountered. Then, updip movement toward the top of the trap begins the filling process. Once a reservoir is filled, oil and gas may leave the trap and migrate into another trap, or a portion of it may leak toward the surface.

Finally, nonhydrocarbon gases are also found in petroleum reservoirs. Carbon dioxide can be a by-product of the oil generation process, but it may also come from thermal decomposition of carbonate cements in argillaceous sandstone reservoirs and from contact metamorphism from igneous intrusion. Nitrogen most likely comes from sedimentary organic matter, especially coals. Hydrogen sulfide in small amounts comes from the sulfur in kerogens or elemental sulfur found in sediments. High concentrations of hydrogen sulfide are typically the product of thermochemical sulfate reduction, where anhydrite reacts with hydrocarbons. Helium is most likely the product of alpha decay of radioactive elements found in crustal rocks. And hydrogen in small amounts is likely derived from hydrocarbon generation, but rare high concentrations of hydrogen can form during serpentinization in the deep crust.

References

Barker, C., 1982. Oil and gas on passive continental margins. In: Watkins, J.S., Drake, C.L. (Eds.), Studies in Continental Margin Geology, vol. 34. American Association of Petroleum Geologists Memoir, pp. 549–565.

Barker, C., Takach, N.E., 1992. Prediction of natural gas composition in ultradeep Sandstone reservoirs. American Association of Petroleum Geologists Bulletin 76, 1859–1873.

Baskin, D.K., 1997. Atomic H/C ratio of kerogen as an estimate of thermal maturity and organic matter conversion. American Association of Petroleum Geologists Bulletin 81, 1437–1450.

Behar, F., Kressmann, S., Rudkiewicz, J.L., Vandenbroucke, M., 1992. Experimental simulation in a confined system and kinetic modeling of kerogen and oil cracking. Organic Geochemistry 19, 173–189.

Behar, F., Loranta, F., Mazeas, L., 2008. Elaboration of a new compositional kinetic schema for oil cracking. Organic Geochemistry 39, 764–782.

Behar, F., Vandenbroucke, M., 1987. Chemical modelling of kerogens. Organic Geochemistry 11, 15–24.

Behar, F., Vandenbroucke, M., Tang, Y., Marquis, F., Espitalie´, J., 1997. Thermal cracking of kerogen in open and closed systems. Organic Geochemistry 26, 321–339.

Bohacs, K.M., Grabbowski Jr., G.J., Carroll, A.R., Mankiewicz, P.J., Miskelll-Gerhardt, K.J., Schwalbach, J.R., Wegner, M.B., Simo, J.A., 2005. Production, destruction, and dilution - the many paths to source-rock development. In: Harris, N.B. (Ed.), The Deposition of Organic-Carbon-Rich Sediments: Models, Mechanisms, and Consequences. SEPM Special Publication No. 82, pp. 61–101.

Braun, R.L., Burnham, A.K., 1994. KINETICS: a computer program to analyze chemical reaction data. Lawrence Livermore National Laboratory Report UCRL-ID-21588, Rev. 2 (September 1994), 14 p.

Braun, R.L., Burnham, A.K., Reynolds, J.G., Clarkson, J.E., 1991. Pyrolysis kinetics for lacustrine and marine source rocks by programmed micropyrolysis. Energy and Fuels 5, 192–204.

Brooks, J.D., Smith, J.W., 1967. The diagenesis of plant lipids during the formation of coal, petroleum and natural gas—I. Changes in the n-paraffin hydrocarbons. Geochimica Cosmochimica Acta 31, 2389–2397.

Burnham, A.K., Braun, R.L., 1985. General kinetic model of oil shale pyrolysis. In Situ 9, 1–23.

Burnham, A.K., Braun, R.L., 1999. Global kinetic analysis of complex materials. Energy and Fuels 13, 1–22.

Calvert, S.E., 1987. Oceanographic controls on the accumulation of organic matter in marine sediments. Special Publication 26. In: Brooks, J., Fleet, A.J. (Eds.), Marine Petroleum Source Rocks. Geological Society of London, pp. 137–151.

Carroll, A.R., Bohacs, K.M., 2001. Lake-type controls on petroleum source rock potential in nonmarine basins. American Association of Petroleum Geologists Bulletin 85, 1033–1053.

Coveney Jr., R.M., Goebel, E.D., Zellar, E.J., Dreschhoff, G.A.M., Angino, E.E., 1987. Serpentinization and the origin of hydrogen gas in Kansas. American Association of Petroleum Geologists Bulletin 71, 39–48.

Demaison, G.J., Hoick, A.J.J., Jones, R.W., Moore, G.T., 1983. Predictive source bed stratigraphy; a guide to regional petroleum occurrence. In: Proceedings of the 11th World Petroleum Congress, vol. 2. John Wiley & Sons, Ltd., London, p. 17.

Demaison, G.J., Moore, G.T., 1980. Anoxic environments and oil source bed genesis. American Association of Petroleum Geologists Bulletin 64, 1179–1209.

Dembicki Jr., H., 2009. Three common source rock evaluation errors made by geologists during prospect or play appraisals. American Association of Petroleum Geologists Bulletin 93, 341–356.

Dembicki Jr., H., 2013. Shale Gas Geochemistry Mythbusting. Oral Presentation at the American Association of Petroleum Geologists 2013 Annual Convention and Exhibition, May 19–22, 2013. American Association of Petroleum Geologists Search and Discovery Article # 80294, Pittsburgh, PA. http://www.searchanddiscovery.com/pdfz/documents/2013/80294dembicki/ndx_dembicki.pdf.html.

Dembicki Jr., H., Anderson, M.J., 1989. Secondary migration of oil: experiments supporting efficient movement of separate, buoyant oil phase along limited conduits. American Association of Petroleum Geologists Bulletin 73, 1018–1021.

Deroo, G., 1976. Correlations between crude oil and source rocks on the scale of sedimentary basins. Research Centre Pau Societe Nationale des Petrole d'Aquitaine Bulletin 10, 317–335.

Didyk, B.M., Simoneit, B.R.T., Brassell, S.C., Eglinton, G., 1978. Organic geochemical indicators of palaeoenvironmental conditions of sedimentation. Nature 272, 216–222.

Dow, W.G., 1978. Petroleum source beds on continental slopes and rises. American Association of Petroleum Geologists Bulletin 62, 1584–1606.

Durand, B., Espitalié, J., 1973. Evolution de la matie're organique au cours de l'enfouissement des sediments. Compte rendus de l'Academie des Sciences (Paris) 276, 2253–2256.

Eglinton, G., 1969. Organic geochemistry: the organic chemist's approach. In: Eglinton, G., Murphy, M.T.J. (Eds.), Organic Geochemistry. Springer-Verlag, New York, pp. 639–675.

England, W.A., 1994. Secondary migration and accumulation of hydrocarbons. In: Magoon, L.B., Dow, W.G. (Eds.), American Association of Petroleum Geologists Memoir 60: The Petroleum System–From Source to Trap, pp. 211–217.

England, W.A., Mackenzie, A.S., Mann, D.M., Quigley, T.M., 1987. The movement and entrapment of petroleum fluids in the subsurface. Journal of the Geological Society of London 144, 327–347.

Erdman, J.G., 1965. Petroleum–its origin in the earth. In: Young, A., Galley, J.E. (Eds.), Fluids in Subsurface Environments, vol. 4. American Association of Petroleum Geologists Memoir, pp. 20–52.

Field, C.B., Behrenfeld, M.J., Randerson, J.T., Falkowski, P., 1998. Primary production of the biosphere: integrating terrestrial and oceanic components. Science 281, 237–240.
Gagosian, R.B., 1983. Processes controlling the distribution of biogenic compounds in recent sediments. In: Meinschein, W.G. (Ed.), Organic Geochemistry of Contemporaneous and Ancient Sediments: Short Course Notes. Society of Economic Paleontologists and Mineralogists, Great Lakes Section, p. 32.
Gagosian, R.B., Peltzer, E.T., 1986. The importance of atmospheric input of terrestrial organic matter to deep sea sediments. Organic Geochemistry 19, 661–669.
Glasby, G.P., 2006. Abiogenic origin of hydrocarbons: an historical overview. Resource Geology 56, 85–98.
Guseva, A.N., Fayngersh, L.A., 1973. Conditions of accumulation of nitrogen in natural gases as illustrated by the Central European and Chu-Sarysu oil-gas basins. Doklady Akademii Nauk SSSR 209, 210–212.
Heilbron, M., Mohriak, W.U., Valeriano, C.M., Milani, E.J., Almeida, J., Tupinambá, M., 2000. From collision to extension: the roots of the southeastern continental margin of Brazil. In: Mohriak, W., Talwani, M. (Eds.), Atlantic Rifts and Continental Margins. American Geophysical Union Geophysical Monograph Series, vol. 115. John Wiley & Sons, Ltd., New York, pp. 1–32.
Horsfield, B., Rullkotter, J., 1994. Diagenesis, catagenesis, and metagenesis of organic matter. In: Magoon, L.B., Dow, W.G. (Eds.), The Petroleum System–From Source to Trap, vol. 60. American Association of Petroleum Geologists Memoir, pp. 189–199.
Horsfield, B., Yordy, K.L., Crelling, J.C., 1988. Determining the petroleum-generating potential of coal using organic geochemistry and organic petrology. Organic Geochemistry 13, 121–129.
Hunt, J.M., 1963. Geochemical data on organic matter in sediments. In: Bese, V. (Ed.), Vortrage der 3. Int. wiss. Konferenz fur Geochemie, Mikrobioloie, und Erdolchemie, Budeapest, vol. 1, pp. 394–412.
Hunt, J.M., 1990. Generation and migration of petroleum from abnormally pressured fluid compartments. American Association of Petroleum Geologists Bulletin 74, 1–12.
Hunt, J., 1991. Generation of gas and oil from coal and other terrestrial organic matter. Organic Geochemistry 17, 673–680.
Hunt, J.M., 1996. Petroleum Geochemistry and Geology, second ed. W. H. Freeman, New York. 743 pp.
Huston, M.A., Wolverton, S., 2009. The global distribution of net primary production: resolving the paradox. Ecological Monographs 79, 343–377.
Ibach, L.E.J., 1982. Relationship between sedimentation rate and total organic carbon content in ancient marine sediments. American Association of Petroleum Geologists Bulletin 66, 170–188.
Jarvie, D., Lundell, L.L., 2001. Kerogen type and thermal transformation of organic matter in the Miocene Monterey Formation. In: Isaacs, C.M., Rullkötte, J. (Eds.), The Monterey Formation: From Rocks to Molecules. Columbia University Press, New York, pp. 268–295.
Karweil, J., 1969. Aktuelle Probleme der Geochemie der Kohle. In: Schenk, P.A., Havenaar, I. (Eds.), Advances in Organic Geochemistry 1968. Pergamon Press, Oxford, pp. 59–84.
Katz, B.J., 1990. Controls on distribution of lacustrine source rocks through time and space. In: Katz, B.J. (Ed.), Lacustrine Basin Exploration: Case Studies and Modern Analogs, vol. 50. American Association of Petroleum Geologists Memoir, pp. 61–75.
Kelts, K., 1988. Environments of deposition of lacustrine petroleum source rocks - an introduction. In: Fleet, A.J., Kelts, K., Talbot, M.R. (Eds.), Lacustrine Petroleum Source Rocks. Geological Society of London Special Publication 40, pp. 3–26.
Kosters, E.C., VanderZwaan, G.J., Jorissen, F.J., 2000. Production, preservation and prediction of source-rock facies in deltaic settings. International Journal of Coal Geology 43, 13–26.
Kudryavtsev, N.A., 1951. Petroleum economy. Neftianoye Khozyaistvo 9, 17–29.
Lehne, E., Dieckmann, V., 2007. Bulk kinetic parameters and structural moieties of asphaltenes and kerogens from a sulphur-rich source rock sequence and related petroleums. Organic Geochemistry 38, 1657–1679.
Le Tran, K.J., 1972. Geochemical study of hydrogen sulfide adsorbed in sediments. In: von Gaertner, H.R., Wehner, H. (Eds.), Advances in Organic Geochemistry 1971. Pergamon Press, Oxford, pp. 717–726.
Lin, C., Eriksson, K.A., Li, S., Wang, Y., Ren, J., Zhang, Y., 2001. Sequence architecture, depositional systems, and controls on development of lacustrine basin fills in part of the Erlian Basin, northeast China. American Association of Petroleum Geologists Bulletin 85 (11), 2017–2043.
Littke, R., Horsfield, B., Leythaeuser, D., 1989. Hydrocarbon distribution in coals and dispersed organic matter of different maceral composition and maturities. Geologische Rundschau 78, 391–410.
Littke, R., Krooss, B., Idiz, E., Frielingsdorf, J., 1995. Molecular nitrogen in natural gas accumulations: generation from sedimentary organic matter at high temperatures. American Association of Petroleum Geologists Bulletin 79, 410–430.

Littke, R., Leythaeuser, D., 1993. Migration of oil and gas in coals. In: Law, B.E., Rice, D.D. (Eds.), American Association of Petroleum Geologists Studies in Geology, vol. 38. Hydrocarbons from Coal, pp. 219–236.

Lopatin, N.V., 1983. Formation of Fossil Fuels. Nedra Press, Moscow. 131 pp.

Louis, M., 1964. Etudes Geochimiques sur les "Schisters cartons" du Toarcian du Basin de Paris. In: Hosbson, G.B., Louis, M.C. (Eds.), Advances in Organic Geochemistry. Pergamon Press, New York, pp. 84–95.

Morley, C.K., 1999. Comparison of hydrocarbon prospectivity in Rift systems. In: Morley, C.K. (Ed.), Geoscience of Rift Systems - Evolution of East Africa. American Association of Petroleum Geologists Studies in Geology No. 44, pp. 233–242.

Orr, W.L., 1974. Changes in sulfur content and sulfur isotope ratios during petroleum maturation—study of Big Horn basin Paleozoic oils. American Association of Petroleum Geologists Bulletin 50, 2295–2318.

Orr, W.L., 1977. Geologic and geochemical controls on the distribution of hydrogen sulfide in natural gas. In: Campos, R., Goni, J. (Eds.), Advances in Organic Geochemistry 1975. Empresa Nacional Adaro de Investigaciones Mineras, Madrid, pp. 571–597.

Orr, W.L., 1986. Kerogen/asphaltene/sulfur relationships in sulfur-rich Monterey oils. Organic Geochemistry 10, 499–516.

Palciauskas, V.V., 1991. Primary migration of petroleum. In: Merrill, R.K. (Ed.), American Association of Petroleum Geologists Treatise of Petroleum Geology: Source and Migration Processes and Evaluation Techniques, pp. 13–22.

Parrish, J.T., 1982. Upwelling and petroleum source beds, with reference to Paleozoic. American Association of Petroleum Geologists Bulletin 66, 750–774.

Pepper, A.S., 1991. Estimating the petroleum expulsion behavior of source rocks: a novel quantitative approach. In: England, W.A., Fleet, A.J. (Eds.), Petroleum Migration: The Geological Society Special Publication, vol. 59, pp. 9–31 London.

Pepper, A.S., Corvi, P.J., 1995. Simple kinetic models of petroleum formation: Part I—oil and gas generation from kerogen. Marine and Petroleum Geology 12, 291–319.

Philippi, G.T., 1965. On the depth, time and mechanism of petroleum generation. Geochimica et Cosmochimica Acta 29, 1021–1049.

Philp, R.P., Gilbert, T.D., 1982. Unusual distribution of biological markers in Australian crude oil. Nature 299, 245–247.

Porfir'ev, V.B., 1974. Inorganic origin of petroleum. American Association of Petroleum Geologists Bulletin 58, 3–33.

Price, L.C., 2000. Organic metamorphism in the California petroleum basins; Chapter B, insights from extractable bitumen and saturated hydrocarbons. U.S. Geological Survey, Bulletin 2174-B, 33.

Revill, A.T., Volkman, J.K., O'Leary, T., Summons, R.E., Boreham, C.J., Banks, M.R., Denwer, K., 1994. Hydrocarbon biomarkers, thermal maturity, and depositional setting of tasmanite oil shales from Tasmania, Australia. Geochimica et Cosmochimica Acta 58, 3803–3822.

Schoell, M., 1983. Genetic characterization of natural gases. Association of Petroleum Geologists Bulletin 67, 2225–2238.

Schowalter, T.T., 1979. Mechanics of secondary hydrocarbon migration and entrapment. American Association of Petroleum Geologists Bulletin 63, 723–760.

Sheng, G., Fu, J., Brassell, S.C., Gowar, A.P., Eglinton, G., Damste', J.S.S., de Leeuw, J.W., Schenk, P.A., 1987. Sulphur containing compounds in sulphur-rich crude oils from hypersaline lake sediments and their geochemical implications. Geochemistry 6, 115–126.

Sinninghe Damste, J.S., de las Heras, F.X.C., van Bergen, P.F., de Leeuw, J.W., 1993. Characterization of Tertiary Catalan lacustrine oil shales: discovery of extremely organic sulfur-rich Type 1 kerogens. Geochimica et Cosmochimica Acta 57, 389–415.

Sinninghe Damsté, J.S., de las Heras, F.X.C., de Leeuw, J.W., 1992. Molecular analysis of sulphur-rich brown coals by flash pyrolysis-gas chromatography-mass spectrometry: the type III-S kerogen. Journal of Chromatography 607, 361–376.

Smith, J.T., Ehrenberg, S.N., 1989. Correlation of carbon dioxide abundance with temperature in clastic hydrocarbon reservoirs: relationship to inorganic chemical equilibrium. Marine and Petroleum Geology 6, 129–135.

Speers, G.C., Whitehead, E.V., 1969. Crude oil. In: Eglinton, G., Murphy, M.T.J. (Eds.), Organic Geochemistry. Springer-Verlag, New York, pp. 20–73.

Swain, F.M., 1970. Non-Marine Organic Geochemistry. Cambridge University Press, London. 445 pp.

Szatmari, P., 1989. Petroleum formation by Fischer-Tropsch synthesis in plate tectonics. American Association of Petroleum Geologists Bulletin 73, 989–998.

Tegelaar, E.W., de Leeuw, J.W., Derenne, S., Largeau, C., 1989. A reappraisal of kerogen formation. Geochimica et Cosmochimica Acta, 53, 3103–3106.

Tegelaar, E.W., Noble, R.A., 1994. Kinetics of hydrocarbon generation as a function of the molecular structure of kerogen as revealed by pyrolysis-gas chromatography. Organic Geochemistry 22, 543–574.

Thrasher, J., Fleet, A.J., 1995. Predicting the risk of carbon dioxide 'pollution' in petroleum reservoirs. In: Grimalt, J.O., Dorronsoro, C. (Eds.), Organic Geochemistry: Developments and Applications to Energy, Climate, Environment and Human History - Selected Papers from the 17th International Meeting on Organic Geochemistry. Iberian Association of Environmental Organic Geochemistry (A.I.G.O.A.), San Sebastian, pp. 1086–1088.

Tissot, B., 1969. Premières données sur les mécanismes et la cinétique de la formation du pétrole dans les bassins sédimentaires. Simulation d'un schema réactionnel sur ordinateur. Revue Institut Francais Petrole 24, 470–501.

Tissot, B., 1987. Migration of hydrocarbons in sedimentary basins: a geological, geochemical, and historical perspective. In: Doligez, B. (Ed.), Migration of Hydrocarbons in Sedimentary Basins. Éditions Technip, Paris, pp. 1–19.

Tissot, B., Durand, B., Espitalie, J., Combaz, A., 1974. Influence of the nature and diagenesis of organic matter in the formation of petroleum. American Association of Petroleum Geologists Bulletin 58, 499–506.

Tissot, B.P., Pelet, R., Ungerer, P.H., 1987. Thermal history of sedimentary basins, maturation indices, and kinetics of oil and gas generation. American Association of Petroleum Geologists Bulletin 71, 1445–1466.

Tissot, B.P., Welte, D.H., 1984. Petroleum Formation and Occurrence. Springer-Verlag, New York. 699 p.

Ungerer, P., Burrus, J., Doligez, B., Chenet, P.Y., Bessis, F., 1990. Basin evaluation by integrated two-dimensional modeling of heat transfer, fluid flow, hydrocarbon generation, and migration. American Association of Petroleum Geologists Bulletin 74, 309–335.

Ungerer, P., Pelet, R., 1987. Extrapolation of oil and gas formation kinetics from laboratory experiments to sedimentary basins. Nature 327, 52–54.

Vandenbroucke, M., Largeau, C., 2007. Kerogen origin, evolution and structure. Organic Geochemistry 38, 719–833.

van Krevelen, D.W., 1961. Coal: Typology-Chemistry-Physics-Constitution. Elsevier, Amsterdam. 514 pp.

Walters, C.C., Qian, K., Wu, C., Mennito, A.S., Wei, Z., 2011. Proto-solid bitumen in petroleum altered by thermochemical sulfate reduction. Organic Geochemistry 42, 999–1006.

Welte, D., 1974. Recent advances in organic geochemistry of humic substances and kerogen: a review. In: Tissot, B., Bienner, F. (Eds.), Advances in Organic Geochemistry 1973. Editions Technip, Paris, pp. 3–13.

Wilkins, R.W.T., George, S.C., 2002. Coal as a source rock for oil: a review. International Journal of Coal Geology 50, 317–361.

Williams, L.A., 1984. Subtidal stromatolites in Monterey formation and other organic-rich rocks as suggested source contributors to petroleum formation. American Association of Petroleum Geologists Bulletin 68, 1879–1893.

Williams, D.F., Lerche, I., 1987. Hydrocarbon production in the Gulf of Mexico region from organic-rich source beds of ancient intraslope basins. Energy, Exploration, and Exploitation 5, 199–218.

Younes, A.I., McClay, K., 2002. Development of accommodation zones in the Gulf of Suez–Red Sea rift, Egypt. American Association of Petroleum Geologists Bulletin 86, 1003–1026.

Ziegler, P.A., Parrish, J.T., Humphreville, R.C., 1979. Paleogeography, upwelling and phosphorites (abs.). In: Cook, P.J., Shergold, J.H. (Eds.), Proterozoic-Cambrian Phosphorites, Project 156 of UNESCO-IUGS. Australian Natl. Univ. Press, Canberra, p. 21.

CHAPTER

3

Source Rock Evaluation

OUTLINE

Definitions and Fundamental Concepts 62

Sample Collection 63
Cuttings 63
Headspace Gas/Canned Cuttings 64
Drilling Mud and Mud Additive Sample 65
Core and Sidewall Cores 65

Total Organic Carbon 66

Rock-Eval Pyrolysis 67
Source Richness Interpretations 68
Thermal Maturity Interpretations 72
Source Quality Interpretations 74

Solvent Extraction, S-A-R-A Analysis, and Extract Data 80
Source Richness and Quality Interpretations 81

Gas Chromatography 82
Source Quality Interpretations 83
Thermal Maturity Interpretations 84
Recognizing Contamination 87

Headspace Gas Analysis 88
Source Richness Interpretations 89
Thermal Maturity Interpretations 90

Pyrolysis–Gas Chromatography 91

Kerogen Isolation 95

Elemental Analysis 96

Vitrinite Reflectance 98
Interferences With Vitrinite Reflectance 99
Lack of Vitrinite 100
Cavings and Reworked Vitrinite 100
Unconformities/Normal Faults 101
Reverse Faults 103
Change in Geothermal Gradient/Heat Flow 103
Igneous Intrusives 104
Suppression and Misidentification 104
Vitrinite Anisotropy 106
Maturity Interpretations With Vitrinite Reflectance 106

Alternative Reflectance Method 109
Liptinite/Sporinite/Exinite Reflectance 109
Solid Bitumen Reflectance 110
Graptolite, Scolecodont, and Chitinozoan Reflectance 110

Visual Kerogen Typing 111

Thermal Alteration Index 113

Kerogen Fluorescence 115

Conodont Alteration Index 117

Wireline Log Interpretations 118
Gamma Ray 119
Formation Density 119
Sonic Transit Time 119
Resistivity 120

Practical Petroleum Geochemistry for Exploration and Production
http://dx.doi.org/10.1016/B978-0-12-803350-0.00003-9

Source Rock Presence Interpretations 120
Source Richness Interpretations 121
Thermal Maturity Interpretations 122
Using Outcrop Samples 125
Strategies in Source Rock Evaluation 126
Background Information 126
Defining the Problem 127
Making Interpretations 127
References 129

DEFINITIONS AND FUNDAMENTAL CONCEPTS

Source rock evaluation is defined as assessing the hydrocarbon-generating potential of sediments by looking at their capacity for hydrocarbon generation, the type of organic matter present, and the thermal maturity of the sediments. The evaluation of a source rock consists of determining the sediments source richness, source quality, and thermal maturity. *Source richness* is a measure of the amount of organic matter in a sediment capable of generating hydrocarbons. It is an indication of the sediment's capacity for hydrocarbon generation and not merely a measure of the amount of organic matter present. *Source quality* is a measure of the types of organic matter present in a sediment, which indicates the type(s) of hydrocarbons the sediment can generate. And the *thermal maturity* is a measure of the degree to which a sediment has been altered by the combined effects of time and temperature. The thermal maturity will indicate what stages of hydrocarbon generation, if any, might have been experienced by the sediment. While source richness and quality may be determined for a single sediment sample at a single depth, individual thermal maturity data have little value. Instead, thermal maturity determinations need a set of data points over a depth range to establish a trend for interpretation. Without the context of a depth trend, it is impossible to establish if individual thermal maturity data points are valid or anomalous.

From a geologic perspective, a source rock should also be described by its geologic age, lithology, depositional setting, areal extent, thickness, and other geologic attributes. These are important characteristics of a source rock and will have some influence on the interpretation of the geochemical data as well as how the data fit into the petroleum system.

A major part of source rock evaluation is determining if a sediment has the potential to be an effective source rock. An *effective source rock* not only has organic matter capable of generating hydrocarbons, but it also has enough organic matter to generate adequate quantities of hydrocarbons to be expelled and contribute to an accumulation. True effective source rocks can be demonstrated to have generated and expelled oil and/or gas, while potential effective source rocks are simply sediments that could generate significant amounts of hydrocarbons. It should also be remembered that source rock evaluation only points to potential effective source rocks. Before a sediment can be called a proven source rock, it must be established that it has contributed to a petroleum accumulation. This can only be established by an oil-to-source rock correlation, as discussed in the next chapter.

A multiple parameter approach is most often adopted for source evaluation studies. Single data types may be in error or influenced by contamination. The multiple parameter approach simply states that more than one data type needs to be used to establish an interpretation. In essence it means that there is a need for corroboration of two or more data types for a high-confidence interpretation to be made.

This chapter will first cover the collection of the main types of geochemical samples used in source rock evaluation projects. This will be followed by sections devoted to the major types of organic geochemical analyses used in these studies. Each section will describe the analysis, explain how the data derived from it are used in making interpretation, and discuss some of the pitfalls that may be encountered. Following the laboratory-based methods will be a brief discussion of the use of wireline log data to provide indirect source rock information. And the final section will suggest some of the strategies to use in source rock evaluation. By understanding the analytical procedures and data used, the role of source evaluation in the exploration process can be better defined, and the data can contribute more insight into the understanding the petroleum system.

SAMPLE COLLECTION

The key to a successful source rock study is the samples. Good quality samples that are representative of the intervals of interest are essential to obtaining good-quality source rock data that can aid an exploration program. The best way to ensure this is to plan a sampling program that will satisfy all the requirements of the study. The following guidelines discuss the different types of samples used in source rock studies, best practices for collection, and cautions that need to be exercised.

Cuttings

Drill cuttings are the most commonly available samples for source rock analysis. Although usually adequate, cuttings samples may have several inherent shortcomings that can influence the data and their interpretation. First, a cuttings sample is collected over a depth interval that is estimated from lag time and is not precisely known. Second, cuttings generally contain some caved material along with the rock from the sampled depth interval. Source rock data can be influenced by this out of place material if enough is present. And finally, because of their relatively high surface area, their small size, and their long exposure to drilling mud, cuttings are also liable to pick up contamination from drilling additives. These shortcomings can be overcome to some degree by careful selection of samples and by following proper collecting procedures.

Cuttings are normally collected as composite samples over 10–30 ft (3–9 m) intervals. When the source rock unit is thin or interbedded with sands, compositing samples over a 10 ft (3 m) interval is recommended if practical. On drilling wells, beginning collecting for a suspected source rock interval at least 250 ft (75 m) above the zone of interest to assure it is sampled. For best results in maturation studies, a series of samples are needed at an interval of every 300–500 ft over several thousand feet to establish a trend of spore-pollen color and/or vitrinite reflectance with depth.

Always collect a larger number of samples than may seem necessary. Not all of the samples collected need to be submitted for organic geochemical analysis. This will, however, ensure sufficient samples will be available to answer any exploration questions that arise.

If you are collecting cuttings from a drilling well, take about two cups (500 cc) of cuttings from the shale shaker, and place them in a clean bucket. Rinse the cuttings several times with clean water to remove the drilling mud. If the sediment is poorly consolidated, rinse as gently as possible to avoid loss. Promptly rinsing off the drilling mud minimizes potential contamination by the mud. Remove any obvious contaminants, apparent cavings, and lost circulation material. Place the cuttings in a mildew-proof cloth sample bag marked with the well name and depth interval, and allow to air dry (no ovens).

If the well has already been drilled and sets of cuttings samples are available, unwashed cuttings are usually preferred over washed cuttings. Washed cuttings are sometimes oven dried thereby potentially altering the organic matter in them. This is especially critical for maturity determinations. One drawback to the unwashed cuttings is that they have had a long exposure to any organic contaminants that may be in the drilling mud. Therefore, be alert for possible contaminants and be cautious during interpretation of the data. Unwashed cuttings usually need to be soaked in clean water prior to rinsing to soften the drilling mud. Again, remove any obvious contaminants, apparent cavings, and lost circulation material and allow to air dry (no ovens) before use. Minimum sample size is about ¼ cup (60 cc) of cuttings. This will usually provide enough material for total organic carbon (TOC), Rock-Eval pyrolysis, and visual kerogen/vitrinite reflectance analysis. If extraction for gas chromatography (GC) analysis is required, more sample material may be needed. When possible, always collect more than the minimum amount of cuttings to ensure enough material will be available if needed.

Headspace Gas/Canned Cuttings

In areas where source rocks maybe prone to gas or gas/condensate production, collecting wet canned cuttings may be appropriate. Canned cuttings samples capture the light hydrocarbons from C_1 up to C_7 outgassing from the sediment's interstitial spaces, and allow them to be analyzed. Because these light hydrocarbons are more similar to gas and condensate, they can add useful information to the evaluation of these types of source rocks. Canned cuttings samples can also be useful supplementary data for evaluation of oil prospects. A major drawback to wet canned cuttings is the increased expense in acquiring, transporting, and analyzing the samples. If this is a factor, consider alternating canned cuttings with dry cuttings samples or limited the canned cuttings samples to certain zones of interest. An additional shortcoming of canned cuttings is that the light hydrocarbons analyzed are very mobile in the subsurface and therefore may not be representative of the interval over which they were collected.

To collect canned cuttings samples, take about two cups (500 cc) of cuttings from the shale shaker, and place them in a clean bucket. Quickly rinse the cuttings several times with clean water to remove the drilling mud. If the sediment is poorly consolidated, rinse as gently as possible to avoid loss. Put the cuttings in an unused clean 1 quart (1 L) size paint can, and cover the cuttings with water leaving at least 1/4 to 1/3 of the air space in the top of the can. Add a few drops of biocide (e.g., glutaraldehyde) to help preserve the sample. Firmly seal the can, and use at least four retaining clips to secure the lid. Label the side of the can with the well name and depth interval with permanent marker. Turn the sealed cans upside down for storage and shipping. If minor leaks occur while inverted, only water is lost and the headspace gas is retained.

An alternate sample container to the 1 quart paint can is the Isojar. It is a high-impact plastic jar with a gas-tight seal in its lid. The jars are labeled with fill lines for cuttings and water, as well as having labels for identifying the well and the depth of the sample. Use of Isojars simplifies the collection of headspace gas/canned cuttings samples and provides a more uniform sample collection process.

Drilling Mud and Mud Additive Sample

It is highly recommended to periodically collect drilling mud samples for reference in the event contamination is encountered in the cuttings. Critical times for mud sample collection are when cuttings sample collection begins, just before and just after a new mud system is being introduced and when the well reaches total depth. Mud samples should be approximately 1/4 cup (50 mL) in size and placed in small screw-top glass bottles marked with the well name and depth. Prior to sealing, add a few drops of biocide to help preserve the mud samples.

Core and Sidewall Cores

Core and sidewall core samples, when available, are an important part of any source rock sampling program. They are usually used to supplement cuttings samples and provide a few distinct advantages. While cuttings samples represent a depth interval in a well, core and sidewall core samples are depth points. As such, they are key samples in maturation studies because they are free of caved material. Core and sidewall core samples are also less likely to be completely contaminated by drilling fluids. The interiors of these samples are often contamination-free and can be used to help assess and compensate for contamination in cuttings samples. When possible, consider adding sidewall coring of suspected source rock intervals to your drilling operations.

On a drilling well, core samples for source rock analysis should be washed with water to remove any drilling mud. A section of core at least 0.5 in. (1.25 cm) thick should provide adequate material for most analyses. Place the core in a sample bag labeled with the well name and depth. Avoid writing directly on the sampled pieces with either felt-tip marker or grease pencil. Sidewall core samples should be rinsed with water to remove any residual drilling mud and packaged in standard sidewall core jars. Label the jars with the well name and depth. As with cuttings, it is highly recommended to collect samples of the drilling mud at the time of coring for reference. If contamination is encountered in the cores or sidewall cores, the drilling mud samples will be needed to help identify the contaminant and compensate for it.

For cores from storage, it is necessary to establish if the core had been wrapped and wax sealed when collected, if it was cut with an oil-cooled saw or drill, or, for older cores, if it has been stored in wax-coated core boxes. All three conditions can adversely affect the geochemical analysis of the rock. Try to sample out of the interior of the core to keep away from surface-related contamination. Also avoid parts of the core marked with felt-tip markers or grease pencils.

TOTAL ORGANIC CARBON

TOC, is the amount of organic carbon present in a source rock expressed as a weight percent. It is a proxy for the total amount of organic matter present in the sediment (Ronov, 1958) and used as an indicator of source richness with respect to how much hydrocarbon the sediment may generate. TOC is determined by taking a portion of the source rock, grinding it to a fine powder, and weighing the sample to be analyzed. The sample is then treated with acid to remove any carbonate minerals to eliminate the carbonate carbon prior to analysis. The analysis consists of heating the carbonate-free sample to a very high temperature in an oxygen-rich atmosphere so that the carbon in the sediment is converted to carbon dioxide. The carbon dioxide is then measured, and the TOC content is calculated. Details of this analysis can be found in Carvajal-Ortiz and Gentzis (2015). While some laboratories use indirect methods of measuring TOC based on pyrolysis product yields, the combustion method is preferred for TOC analysis.

Source richness interpretations of TOC usually use a semiquantitative scale (Peters, 1986; Jarvie, 1991), such as shown in Fig. 3.1. While this scale is broadly used, most petroleum geochemists believe a sediment will need at least 2.0% original TOC to be an effective source rock, assuming a large proportion of the organic matter is reactive.

Although TOC is a good starting point for source rock evaluation, it should be used cautiously on its own. As shown in Chapter 2, all of the organic matter in sediments is not the same. Some will be capable of generating hydrocarbons while some will be inert and generate nothing (Tissot et al., 1974). TOC measures all the organic matter in a sediment and does not differentiate between the two.

Another problem with TOC is that it is maturity sensitive. Daly and Edman (1987) observed the TOC in a source rock decreases as hydrocarbons are generated and expelled. In addition to the amount of TOC decreasing with increasing maturity, the nature of the organic matter also changes. As the organic matter goes from the immature to the overmature state, the relative inert kerogen content of the sedimentary organic matter increases as the reactive organic matter is consumed in hydrocarbon generation. The TOC is, therefore, less accurate as an indicator of source richness as the source rock becomes increasingly mature. As a result, it is essential that the sediment's maturity be accounted for in making interpretations of TOC, and it is better to use TOC in conjunction with other data to obtain a more complete picture of source rock richness.

Contamination can also interfere with interpretation of TOC data. Specifically, oil-based drilling muds and organic drilling mud additives can increase the apparent TOC content of a sediment (Carvajal-Ortiz and Gentzis, 2015). It is critical to check on what drilling mud additives have been used as well as the type of drilling mud as a quality control check for TOC data. In addition, check the pyrograms from the Rock-Eval analysis (discussed later) to look for anomalies that may be related to drilling mud contamination. If necessary, a preanalysis

Richness	TOC
Poor	0.0-0.5
Fair	0.5-1.0
Good	1.0-2.0
Very Good	>2.0

FIGURE 3.1 Semiquantitative source richness interpretation of total organic carbon (TOC). *After Peters, K.E., 1986. Guidelines for evaluating petroleum source rock using programmed pyrolysis. American Association of Petroleum Geologists Bulletin 70, 318–329; Jarvie, D.M., 1991. Total organic carbon (TOC) analysis. In: American Association of Petroleum Geologists Treatise of Petroleum Geology: Source and Migration Processes and Evaluation Techniques, 113–118.*

solvent extraction of the sediment may be necessary to remove these contaminants. In this case, the bitumen is removed with the contaminants, and the TOC will only reflect the total organic matter in the insoluble material in the sediment.

ROCK-EVAL PYROLYSIS

Rock-Eval pyrolysis is a quick, inexpensive analysis that can provide information about a sediment's hydrocarbon generating potential, hydrocarbon generation type, and maturity. This laboratory pyrolysis technique has been a petroleum industry standard analysis for source rocks since the early 1980s and continues to be a mainstay of source rock analysis today. Pyrolysis is the thermal decomposition of organic matter in an inert atmosphere. It is relevant to source rock evaluation in that this thermal decomposition is roughly analogous to the maturation process kerogen undergoes during hydrocarbon generation. Although the heating rate is much higher in the laboratory, much can be learned from how source rocks behave during the Rock-Eval analysis.

The results from a Rock-Eval analysis are shown in Fig. 3.2. The data produced as well as the analysis have been discussed in detail by Espitalie et al. (1986), Peters (1986) and

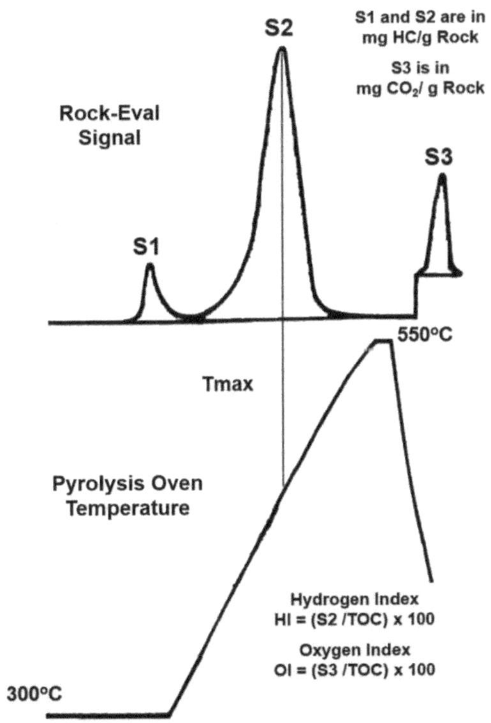

FIGURE 3.2 Schematic of the Rock-Eval pyrolysis analysis showing the data recorded and the pyrolysis oven heating profile. *Modified after Peters, K.E., 1986. Guidelines for evaluating petroleum source rock using programmed pyrolysis. American Association of Petroleum Geologists Bulletin 70, 318–329.*

others. The pyrogram in the upper part of Fig. 3.2 is the recording of the detectors used in the instrument. The temperature profile used during the analysis is shown in the lower part of Fig. 3.2. Often, the temperature profile is overlain on the pyrogram in a single plot when data is received from contract laboratories.

The analysis consists of weighing a sample of ground rock, typically 50–100 mg, and placing it into a crucible that goes into the Rock-Eval's pyrolysis oven. The rock sample is rapidly heated in a helium atmosphere to 300°C and held isothermally for up to 5 min. During this time the volatile organic material that evolves from the sample is measured by a flame ionization detector and is recorded as the S1 peak. The S1, expressed in milligrams hydrocarbon/gram rock or mg HC/g rock, is usually characterized as the hydrocarbons that have already been generated, but it also will include any preserved bitumen from the depositional environment. The sample is then heated from 300 to 550°C at a rate of 25°C/minute in the standard experiment, and the volatile organic material that evolves is measured and recorded as the S2 peak. The S2, also expressed in mg HC/g rock, is considered the hydrocarbon generating potential remaining in the sediment's kerogen. It is used as an indicator of the amount of hydrogen associated with the kerogen in the sediment. During the heating of the sample from 300 to 390°C, the carbon dioxide that evolves from the sample is measured and recorded as the S3 peak. The S3, expressed in milligrams carbon dioxide/gram rock or mg CO_2/g rock, is used as an indicator of the amount of oxygen associated with the kerogen in the sediment. In addition to these measurements of evolved volatile organic matter and carbon dioxide, the temperature at which the S2 peak reaches a maximum is recorded as T_{max}. Total analysis time is approximately 20 min. Most geochemists will limit the sediments analyzed by Rock-Eval to those containing at least 1.0% TOC in an attempt to insure meaningful results are obtained.

Three additional Rock-Eval parameters are derived from these measurements and the weight percent TOC measurement for the rock sample. The hydrogen index, HI, is the S2/TOCx100 which represents the amount of hydrocarbon that can be generated relative to the amount of organic matter in the source rock. The HI is expressed as milligrams hydrocarbon per gram TOC or mg HC/g TOC. The oxygen index, OI, is the S3/TOCx100 which represents the amount of CO_2 that can be generated relative to the amount of organic matter in the source rock. The OI is expressed as milligrams carbon dioxide per gram TOC or mg CO_2/g TOC. And the production index, PI, is S1/(S1 + S2) which represents the amount of hydrocarbon already generated relative to the total amount of hydrocarbon that could be generated.

Initially, some of the Rock-Eval parameters are usually plotted versus depth, as shown in Fig. 3.3. This helps to place the data in a stratigraphic context and provides a quick look to identify zones of interest.

Source Richness Interpretations

Initially, source rock evaluation programs utilized the Rock-Eval S1 and S2 parameters to provide information about source richness, with S1 representing the hydrocarbons that have already been generated, while the S2 represents the hydrocarbons generating potential remaining in the sediment's kerogen. The semiquantitative scales shown in Fig. 3.4 from Peters (1986) are in general use. At the time these scales were introduced, most wells were drilled with water-based drilling mud. Current drilling practices rely heavily on oil-based drilling mud and a large variety of organic drilling mud additives. These materials can

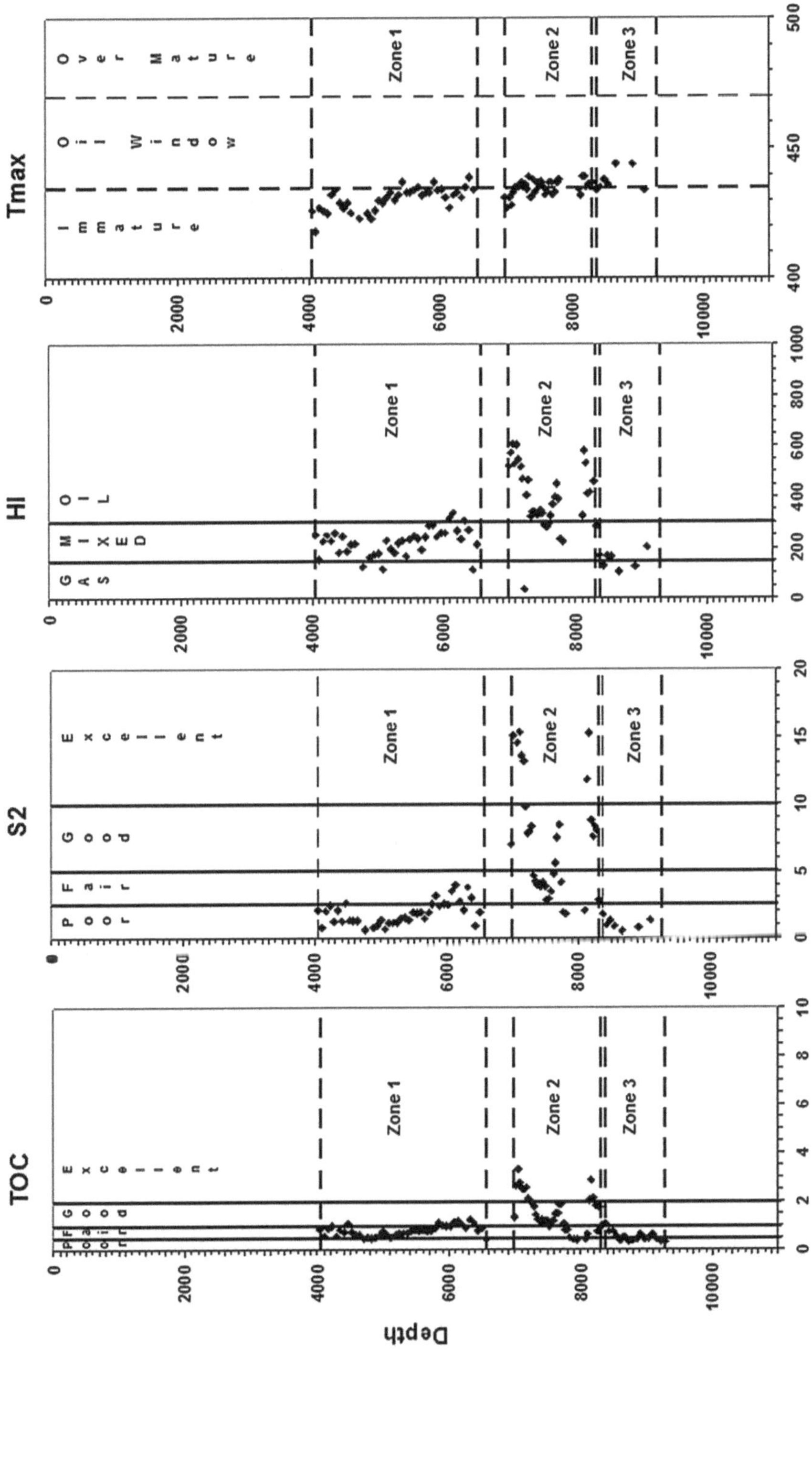

FIGURE 3.3 The main Rock-Eval pyrolysis parameters plotted versus depth to place the data in a stratigraphic context.

Richness	S1	S2
Poor	0.0-0.5	0.0-2.5
Fair	0.5-1.0	2.5-5.0
Good	1.0-2.0	5.0-10.0
Very Good	>2.0	>10.0

FIGURE 3.4 Semiquantitative source richness interpretation of the Rock-Eval S1 and S2 parameters. *After Peters, K.E., 1986. Guidelines for evaluating petroleum source rock using programmed pyrolysis. American Association of Petroleum Geologists Bulletin 70, 318–329.*

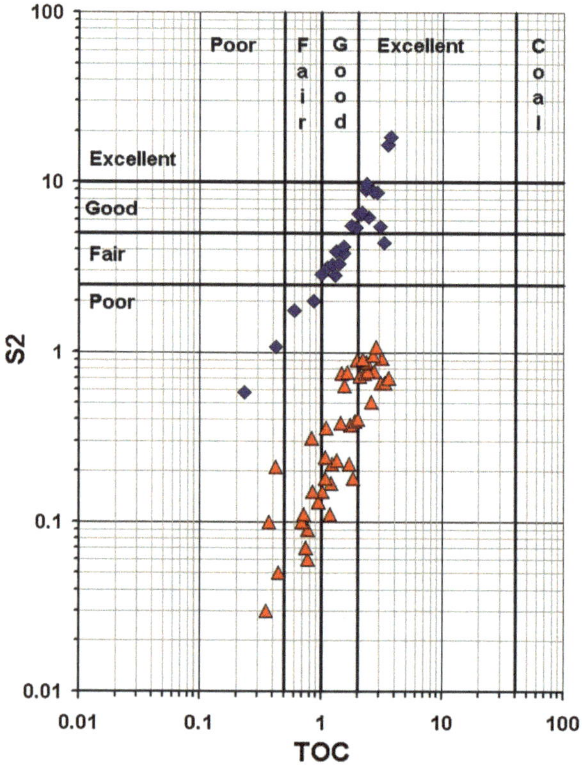

FIGURE 3.5 The source richness total organic carbon (TOC)-S2 cross-plot showing two populations of source rocks with similar TOC values but different S2 values. The population with the higher S2 values shows greater hydrocarbon-generating potential. *Modified from Dembicki Jr., H., 2009. Three common source rock evaluation errors made by geologists during prospect or play appraisals. American Association of Petroleum Geologists Bulletin 93, 341–356.*

contribute significant amounts of material to the S1 peak and obscure its true value. As a result, the S1 peak has fallen out of favor as a source richness indicator.

Most current source evaluation programs utilized the S2 parameter as the principle source richness indicator. In addition to using the simple cutoff values, S2 is often combined with TOC data in a cross-plot, as shown in Fig. 3.5. The two data sets shown clearly illustrate why TOC should not be used alone as an indicator of source richness. Both data sets have essentially the same TOC distributions but very different S2 values. The source rock represented by the diamond symbols has higher S2 values and therefore more hydrocarbon

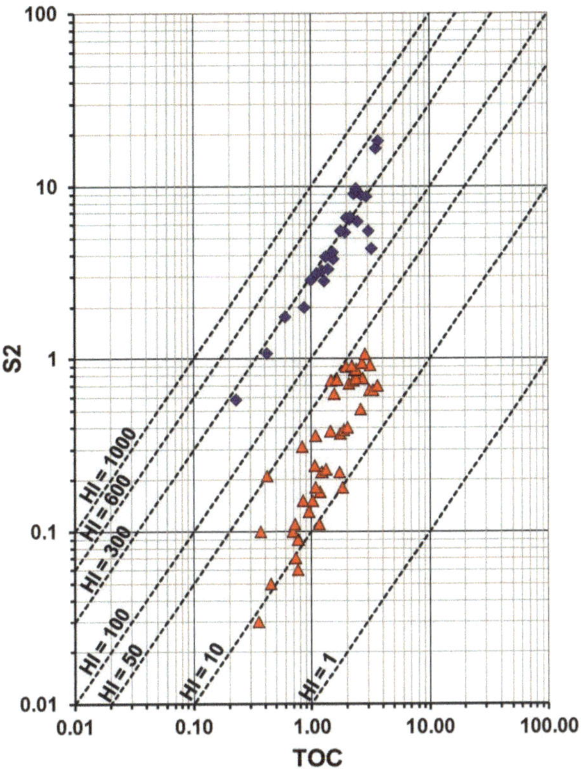

FIGURE 3.6 Total organic carbon (TOC)-S2 cross-plot showing isohydrogen index lines.

generating potential and better source richness than the source rock plotted with the triangles. The two sets of data are also observed to form two roughly parallel trends in the cross-plot. Assuming they have a similar thermal maturity, this likely indicates the two source rocks have two difference kerogen type mixtures. This is illustrated in Fig. 3.6 where isohydrogen index lines have replaced the cutoff value lines on the S2 and TOC cross-plot. The two parallel data trends have different average hydrogen indices which suggests different mixes of kerogens: the source rock represented by the diamond symbols being more oil prone, and the source rock plotted with the triangles being gas prone-to-inert.

While the S2 is supposed to represent the hydrocarbons generating potential remaining in the sediment's kerogen, there are instances when some of the material that should be confined to the S1 are carried over and included in the S2. This can occur when a high proportion of the generated material consists mainly of resins and asphaltenes that may not be volatile at 300°C. These materials would be volatilized at higher temperatures and thereby be included in the S2 peaks. A similar situation can occur with contamination of the sediments by organic drilling mud additives. Some of these additives may contain or consist of processed asphalt or Gilsonite that will contribute to the S2 peak. To recognize the presence of resins and asphaltenes in the S2 peak either from the bitumen or drilling mud additives, it is necessary to examine the pyrograms from the Rock-Eval analysis. These contributions to the S2 peak can be recognized as an asymmetry in the peak on the low-temperature side.

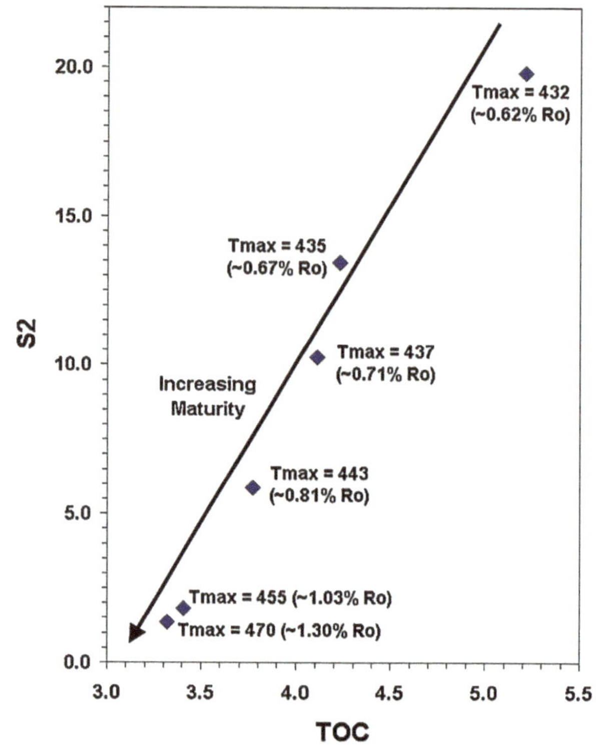

FIGURE 3.7 As a source rock generates and hydrocarbons migrate off, the amount of organic matter in the source rock will decrease with a corresponding decrease in TOC, and the amount of reactive kerogen will decrease (the amount of hydrogen will decrease) resulting in a decrease in Rock-Eval S2. *Modified from Dembicki Jr., H., 2009. Three common source rock evaluation errors made by geologists during prospect or play appraisals. American Association of Petroleum Geologists Bulletin 93, 341–356.*

If the amount of material is substantial, a preanalysis solvent extraction of the sediment may be necessary to remove the bitumen or contaminants to obtain a more accurate S2 measurement.

Another caution in using Rock-Eval S2 for a source richness indicator concerns what happens to S2 as the source rock matures. As previously stated, the amount of organic matter in the source rock, measured by the TOC, will decrease with increasing maturation (Daly and Edman, 1987) as the amount of reactive kerogen, measured by the S2, gets consumed (Espitalie et al., 1977b). This is illustrated in the source rock maturation series in Fig. 3.7. With increasing thermal maturity, a source rock will gradually look less like a source rock and sediments that may have been oil prone in their immature state may appear gas prone. Because of this, it is necessary to know what the maturity is for the sediments being evaluated to avoid misinterpretation (Dembicki, 2009).

Thermal Maturity Interpretations

T_{max} is the primary Rock-Eval parameter for assessing thermal maturity. It was recognized early that T_{max} would increase with increasing depth (Espitalie et al., 1977a,b) and was eventually related to other maturity indicators such as vitrinite reflectance (as discussed later). Typical T_{max} values for the top and bottom of the oil window are shown in Fig. 3.8.

Maturation	Tmax
Top oil window	~435-445
Bottom oil window	~470

FIGURE 3.8 Thermal maturity interpretations for Rock-Eval T_{max}. *After Espitalie, J., 1986. Use of Tmax as a maturation index for different types of organic matter: comparison with vitrinite reflectance. In: Burrus, J. (Ed.), Thermal Modeling in Sedimentary Basins. Proceedings of Meeting: 1st IFP Exploration Research Conference, Carcans, France, June 3–7, 1985, vol. 44. Editions Technip, Paris, pp. 475–496; Peters, K.E., 1986. Guidelines for evaluating petroleum source rock using programmed pyrolysis. American Association of Petroleum Geologists Bulletin 70, 318–329.*

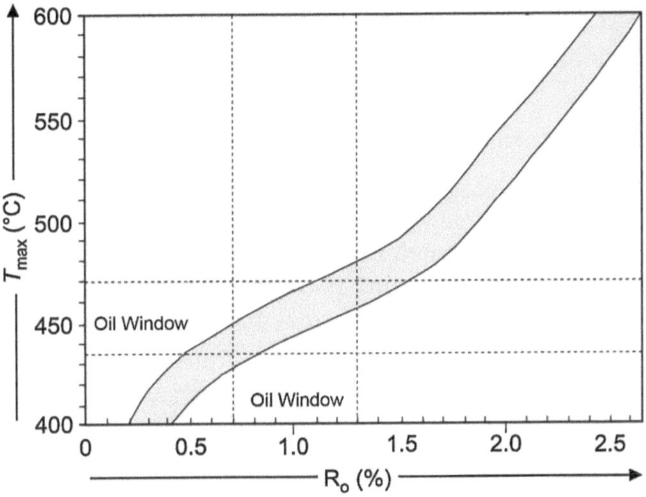

FIGURE 3.9 Observed relationship of T_{max} with vitrinite reflectance. The *gray band* on the diagram illustrates the variation observed in the data. *Modified after Espitalie, J., Deroo, G. Marquis, F., 1985. La pyrolyse Rock-Eval et ses applications. Deuxième partie. Oil & Gas Science and Technology, l'Institut Francais du Petrole Energies nouvelles 40, 755–784.*

The temperature range for the top of the oil window reflects that the absolute changes in T_{max} vary with kerogen type as well as maturity (Espitalie et al., 1985; Espitalie, 1986; Peters, 1986). While T_{max} values for Type II kerogen will be approximately 435°C at the top of the oil window, Type I kerogen will have T_{max} values of about 440°C. Significant hydrocarbon generation in Type III will be indicated by T_{max} values of about 445°C. As maturity progresses, the kerogen type influence diminishes, and all three kerogen types have T_{max} values of approximately 470°C at the bottom of the oil window.

Unfortunately, using T_{max} as a maturity indicator is not that simple. As with all maturity indicators, individual T_{max} values have little intrinsic value. Taken individually, a T_{max} value may be a valid indication of maturity, or it may be a spurious point. However, a set of T_{max} values over a depth range, as shown in Fig. 3.3, can help establish a trend which aids in deciphering maturity.

In addition, the observed relationship between T_{max} and vitrinite reflectance, shown in Fig. 3.9, displays a range of variation in T_{max} at any given maturity level. Some of this variation can be attributed to kerogen mixtures in the source rock but only for the immature and low-maturity samples. Espitalie (1986) felt that minor variations in the instrument performance, mineral matrix effects, and retention of heavy resin and asphaltene compounds from the bitumen influenced the T_{max} measurement that resulted in this variation.

Further complicating the problem, the S2 peak can occasionally have multiple maxima, shown in the pyrogram in Fig. 3.10, making it difficult to pick an appropriate T_{max} for the analysis. In contrast, when the S2 peak is small, <0.2 mg HC/g rock, inaccurate T_{max} values are often selected during the Rock-Eval analysis (Peters, 1986). And S2 asymmetry that can shift the T_{max} can also be caused by contamination by some oil-based drilling muds and certain drilling mud additives (Carvajal-Ortiz and Gentzis, 2015).

To guard against making poor maturity interpretations with T_{max}, remember it is a trend tool so data over a large depth range is needed to make valid observations. Examination of the pyrograms is essential to look for anomalies that may call some of the T_{max} data into question. And consider that there is natural scatter in the T_{max} data that must be factored in when making the interpretations.

Source Quality Interpretations

Most source rock evaluation programs rely on Rock-Eval pyrolysis to provide information about kerogen type. A simple approach to this interpretation is to use HI or the S2/S3 ratio, as shown in Fig. 3.11. While these parameters may provide a first approximation for discerning oil-prone, gas-prone, and mixed generating potential, it is difficult to place these interpretations in the proper context with respect to thermal maturity.

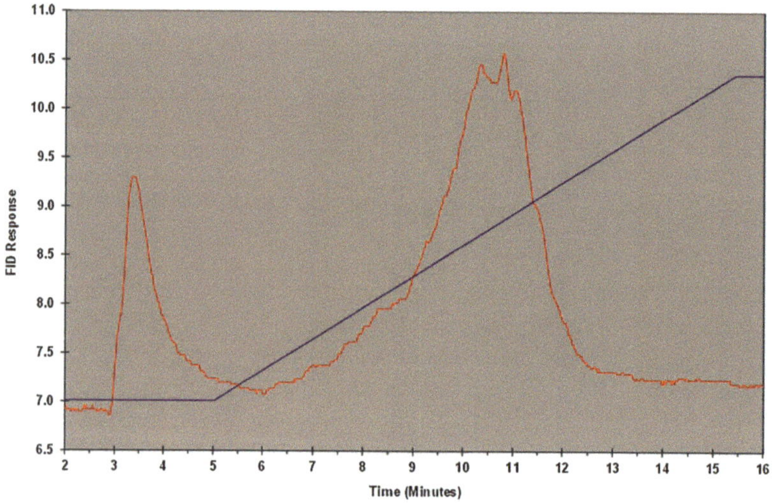

FIGURE 3.10 A Rock-Eval analysis pyrogram showing multiple maxima on the S2 peak.

Type	HI	S2/S3
Gas	0-150	0-3
Gas and oil	150-300	3-5
Oil	>300	>5

FIGURE 3.11 Source quality interpretation based on the hydrogen index, HI, and the S2/S3 ratio. *From Peters, K.E., 1986. Guidelines for evaluating petroleum source rock using programmed pyrolysis. American Association of Petroleum Geologists Bulletin 70, 318–329.*

The main Rock-Eval source quality interpretation scheme utilizes the HI and OI plotted on a pseudo-van Krevelen diagram. HI and OI are approximations for the H/C atomic ratio and O/C atomic ratio, respectively, from kerogen elemental analysis (Espitalie et al., 1977a; Peters, 1986; Baskin, 1997), as shown in Fig. 3.12. A typical pseudo-van Krevelen diagram, shown in Fig. 3.13, has the Type I, Type II, and Type III kerogen trends. These trends start as immature kerogen on the right side of the diagram and converge in the overmature state near the origin. Type II-S kerogen plots along the Type I trend (Williams, 1984) and Type IV kerogen plots below the Type III trend just above the OI axis (Peters, 1986). While Type I and Type II-S kerogens plot along the same trend on a pseudo-van Krevelen diagrams, they can be differentiated based on the depositional environment of the source rock. Type II-S kerogen is found in marine source rocks, while Type I kerogen is associated mainly with lacustrine sediments.

If a source rock contained only one kerogen type, the interpretation of pseudo-van Krevelen diagrams would be straight forward. However, as mentioned in Chapter 2, very few source rocks contain only one type of kerogen, and mixtures of kerogen types make interpreting Rock-Eval data on pseudo-van Krevelen diagrams more challenging. Looking at the data plotted on the pseudo-van Krevelen diagram in Fig. 3.13, most of the data points fall on a trend between the Type II and Type III curves. An obvious assumption would be that this a simple mixing trend for source rock samples that contain mainly Type II on one end and most Type III on the other. This may be the case, but there are other possible interpretations. To investigate this further, Dembicki (2009) used binary mixtures of Type III kerogen with Type I and Type II and binary mixtures of Type IV kerogen with Type I, Type II, and Type III to simulate how these combinations would appear on pseudo-van Krevelen diagrams.

The HI and OI of mixtures of Types I and II with Type III are plotted on a pseudo-van Krevelen diagram in Fig. 3.14. The Type II–Type III mixtures appear to accurately represent

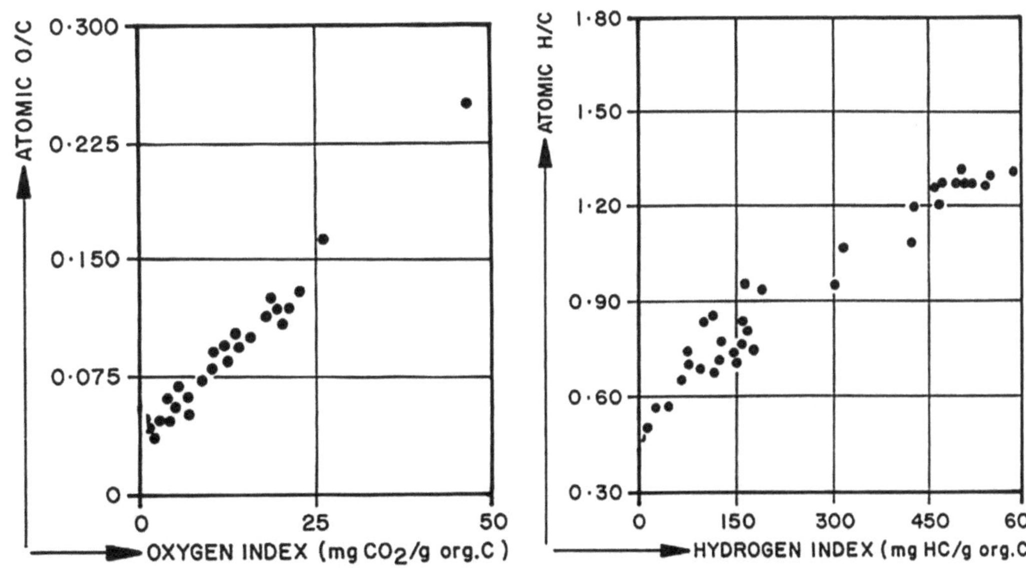

FIGURE 3.12 The relationship between oxygen index and atomic O/C ratio and hydrogen index and atomic H/C ratio. *From Tissot, B.P., Welte, D.H., 1984. Petroleum Formation and Occurrence. Springer-Verlag, New York, 699 p.*

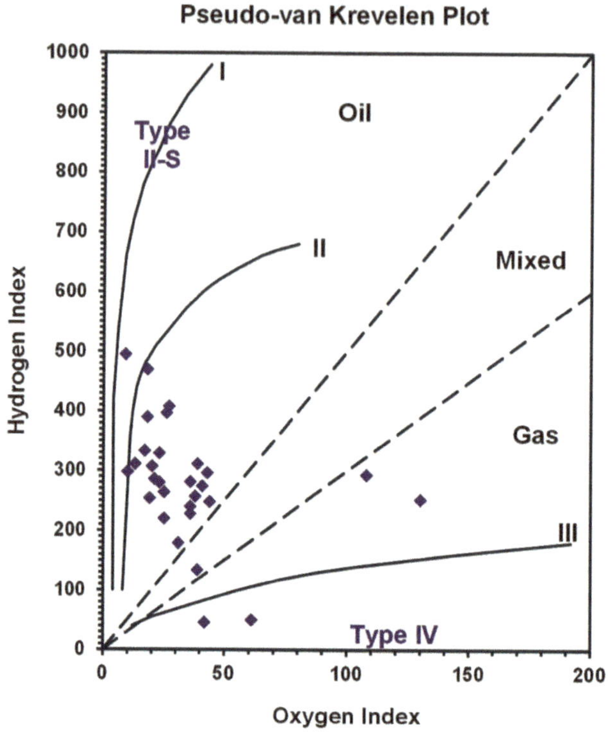

FIGURE 3.13 Pseudo-van Krevelen plot of hydrogen index versus oxygen index. *Modified from Dembicki Jr., H., 2009. Three common source rock evaluation errors made by geologists during prospect or play appraisals. American Association of Petroleum Geologists Bulletin 93, 341–356.*

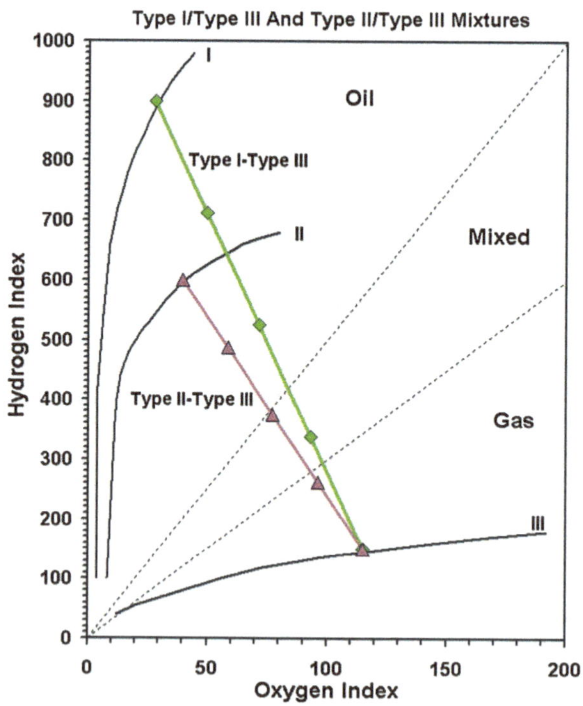

FIGURE 3.14 Simulation of the mixing Type I and Type II kerogens with Type III kerogen plotted on a Pseudo-van Krevelen diagram. Kerogen mixtures shown by the trends are in the following proportions 100/0, 75/25, 50/50, 25/75, 0/100. *Modified from Dembicki Jr., H., 2009. Three common source rock evaluation errors made by geologists during prospect or play appraisals. American Association of Petroleum Geologists Bulletin 93, 341–356.*

the simple interpretation of the kerogen types present. However, the Type I–Type III mixtures show overlap with the Type II–Type III mixing zone on the pseudo-van Krevelen diagram, and interpretations can then become unclear (Dembicki, 2009). The 75% Type I/25% Type III mixture, for example, could easily be interpreted to contain 100% Type II kerogen. With additional Type III kerogen in the mixture, the data points could be interpreted as Type II–Type III mixtures instead of Type I–Type III mixtures. This confusion could lead to erroneous conclusions about the hydrocarbon-generating capacity of the source rock. And because Type I kerogen is usually lacustrine and Type II kerogen is marine, inferences about the depositional setting of the source rock based on kerogen type could also be incorrect (Dembicki, 2009).

When mixing of Types I, II, and Ill kerogens with Type IV, as shown in Fig. 3.15, the pseudo-van Krevelen diagram again does not always accurately represent the type of hydrocarbon generation possible from the source rocks. Because Type IV kerogen is essentially inert, it will not contribute significantly to the S2 measurement. Instead it will only contribute to the TOC which in turn will lower the HI (Dembicki, 2009), thereby making the source rock appear more gas prone. While the mixtures of Type I and Type IV kerogen suggest the presence of Type II kerogen or mixtures of Type II and III kerogens, Type II and IV kerogen mixtures appear as mixtures of Types II and III. The only accurately represented source rocks are the mixtures of Type III and IV kerogens (Dembicki, 2009).

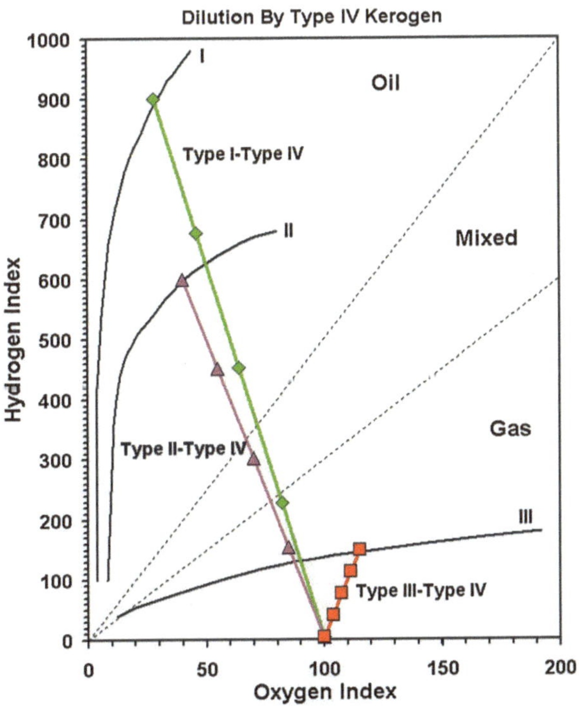

FIGURE 3.15 Simulation of the diluting Type I, Type II, and Type III kerogens with Type IV kerogen plotted on a Pseudo-van Krevelen diagram. Kerogen mixtures shown by the trends are in the following proportions 100/0, 75/25, 50/50, 25/75, 0/100. *Modified from Dembicki Jr., H., 2009. Three common source rock evaluation errors made by geologists during prospect or play appraisals. American Association of Petroleum Geologists Bulletin 93, 341–356.*

In addition to kerogen mixtures, the mineral matrix of the source rock may also have some influence on interpreting the data on pseudo-van Krevelen diagrams. Espitalie et al. (1980) and Katz (1983) demonstrated that sediments with TOC contents of less than 2.0% can exhibit hydrocarbon retention on mineral grains that result in significant reductions in the HI. There is also the potential for thermal decomposition of small amounts of carbonate minerals during the Rock-Eval analysis that can contribute carbon dioxide to the S3 peak and increase the OI (Katz, 1983). And contamination may impact S2 (as discussed earlier), which in turn influences HI leading to inaccurate interpretations.

There are two additional graphical methods for characterizing kerogen type and the type of hydrocarbon that may be generated using Rock-Eval data. The first is the HI versus T_{max} cross-plot shown in Fig. 3.16. It shows trends for Type I, II, and III kerogens, similar to the pseudo-van Krevelen diagram, with the immature end of the trends on the left side of the plot with the trends converging as they go to overmature state on the right side. These diagrams were introduced as a means of discerning kerogen type for the Rock-Eval III Oil Show Analyzer (Espitalie et al., 1984). This version of the instrument was designed for well

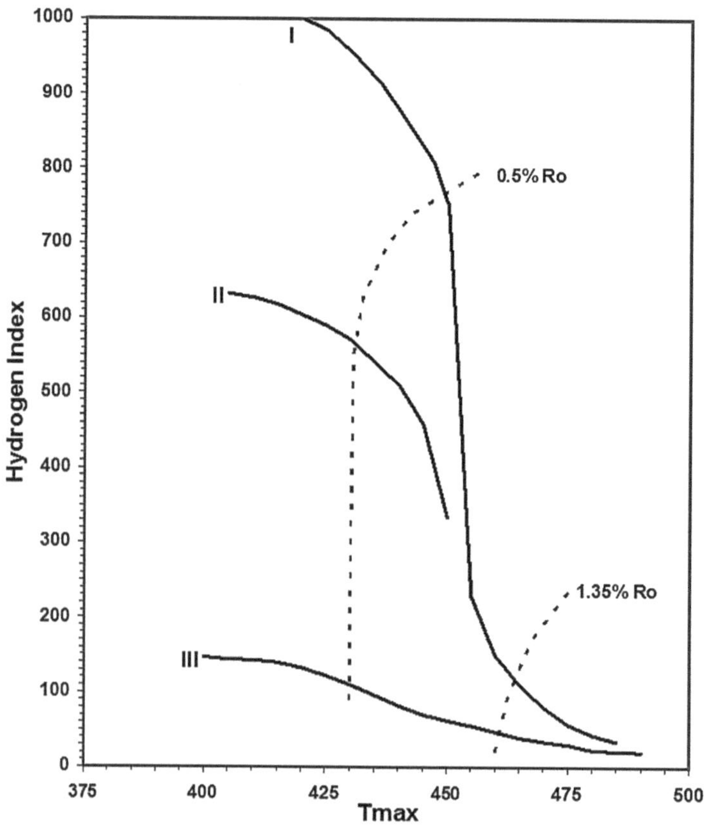

FIGURE 3.16 An alternate source quality interpretation diagram based on hydrogen index and T_{max}. *After Espitalié, J., Marquis, F., Barsony, I., 1984. Geochemical logging. In: Voorhees, K.J. (Ed.), Analytical Pyrolysis–Techniques and Applications. Butterworth, Boston, pp. 276–304.*

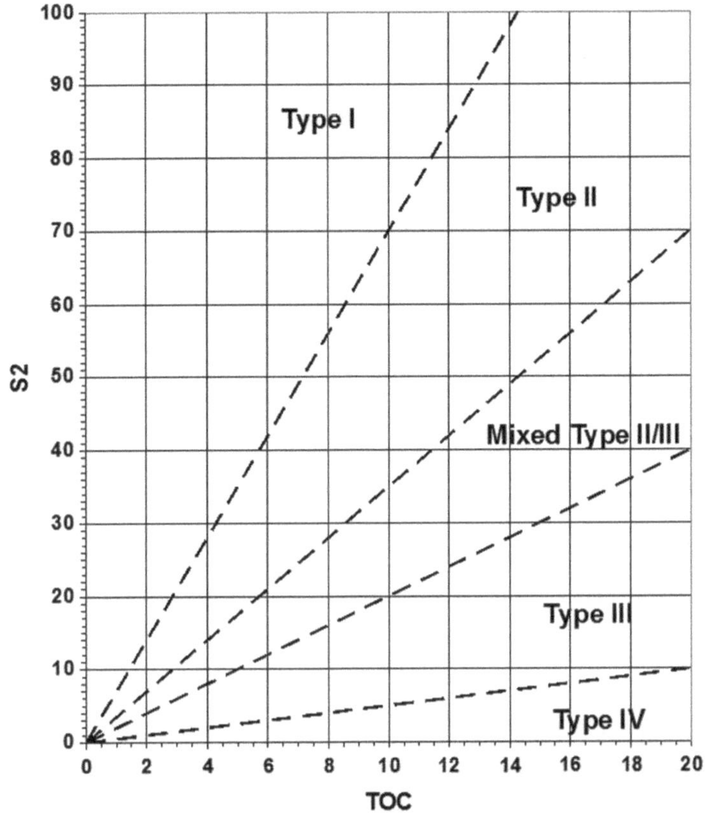

FIGURE 3.17 Total organic carbon (TOC)-S2 cross-plot for characterizing kerogen type and the type of hydrocarbon generation. The *dashed diagonal lines* separating the kerogen type regions are isohydrogen index lines for hydrogen indices of 700, 350, 200, and 50, from left to right, respectively. *After Katz, B.J., and Elrod, L.W., 1983. Organic geochemistry of DSDP Site 467, offshore California, Middle Miocene to Lower Pliocene strata: Geochimica et Cosmochimica Acta 47, 389–396.*

site use and had converted the thermal conductivity detector for measuring S3 into a means of determining TOC. Studies by Espitalie et al. (1986) showed that the HI-T_{max} plot could be used in place of the pseudo-van Krevelen diagram for preliminary kerogen typing. The diagram is preferred by some geochemists because it combines an indicator of kerogen type, HI, with a maturity indicator, T_{max}. However, the HI in this plot is still subject to the problem of mixed kerogen types observed with the pseudo-van Krevelen diagram, as well as variations of T_{max} with both maturity and kerogen type, as mentioned earlier. Therefore, caution should be used when using the HI-T_{max} cross-plot alone.

The second additional method is a simple TOC-S2 cross-plot using linear scales, as shown in Fig. 3.17. This method is based on the work of Katz and Elrod (1983) and further supported by Dahl et al. (2004). The diagonal lines separating the kerogen type regions are isohydrogen index lines. This approach is also subject to the problems of mixed kerogen types observed with the pseudo-van Krevelen diagram and the HI-T_{max} cross-plot and contamination.

Because of kerogen mixtures and mineral matrix effects, Rock-Eval data are usually inadequate by themselves to accurately determine what kinds of kerogen that are present and what kind of hydrocarbons they may generate. For proper assessments, think about the potential for kerogen mixtures, know something about the depositional setting, and factor in the maturity of the sediments. Most importantly, use supplemental geochemical data such as pyrolysis–gas chromatography (PGC) (discussed later) to corroborate and refine the interpretations made.

SOLVENT EXTRACTION, S-A-R-A ANALYSIS, AND EXTRACT DATA

The extractable organic matter, the bitumen, can be removed from the sediments with solvents typically using a Soxhlet extraction apparatus similar to the one shown in Fig. 3.18. The sediment is dried, ground to a fine powder, weighed, and placed inside the thimble in the central part of the apparatus. The thimble is usually made from filter paper or porous ceramic and precleaned prior to use. The solvent in the flask, often dichloromethane, is heated causing it to evaporate. The hot solvent vapor travels through the bypass sidearm up to the condenser, where it cools, condenses back into a liquid, and drips down onto the sediment in the thimble. The chamber containing the thimble will slowly fill with warm solvent until it is almost full. It then empties by siphon action and returns back down to the flask. This cycle will be allowed to repeat many times, usually for 24–48h. During each cycle, a portion

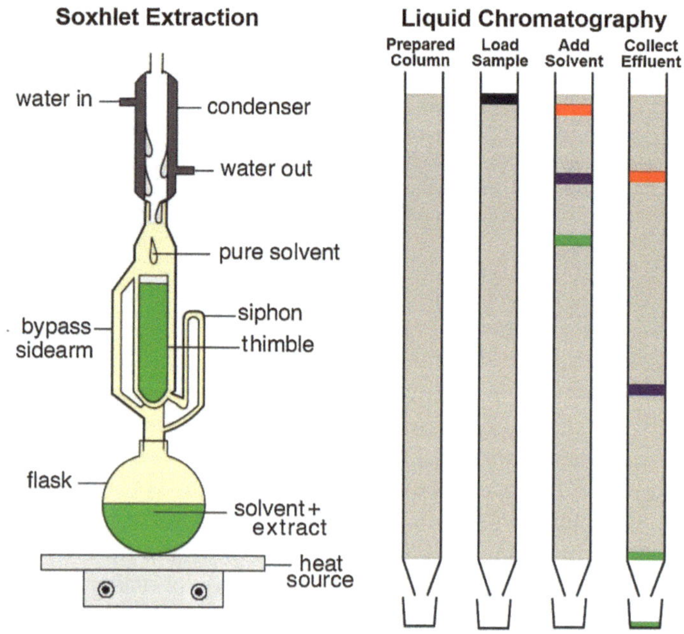

FIGURE 3.18 Schematics of a Soxhlet extraction apparatus used in solvent extraction of the bitumen from source rocks and liquid column chromatography used to separate the extracted bitumen into saturate, aromatic, and N-S-O fractions.

of the bitumen dissolves in the solvent and is eventually isolated in the solvent in the flask. Only clean warm solvent is vaporized and recycled to extract the sediment. At the end of an extraction, the excess solvent is removed by evaporation, leaving behind only the extracted bitumen. The bitumen is weighed, and the amount of extractable organic matter in parts per million (ppm) is calculated. This recovered material is often referred to as the C_{15}+ extractable organic material. This is because compounds lighter than approximately C_{15} will be partially or completely lost during the extract and evaporation processes.

While some of the extractable organic matter may be used as is for gas chromatographic analysis (discussed later), most often it is separated by some form of chromatography into saturated, aromatic, resin (N-S-O), and asphaltene fractions. This S-A-R-A (Saturates-Aromatics-Resins-Asphaltenes) can be very rigorous or cursory depending on the data application. For source rock evaluation purpose, the separations are typically not rigorous. As noted in the next chapter, the S-A-R-A analysis is often applied to crude oils as well. To illustrate the separation used in the S-A-R-A analysis, a generic liquid column chromatography analysis method will be described.

At the start of the analysis, the amount of bitumen to be separated is weighed. The first step in the S-A-R-A analysis is the removal of the asphaltenes. This is accomplished by precipitating them out of the source rock extract (or crude oil) by adding n-heptane which causes the individual asphaltenes molecules to form aggregates that are insoluble. The material that remains in solution contains the hydrocarbons and N-S-O fractions is then separated on a liquid chromatography column. The column holds a stationary phase, typically silica gel or alumina, in equilibrium with a solvent. The sample material (either deasphalted bitumen or oil) is loaded onto the top of the column. The different groups of compounds in the sample material pass through the column at different rates due to differences in their partitioning behavior between the mobile liquid phase (the solvent) and the stationary phase (the silica gel or alumina). The solvent used for the mobile liquid phase can be changed to alter the partitioning behavior of the sample components. The separated groups of compounds are recovered by collecting aliquots of the column effluent as a function of time, solvent type, and solvent volume. The material recovered in the asphaltene precipitation and the material recovered in each of the fractions collected are taken to dryness and weighed.

Saturate, aromatic, and N-S-O, or resin, fractions are collected from the column. The saturate fraction contains primarily saturated hydrocarbon compounds. The aromatic fraction contains aromatic compounds, mostly hydrocarbons, but also including aromatic sulfur compounds. The N-S-O or resin fraction contains the bulk of the nitrogen, sulfur, and oxygen (NSO) bearing compounds. The saturate fraction is often used in gas chromatographic analysis (see later discussion), and both the saturate and aromatic fractions can be used in biomarker analysis, as discussed in Chapter 4.

Source Richness and Quality Interpretations

Prior to Rock-Eval pyrolysis data becoming an industry standard for source rock evaluation, most source richness and quality determinations were made using TOC and extractable organic matter data. The amount of C_{15}+ extract and C_{15}+ hydrocarbons in the extract (saturate + aromatic fractions), both expressed in ppm, were often used as an indicator of source rock richness. The ratio of ppm C_{15}+ extract and/or hydrocarbons expressed as a percent of the TOC were often used to indicate source rock quality. Some examples of how this data were used can be found in Baker (1960), Fuloria (1967), and Claypool et al. (1978).

There were a few shortcomings to these methods. Using extract data as source richness and quality indicators are biased toward oil-prone source rocks. The source rock also had to be mature enough to have generated some bitumen to be recognized as a source rock. However, the biggest problem stems from the analytical methods themselves. In the era before Rock-Eval, most of the analytical work was done by internal laboratories. The exact details about the solvents used during extraction, the separation methods, and the cutoff values applied were usually proprietary and did not appear in many published reports. As a result, it was very difficult to utilize outside extract data in source rock evaluations. With the standardization of analysis, data measured, and general interpretation from Rock–Eval pyrolysis, extract data were quickly abandoned.

GAS CHROMATOGRAPHY

While Rock-Eval pyrolysis and extract data deal with bulk properties of the organic matter in sediments, GC provides a means of determining the distribution of different compound types found in the bitumen. As the name implies, GC is a separation technique. But instead of the coarse separation obtained by the liquid column chromatography described earlier, GC provides a higher degree of separation to resolve more detail about the composition of the bitumen.

A schematic of a gas chromatograph is shown in Fig. 3.19. The analysis begins with a small sample of bitumen or the saturate fraction dissolved in a solvent, which is then introduced into the injection port via a syringe. The injected sample is vaporized onto the head of the gas chromatographic column. The column is a very small diameter tube, usually made of fused silica glass. The interior of the tube is coated with the stationary phase. As the vaporized sample is transported through the column by the flow of an inert carrier gas, separation occurs as a result of differences in the vapor pressure of individual compounds in the sample

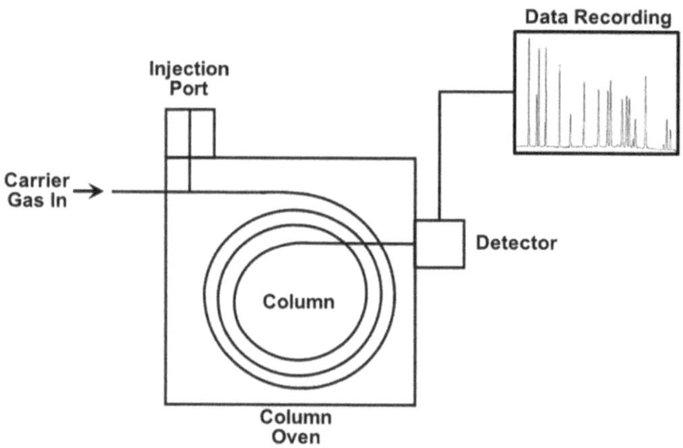

FIGURE 3.19 A schematic of a typical gas chromatograph used in the analysis of source rock extracts, crude oils, and natural gases.

and the partitioning of the compounds between the stationary phase and the carrier gas. The gas chromatographic column is usually housed in an oven which can be slowly heated to aid in the separation.

As compounds are eluted from the column, they are carried to a detector, such as a flame ionization detector or mass spectrometer, where the compounds in the carrier gas are sensed, and their relative concentrations are measured. These relative concentrations are plotted versus time as gas chromatograms, as shown in Fig. 3.20. Gas chromatograms consist of a series of peaks, representing the compounds eluting off the column. The higher the resolution of the analysis, the sharper the peaks will appear. Peaks are often identified by their retention time on the column during analysis. The retention time for specific compounds are usually established by running a standard mixture of reference compounds prior to analysis of an unknown sample.

In the example chromatogram in Fig. 3.20, the large peaks are normal paraffin peaks. Two peaks closely follow the n-C_{17} and n-C_{18} normal paraffin peaks. These are the isoprenoids pristane and phytane, respectively. Because pristane and phytane are almost always present in source rock extracts, they can be used to identify the n-C_{17} and n-C_{18} normal paraffin peaks, which in turn allows the other normal paraffin peaks to be identified. Besides the peaks in the chromatogram, other useful features include the shape of the peak envelope and the amount and shape of the unresolved material below the peaks.

Whole extract (bitumen) and saturate fraction chromatograms can provide insight into the type of hydrocarbons the source rock might generate and the maturity of the source rock.

Source Quality Interpretations

Clues to the kerogen type in a source rock can be obtained from the whole extract (bitumen) or saturate fraction gas chromatograms. The gas chromatograms are evaluated qualitatively

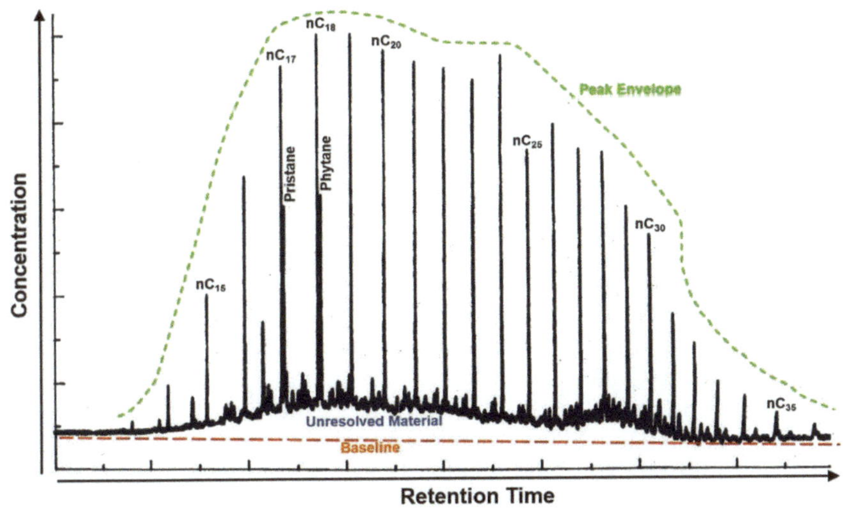

FIGURE 3.20 An example of a gas chromatogram of a saturate fraction from a source rock extract.

to determine the potential of a rock to generate oil versus gas, as well as identify the dominate kerogen type. Some example chromatograms are shown in Fig. 3.21.

Chromatograms of oil-prone rocks, such as Fig. 3.21A and B, exhibit prominent n-paraffin peaks with the bulk of the material in the greater than C_{15} carbon number range. These source rock saturate fraction chromatograms resemble saturate fraction chromatograms of crude oils. Chromatogram A shows little unresolved material below the peaks and predominantly large n-paraffin peaks extending out to the C_{30} range, which is characteristic of Type I kerogen. Chromatogram B shows abundant unresolved material below the peaks with n-paraffins decreasing in height in the C_{15}–C_{30} range. This is the characteristic of source rocks dominated by Type II kerogen. Chromatograms of gas-prone source rocks are typically dominated by shorter-chain carbon compounds, with the bulk of the material in the less than C_{20} range. These characteristics, as shown in chromatogram C in Fig. 3.21, are more typical of source rocks containing mostly Type III kerogen. It should be noted that it usually requires some experience with viewing and working with gas chromatograms to become proficient with interpreting them properly. It is also best to utilize other data such as Rock-Eval or PGC data to corroborate any interpretations made.

Thermal Maturity Interpretations

Compositional changes in gas chromatograms with increasing thermal maturity were discussed in detail in Chapter 2. As a brief review here, an example of changes in the gas chromatograms of saturated hydrocarbons with maturity is shown in Fig. 3.22. The immature stage shows a bimodal envelope of unresolved material beneath the peaks, and there is an odd carbon number predominance in the C_{25}–C_{35} n-paraffins. In the moderately mature sample, there is a noticeable reduction in the higher molecular weight "hump" of unresolved material, and the odd carbon number predominance is gone. In the chromatogram of the mature sample, the higher molecular weight "hump" of unresolved material has been eliminated and there is a substantial increase in the n-paraffins in the C_{15}–C_{18} range making the saturate fraction look more like crude oil.

In addition to looking at the gas chromatograms, the height of the n-paraffin peaks can also be used to track changes with maturity. One method is to calculate the Carbon Preference Index, or CPI, as shown in Fig. 3.23. The CPI was developed by Bray and Evans (1961) to help distinguish immature n-paraffin distributions from more mature n-paraffin distributions in source rock and crude oils. Most immature source rocks exhibit CPIs greater than 1.0, while mature source rocks and crude oil have CPIs of about 1.0 (Bray and Evans, 1965).

Another way of using the peak heights of the n-paraffin to track changes with maturity is to plot the peak heights versus carbon number, as shown in Fig. 3.24. The shallow, immature n-paraffin distribution exhibits a saw-toothed pattern with the odd carbon number n-paraffins dominating the C_{25}–C_{33} range. As the sediments get deeper and more mature, the saw-toothed pattern diminishes and the maximum of the distribution of n-paraffins shifts down to the lower molecular weight range, becoming more like crude oil. CPI values for these n-paraffin distributions are also included in Fig. 3.24 to demonstrate how they change with increasing maturity.

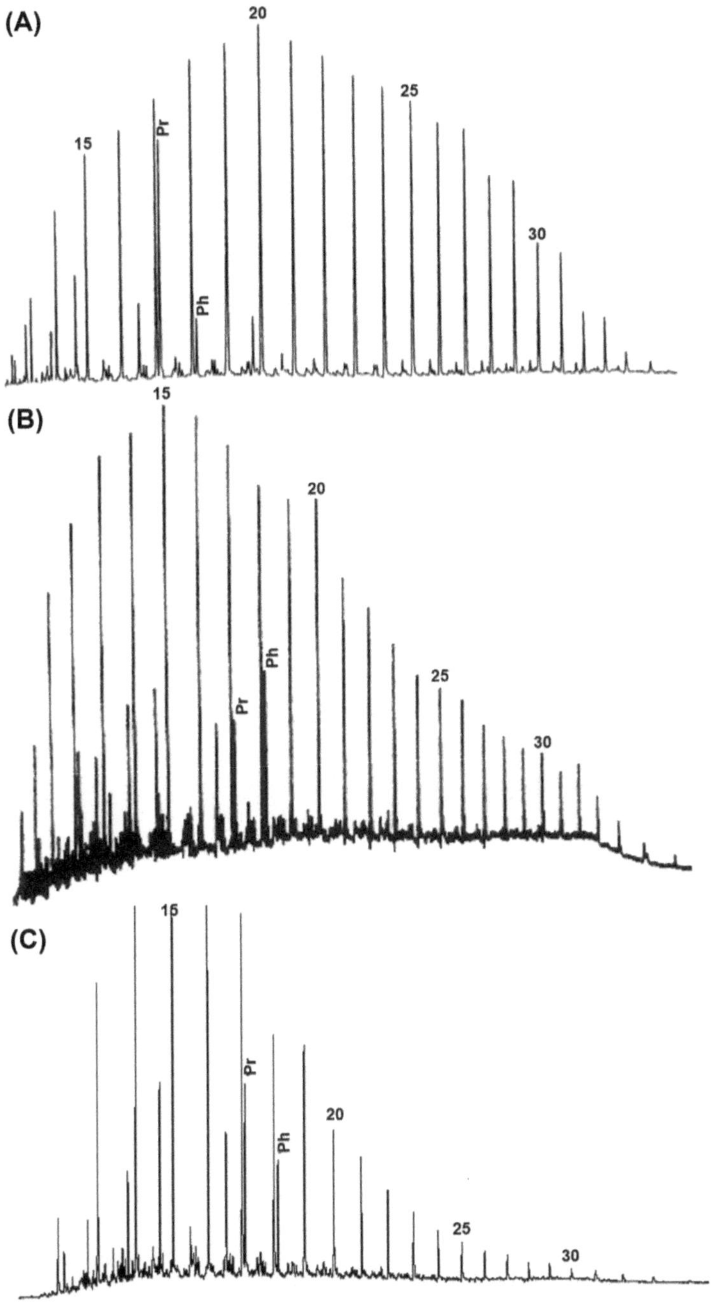

FIGURE 3.21 Rock extract gas chromatograms from source rock with different kerogen types. Type I, Type II, and Type III are the dominant kerogen types in the extracts shown in chromatograms (A), (B), and (C), respectively.

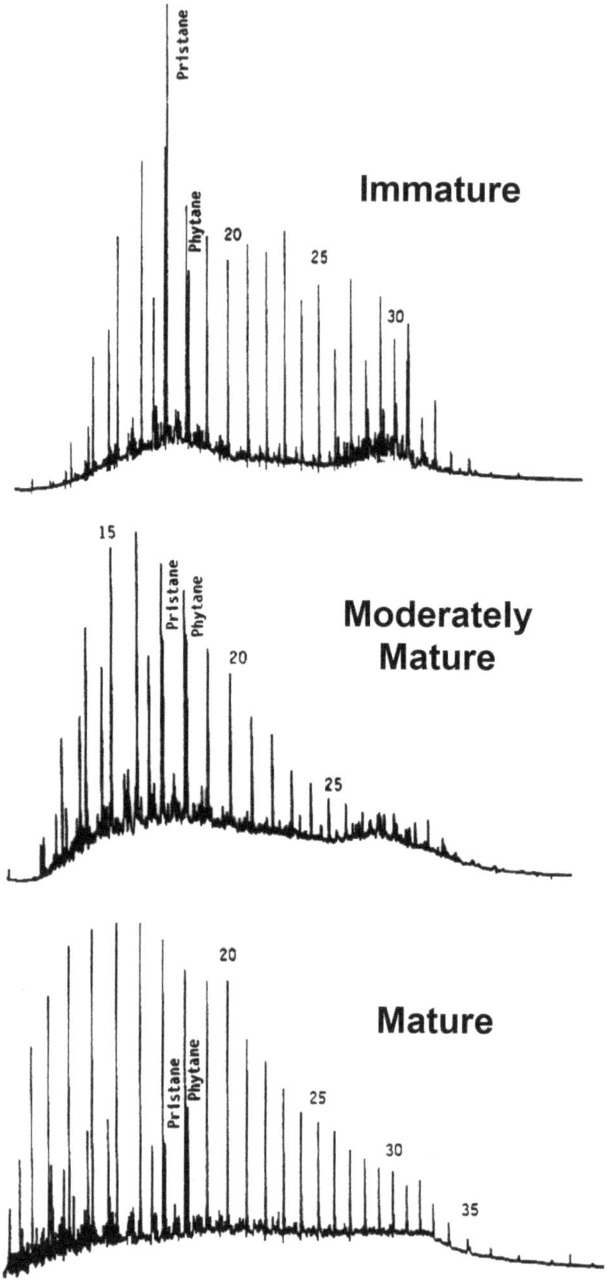

FIGURE 3.22 Progressive changes in the source rock extract gas chromatograms due to increasing maturity from a series of samples with Type II dominated kerogen.

$$\text{CPI} = 0.5 \times \left[\frac{\Sigma\, C_{21} \text{ to } C_{31} \text{ Odd n-Paraffins}}{\Sigma\, C_{22} \text{ to } C_{32} \text{ Even n-Paraffins}} + \frac{\Sigma\, C_{21} \text{ to } C_{31} \text{ Odd n-Paraffins}}{\Sigma\, C_{20} \text{ to } C_{30} \text{ Even n-Paraffins}} \right]$$

FIGURE 3.23 Equation for the calculation of CPI as defined by Bray and Evans (1961).

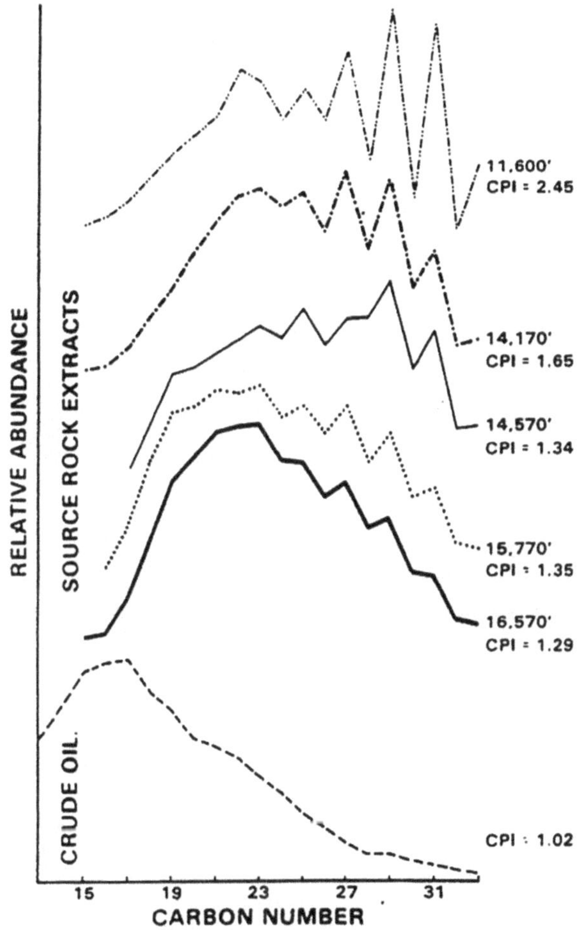

FIGURE 3.24 Progressive changes the n-paraffin distributions and CPI due to increasing maturity for a series of source rocks samples with similar kerogen type.

Recognizing Contamination

Contamination of cutting and core samples, especially from oil-based drilling mud and drilling mud additive, is often recognized in whole extract and saturate fraction gas chromatograms. An example, in Fig. 3.25, shows the distinctive pattern for Novaplus, an olefin-based synthetic drilling mud base oil in a source rock extract. The characteristic clusters of olefin peaks around C_{16} and C_{18} dominate the chromatogram even at contamination levels

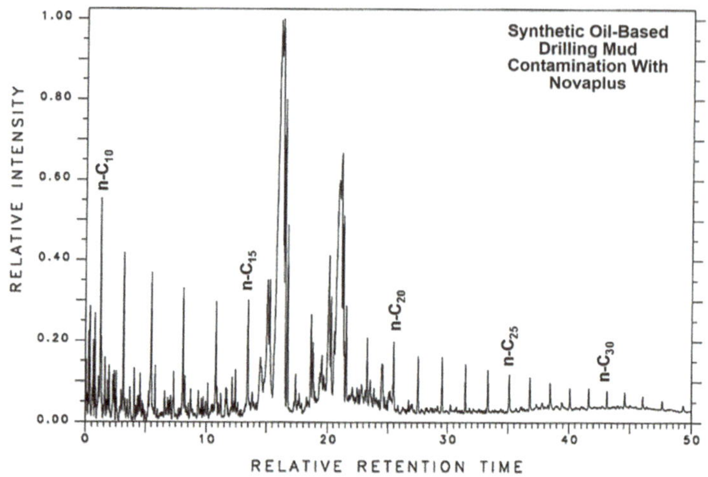

FIGURE 3.25 Source extract gas chromatogram showing contamination of Novaplus, a synthetic oil-based drilling mud.

of about 1–2% by volume. Other contamination such as diesel, grease pencil, and processed asphalt mud additives typically stand out from the bitumen indigenous to the source rock. To assist in the recognition of contamination from oil-based drilling mud and drilling mud additive, obtain the mud report for the well listing the type of mud and all additives used. Reference gas chromatograms are often available from the manufacturer for comparison. As mentioned earlier in the section on sample collection, drilling mud and mud additive sample should be collection during the drilling of the well. These samples can also be used as reference materials to help recognize possible contamination.

HEADSPACE GAS ANALYSIS

Headspace gas data have applications in both source rock evaluation and reservoir geochemistry. The discussion here will emphasize utilizing the data in source evaluation. Discussion of the reservoir applications will be covered in Chapter 5.

Although beneficial in oil plays, headspace gas data from canned cuttings are especially useful in areas where source rocks maybe prone to gas or gas/condensate production. The headspace gas captures the light hydrocarbons from C_1 up to C_7 outgassing from the sediments interstitial spaces, as shown in Fig. 3.26 and allow them to be analyzed. Because of difficulties in obtaining quantitatively reproducible results from C_5 to C_7, the analysis is usually confined to the C_1 to C_4 light hydrocarbons.

If quart size metal paint cans were used in the collection of the canned cuttings, the analytical procedure begins with a septum being glued to the top of the can. The septum provides a leak-free seal that can be pieced by a syringe in order that the headspace gas can be sampled. If the Isojars were used, a septum is already incorporated into the lid. The cans or the Isojars are brought to a constant temperature, usually in a water bath, prior to analysis to allow

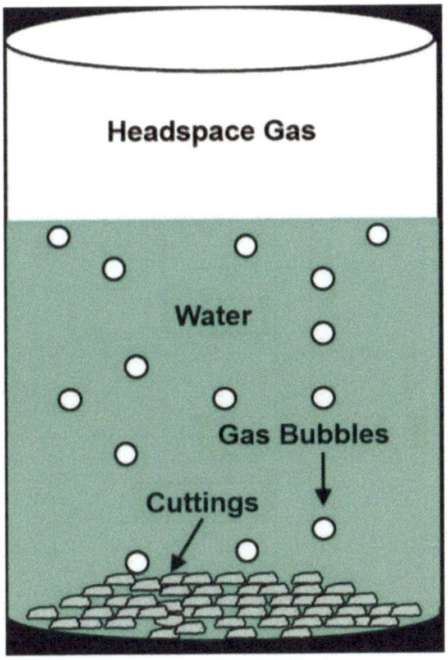

FIGURE 3.26 A schematic of a canned cuttings sample used in headspace gas analysis.

the partitioning of the light hydrocarbons gases between the headspace and the water to equilibrate. A fixed sized sample of the headspace gas is then taken with a gas-tight syringe and injected into a gas chromatograph for analysis. The gas chromatograph separates the headspace gas into its component parts and allows the measurement of the contribution of each component to the gas. After correcting for the air in the original headspace of the can or Isojar, the concentrations of methane (C_1), ethane (C_2), propane (C_3), n-butane(nC_4), and isobutane(iC_4) are calculated. The headspace gas data are usually reported as the concentration of C_1–C_4 hydrocarbons in ppm, the % wet gas (C_2–C_4) and the iC_4/nC_4 ratio, and are often plotted versus depth, as shown in Fig. 3.27.

In addition to the gas chromatographic analysis, the headspace gas is occasionally used for stable carbon isotope determination. This is more typically done on headspace gas in reservoir intervals and as such will be covered in the Chapter 5.

Source Richness Interpretations

The concentration of the headspace gas is the main indicator of source richness. As with extraction data, minor variations in laboratory protocols often make it difficult to compare the results from one laboratory to the other. This has resulted in cutoff values for source richness assessments being laboratory specific because of these variations in methods. However, Noble (1991) offers some general guidelines for the concentration of total gas, C_1–C_4, expressed in ppm. Total gas concentrations of less than 100 ppm are background, while 1000–10,000 ppm indicate rich source rocks and above 10,000 ppm would suggest a very rich source.

Care must be taken with applying these interpretations. Kerogen type may influence the amount of light hydrocarbons that is generated. There is also the potential for these light

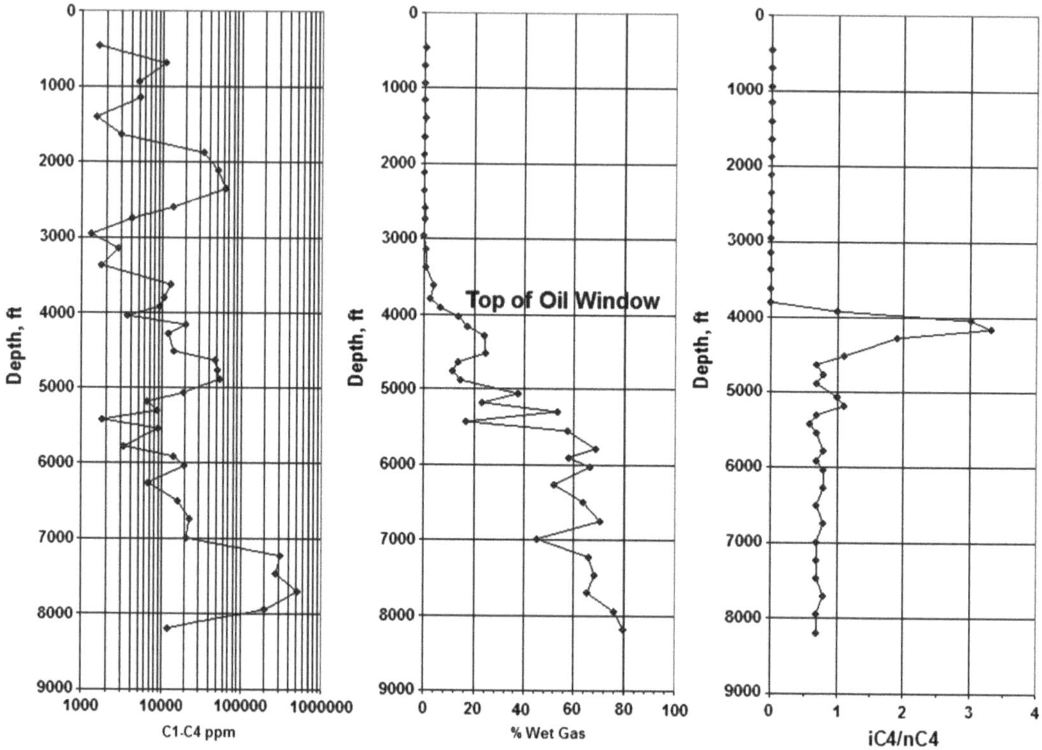

FIGURE 3.27 A plot of headspace gas parameters versus depth.

hydrocarbons to more easily migrate out of the source rock resulting in indications of a less rich interval. If oil-based drilling mud was used, the oil can act as a solvent and hold a portion of the light hydrocarbons in solution, giving lower total concentrations.

Thermal Maturity Interpretations

There are two main ways to interpret maturity from headspace gas data. The first utilizes the % wet gas as an indicator of maturity, with the top of the oil window usually placed at about 10%. The wet gas will continue to increase until peak generation and then begin to decrease. The bottom of oil window is indicated when the % wet gas decreases to about 20%. The second uses the isobutane/n-butane (iC_4/nC_4) ratio. When the iC_4/nC_4 is less than 0.5, the sediments are likely immature, ratios between 0.5 and 1.0 are considered mature, and when the iC_4/nC_4 ratio is greater than 1.0 the sediments are postmature.

These maturity interpretations should be considered tentative and require corroboration from other maturity indicators. This is due to the influence of kerogen type on the light hydrocarbons generated, as well as the potential for these light hydrocarbons to more easily migrate in the subsurface. There are also potential problems in interpretation if the well was drilled with oil-based drilling mud. If not rinsed off the cuttings at the time of collection, oil-based mud can act as solvents and hold a portion of the light hydrocarbons in solution which may alter the % wet gas and the iC_4/nC_4 ratio. Caution is urged when utilizing these data.

PYROLYSIS–GAS CHROMATOGRAPHY

PGC is essentially combining Rock-Eval pyrolysis and GC to reveal the composition of the hydrocarbons and other compounds that make up the S2 peak. Knowing what compounds make up the S2 peak provides a direct indicator of the kerogen type in the source rock and the types of hydrocarbons that can be generated by the kerogen (Giraud, 1970; Larter and Douglas, 1980; Dembicki et al., 1983; Horsfield, 1989). It is used to augment Rock-Eval data to provide more accurate assessments of kerogen type (Dembicki, 1993, 2009).

PGC analysis will use an instrument similar to the one shown schematically in Fig. 3.28. A source rock sample is sealed in the pyrolysis oven, heated to 300°C, and held at that temperature for 3–5 min to remove the S1 peak. The material from the S1 peak is collected in one of the traps. After the collection of the S1, the pyrolysis oven is ramped up from 300°C to about 550°C at a fixed heating rate to collect the S2 material in the second trap. The material collected in the traps can be liberated by heating to about 300–325°C. The collected S1 peak material may either be vented or analyzed by GC. The GC analysis of the S1 peak is often referred to as thermal extraction-GC and is useful for characterizing the bitumen already generated, identifying possible contamination, and in reservoir geochemistry applications, as discussed in Chapter 5. The gas chromatographic analysis of the collected S2 peak material provides a qualitative and quantitative analysis of the chemistry of the thermal decomposition products of the kerogen.

PGC assists in kerogen type assessment by providing details of the contents of the equivalent of the Rock-Eval S2 peak. There is some variation in the data due to differences in the

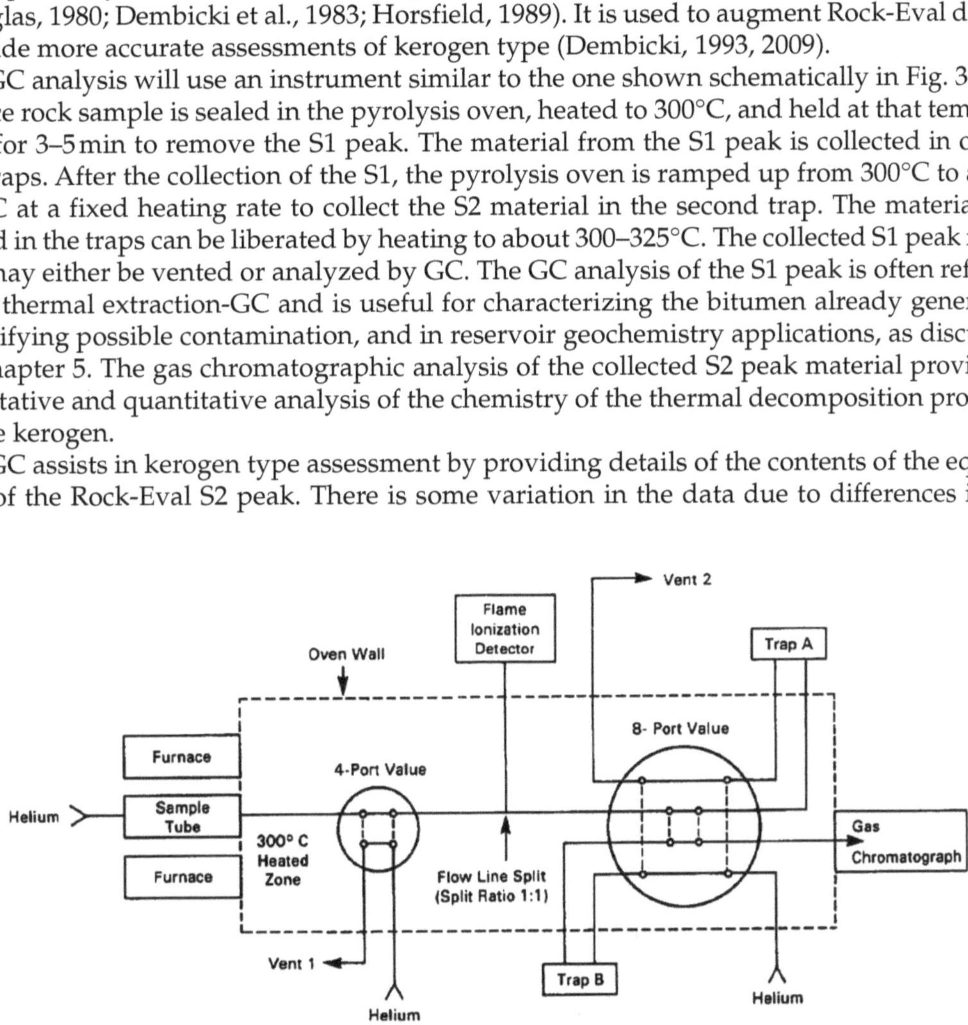

FIGURE 3.28 A schematic of a typical pyrolysis-gas chromatograph used in source rock analysis. *From Dembicki Jr., H., Horsfield, B., Ho, T.T.Y., 1983. Source rock evaluation by pyrolysis-gas chromatography. American Association of Petroleum Geologists Bulletin 67, 1094–1103.*

design of the PGC instrument and the operating conditions used. For this discussion, the results from the study done by Dembicki (2009) will be used as an example of how PGC improves kerogen-type assessments. Other interpretation schemes, such as Larter and Douglas (1980) and Horsfield (1989), use slightly different approaches but can be applied with equally satisfactory results.

Fig. 3.29 shows typical pyrolysis-gas chromatograms of immature source rocks containing predominantly Type I, Type II, and Type III kerogen. The oil-prone Type I and Type

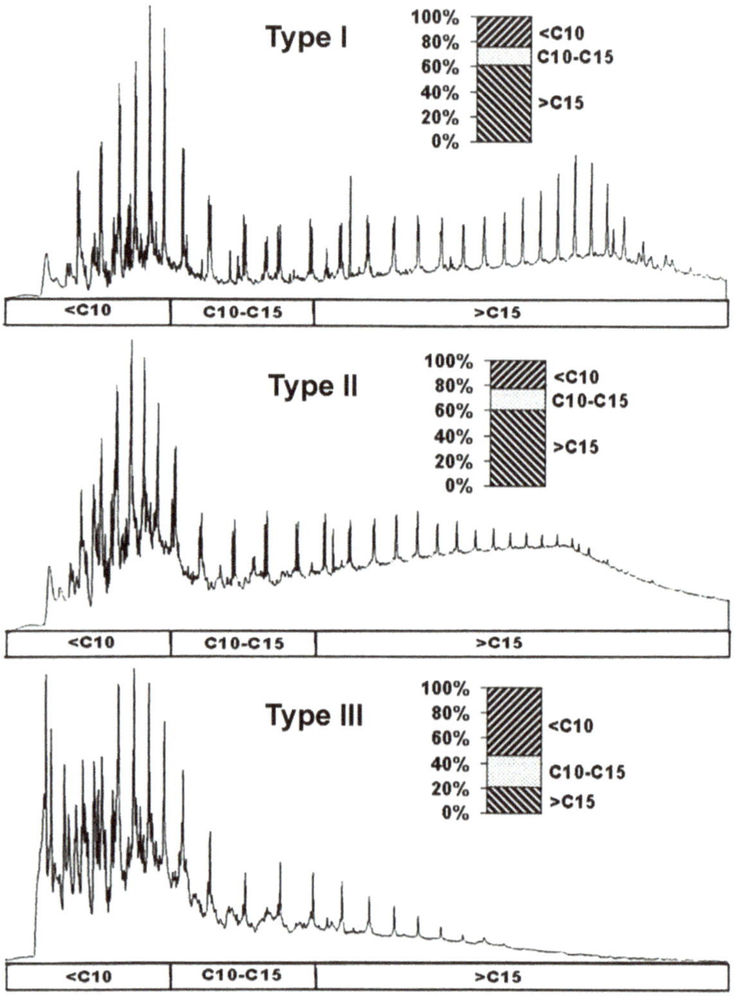

FIGURE 3.29 Whole rock pyrolysis–gas chromatograms of source rocks containing Type I, Type II, and Type III kerogen. Also included is the amount of material in each chromatogram in the $<C_{10}$, C_{10}–C_{15}, and $>C_{15}$ carbon number ranges for each kerogen type. The Type I source rock is the Eocene age Green River Shale from the Green River Basin, the Type II source rock is the Jurassic age Kimmeridge Clay from Dorset, England, and the Type III source rock is a Tertiary age deltaic sediment from Southeast Asia. *From Dembicki Jr., H., 2009. Three common source rock evaluation errors made by geologists during prospect or play appraisals. American Association of Petroleum Geologists Bulletin 93, 341–356.*

II kerogens give pyrolysis products that extend out to the high molecular weight range (C_{30}+). While Type I kerogen produces abundant long-chained n-alkanes and n-alkenes, as shown by the prominent peaks in the chromatogram in the >C_{15} range, the pyrolysis products from Type II kerogen consist of more naphthenic compounds represented by the large unresolved "hump" of material in the >C_{15} range. The Type III gas-prone kerogen is significantly different, showing the bulk of the pyrolysis products confined to the <C_{10} portion of the chromatogram. These differences are also reflected in the amount of material in the <C_{10}, C_{10}–C_{15}, and >C_{15} carbon number ranges for each kerogen type, shown in bar diagrams in Fig. 3.29. High concentrations of the >C_{15} compounds are characteristic of the Type I and Type II kerogens, reflecting the oil-prone nature of these kerogens. In contrast, the Type III bar diagram shows most of the products are in the <C_{10} fraction, as expected for gas-prone kerogens.

In addition to Types I, II, and III, Type II-S and Type IV can be included in this classification scheme. With respect to their pyrolysis–gas chromatograms, Type II-S kerogens are virtually identical to Type II. As a result, Type II-S would be distinguished as having a Type II PGC signature while plotting along the Type I trend on a Rock-Eval pseudo-van Krevelen diagram. And due to their essentially inert nature, Type IV kerogens would not yield any significant amount of hydrocarbons upon pyrolysis and little or no material would be observed in a pyrolysis–gas chromatogram.

When dealing with mixtures of kerogen types in a source rock, PGC results can give a clear picture of the kerogen types present. Using the results from the study done by Dembicki (2009), pyrolysis–gas chromatograms are shown in Fig. 3.30 of mixtures of Type I and III and Type II and III. The mixtures are varied from 0% to 100% Type III in 25% increments. PGC results produced by these mixtures are intermediate between the end-point kerogen types. The changes in the chromatograms are progressive with the amount of pyrolysis products in the >C_{15} fraction decreasing in both the Type I–III and Type II–III mixtures as the amount of material in <C_{10} fraction increases. Features of the >C_{15} region of the pyrolysis–gas chromatograms, allow the differentiation of mixtures with Type I and II kerogens with up to 75% Type III kerogen. The Type I mixtures exhibit more prominent peaks, while the Type II mixtures exhibit a more prominent naphthenic unresolved "hump." While some of these changes in the chromatograms may appear to be subtle at first glance, experienced interpreters of the data can accurately estimate the types of kerogens present.

These progressive changes are also reflected in the bar diagrams of the <C_{10}, C_{10}–C_{15}, and >C_{15} carbon number ranges in Fig. 3.31. The amounts of material in the <C_{10}, C_{10}–C_{15}, and >C_{15} carbon number ranges are essentially the same for Type I and Type II kerogen. Therefore, interpretations using the quantified data will only distinguish between oil-prone and gas-prone kerogen and not the individual kerogen types.

Maturation, as with Rock-Eval data, can have a profound impact on how PGC data are interpreted. The data discussed earlier came from immature source rock samples, and the interpretation methods described are valid for samples that range from immature to near peak oil generation. After peak oil generation, Type I and Type II kerogen have been significantly depleted in the larger chemical moieties that generate oil (>C_{15}), and their pyrolysis products begin to appear more like Type III. With increasing maturity, eventually the pyrolysis products of all the reactive kerogen types begin to look the same. This is similar to the convergence of the kerogen type trends on the Rock-Eval pseudo-van Krevelen diagram.

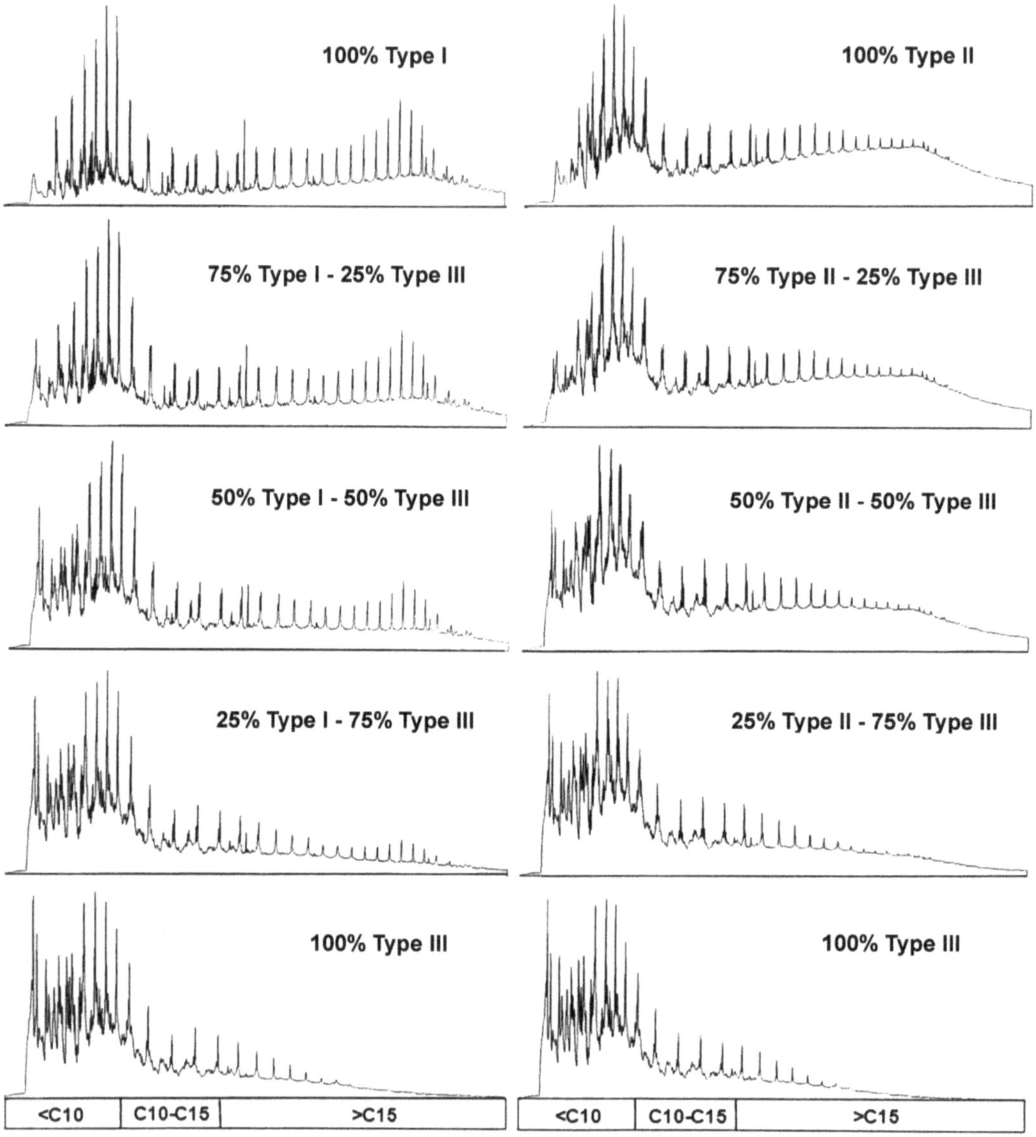

FIGURE 3.30 Pyrolysis–gas chromatographic results showing the progressive changes resulting from the mixing Type I and Type II kerogen with Type III kerogen. *From Dembicki Jr., H., 2009. Three common source rock evaluation errors made by geologists during prospect or play appraisals. American Association of Petroleum Geologists Bulletin 93, 341–356.*

Therefore caution must be exercised when attempting to use PGC on source rock samples at maturities higher than peak oil generation.

Contamination of the source rock sample can also have an impact on PGC data. This may be in the form of bitumen rich in resins and asphaltenes or organic drilling mud additives, especially those consisting of processed asphalt or Gilsonite. These materials may not be volatile at 300°C during the thermal extraction step. As with Rock-Eval data, it is necessary to

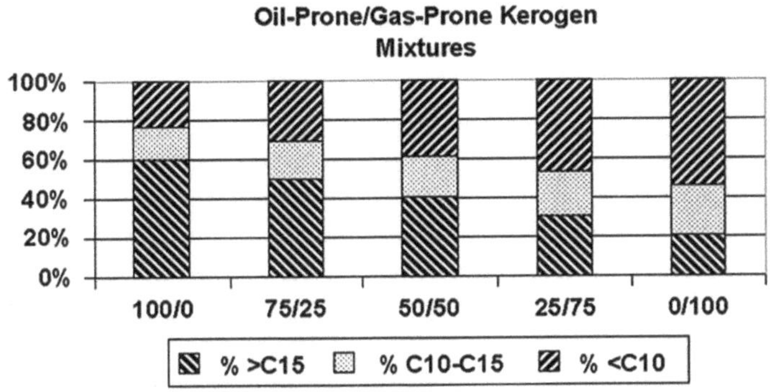

FIGURE 3.31 Pyrolysis–gas chromatographic results showing the progressive changes resulting from the mixing oil-prone Type I or Type II kerogen with gas-prone Type III kerogen quantified in bar diagrams of the $<C_{10}$, C_{10}–C_{15}, and $>C_{15}$ carbon number ranges. *From Dembicki Jr., H., 2009. Three common source rock evaluation errors made by geologists during prospect or play appraisals. American Association of Petroleum Geologists Bulletin 93, 341–356.*

inspect the pyrograms to look for asymmetry in the S2 peak on the low-temperature side. If the amount of material is substantial, a preanalysis solvent extraction of the sediment may be necessary to remove the bitumen or contaminants to obtain PGC data more representative of the kerogen.

KEROGEN ISOLATION

Kerogen isolation is not in itself an analytical procedure. It is a preparatory step for elemental analysis and the optical kerogen analysis methods discussed later. The first step is a coarse grinding of the source rock sample to not smaller than very coarse sand size (1–2 mm). Grinding to a finer size risks breaking up kerogen particles and obscuring their structure. The rock is then treated with hydrochloric acid (HCl) to remove carbonates. Because the reaction of HCl with carbonates can be violent to the point of break up of kerogen particles, one approach is to do the HCl treatment several times starting with a dilute solution and increasing the concentration with each subsequent step up to 6 N HCl. The residue from the HCl treatment must be rinsed thoroughly with distilled water prior to proceeding to remove any residual calcium ions to avoid formation of insoluble calcium fluoride.

The next step is treatment with 40% hydrofluoric acid (HF) to remove silicates including quartz. Often, the HF is mixed with some 6 N HCl to facilitate the reactions. HF is a very hazardous material and requires a special chemical exhaust hood and extra personal protection equipment for safety. The HF treatment may be repeated to insure complete dissolution of any reactive silicate mineral present in the rock sample. Again, the residue must be rinsed thoroughly with distilled water prior to proceeding to remove any residual HF in the sample.

At this stage, the remaining material will consist of organic matter and refractory mineral including pyrite and heavy oxides such as zircon, rutile, and anatase (Vandenbroucke and Largeau, 2007). In addition, some fluorides may form which may be difficult to remove (Durand and Niacaise, 1980). A second treatment with 6 N HCl may be needed to facilitate their removal.

This kerogen concentrate with refractory minerals is often used to prepare strewn slide for transmitted light microscopy for visual kerogen analysis and assessment of the thermal alteration index (TAI) discussed later. However, for use in vitrinite reflectance or elemental analysis, the kerogen concentrate requires the bulk of the refractory minerals be removed by heavy liquid separation, typically using zinc bromide ($ZnBr_2$) at a specific gravity 2.1 g/cm^3. The "kerogen" floats on the $ZnBr_2$ and is removed, rinsed, and dried.

Occasionally, contamination by oil-based drilling mud interferes with the kerogen isolation procedure. The oil coats the mineral surfaces inhibiting or preventing their reaction with the acids. To remedy this, samples may be solvent extracted to remove the oil prior to kerogen isolation.

While a highly concentrated kerogen fraction can be obtained by these methods, the complete isolation of the kerogen from its mineral matrix is impossible. Some loss of kerogen is experienced at every step. The most common mode of kerogen loss in the process is through rinsing the rock sample after acid digestion. Small, mainly unstructured kerogen particles are often liberated during these reactions and washed away during rinsing. In addition, hydrolysis reactions involving the kerogen due to exposure to HCl and HF and oxidation promoted by the exothermal nature of the mineral dissolution reactions can change the composition of the kerogen (Durand and Niacaise, 1980). And not all the minerals have been removed. The kerogen is often observed to clump or cluster around pyrite allowing it to float during the heavy liquid separation (Saxby, 1970).

ELEMENTAL ANALYSIS

Elemental analysis for carbon, hydrogen, oxygen, and nitrogen is done on kerogen isolated from the source rock. The first step in the analysis is to weigh a sample of the kerogen concentrate and then burn it in an oxygen atmosphere. Water (H_2O), carbon dioxide (CO_2), and nitrous oxides are formed from the organic matter. The H_2O and CO_2 are measured to calculate the hydrogen and carbon contents. The nitrous oxides are first reduced to nitrogen (N_2) with a catalyst, and the N_2 is then measured to calculate the nitrogen content. In the second step, oxygen is determined by pyrolyzing some of the kerogen so that all the oxygen in the sample is released as either carbon dioxide (CO_2) or carbon monoxide (CO). The CO is then converted to CO_2 using a catalyst and all the CO_2 is measured to calculate the oxygen content. The weight percent H, C, O, and N are then divided by their atomic weight and H/C, O/C, and N/C atomic ratios are calculated.

The elemental composition of kerogen will depend on both its original elemental composition and the maturation level it has achieved. If you know the kerogen type or the maturity of the source rock, it should be possible to infer the other information using elemental analysis (Baskin, 1997). In practice, elemental analysis data is most often used as an indicator of kerogen type using an independent method for determining maturity (Tissot et al., 1974). The main interpretation tool used with elemental analysis data is the van Krevelen diagram (Tissot and Welt, 1984), shown in Fig. 3.32. As discussed in Chapter 2, the van Krevelen diagram plots the H/C atomic ratio of kerogen versus the O/C atomic ratio and the pathways for Type I, II, and III. These kerogen pathways begin separately while they are in the diagenesis/immature zone. The board bands for the composition of each kerogen type reflects variation

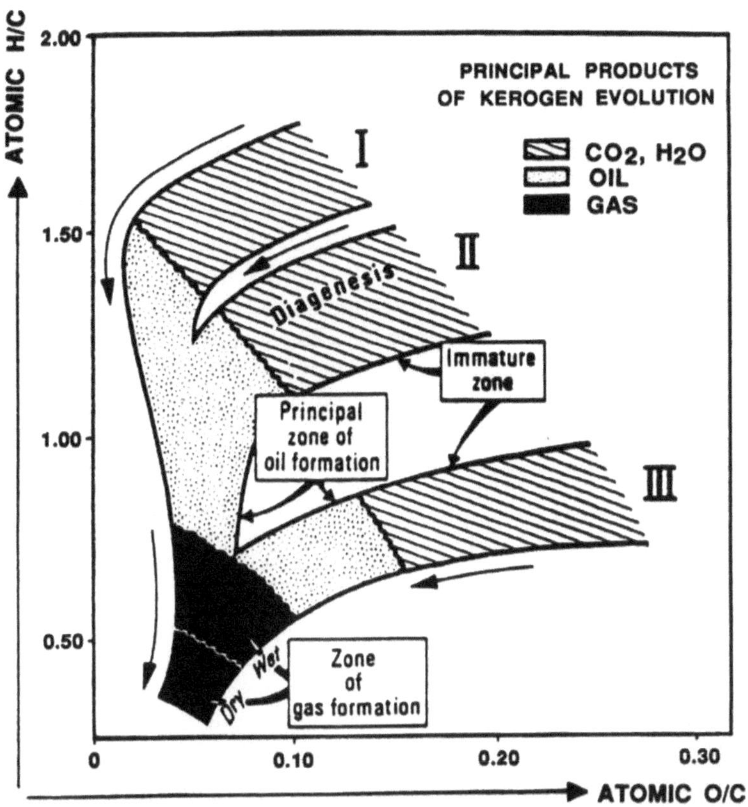

FIGURE 3.32 The van Krevelen diagram plotting atomic H/C ratios versus atomic O/C ratios adapted for kerogen. *From Tissot, B.P., Welte, D.H., 1984. Petroleum Formation and Occurrence. Springer-Verlag, New York, 699 p.*

in composition mostly likely due to mixtures of kerogen types present in the samples used, as well as compositional variation in the actual end-member kerogens. As the kerogens evolve with thermal maturity, they enter the principal zone of oil formation. The Type I and II kerogen trends begin to merge as this stage progresses. These basic trends document that hydrogen and oxygen are being depleted in all three kerogens, while the carbon content is being enriched. Finally as all three kerogen types enter the zone of gas formation, their elemental compositions become virtually indistinguishable from each other.

If the kerogen type in a source rock is known from other data, the elemental analysis results can be used to estimate maturity. Average values of atomic H/C and O/C ratios relative to vitrinite reflectance, shown in Fig. 3.33, have been worked out by Tissot and Welte (1984) for this purpose. Note that due to kerogen mixtures, these values are only approximate and should only be used in conjunction with other maturity data.

Elemental analysis was used extensively prior to 1980 and can be found in many source rock data compilations from that time. However, elemental analysis is seldom used today as a routine source evaluation tool. This is mainly due to the time required for proper sample preparation for the analysis. As shown in the earlier discussion, it is difficult to isolate a clean kerogen sample

	Type I		Type II		Type III	
% Ro	H/C	O/C	H/C	O/C	H/C	O/C
0.6	1.45	0.05	1.25	0.08	0.80	0.18
1.3	0.70	0.05	0.70	0.05	0.60	0.08
2.0	0.50	0.05	0.50	0.05	0.50	0.05

FIGURE 3.33 Average values of atomic H/C and O/C ratios. *Modified from Tissot, B.P., Welte, D.H., 1984. Petroleum Formation and Occurrence. Springer-Verlag, New York, 699 p.*

from a potential source rock. There are always refractory minerals associated with the isolated kerogen, and these refractory minerals can contribute hydrogen and oxygen to the analysis (Durand and Monin, 1980). To assure that the kerogen is relatively free of refractory minerals, the ash content of the kerogen should be less than 10%. Kerogen loss during the rinsing and heavy liquid separation during kerogen isolation may also result in the final product not being representative of the kerogen in the source rock under analysis. And finally, elemental analysis data is subject to misinterpretation due to the presence of mixtures of kerogen in the source rock, similar to the mixed kerogen problem discussed with the Rock-Eval pseudo-van Krevelen diagram.

VITRINITE REFLECTANCE

Vitrinite is a type of kerogen particle, or maceral, formed from humic gels that are thought to be derived from the lignin-cellulose cell walls of higher plants (Teichmuller, 1989). It is a common component of coals and source rocks. The increase in reflectance of vitrinite particles with increasing time and temperature was first observed in coals and used as a means of determining a coal's rank or thermal maturity (Teichmuller, 1982a). After vitrinite was recognized as a component in source rock kerogens, this systematic increase in vitrinite reflectance was related to the hydrocarbon generation history of sediments and adopted as a maturity indicator for source rock evaluation.

Vitrinite reflectance is measured on populations of randomly oriented particles in source rock samples. Measurements can be made using either a polished whole rock mount, often consisting of cutting or core chips embedded in epoxy, or using a kerogen concentrate embedded in epoxy. The epoxy mounts are ground flat and polished for microscopic examination. The whole rock mounts maintain the relationship between the kerogen particles and the mineral matrix providing some clues to whether the vitrinite is in situ or reworked. Whole rock mounts are best used with organic-rich source rocks. In contrast, the kerogen concentrate mounts increase the probability of finding vitrinite in the samples where the organic matter content is not particularly high.

Observations are made under oil immersion using a reflected light microscope. A calibrated light beam is directed at a vitrinite particle, and the amount of light reflected from the surface is then measured by a sensor, usually a photomultiplier tube, as shown in Fig. 3.34. The microscope systems are calibrated with a set of known reflectance standards at regular intervals throughout the day to detect and correct for any changes in the light source or instrument drift in the sensor that may occur. Barker and Pawlewicz (1993) suggest that 20–30 measurements be made on each sample, when possible, to obtain statistically reliable results. More than one population of reflectance values may be found in a single sample. Mean values and standard deviations are calculated for each of the populations of vitrinite particles from each sample and reported as percent reflectance in oil immersion, % Ro. The mean values,

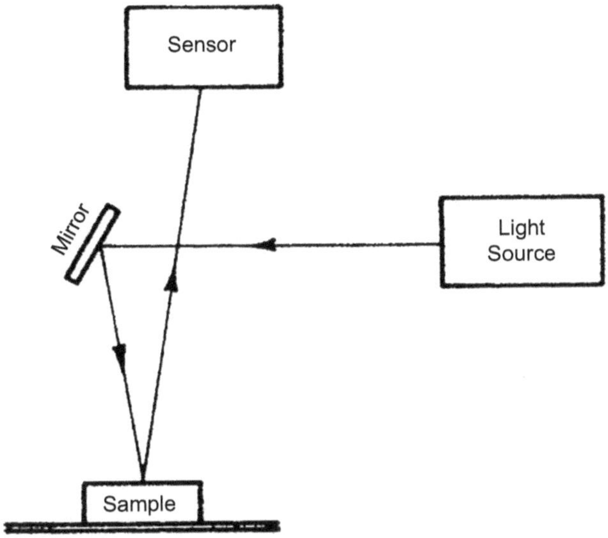

FIGURE 3.34 A schematic of the measurement of vitrinite reflectance.

often with standard deviation error bars, are plotted on a log scale versus depth on a linear scale. A good overview of the vitrinite reflectance method can be found in Cardott (2012).

Recently, ASTEM method D7708e11 (ASTM, 2014) was developed to standardize the measurement of vitrinite reflectance in sedimentary rocks. This method specifies the use of whole rock samples and sets down reporting standards for the observations made. While this is a step in the right direction toward improved vitrinite reflectance measurements and better agreement between laboratories, problems still exist. Hackley et al. (2015) reported on an interlaboratory comparison using the ASTM method and a common set of samples. Organically lean and higher maturity samples were found to give lower repeatability and reproducibility of the measurements. Hackley et al. (2015) also found that misidentification of other macerals, especially solid bitumen, as vitrinite contributed to poor reproducibility from lab to lab. These results point out the need for additional work on this method as well as the need for more training of the operators making these measurements.

Ideally, the vitrinite reflectance trend should show the data increases with increasing depth as a straight line with a surface intercept for the trend between 0.20% and 0.23% Ro, as shown in Fig. 3.35. Unfortunately, this is not always the case. Vitrinite reflectance data do not always increase with increasing depth. The plot of the log vitrinite reflectance versus linear depth does not always plot as a straight line. And the surface intercept of the trend may not be between 0.20% and 0.23% Ro. It is therefore necessary to investigate some of the causes of these deviations from the ideal to make proper and accurate interpretations of vitrinite reflectance data, as discussed below.

Interferences With Vitrinite Reflectance

As stated above, vitrinite reflectance trends with depth do not always exhibit the characteristics of the "ideal" trend. There are many possible interferences with the vitrinite trend that

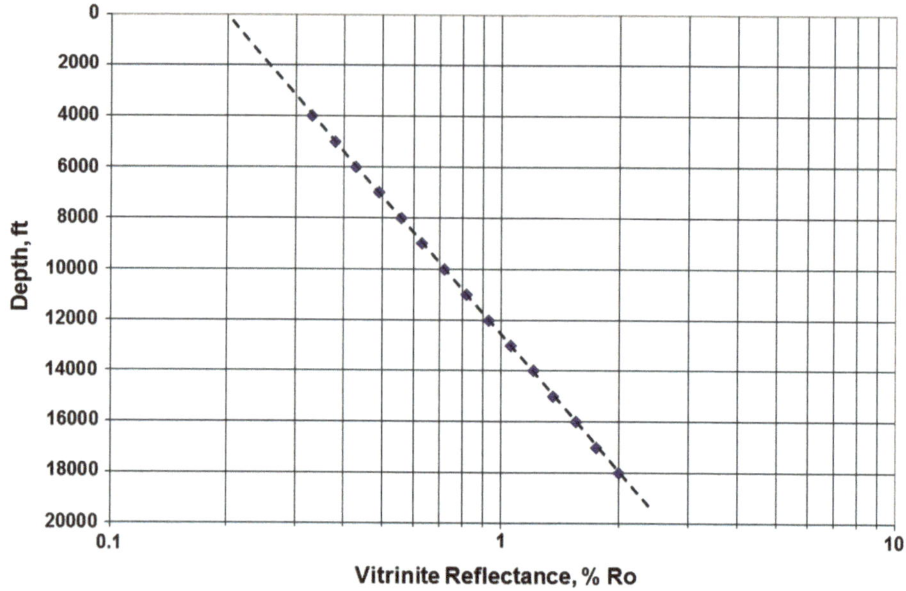

FIGURE 3.35 An ideal vitrinite reflectance trend with depth.

can complicate the interpretation of the data. These include lack of vitrinite, caved sediments, reworked organic material, faults, unconformities, thermal events, suppression/misidentification, and anisotropy. Often, more than one of these interferences can be encountered in the same data set. The following discussion will describe each of these interferences, give an example when possible, and suggest a way to compensate for them during the maturity interpretation process.

Lack of Vitrinite

Not all sediments contain vitrinite. Some depositional settings may simply not receive significant contributions from higher plants, resulting in little or no indigenous vitrinite. Because of its higher plant origins, vitrinite should not exist in pre-Devonian sediments in the lower Paleozoic and Precambrian. If there is no vitrinite present, its reflectance cannot be measured, but this sometimes leads to measurements being made on kerogen particles with "vitrinite-like" appearances (more about misidentification of vitrinite later).

Cavings and Reworked Vitrinite

During the drilling of a well, up-hole sections may become progressively unstable due to physical and mineralogical characteristics of the rock and the rock's interaction with the drilling fluid. This can result in up-hole intervals caving, or sloughing off, into the borehole, contributing lower maturity vitrinite particles to a cuttings sample. In the case of reworked vitrinite, preexisting source rocks can be eroded and redeposited, thereby contributing potentially higher maturity vitrinite particles to a sediment.

Caved and reworked vitrinite can often be detected in the histograms of vitrinite measurements, as shown in Fig. 3.36. Instead of having one population of reflectance measurements,

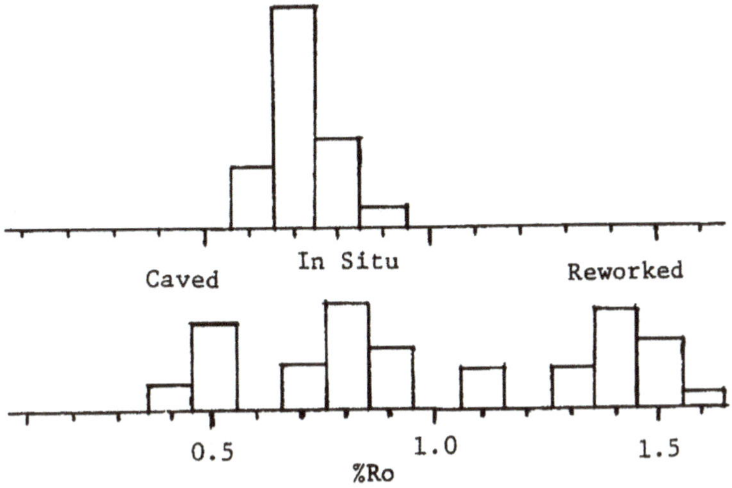

FIGURE 3.36 Histograms of vitrinite reflectance measurements. The top histogram is an "ideal" data set of a single in situ population. The lower histogram shows a more typical histogram with caved, in situ, and reworked populations.

two or more exist. If multiple samples in the well contain potential caved or reworked vitrinite, plotting all the populations for each sample versus depth can help to decipher which are caved, which are reworked, and which are in situ, as shown in Fig. 3.37. It also helps to have some core and/or sidewall core samples as part of the data set. Core and sidewall core will not contain any caved material. It is usually assumed that the lowest reflectance population in a core or sidewall core sample is likely the in situ vitrinite. If core or sidewall core material are not available, cuttings samples collected below just casing points are often used instead. Casing cuts off any potential caving from above making samples just below the casing point desirable for sorting out caving problems.

Unconformities/Normal Faults

Loss of sediment at erosional unconformities can result in an offset in the vitrinite reflectance trend toward higher maturity, as shown in Fig. 3.38. The loss of section brings higher maturity sediments to be juxtaposed to lower maturity sediments at the unconformity surface (Dow, 1977). Similar patterns are observed with offsets in vitrinite reflectance trends due to normal faulting. Some have suggested that by extending the deeper trend up to the point of overlap with the shallower tend, an estimate of the amount of sediment lost at an unconformity, or offset of the fault, can be made, as shown in Fig. 3.38 (Dow, 1977). However, the estimated offset should only be considered a minimum. With time and additional heating, the offset at the unconformity/normal fault can be reduced or eliminated by a process called annealing (Katz et al., 1988). Essentially, the vitrinite in the less mature shallow section continues to mature until it catches up to the more mature deeper section. As this progresses, the offset between the two trends diminishes. In many instances, little or no offset in the vitrinite reflectance trend may be observed at unconformities or normal faults, given enough geologic times and additional burial.

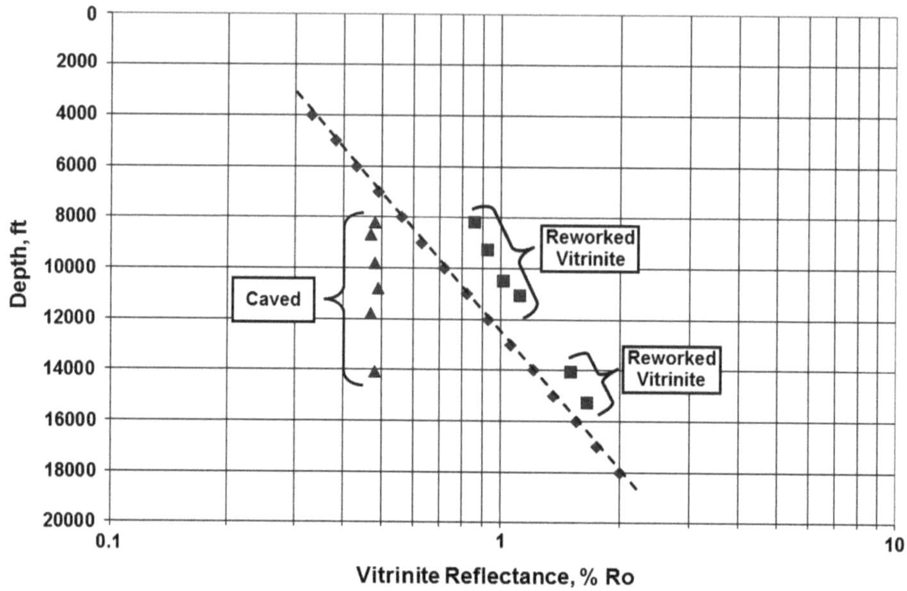

FIGURE 3.37 A vitrinite reflectance trend showing the effects of caved and reworked vitrinite. *After Dow, W., 1977. Kerogen studies and geological interpretations. Journal of Geochemical Exploration 7, 79–99.*

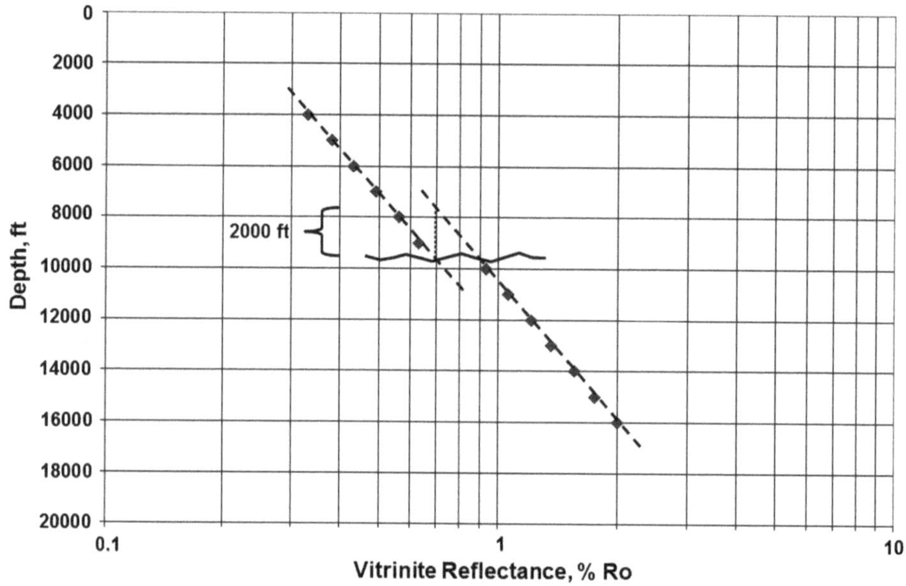

FIGURE 3.38 A vitrinite reflectance trend showing the effects of an erosional unconformity resulting in an offset in the trend to higher maturity values. A similar offset can be observed with normal faults. By matching the values on the two trends, an estimated of the minimum amount of sediment lost, or fault offset, can be estimated. *After Dow, W., 1977. Kerogen studies and geological interpretations. Journal of Geochemical Exploration 7, 79–99.*

Unconformities that occur at the surface will also impact the vitrinite reflectance trend. With loss of sediment at the surface due to erosion, the surface intercept of the vitrinite reflectance trend will not be in the vicinity of 0.20–0.23% Ro, but rather at a higher value. By extending the vitrinite reflectance trend above the ground surface to a point in the 0.20–0.23% Ro range, an estimate of the amount of sediment lost at the unconformity can be made. Again, caution should be used with these estimates of lost section.

Reverse Faults

Displacement of sediments by a reverse, or thrust, fault will move higher maturity sediments up and over lower maturity sediments (Dow, 1977). This will result in an offset in the vitrinite reflectance trend as shown in Fig. 3.39. As with unconformities and normal faults, some workers have suggested that offset, or vertical displacement, of the fault can be estimated from the vitrinite reflectance trend. In this case, the reflectance value of the lower maturity subthrust tend at the fault scarp is projected up to the overthrust vitrinite trend to estimate of the amount of vertical displacement, as shown in Fig. 3.39 (Dow, 1977). However, the estimated offset should again only be considered a minimum due to the process of annealing, as with unconformities and normal faults discussed earlier.

Change in Geothermal Gradient/Heat Flow

An increase in the vitrinite reflectance gradient at depth suggests there has been a higher geothermal gradient or heat flow in the past. Reduction in heat flow may be the result of the cooling of the lower crust. A reduction in heat flow may also be from a change in the thermal

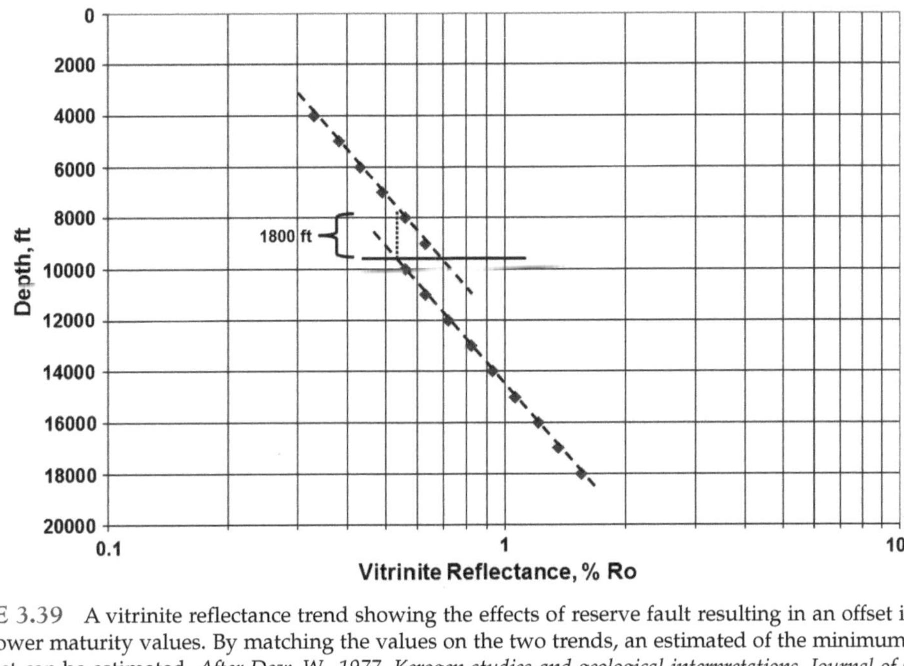

FIGURE 3.39 A vitrinite reflectance trend showing the effects of reserve fault resulting in an offset in the trend toward lower maturity values. By matching the values on the two trends, an estimated of the minimum amount of fault offset can be estimated. *After Dow, W., 1977. Kerogen studies and geological interpretations. Journal of Geochemical Exploration 7, 79–99.*

properties of the sediments caused by overpressure (Hunt, 1996). Excess fluids confined in the pore spaces during periods of overpressure can lower the thermal conductivity of the sediments and retain more heat below the overpressure. These reductions in heat flow result in a kink, or dog-leg, in the vitrinite reflectance trend with depth showing a more rapidly increasing thermal maturity in the deeper section, as shown in Fig. 3.40. Law et al. (1989) documented more than one "kink" in the trend can occur if multiple overpressure zone develop during the history of the sediments.

Igneous Intrusives

Localized high-temperature heating by an igneous intrusive can cause a high-maturity excursion in the vitrinite trend, as shown in Fig. 3.41. Thermal effects of an igneous intrusive typically extend out to a distance equal to about twice the thickness of the intrusive on both sides of the igneous body (Dow, 1977). Occasionally, a similar pattern in the vitrinite trend is observed where no igneous body is encountered in the well bore. These are most likely the results of thermal alterations by hot fluids from nearby igneous bodies moving out along bedding plane.

Suppression and Misidentification

Occasionally, there is a localized decrease in reflectance values in a vitrinite trend with depth, as shown in Fig. 3.42. This decrease is usually confined to an interval corresponding to a rich oil-prone source rock. One explanation for this phenomenon is called suppression. This occurs when oil generated in rich oil-prone source rocks supposedly invades vitrinite particles under subsurface pressures and lower their reflectivity (Carr, 2000). Corrections for

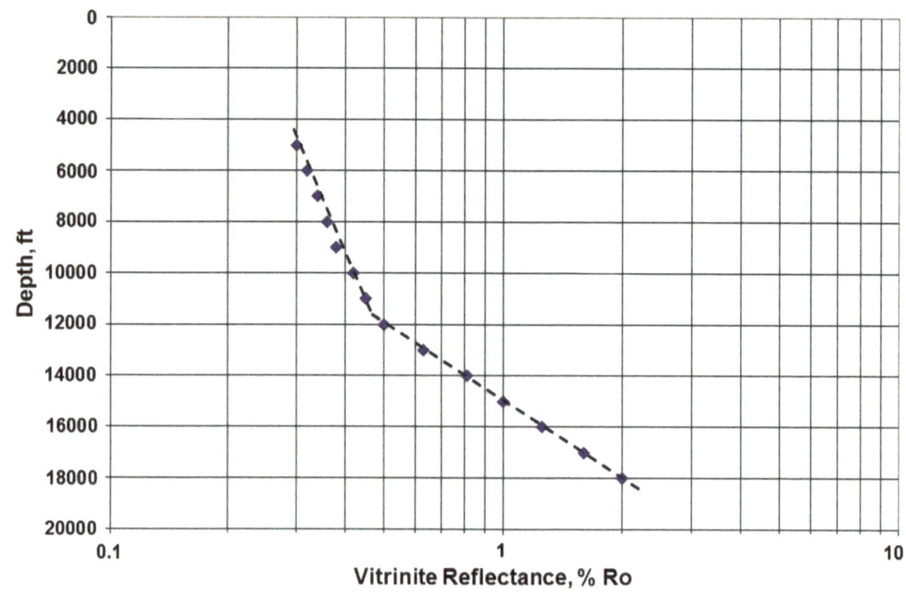

FIGURE 3.40 A vitrinite reflectance trend showing the effects of a decrease in heat flow/geothermal gradient going from older to younger sediments.

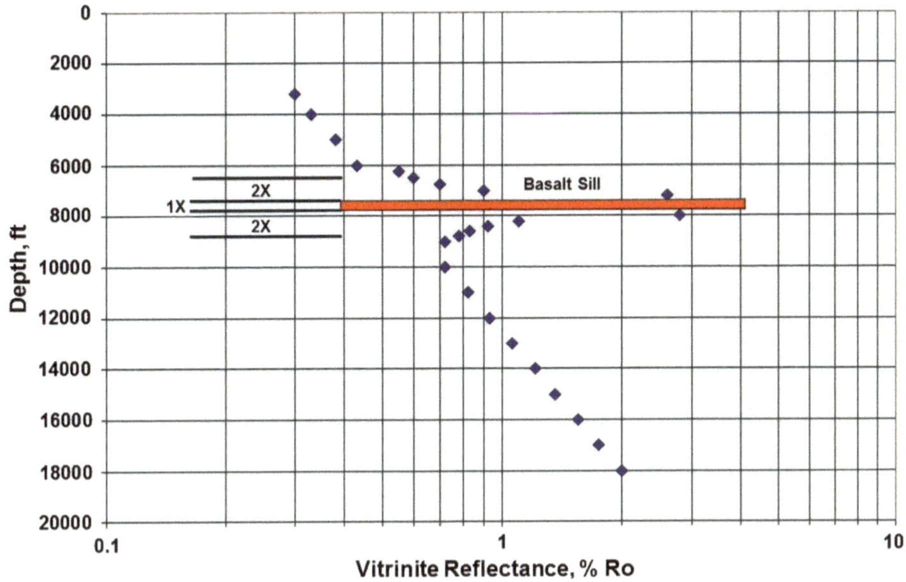

FIGURE 3.41 A vitrinite reflectance trend showing the effects of metamorphism induced by local igneous intrusion. *After Dow, W., 1977. Kerogen studies and geological interpretations. Journal of Geochemical Exploration 7, 79–99.*

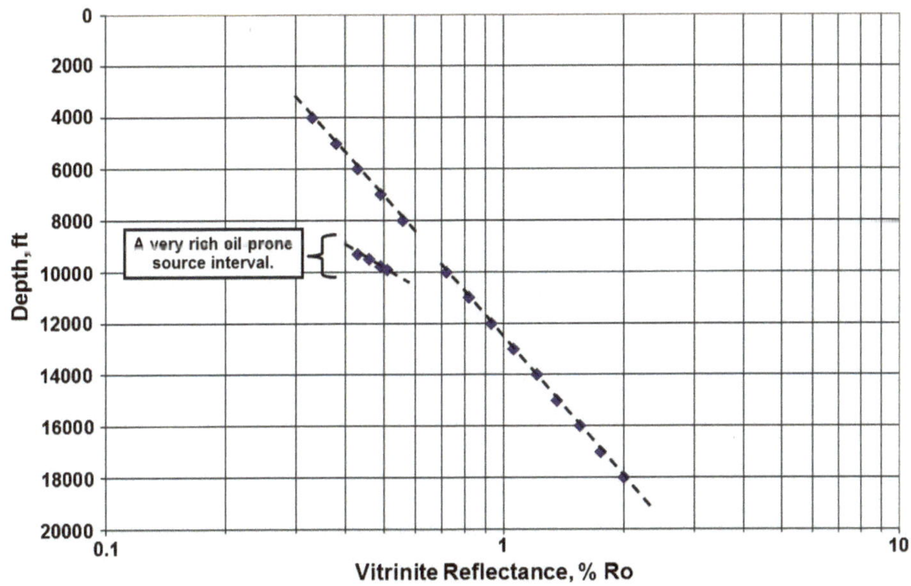

FIGURE 3.42 The effects of "suppression" or misidentification of liptinite or solid bitumen particle as vitrinite.

the "suppressed" vitrinite related to the HI of the source rock have been proposed (e.g., Lo, 1993). However, because the decrease in reflectance is localized, the remainder of the vitrinite trend above and below the "suppressed" zone should be sufficient for any correction.

While suppression may affect sedimentary organic matter, misidentification of vitrinite-like particles should not be overlooked as a possible cause for localized lower than expected reflectance values. Rich oil-prone source rocks often contain little or no indigenous vitrinite, and this lack of vitrinite may lead to misidentification of other kerogen particles with vitrinite-like appearances, such as solid bitumen. The problem of misidentification of vitrinite-like particles is widespread. Hackley et al. (2015) have stated that the lack of petrographic distinction of solid bitumens and low-reflecting inert macerals from vitrinite is the most difficult problem encountered during vitrinite reflectance measurements. The lower reflecting solid bitumen (see later discussion) would be abundant in rich oil-prone source rocks and misidentification might explain some of the suppression phenomenon.

Whatever the cause of the localized decreases in reflectance observed in vitrinite trends, the data above and below the affected zone should allow a trend to be established and maturity interpretations to be made.

Vitrinite Anisotropy

Teichmuller (1982a) observed that vitrinite becomes anisotropic at reflectance values above 1.0% Ro. However, Houseknecht and Weesner (1997) determined that below vitrinite reflectance values of 2.0% Ro, the anisotropy is low, and the error introduced is minimal. Above 2.0%, the anisotropy increases rapidly, and the error can be significant without proper measurement. In these cases, rotation of the vitrinite particles is needed to determine the maximum reflectance for measurement. If not done properly, measurements on randomly oriented particles will likely result in lower than actual reflectance values being reported.

Maturity Interpretations With Vitrinite Reflectance

Vitrinite reflectance, like all maturity indicators, is a trend tool. A valid data set should consist of 10–20 vitrinite reflectance sample analyses spanning a depth range of a minimum of 4000–5000 ft (approximately 1200–1500 m). Mean reflectance values of the in situ vitrinite population from each sample should be plotted on a log scale versus depth on a linear scale. Next, a straight line should be drawn through the data points that appear to fall on trend, while looking for evidence of the potential interferences discussed earlier. Resist the use of regression to draw the trend line. At this stage, reworked vitrinite, caved vitrinite, or other spurious data may still be included in the data set and influence the regression. It may be necessary to go back to the raw data and histograms to reassess which population of reflectance values may be in situ before a final trend is arrived at.

Once a reasonable trend has been established, the standard interpretations of the stages of hydrocarbon generation of a source rock, shown in Fig. 3.43, can be applied to estimate depth to the onset of hydrocarbon generation, peak generation, and so on. These threshold values are based on the work of Dow (1977), Senftle and Landis (1991), and others. The threshold values for oil generation are for an average marine oil-prone source rock. The gas generation threshold values are more generic.

Oil-Prone Generation		Gas-Prone Generation	
Generation Stage	R_o(%)	Generation Stage	R_o(%)
Immature	<0.6	Immature	<0.8
Early oil	0.6–0.8	Early gas	0.8–1.2
Peak oil	0.8–1.0	Peak gas	1.2–2.0
Late oil	1.0–1.35	Late gas	>2.0
Wet gas	1.35–2.0		
Dry gas	>2.0		

FIGURE 3.43 General interpretation scheme for vitrinite reflectance for the main stages of oil and gas generation. *From Dembicki Jr., H., 2009. Three common source rock evaluation errors made by geologists during prospect or play appraisals. American Association of Petroleum Geologists Bulletin 93, 341–356, based on Dow, W., 1977. Kerogen studies and geological interpretations. Journal of Geochemical Exploration 7, 79–99, Senftle, J.T., Landis, C.R., 1991. Vitrinite reflectance as a tool to assess thermal maturity. In: American Association of Petroleum Geologists Treatise of Petroleum Geology: Source and Migration Processes and Evaluation Techniques 119–125, and others.*

Kerogen Type	R_o(%)
Type I	0.7
Type II	0.6
Type IIS	0.45-0.5
Type III	0.8

FIGURE 3.44 Adjustment for vitrinite reflectance value for the onset of hydrocarbon generation for kerogen different kerogen types. *From Dembicki Jr., H., 2009. Three common source rock evaluation errors made by geologists during prospect or play appraisals. American Association of Petroleum Geologists Bulletin 93, 341–356, based on Tissot, B., 1984. Recent advances in petroleum geochemistry applied to hydrocarbon exploration. American Association of Petroleum Geologists Bulletin 68, 546–563, Petersen, N.F., Hickey, P.J., 1987. California Plio-Miocene oils: evidence of early generation. In: Meyer, R.F. (Ed.), Exploration for Heavy Crude Oil and Natural Bitumen: American Association of Petroleum Geologists Studies in Geology, vol. 25. pp. 351–359, Sweeney, J.J., Burnham, A.K., Braun, R.L., 1987. A model of hydrocarbon generation from Type I kerogen: application to Uinta Basin, Utah. American Association of Petroleum Geologists Bulletin 71, 967–985, Tissot, B., Pelet, R., Ungerer, P., 1987. Thermal history of sedimentary basins, maturation indices, and kinetic of oil and gas generation. American Association of Petroleum Geologists Bulletin 71, 1445–1466, and others.*

Over the years, field observations and kinetics studies such as Tissot (1984), Petersen and Hickey (1987), Sweeney et al. (1987), and Tissot et al. (1987) have indicated that the onset of significant hydrocarbon generation can vary depending on the predominant kerogen type in the source rock. For example, as discussed in Chapter 2, Type IIS kerogen is likely to begin hydrocarbon generation earlier than Type II, and Type I kerogen is likely to begin hydrocarbon generation slightly later than Type II. These kerogen type–based adjustments to the onset of significant hydrocarbon generation are summarized in Fig. 3.44.

After establishing the vitrinite reflectance maturity trend, it is vital these data be placed into a geologic context for proper interpretation. This is easily accomplished using the burial history of the stratigraphic sequence of the well (e.g., Katz et al., 1988; Law et al., 1989;

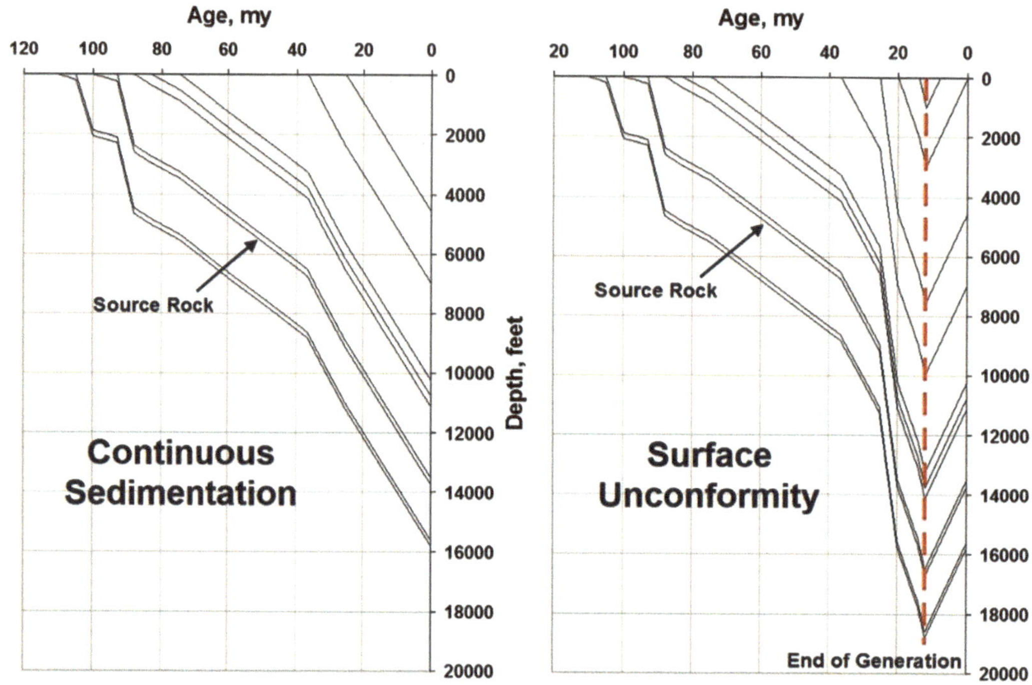

FIGURE 3.45 Two possible burial histories for the same vitrinite trend. In the one on the left, burial is fairly continuous to present day, while the one on the right contains a surface unconformity. *From Dembicki Jr., H., 2009. Three common source rock evaluation errors made by geologists during prospect or play appraisals. American Association of Petroleum Geologists Bulletin 93, 341–356.*

Dembicki, 2009). Using an example from Dembicki (2009), if sedimentation is fairly continuous, as shown in burial history on the left side of Fig. 3.45, the vitrinite reflectance trend should indicate the current maturity. But when a surface unconformity is present, as shown in burial history on the right side of Fig. 3.45, the vitrinite reflectance trend indicates the maturity level that was reached prior to the uplift and erosion. The depth to the top of the oil window is actually deeper than the vitrinite trend indicates and needs to be adjusted for the amount of lost sediment.

Recent trends in source rock evaluation projects, especially with unconventional resources, have tended to focus on sampling only in the interval of interest. In these instances, only a small number of samples are being collected and analyzed over a 200–300 ft (approximately 60–90 m) interval, including maturity data. This practice can result in erroneous conclusion about the maturity of the interval. Considering good oil-prone source rocks often have little or no vitrinite in their kerogen, by focusing on a source rock interval, little or no vitrinite may be encountered. Without whole well data, there may be no indication of the maturity in the interval of interest. But what if vitrinite is found? With data from only a short depth interval, there is no context for interpreting data. If cuttings are used, the vitrinite observed may be from caved material and gives a lower than actual maturity. Or, it could be reworked vitrinite that will give an indication of higher maturity than actually exists. And there is always the possibility that other kerogen particles may be misidentified as vitrinite and give an inaccurate maturity. Without the context of whole well data, it may not be possible to properly assess the validity of these samples. However, with whole well data, a trend can be established that can be used to evaluate any data collected from the interval of interest. And if no data were collected from the interval of interest or the data are suspect, the whole well trend can be used to determine what the maturity of the source rock should be.

Vitrinite reflectance is often used to make inferences about hydrocarbon generation and migration. Hydrocarbon generation depends on the kerogen type in the source rock as well as its time–temperature history. While vitrinite reflectance can indicate if hydrocarbon generation is possible and suggest what type(s) of hydrocarbon could be formed, it cannot directly indicate when hydrocarbon generation might have started or tell how much hydrocarbon has been generated. With respect to migration, maturity has a role in determining when the expulsion of hydrocarbons occurs. But expulsion is more directly related to the richness of the source rock, its porosity and permeability, and the amount of hydrocarbon that has been generated (England et al., 1991; Palciauskas, 1991). While one source rock might begin expulsion at 0.8% Ro, a richer source rock might start expelling at 0.7% Ro. As a result, vitrinite reflectance cannot directly indicate when migration occurs. It can only suggest when expulsion might be possible.

How should vitrinite reflectance data be used? First and foremost, it is an indicator of a sediment's thermal maturity which is the product of its cumulative time–temperature history. It can indicate if generation could have taken place and suggest the types of hydrocarbons that may have been formed, thereby identifying which source rocks are potential contributors to a petroleum system. And vitrinite reflectance data can be used to help constrain and validate basin models, as discussed in Chapter 8.

ALTERNATIVE REFLECTANCE METHOD

In addition to vitrinite reflectance, there are other organic particle reflectance measurements that are used for maturation analysis. These include the liptinite/sporinite/exinite, solid bitumen, graptolites, scolecodonts, and chitinozoans.

Liptinite/Sporinite/Exinite Reflectance

Liptinite is a maceral group in coal petrology that was previously referred to as the exinite group. It consists of the sporinite, cutinite, alginite, and resinite maceral types. Typically, the sporinite maceral is the focus of reflectance measures, although references in published literature will also refer to liptinite or exinite reflectance. Early work in coal petrology recognized that the reflectance of both liptinite and vitrinite would increase with increasing thermal maturity in a predictable way (Alpern, 1970; Alpern et al., 1978; Stach et al., 1982). As shown by the comparison of vitrinite reflectance to liptinite/sporinite reflectance in Fig. 3.46, liptinite reflectance is lower than the equivalent vitrinite reflectance up to about 1.7% Ro vitrinite reflectance. Above that point, the trends merge. While showing promise as an alternate maturity indicator in the absence of vitrinite, use of liptinite/sporinite reflectance is not widespread. Few contract laboratories offer it as a service, but it does occasionally appear in older source rock data compilations.

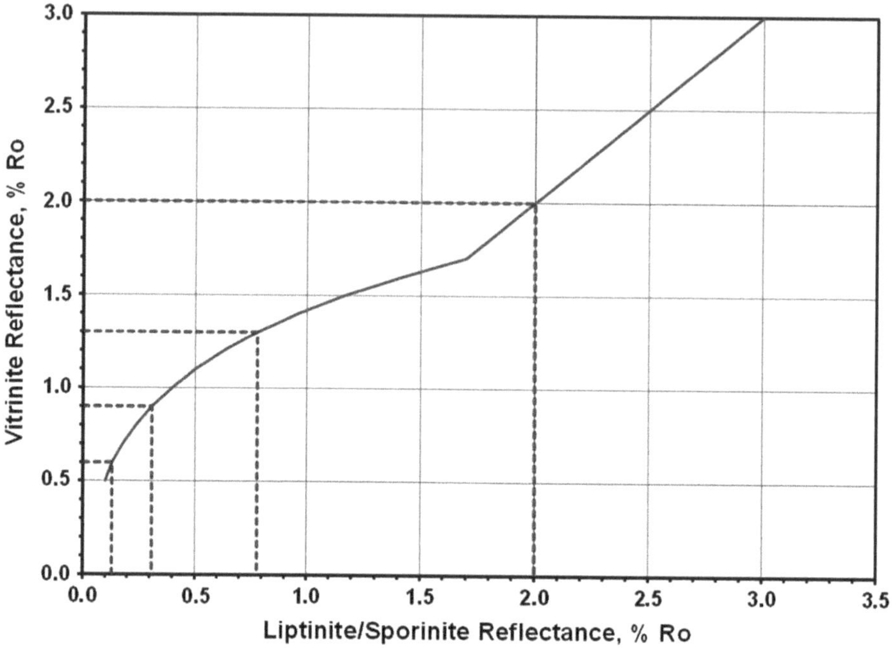

FIGURE 3.46 Liptinite/Sporinite reflectance as compared to vitrinite reflectance. *Based on data found in Bertrand, R., 1990. Correlations among the reflectances of vitrinite, chitinozoans, graptolites and scolecodonts. Organic Geochemistry 15, 565–574.*

Solid Bitumen Reflectance

Solid bitumen, also referred to as solid hydrocarbon, is frequently observed during microscopic examination of kerogen. In contrast to vitrinite and liptinite, solid bitumen is not a kerogen component but rather a product of generation from the kerogen that has flowed into pore spaces within mineral grains (Thompson-Rizer, 1987; Jacobs, 1989). Robert (1973) observed that the reflectivity of solid bitumen increased with increasing maturity similar to vitrinite reflectivity. While a number of workers in the field have proposed relationships to convert solid bitumen reflectance to equivalent vitrinite reflectance, including Jacob (1985), Bertrand (1993), and Landis and Castaño (1995), no unified method has been discovered. The differences between these relationships are thought to be due to variations in optical properties of solid bitumen derived from different kerogen types, as well as differences observed in solid bitumen in source rocks versus reservoir rocks. If local relationship can be established between solid bitumen reflectivity and other maturity indicators (e.g., Gentzis and Goodarzi, 1990), the method can provide useful information. Caution is urge in applying solid bitumen reflectivity.

Graptolite, Scolecodont, and Chitinozoan Reflectance

The need for Lower Paleozoic maturity indicators has lead several workers in the field to investigate the use of the reflectivity of graptolites, scolecodonts, and chitinozoans. Graptolites

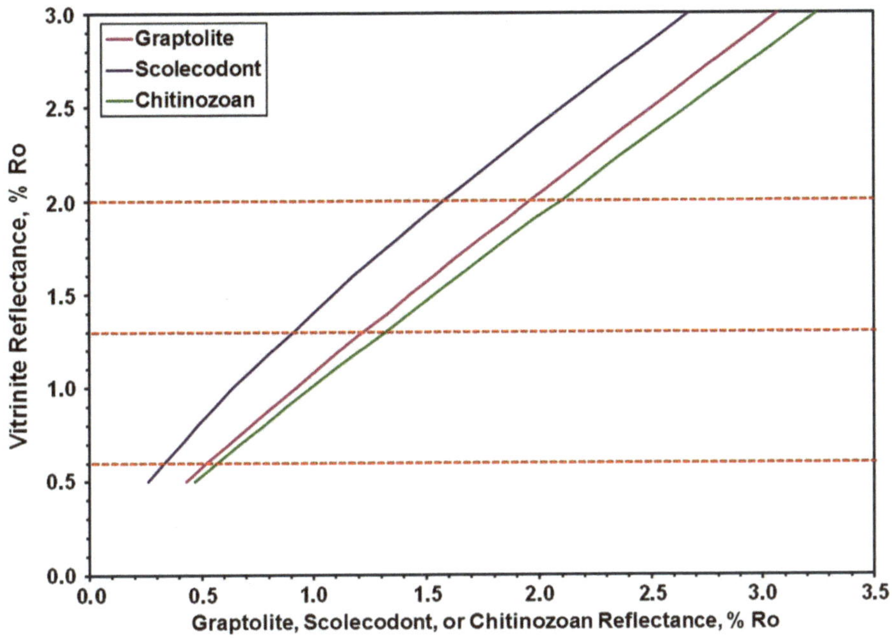

FIGURE 3.47 Graptolite, scolecodont, and chitinozoan reflectance as compared to vitrinite reflectance. *Based on data found in Bertrand, R., 1990. Correlations among the reflectances of vitrinite, chitinozoans, graptolites and scolecodonts. Organic Geochemistry 15, 565–574.*

are colonial animals likely related to hydrozoans, commonly dendritic or branching saw-blade like, or "tuning fork" shaped preserved as a black carbonized film. They range from Upper Cambrian through Lower Carboniferous (Mississippian). Scolecodonts are the chitinous jaws of certain polychaete annelids. They can be found in sediments from the Cambrian to the present but are most common in Ordovician, Silurian, and Devonian marine deposits. The chitinozoans are organic-walled microfossils with unknown biological affinities exhibiting a vesicular shape with one open end and a body cavity. Their range in the fossil record is from the lowermost Ordovician to the uppermost Devonian/lowermost Carboniferous. Bertrand (1990) has summarized the reflectivity of these zooclasts as compared to vitrinite reflectance. The results of these comparisons are shown in Fig. 3.47. While not as well studied as vitrinite reflectance, these zooclast reflectivities show promise as Lower Paleozoic maturity indicators. The main problem with using these reflectivities is that these data are typically not available through contract laboratories, and there are only a few practitioners in academic and government laboratories. This may change as more emphasis is placed on Lower Paleozoic plays, especially in unconventional resources.

VISUAL KEROGEN TYPING

Visual kerogen typing, a form of organic petrography, is the microscopic method for the examination of kerogen. It is based on the premise that optically classified kerogen particles

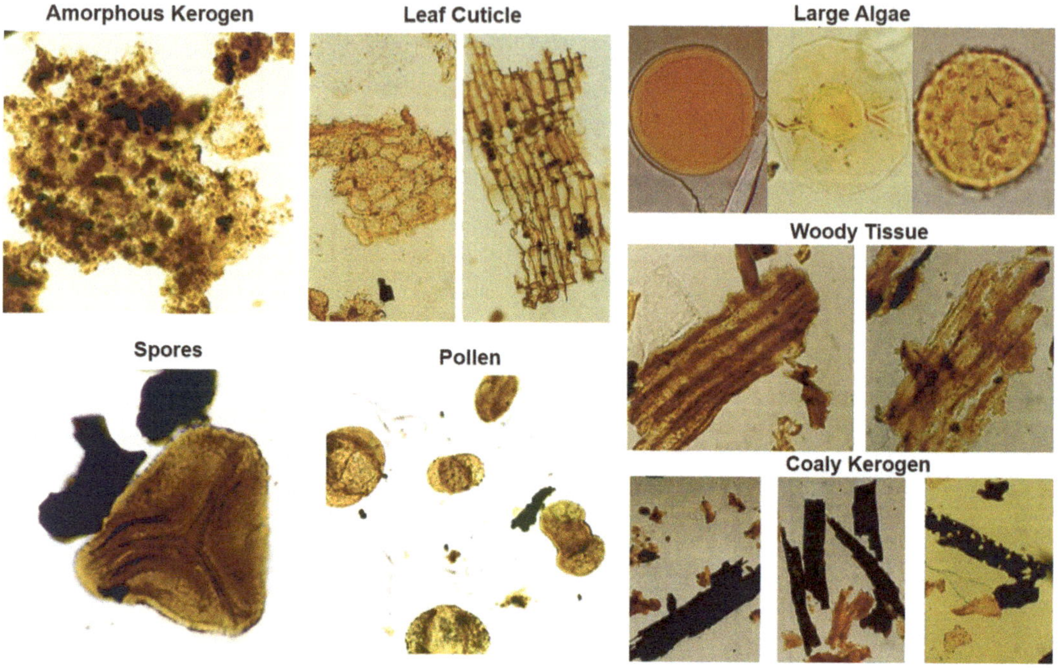

FIGURE 3.48 Some kerogen particle types observed using transmitted light microscopy.

can be related to hydrocarbon-generating potential (Staplin, 1969) of a source rock. These microscopic observations are usually done using a kerogen concentrate with refractory minerals prepared as a strewn slide. The analysis is most often done in transmitted light, typically in conjunction with the assessment of the TAI discussed later. Some typical kerogen particle types observed using transmitted light microscopy are shown in Fig. 3.48. The analysis consists of classifying the kerogen particles into a series of types and estimating the contribution of each particle type to the total kerogen. Most classification schemes utilize terminology derived from transmitted light descriptions used in palynology, some use terminology derived from coal petrology, and a few used a hybrid scheme using elements of both. Examples of transmitted and reflected light classification schemes are shown in Fig. 3.49. Both have an amorphous category. Herbaceous, woody, and coaly categories in the transmitted light scheme are essentially equivalent to the liptinite/exinite, vitrinite, and inertinite categories in the reflected light schemes, respectively. Classification schemes can vary from lab to lab or analyst to analyst, sometimes using the same particle names to describe different materials. It is important to know the definition of the categories in the classification scheme used when interpreting a data set to avoid any confusion or misinterpretation.

Interpretation of visual kerogen data is based on each kerogen particle type having a specific hydrocarbon-generating capacity. Particles including spores, pollen, leaf cuticle, and thin cell wall structures were thought to be hydrogen rich and should be oil prone, while woody tissue, thick cell wall structures, and vitrinite were thought to be lignin rich and should be gas prone (Staplin, 1969). Coaly kerogen and inertinite are thought to be inert and have no real

Transmitted Light Classification		Reflected Light Classification		
Particle Type	Description	Particle Type	Description	Interpretation
Amorphous	Fluffy, structureless material	Amorphous	Structureless material	Oil-Prone
Herbaceous	Spores, pollen, leaf cuticle, membraneous and thin cell wall structures	Liptinite/ Exinite	Algae, spores, pollen, leaf cuticle, resins	Oil-Prone
Woody	Vascular plant tissue, fibrous, and thick cell wall structures	Vitrinite	Humic material with obscured cell structure and light gray color	Gas-Prone
Coaly	Black opaque organic material	Inertinite	Highly reflective, charcoal-like, may or may not exhibit cell structure	Inert

FIGURE 3.49 Examples of transmitted and reflected light classification schemes used in visual kerogen typing.

hydrocarbon-generating potential. Amorphous kerogen was originally thought to be derived almost exclusively from algae and therefore oil prone (Staplin, 1969; Harwood, 1977; Tissot, 1984).

In practice, however, there is a poor correlation between interpreted hydrocarbon-generating potential based on visual kerogen typing and interpretation based on chemical data. This lack of correlation arises primarily from little chemical data existing that confirms the hydrocarbon generating potential of the classes of kerogen particles, particularly in the amorphous kerogen, in the classification schemes. In the case of amorphous kerogen, studies have shown that it can be derived from a variety of organic materials and can be both oil prone and gas prone (e.g., Jones and Edison, 1978; Powell et al., 1982). Optical methods have been developed to distinguish between oil- and gas-prone amorphous kerogens (e.g., Massoud and Kinghorn, 1985; Mukhopadhyay et al., 1986; Thompson and Dembicki, 1986), but they are not in general use.

Other problems may also contribute to the lack of correlation between visual and chemical hydrocarbon-generating potential assessments. As noted earlier in the section on kerogen isolation, the loss of both small less dense and large heavy particles during the kerogen isolation procedures can introduce bias into the kerogen sample observed. In addition, estimates of the concentration of the kerogen particle types are subjective. And, Powell et al. (1982) found that microscopists with a palynological background tend to overestimate the concentration of structured particles and nearly all observers overestimate the amount of amorphous material kerogen present. To remedy this, Kuncheva et al. (2008) developed an image analysis–based system to automated kerogen classification in microscope images of dispersed kerogen to give better estimates of the amount of each kerogen type present. However, this system is not used by microscopists doing this analysis.

THERMAL ALTERATION INDEX

TAI is a maturity indicator based on observations of the progressive change in the color of spore and pollen particles in kerogen with increasing maturity (e.g., Gutjahr, 1966; Correia, 1969; Staplin, 1969). The first formal scale was developed by Staplin (1969), and it used a 1–5 scale employing + and – notations to signal intermediate steps, as shown in Fig. 3.50. The

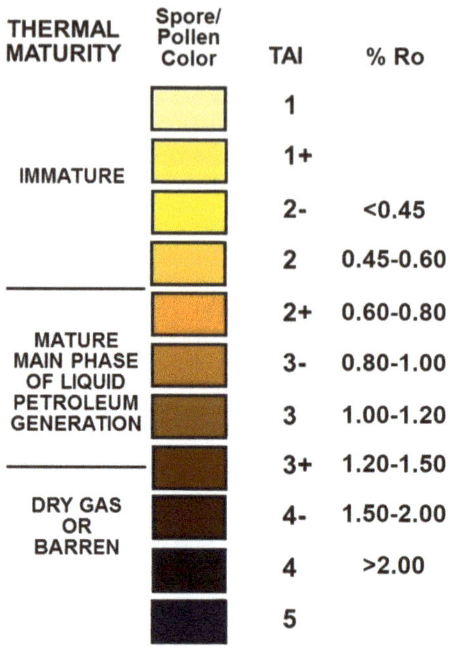

FIGURE 3.50 The thermal alteration index of Staplin (1969) as interpreted by Pearson (1984).

analysis is performed by observing the color of spores and pollen in transmitted light using a strewn slide of a kerogen concentrate with refractory minerals. The analysis is most often done in conjunction with visual kerogen typing (discussed earlier). Ideally, a large number of observations of spores and/or pollen color are made on the kerogen slide, and a representative TAI is determined based on these observations.

TAI, like all maturity indicators, is a trend tool. A set of observations from samples taken over a large depth interval are required to establish a trend. The original work by Staplin (1969) did not assign equivalent vitrinite reflectance ranges to the TAI values but simply related the TAI to stages of hydrocarbon generation. Since then, many individual labs and analysts have developed their own internal scales for relating TAI to vitrinite reflectance. While most analysts use a 1–5 scale similar to Staplin, others use a 1–5 scale with decimal steps instead of the +/− notation, and a few use a 1–10 scale. As a result, it is important to obtain a copy of the scale used by the analyst as well as the vitrinite reflectance equivalence of each step on the scale for proper interpretation of the data.

Difficulties using TAI can arise because the determination of spore/pollen color is a subjective observation. Even with color standards such as Pearson (1984), the interpretation of color by TAI analysts is based on the individual's perception of what is seen down the microscope (Gutjahr, 1966). In addition, particle thickness can influence the perception of color. An intact spore (double thickness) viewed in the kerogen slide may not appear to have the same color as a broken spore (single thickness) (Jones and Edison, 1978). Thickness difference can also be observed from species to species. To avoid this thickness problem, Jones and Edison (1978) suggest observing only intact spore or pollen grain in the same taxon when possible.

To eliminate the subjectivity in the TAI analysis, Marshall (1991) developed an image analysis–based quantitative color assessment for spore and pollen grains. However, this system is not used by microscopists doing this analysis.

Because it is based on spores and pollen, TAI is limited to use in mid-Paleozoic and younger sediments. It is also subject to interferences by caved and reworked material, similar to vitrinite reflectance.

When spore and pollen particles are absent, Peters et al. (1977) suggested that the color of other kerogen particles, including amorphous, can be used to make TAI observations. This can be very problematic. Spore/pollen color changes are well documented by Gutjahr (1966) and Staplin (1969) and reflect the change in chemical structure in these particles with increasing maturity. While other kerogen particles may change color with maturity, they may not be reflecting the same structural changes taking place in spore and pollen grains that TAI is based on. Therefore, using the TAI scale for kerogen particles other than spores or pollen may lead to erroneous TAI assessments. When reviewing reports of TAI data, it is imperative to determine if the TAI measurements were made on spores, pollen, or other particles. Use data not derived from spores or pollen very cautiously.

KEROGEN FLUORESCENCE

When exposed to incident ultraviolet (UV) light or blue light, some of the aromatic structures in the kerogen absorb part of the light energy and excite electrons to go to a higher energy state. As the electrons return to their original lower energy state, they fluoresce and emit visible and UV light with longer wavelengths than the original absorbed light. In source rock evaluation, kerogen fluorescence is used mainly to estimate the thermal maturity of a source rock. The technique is based on observed changes in the color and intensity of kerogen fluorescence with increasing maturity.

Analysis of kerogen fluorescence can be done either as a qualitative assessment or as quantitative spectral analysis. In either case the analysis is performed using a strewn slide of a kerogen concentrate. For the qualitative assessment, observations are made of the color of fluorescing kerogen in incident UV or blue light. The usual excitation wavelength of the UV light is 365 nm, and if blue light is used, the excitation wavelength is 477 nm (van Gijzel, 1979). The qualitative assessment is most often done in conjunction with the visual kerogen typing and TAI determination, discussed earlier. Ideally, a large number of observations of fluorescence color are made on the kerogen slide, and a representative color is determined based on these observations. The representative color of the fluorescence can then be related to the maturity using the conversion to a vitrinite reflectance range shown in Fig. 3.51. Maturity based on this qualitative assessment of the color of kerogen fluorescence is not a precise indicator. It is typically used as corroboration for observation made with vitrinite reflectance and TAI.

More precision can be obtained doing a quantitative spectral fluorescence analysis of the kerogen. This is done in incident UV light with an excitation wavelength of 365 nm. Instead of simply observing the color of the emitted fluorescence of the kerogen, the light is measured by a spectrometer between the wavelengths of 400 and 700 nm (Thompson-Rizer and Woods, 1987). The main measurements made on the recorded spectra are the wavelength of

Fluorescence Color	Vitrinite Reflectance
Yellow-Green	<0.4 % Ro
Yellow	~0.4-0.5 % Ro
Orange	~0.5-0.7 % Ro
Orange-Brown	~0.7-1.0 % Ro
None	>1.0 % Ro

FIGURE 3.51 Approximate vitrinite reflectance ranges for the fluorescence color of kerogen. *Adapted from the observations of van Gijzel, P., 1979. Manual of the Techniques and Some Geological Applications of Fluorescence Microscopy: American Association of Stratigraphic Palynologists, 12 Annual Meeting Workshop, Dallas, 55 p.; Hagemann, H., Hollerbach, A., 1981. Spectral fluorometric analysis of extracts a new method for the determination of the degree of maturity of organic matter in sedimentary rocks. Bulletin Des Centres De Recherchese Exploration-Production Elf-Aquitaine 5 (2), 635–650; Stach, E., Mackowsky, M.-Th., Teichmüller, M., Taylor, G.H., Chandra, D., Teichmüller, R., 1982. Stach's Textbook of Coal Petrology, second ed. Gebrüder Borntraeger, Berlin. 535 p.; Teichmüller, M., 1982b. Fluorescence Microscopical Changes of Liptinites and Vitrinites during Coalification and Their Relationship to Bitumen Generation and Coking Behaviour: The Society for Organic Petrology. Special Publication No. 1, 74 p.; Thompson-Rizer, C.L., Woods, R.A., 1987. Microspectrofluorescence measurements of coals and petroleum source rocks. International Journal of Coal Geology 7, 85–104.*

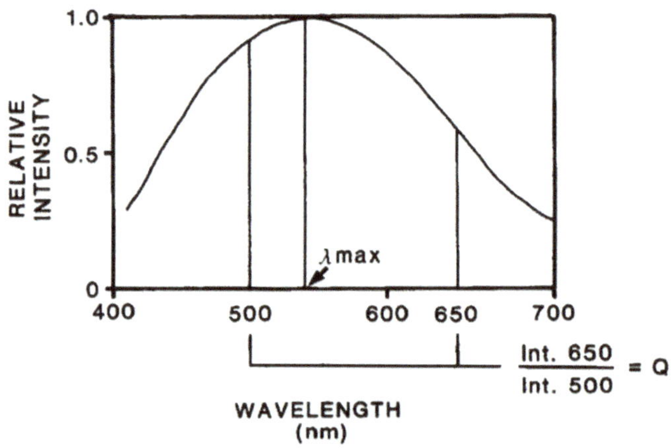

FIGURE 3.52 An example of a UV fluorescence spectra showing the wavelength of maximum intensity, λmax, and derivation of Q, the red/green quotient (relative intense at 650 nm/relative intensity at 500 nm). *Taken from Stasiuk, L.D., Osadetz, K.G., Potter, J., 1990. Fluorescence spectral analysis and hydrocarbon exploration: examples from Paleozoic source rocks. In: Beck, L.S., Harper, C.T. (Eds.), Innovative Exploration Techniques, Saskatchewan Geological Society, Special Publication No. 12, pp. 242–251.*

maximum fluorescence, λmax, and Q, the red/green quotient (relative intense at 650 nm/relative intensity at 500 nm) (Teichmuller and Durand, 1983; Stasiuk et al., 1990; Bertrand et al., 1993), as shown in Fig. 3.52. For consistency, one kerogen particle type, usually liptinite, is used for a series of samples over a depth range. After a number of observations are made for each sample, an average λmax and Q are calculated and plotted versus depth. λmax and Q can also be converted into equivalent vitrinite reflectance values using the relationships shown in Fig. 3.53.

Kerogen fluorescence, like all maturity indicators, is a trend tool, and a set of observations from samples taken over a large depth interval are required to establish a trend. It is also subject to the effects of caved sediments and contamination by fluorescing organic drilling mud additives.

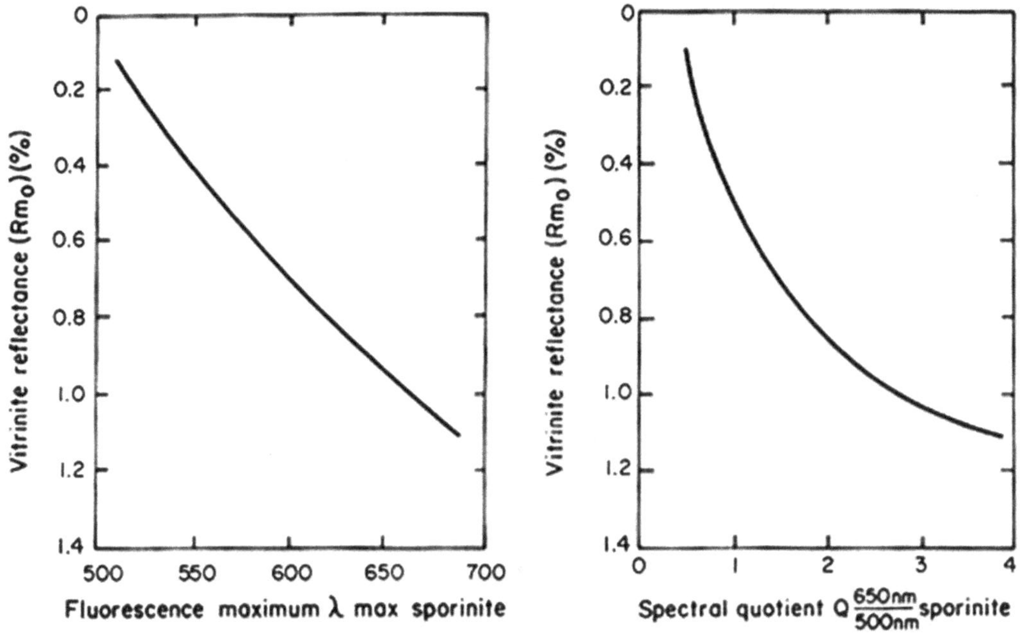

FIGURE 3.53 The observed relationship between vitrinite reflectance and the kerogen fluorescence parameter lmax and Q. *From Teichmüller, M., Durand, B., 1983. Fluorescence microscopical rank studies on liptinites and vitrinites in peat and coals, and comparison with results of the Rock-Eval pyrolysis. International Journal of Coal Geology 2, 197–230.*

While the qualitative assessment of kerogen fluorescence is commonly provided during visual kerogen analysis by contract laboratories, quantitative spectral fluorescence is rarely done in contract laboratories and is confined mainly to academic institutions. Quantitative spectral fluorescence was, however, more commonly utilized in the past and may be found in older source rock evaluation reports and data compilations.

CONODONT ALTERATION INDEX

Conodonts are microscopic jaw parts composed of apatite from an extinct group of early chordates that existed from Cambrian to Triassic times. Epstein et al. (1977) observed that conodonts undergo systematic color changes that are related to thermal maturity in both natural samples and laboratory-heated sediments. Based on this observation, they developed the Conodont Alteration Index, CAI, for use as a thermal maturity indicator, as shown in Fig. 3.54. The equivalent vitrinite reflectance ranges for each step on the CAI scale as established by Harris et al. (1978) are also shown in Fig. 3.54. The CAI 1–5 scale is similar in principle to Staplin's TAI, with 1 being the lower maturity light color and 5 being the more mature darker color. Although not as discerning as TAI in the lower thermal maturity range, CAI can help to place sediments within the major phases of hydrocarbon generation. The importance of the CAI is that it can be used in Lower Paleozoic sediments where vitrinite, spores, and pollen are not available. CAI is typically not available through contract laboratories and must

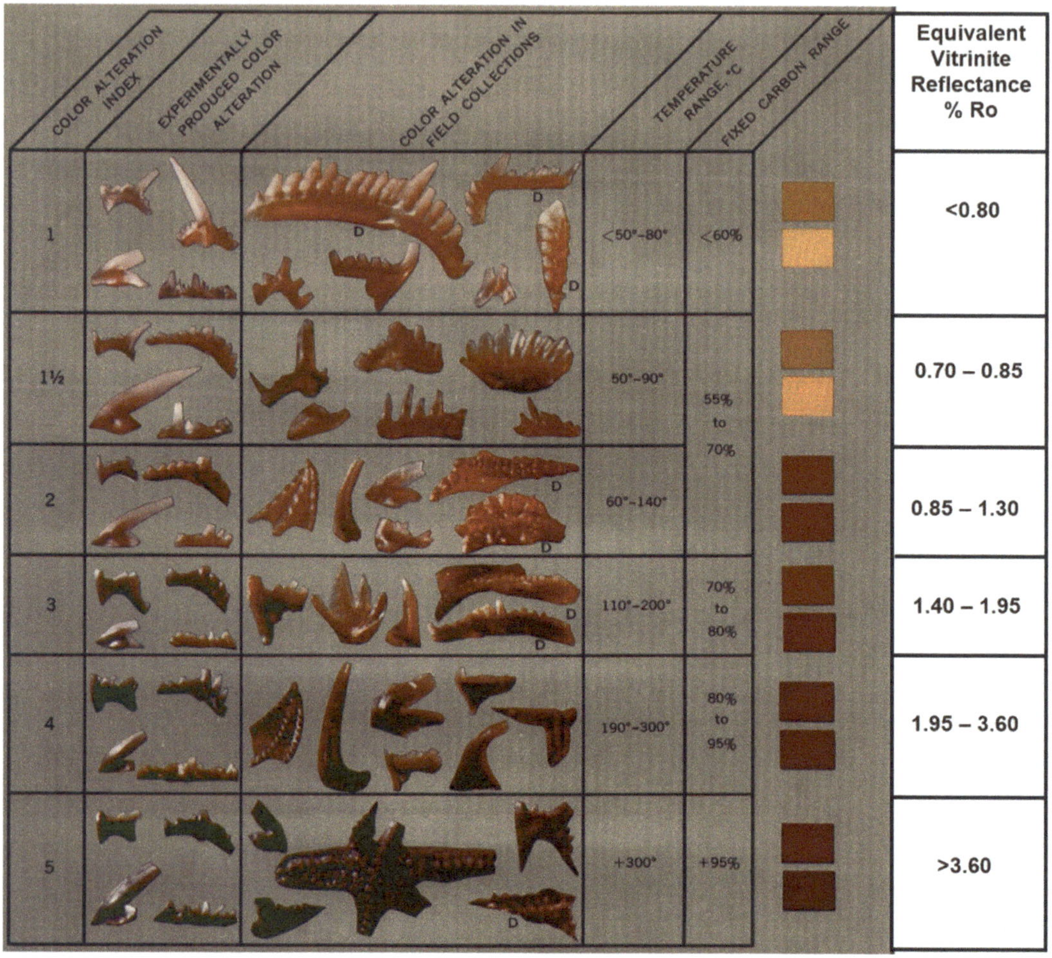

FIGURE 3.54 Changes in conodont color from both heating experiments and filed examples establishing the basis for the conodont color alteration index, CAI *(Modified from Epstein, A.G., Epstein, J.B., Harris, L.D., 1977. Conodont color alteration; an index to organic metamorphism. United States Geological Survey Professional Paper 995, 27.)*, with the conversion of CAI to vitrinite reflectance *(After Harris, A.G., Harris, L.D., Epstein, J.B., 1978. Oil and gas data from Paleozoic rocks in the Appalachian basin; maps for assessing hydrocarbon potential and thermal maturity (conodont color alteration isograds and overburden isopachs): U.S. Geological Survey Miscellaneous Investigations Series Map I–917–E, 4 sheets, scale 1:2,500,000.)*.

be obtained through academic and government laboratories with conodont research groups. This may change as more emphasis is placed on Lower Paleozoic plays, especially in unconventional resources.

WIRELINE LOG INTERPRETATIONS

There are times when samples of potential source rocks are not available for an exploration area, and no published source rock data exist. In these instances, wireline log data are usually available and may be used to make some preliminary assessments of some source rock characteristics. The ability to discern the presence of potential source rocks and some of their geochemical characteristics using conventional wireline log data is dependent on the way some logging tools respond to the presence of organic matter in sediments. Typically, source rock studies using wireline logging focus on four common tools: gamma ray, formation density, sonic, and resistivity.

Gamma Ray

The gamma ray signal detected in wireline logs is derived primarily from the naturally occurring radioactive elements uranium, potassium-40, and thorium. It has been long known that source rocks are often "hot" zones on gamma ray logs. The high gamma ray response associated with source rocks comes from the uranium content of the sediments. Uranium compounds are water soluble when the sediment's Eh is oxidative, but under reducing conditions, uranium compounds precipitate and become fixed in the sediments. Organic matter in sediments can cause reducing conditions to develop which in turn precipitates uranium from natural waters (Fertl and Chilingar, 1988), and the amount of uranium in the sediments is often directly proportional to the amount of organic matter present. Increases in gamma ray response with organic matter content are frequently detected in the bulk gamma ray signal. However, an increase in bulk gamma ray may also be in response to increases in potassium-40 and/or thorium in the sediment. A better indicator of increased organic matter content with increasing gamma ray signal can be obtained using only the uranium signal derived from spectral gamma ray logs (Fertl and Rieke, 1980). Gamma ray logs used for source rock studies must be corrected for borehole size.

Formation Density

The density of a shale's mineral matrix is about 2.25 g/cc and greater, while the density of organic matter is about 0.9–1.1 g/cc (Meyer and Nederlof, 1984). Because of this disparity in density, density logs may be used to indicate the amount of organic matter present in sediments (Schmoker, 1979). After density logs are corrected for porosity, increases in organic matter content should result in decreases in the bulk formation density. One potential interference with this response is the presence of large amounts of pyrite (density 5.0 g/cc) in the sediment. Schmoker (1979) suggested that the logs could be compensated for the pyrite by assuming the amount of pyrite present was proportional to the organic matter content.

Sonic Transit Time

Organic matter is a poor conductor of acoustic energy. It also lowers the bulk density of the sediment, which in turn can decrease sonic velocity. Therefore, as the amount of organic matter in sediments increases, the sonic log's interval transit time should increase (decreasing sonic velocity) (Herron, 1991). While this relationship seems straight forward, there are many other influences on the interval transit time including mineral composition of the matrix, grain-to-grain pressure, and the water–organic matter ratio (Meyer and Nederlof, 1984). As a result, the sonic log response is not a stand-alone indicator and needs be used with other wireline log responses.

Interval transit time has also been suggested to be related to the thermal maturity of sediments. This relationship is not directly related to the organic matter in the sediments. Instead, it is linked to the compaction and diagenesis of the sediments that parallels the thermal maturity of the sediments.

Resistivity

Organic matter in sediments is nonconductive. As kerogen in a source rock, the organic matter has little impact on wireline log resistivity measurements. It is the pore water that has the greatest influence on resistivity measurements. But with increasing burial, hydrocarbon generation will occur. These hydrocarbons move out into the pore space, displacing conductive pore fluid thereby increasing the formation resistivity (Meissner, 1978). Whereas the gamma ray, density, and sonic logs may be used to indicate the presence of organic matter in sediments, the resistivity log is better used as an indicator of thermal maturity and hydrocarbon generation (Herron, 1991). Resistivity logs used in source rock studies must be corrected for temperature.

Source Rock Presence Interpretations

Identifying intervals of potential source rocks in a stratigraphic sequence is an important first step in any source rock evaluation program. This information can be used to direct sampling programs in wells by indicating where the zones of highest source potential are. Using wireline log data for this task can be as simple as look for "hot shale" zones in the gamma ray logs. However, a more integrated approach is often more effective.

One method is to basically look at a composite log consisting of gamma ray, density, sonic, and resistivity logs, such as the North Sea example in Fig. 3.55. Typically, source rock intervals will show an increase in gamma ray response, a decrease in formation density, and an increase in sonic transit time, as shown in the example. Resistivity may or may not show an increase depending on the maturity. In the case of the example, only a moderate increase in resistivity is observed suggesting a relatively low thermal maturity. In lacustrine sediments, gamma ray response may not be significant due to lower inherent uranium content in lacustrine waters (Meyer and Nederlof, 1984). The other log responses should be similar in lacustrine sediments.

Another method utilizes cross-plots of density/resistivity and sonic transit time/resistivity as described by Meyer and Nederlof (1984). An example of a density/resistivity cross-plot is shown in Fig. 3.56. Discriminant function analysis was done using gamma ray, sonic, density, and resistivity log data to establish the boundary conditions that indicate source rocks. While the discriminant functions determined by Meyer and Nederlof (1984) are a good starting point for applying these methods, a locally derived discriminant function may be needed in some cases.

A third method for source rock recognition is the Δlog R overlay method of Passey et al. (1990). The method plots porosity and resistivity on the same axis versus depth and looks for the separation between the curves, as shown in Fig. 3.57. It is often useful to plot the gamma ray curve parallel to the overlay plot to help confirm the presence of potential source rock intervals. Note that coals give similar responses to source rocks using this method.

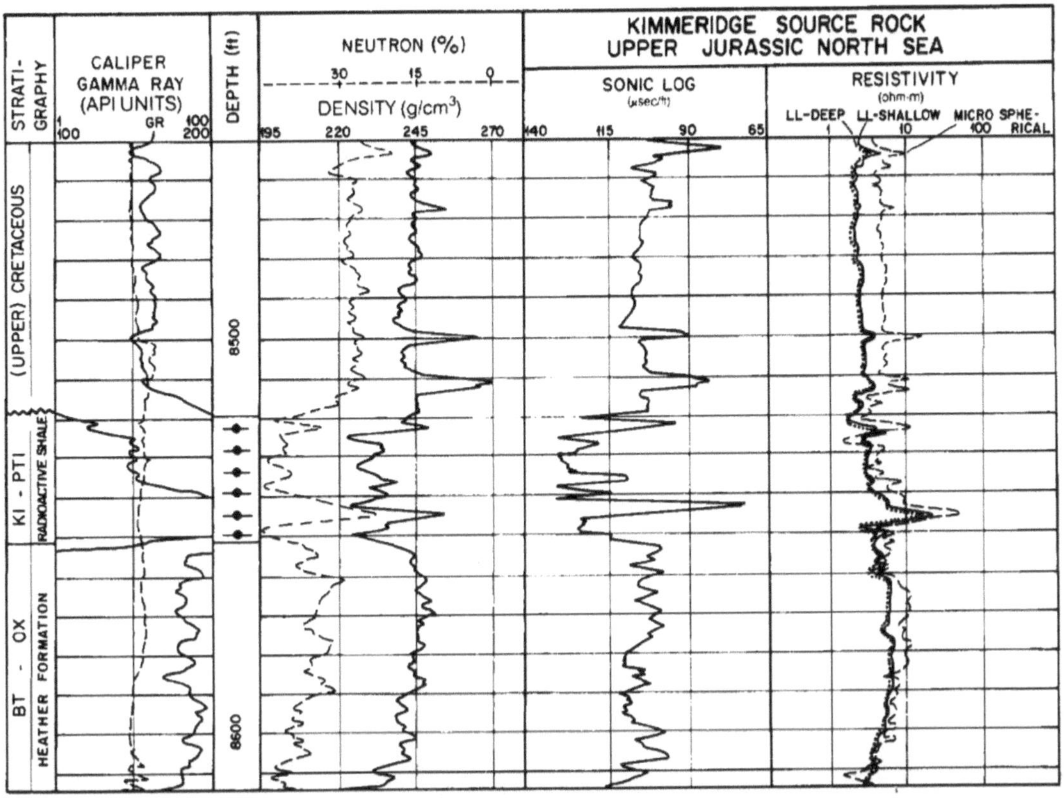

FIGURE 3.55 A composite log of Kimmeridgian source rock from the North Sea. The source rock is the stratigraphic interval KI-PTI radioactive shale from about 8519 to 8552 ft depth. *From Meyer, B.L., Nederlof, M.H., 1984. Identification of source rocks on wireline logs by density/resistivity and sonic transit time/resistivity crossplots. American Association of Petroleum Geologists Bulletin 68, 121–129.*

Source Richness Interpretations

The most common use of wireline logs in source rock interpretations is the prediction of % TOC. Early methods used either gamma ray (Schmoker, 1981) or formation density (Schmoker, 1979) to estimate the % TOC in source rocks, as shown in Figs. 3.58 and 3.59, respectively. Because of variation in lithology, compaction, and other geologic factors, no universal conversion of log response to % TOC exists for gamma ray and formation density. Instead, both of these methods require a set of laboratory-derived TOC values from a single source rock interval from a series of wells with corresponding wireline log data to establish a correlation of % TOC with log response. Once these correlations are established, measured TOC data can be supplemented by log-derived estimations.

A more recent method uses an extension of the Δlog R overlay method of Passey et al. (1990) to estimate % TOC from wireline logs. Quantitative estimates of % TOC are based on the amount of separation observed between the resistivity and sonic transit time curves with adjustments for the maturity of the source rock expressed as Level of Organic Metamorphism

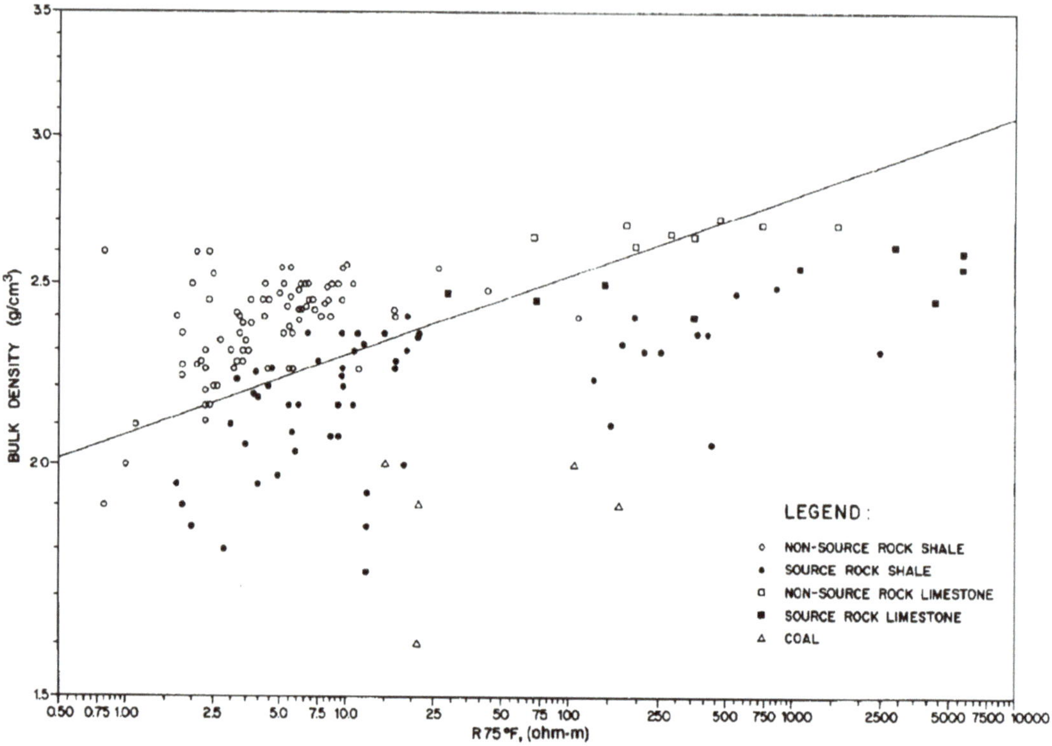

FIGURE 3.56 The density/resistivity cross-plot quick-look method. Potential source rocks should plot along below the line base case line determined by discriminant function analysis using gamma ray, sonic, density, and resistivity log data. *From Meyer, B.L., Nederlof, M.H., 1984. Identification of source rocks on wireline logs by density/resistivity and sonic transit time/resistivity crossplots. American Association of Petroleum Geologists Bulletin 68, 121–129.*

(LOM, after Hood et al., 1975). It has a fairly robust universal % TOC predictor for this combined log response at maturity levels less than about LOM 10.5 (vitrinite reflectance of about 0.9% Ro), but the method underestimates % TOC at higher maturities (Passey et al., 2010). Sondergeld et al. (2010) has suggested a correction to the conversion equation to remedy this problem.

Thermal Maturity Interpretations

There are two methods for using wireline log data to predict thermal maturity in source rock interval. The first method utilizes resistivity data, while the second is based on the sonic log's interval transit time.

To use resistivity as a qualitative maturity indicator, the work of Meissener (1978) should be used as the primary example. The resistivity of the source rock interval, in this case the Bakken of the Williston Basin, was observed to make a substantial and rapid increase in what was interpreted as the top of the oil window. Meissener (1978) attributed the increase in resistivity to generated hydrocarbons moving out of the kerogen, filling pore spaces, and

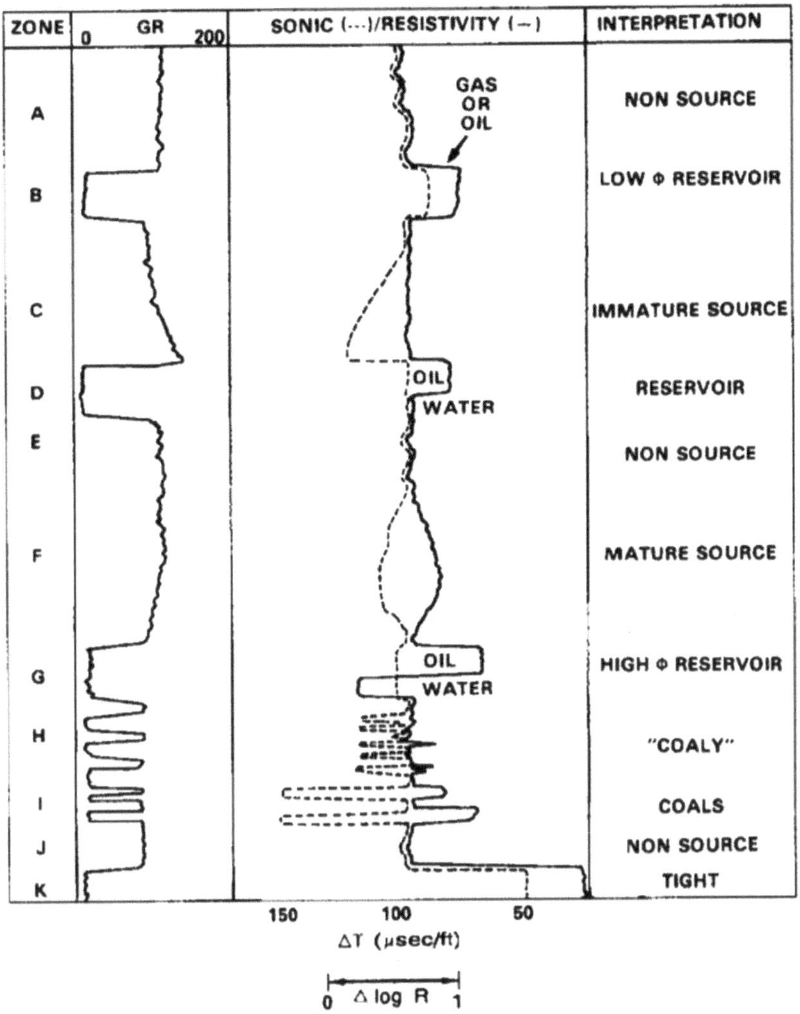

FIGURE 3.57 Schematic guide for the interpretation of the quick look Δlog R method of Passey et al. (1990) combining the overlay of the sonic and resistivity log responses with gamma ray (GR).

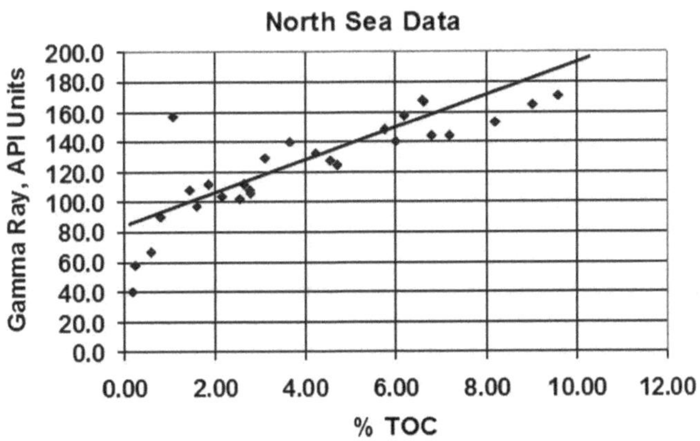

FIGURE 3.58 Calibrated response of gamma ray with % total organic carbon (TOC) from a North Sea data set. Note the lack of correlation below about 2.0% TOC.

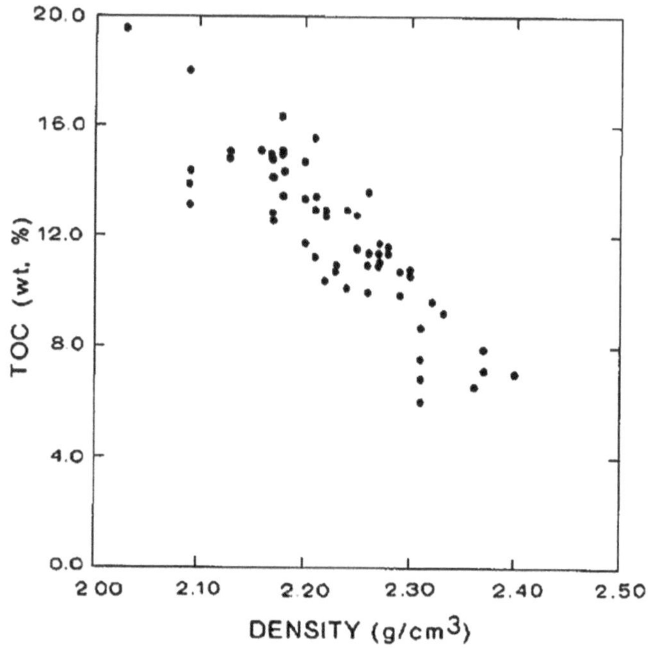

FIGURE 3.59 Calibrated response of formation density with % TOC. *From Schmoker, J.W., Hester, T.C., 1983. Organic carbon in Bakken formation, United States portion of Williston Basin. American Association of Petroleum Geologists Bulletin 67, 2165–2174.*

displacing conductive formation waters. By mapping the resistivity, Meissener (1978) was able to determine where the Bakken was mature and generating and where it was immature.

A more quantitative approach to using resistivity as a maturity indicator can be found in Smagala et al. (1984). This work developed a correlation between measured vitrinite reflectance in the Niobrara Formation in the Denver Basin and the wireline log resistivity of a specific source rock interval within the Niobrara, shown in Fig. 3.60. The relationship was then used to predict maturity in other parts of the basin where only wireline line resistivity logs were available. Schmoker and Hester (1990) used this same technique for the Bakken shale in the Williston Basin and the Woodford shale in the Anadarko basin. Because of the dependence of resistivity in sediments on composition of the pore waters in a formation, compaction, and lithology, no universal predictors can be derived for this method. Instead, local relationships need to be established between vitrinite reflectance and resistivity for specific source rock intervals to apply this method.

A similar approach was proposed by Lang (1994) to predict thermal maturity based on the sonic log's interval transit time. A correlation was developed between measured vitrinite reflectance and log-derived interval transit time of source rocks in the North Slope and Colville basin areas of Alaska. McTavish (1998) conducted a similar study in the North Sea, developing correlations of vitrinite reflectance with interval transit times in Paleozoic carbonaceous shales and post-Paleozoic argillaceous shale. Because of the dependence of interval transit times in sediments on both the lithology and compaction, no universal predictors can

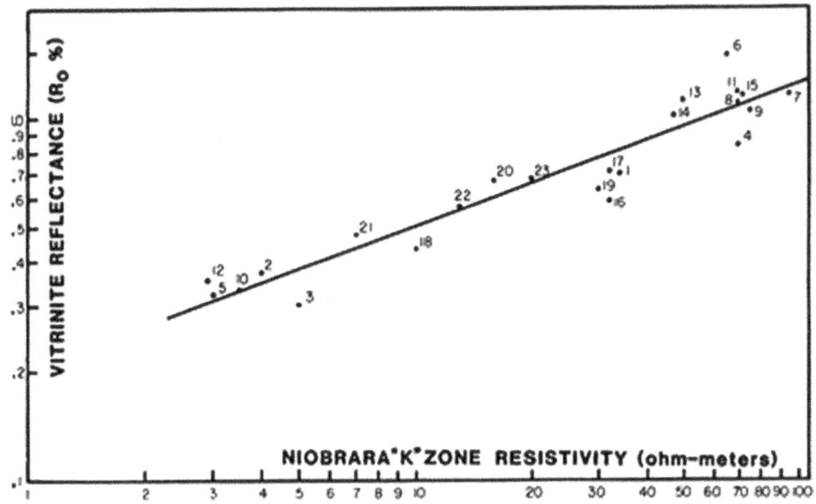

FIGURE 3.60 Calibrated response of vitrinite reflectance with resistivity for a specific zone in the Niobrara Formation in the Denver Basin. *From Smagala, T.M., Brown, C.A., Nydegger, G.L., 1984. Log-derived indicator of thermal maturity, Niobrara Formation, Denver Basin, Colorado, Nebraska, Wyoming. In: Woodward, J., Meissner, F.F., Clayton, J.L., (Eds.), Hydrocarbon Source Rocks of the Greater Rocky Mountain Region: Rocky Mountain Association of Geologists, pp. 355–363.*

be derived for this method. Instead, local relationships need to be established between vitrinite reflectance and interval transit time to be able predict thermal maturity.

USING OUTCROP SAMPLES

Most source rock evaluation programs are carried out using subsurface samples, but occasionally it is necessary or desirable to include outcrop samples. Outcrop samples differ primarily from subsurface samples because outcrop samples are subject to weathering of both the inorganic and organic components of the sediment. While the weathering of the inorganic components of the sediments does not directly impact the evaluation of an outcrop sample's source potential, weathering of the organic component certainly will.

When the organic matter weathers, several things can happen to the source potential indicators. First, the amount of organic matter present decreases making the sediment appear less organic rich. Second, the hydrogen content of the kerogen usually decreases making the sediment appear less capable of generating hydrocarbons and appear less oil prone and more-gas prone. And third, maturity indicators can be affected. Spores and pollen may darken and vitrinite reflectivity may increase due to oxidation, making the sediment appear more mature. The result is that outcrop samples usually appear less rich, less oil prone, and more mature than equivalent subsurface samples would.

The key to successfully using outcrop data is to be cautious. Organic matter in a sediment weathers at faster rates than does the mineral components. Therefore, when you observe weathering of the sediment, it is almost assured that the organic matter has been altered. It is best to try to use samples that exhibit minimal alteration. This may mean digging or drilling into the outcrop to collected as far below the surface as possible. When submitting the samples to a laboratory for analysis, clearly indicate they are outcrop samples so the analysts are aware and can look for the effects of weathering in the organic matter. And remember, the results obtained are likely more pessimistic than they would be from a subsurface sample, so a restrained interpretation of the data may be appropriate.

STRATEGIES IN SOURCE ROCK EVALUATION

Background Information

Prior to beginning to make interpretations of a source rock data set, there are certain pieces of essential information that should be gathered. Stratigraphy and geological age of the sediments are important to the study. If wireline logs have been run and formation tops picked, use this information as a guideline for the wells stratigraphy. Note any unconformities, faults, igneous intrusions, and other geologic features that may influence the geochemical data or interpretations.

The sediments lithology is also crucial information. Use the lithology log as a guide. Look for dark gray, dark brown, and black shales and limestone. Note the location of any coals encountered. Record any mineralized zones; they are possible indicators of hydrothermal alteration of the sediments. Document any oil stained intervals (including the cut and fluorescence), gas kicks, or the mention of bitumen in the sediments.

Note any casing points in the well. Remember, cavings are minimized just below casing points, and this can help with sorting out in situ vitrinite and/or spores/pollen for maturity studies.

Wireline log data are also very useful. Washouts on caliper log can indicate zones of caving sediments. Gamma ray log may help point toward potential source rock intervals. Look for gamma ray values above 80 API units. Collect any bottomhole temperature and pressure measurements that have been taken in conjunction with the logging runs. And, if possible, try the Δlog R overlay method of Passey et al. (1990). This sonic transit time/resistivity log overlay is a quick way to help identify potential source rock intervals.

The drilling mud composition is also very important. Determine whether the drilling mud was water based or oil based. If oil based, was a synthetic oil used, and if so which one? Note all drilling additives and when they were added to the drilling mud. And find out if drilling mud samples have been collected and if they are available for analysis if needed.

And finally, check on the drill bit used. Polycrystalline diamond compact, PDC, bits have an adverse effect on drill cuts, especially when used in conjunction with oil-based drilling muds. The extreme localized heat and pressure exerted on the sediment by these drill bits can induce physical and chemical changes in the cuttings referred to as bit metamorphism (Graves, 1986). The result is a characteristics texture called PDC-bit platelets (Graves, 1986) that indicate alteration of the geochemical properties of the sediment is likely (Wenger et al., 2009). In addition to the heating, bit metamorphism can incorporate drilling mud into the cuttings. The overall effects of bit metamorphism are reduced TOC and HI values, increases in the vitrinite reflectance, and darken spores and pollen grains in the sediments.

Defining the Problem

Before a source evaluation study begins, it is important to clearly define what you hope to learn from the data. Are you trying to identify source rocks or confirming source potential of known source rocks? If you know the source rocks are there, is determining maturity levels the main goal? Setting down the objectives of the study will guide you through the process and help focus your efforts. You will also need to assess whether you have the appropriate samples to answer your questions and if the appropriate geochemical analyses have been done. Recognizing that more samples or additional analytical work are needed early in the process is more efficient than discovering it after you have been through the evaluation process.

Making Interpretations

Once all the background information and geochemical data have been gathered, the process of interpretation can actually begin. A typical work flow is shown in Fig. 3.61. The first step is usually a review of the data for indications of any contamination in the samples. Recognizing contamination early prevents erroneous conclusions from being drawn. Begin with a review of the drilling mud information to look for indications of oil-based muds or

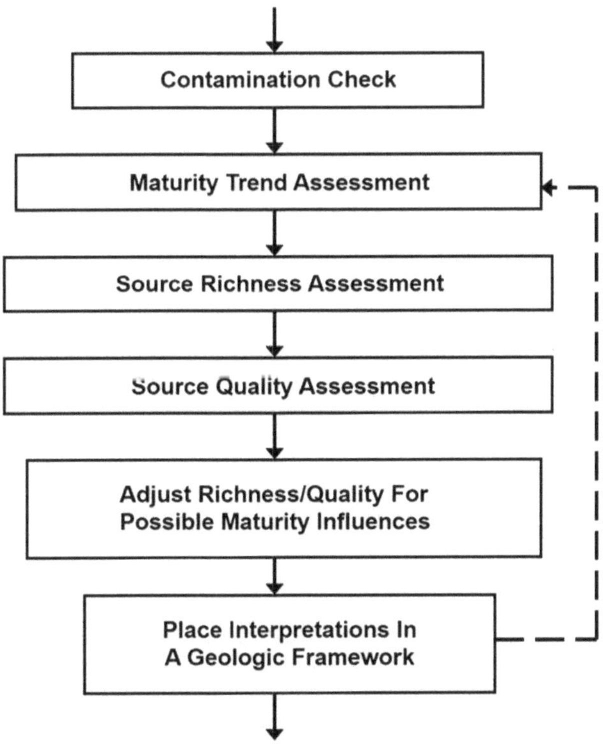

FIGURE 3.61 Flow chart for a source rock evaluation project as a component of a part of a comprehensive petroleum system analysis.

organic drilling additives. Then look at Rock-Eval S1 values, as well as any gas chromatographic data from whole rock extracts or saturate fractions. If the data look suspect, a discussion with the laboratory that provided the data is recommended. The lab analysts can help explain the inconsistencies in the data and may suggest way to mitigate any problem areas that may exist.

Once the contamination issue has been addressed, it is recommended that the thermal maturity data set be examined. Because maturity has an impact on how most source rock richness and quality data are interpreted, it is essential to understand the maturity of the source rocks early. Use all the possible maturity indicators available, weighing heavily on vitrinite reflectance and TAI. Look to T_{max}, % wet gas from headspace gas, fluorescence, and other indicators to corroborate the maturity trend. Remember to factor in any known igneous or hydrothermal activity, as well as investigating if PDC drill bits were used in the well.

When a maturity trend has been established, it is time to address source richness and quality. Usually source richness is evaluated first. Remember that TOC and S2 will decrease with increasing maturity. Higher maturity samples need to be recognized when richness assessments are being made. Although there are published methods to estimate the original TOC and S2 from samples with advanced maturity, these methods are based on making some assumptions about kerogen type and HI and usually result in biased guesses. Similarly, source quality assessments can be impacted by maturity. As a source rock increases in maturity, HI and OI decrease, H/C and O/C decrease, and PGC data for oil-prone kerogen look more gas prone. All these changes are progressive and influenced by changes in the kerogen due to hydrocarbon generation and migration. The bottom line is that as a source rock matures, it will look less like a source rock. A rich, oil-prone immature sediment could very well look like a lean, gas-prone sediment when it reaches the main stage of gas generation. Remember to account for maturity when making source richness and quality interpretations. An overmature sediment today should not be condemned because we cannot know what it started out as.

When the maturity, richness, and quality interpretations have been made, it is time to place your geochemical interpretation in a geologic context. The following are some of the many questions that should be asked to determine if the geochemical data is consistent with the geologic setting. Are the source rocks in a position to charge reservoirs? If the source rocks are immature at the well location, could they be mature down dip? Based on maturity, are the source rocks deeper in the section than the total depth of the well and therefore have not been sampled? If you are looking at a play adjacent to the well sampled, are the source rocks laterally consistent or could they be subject to facies change? Are the source rocks appropriate for their depositional setting, i.e., do you have marine kerogen in marine sediments?

The geochemical interpretations cannot stand alone. They need to make geologic sense, or they have little value. If there are inconsistencies in the geologic setting and/or geochemical interpretation, you may have to reassess the interpretation and begin the process again. Also consider that there may be errors in the geologic data. Early assignments of geologic age or depositional setting may be tentative and corrected by later more in depth study.

Source rock evaluation is but a part of a comprehensive petroleum system analysis, albeit an important part. A good unbiased effort in defining the source rock's characteristics and thermal maturity will pay large dividends in an exploration program. Take the time to understand the data and apply them properly.

References

Alpern, B., 1970. Classification petrographique des constituants organique fossils des roches sedimentaires. Revue de l'Institut Francais du Petrole 25, 1233–1267.

Alpern, B., Durand, B., Durand-Souron, C., 1978. Proprieties optique de residus de la pryrolyse des kerogenes. Revue de l'Institut Francais du Petrole 33, 867–890.

ASTM, 2014. D7708e11 Standard test method for microscopical determination of the reflectance of vitrinite dispersed in sedimentary rocks. Sec. 5, v. 05.06. In: Annual Book of ASTM Standards: Petroleum Products, Lubricants, and Fossil Fuels; Gaseous Fuels; Coal and Coke. ASTM International, West Conshohocken, PA. 10 p.

Baker, D.R., 1960. Organic geochemistry of Cherokee Group in southeastern Kansas and northeastern Oklahoma. American Association of Petroleum Geologists Bulletin 42, 1621–1642.

Barker, C.E., Pawlewicz, M.J., 1993. An empirical determination of the minimum number of measurements needed to estimate the mean random vitrinite reflectance of disseminated organic matter. Organic Geochemistry 20, 643–651.

Baskin, D.K., 1997. Atomic H/C ratio of kerogen as an estimate of thermal maturity and organic matter conversion. American Association of Petroleum Geologists Bulletin 81, 1437–1450.

Bertrand, R., 1990. Correlations among the reflectances of vitrinite, chitinozoans, graptolites and scolecodonts. Organic Geochemistry 15, 565–574.

Bertrand, R., 1993. Standardization of solid bitumen reflectance to vitrinite in some Paleozoic sequences of Canada. In: Goodarzi, F., Macqueen, R.W. (Eds.), Geochemistry and Petrology of Bitumen with Respect to Hydrocarbon Generation and Mineralization: Energy Sources, vol. 15, pp. 269–287.

Bertrand, Ph, Bordenave, M.L., Brosse, E., Espitalié, J., Houzay, J.P., Pradier, B., Vandenbroucke, M., Walgenwitz, F., 1993. Other methods and tools for source rock appraisal. In: Bordenave, M.L. (Ed.), Applied Petroleum Geochemistry. Editions Technip, Paris, vol. 3, pp. 279–371 (Chapter II).

Bray, E.E., Evans, E.D., 1961. Distribution of n-paraffins as a clue to recognition of source beds. Geochimica et Cosmoschimica Acta 22, 2–9.

Bray, E.E., Evans, E.D., 1965. Hydrocarbons in non-reservoir-rock source beds. American Association of Petroleum Geologists Bulletin 49, 248–257.

Cardott, B.J., 2012. Introduction to vitrinite reflectance as a thermal maturity indicator. In: Adapted from an Oral Presentation at Tulsa Geological Society Luncheon, May 8, 2012. American Association of Petroleum Geologists Search and Discovery. Article #40928 http://www.searchanddiscovery.com/documents/2012/40928cardott/ndx_cardott.pdf.

Carr, A.D., 2000. Suppression and retardation of vitrinite reflectance, Part I. Formation and significance for hydrocarbon generation. Journal of Petroleum Geology 23, 313–343.

Carvajal-Ortiz, H., Gentzis, T., 2015. Critical considerations when assessing hydrocarbon plays using Rock-Eval pyrolysis and organic petrology data: data quality revisited. International Journal of Coal Geology 152, 113–122.

Claypool, G.E., Love, A.H., Maughan, E.K., 1978. Organic geochemistry, incipient metamorphism, and oil generation in black shale members of Phosphoria Formation, western interior United States. American Association of Petroleum Geologists Bulletin 62, 98–120.

Correia, M., 1969. Contribution a la recherche de zones favorable a la genese du petrole par l'observation microscopique de la matiere organique figure. Revue de l'Institut Francais du Petrole 24, 1417–1454.

Dahl, B., Bojesen-Koefoed, J., Holm, A., Justwan, H., Rasmussen, E., Thomsen, E., 2004. A new approach to interpreting Rock-Eval S2 and TOC data for kerogen quality assessment. Organic Geochemistry 35, 1461–1477.

Daly, A.R., Edman, J.D., 1987. Loss of organic carbon from source rocks during thermal maturation. American Association of Petroleum Geologists Bulletin 71, 546.

Dembicki Jr., H., 1993. Improved determination of source quality and kerogen type by combining Rock Eval and pyrolysis-gas chromatography results. American Association of Petroleum Geologists 1993 Annual Convention Program 90.

Dembicki Jr., H., 2009. Three common source rock evaluation errors made by geologists during prospect or play appraisals. American Association of Petroleum Geologists Bulletin 93, 341–356.

Dembicki Jr., H., Horsfield, B., Ho, T.T.Y., 1983. Source rock evaluation by pyrolysis-gas chromatography. American Association of Petroleum Geologists Bulletin 67, 1094–1103.

Dow, W., 1977. Kerogen studies and geological interpretations. Journal of Geochemical Exploration 7, 79–99.

Durand, B., Monin, J.C., 1980. Elemental analysis of kerogens (C, H, O, N, S, Fe). In: Durand, B. (Ed.), Kerogen. Editions Technip, Paris, pp. 113–142.

Durand, B., Nicaise, G., 1980. Procedures for kerogen isolation. In: Durand, B. (Ed.), Kerogen. Editions Technip, Paris, pp. 35–53.
England, W.A., Mann, A.L., Mann, D.M., 1991. Migration from source to trap. In: American Association of Petroleum Geologists Treatise of Petroleum Geology: Source and Migration Processes and Evaluation Techniques, pp. 23–46.
Epstein, A.G., Epstein, J.B., Harris, L.D., 1977. Conodont color alteration; an index to organic metamorphism. United States Geological Survey Professional Paper 995, 27.
Espitalie, J., 1986. Use of Tmax as a maturation index for different types of organic matter: comparison with vitrinite reflectance. In: Burrus, J. (Ed.), Thermal Modeling in Sedimentary Basins. Proceedings of Meeting: 1st IFP Exploration Research Conference, Carcans, France, June 3–7, 1985, vol. 44. Editions Technip, Paris, pp. 475–496.
Espitalie, J., Laporte, J.L., Madec, M., Marquis, F., LePlat, P., Paulet, J., Boutefeu, A., 1977a. Methode rapide de caracterisation des roches meres de leur potentiel petrolier et de leur degre d'evolution. Revue Institut Francais Petrole 32, 23–42.
Espitalie, J., Madec, M., Tissot, B., Mennig, J.J., Leplat, P., 1977b. Source rock characterization method for petroleum exploration. In: Proceedings of the 9th Annual Offshore Technology Conference, v. 3, pp. 439–448.
Espitalie, J., Madec, M., Tissot, B., 1980. Role of mineral matrix in kerogen pyrolysis: influence on petroleum generation and migration. American Association of Petroleum Geologists Bulletin 64, 59–66.
Espitalié, J., Marquis, F., Barsony, I., 1984. Geochemical logging. In: Voorhees, K.J. (Ed.), Analytical Pyrolysis–Techniques and Applications. Butterworth, Boston, pp. 276–304.
Espitalie, J., Deroo, G., Marquis, F., 1985. La pyrolyse Rock-Eval et ses applications. Deuxième partie. Oil & Gas Science and Technology. l'Institut Francais du Petrole Energies nouvelles 40, 755–784.
Espitalie, J., Deroo, G., Marquis, F., 1986. La pyrolyse Rock-Eval et ses applications. Troisième partie. Oil & Gas Science and Technology. l'Institut Francais du Petrole Energies nouvelles 41, 73–89.
Fertl, W.H., Chilingar, G.V., 1988. Total organic carbon content determined from well logs. SPE Formation Evaluation 3, 407–419.
Fertl, W.H., Rieke, H.H., 1980. Gama-ray spectral evaluation techniques identify fractured shale reservoirs and source-rock characteristics. Journal of Petroleum Technology 31, 2053–2062.
Fuloria, R.C., 1967. Source rocks and criteria for their recognition. American Association of Petroleum Geologists Bulletin 51, 842–848.
Giraud, A., 1970. Application of pyrolysis and gas chromatography to geochemical characterization of kerogen in sedimentary rocks. American Association of Petroleum Geologists Bulletin 54, 439–451.
Gentzis, T., Goodarzi, F., 1990. A review of the use of bitumen reflectance in hydrocarbon exploration with examples from Melville Island, Arctic Canada. In: Nuccio, V.F., Barker, C.E. (Eds.), Applications of Thermal Maturity Studies to Energy Exploration. Rocky Mountain Section Society of Economic Paleotologists and Mineralogists, pp. 23–36.
Graves, W., 1986. Bit-generated rock textures and their effect on evaluation of lithology, porosity, and shows in drill-cutting samples. American Association of Petroleum Geologists Bulletin 70, 1129–1135.
Gutjahr, C.C.M., 1966. Carbonization measurements of pollen-grains and spores and their application. Leidse Geologische Mededelingen 38, 1–29.
Hackley, P.C., Araujo, C.V., Borrego, A.G., Bouzinos, A., Cardott, B.J., Cook, A.C., Eble, C., Flores, D., Gentzis, T., Gonçalves, P.A., Filho, J.G.M., Hámor-Vidó, M., Jelonek, I., Kommeren, K., Knowles, W., Kus, J., Mastalerz, M., Menezes, T.R., Newman, J., Oikonomopoulos, I.K., Pawlewicz, M., Pickel, W., Potter, J., Ranasinghe, P., Read, H., Reyes, J., De La Rosa Rodriguez, G., de Souza, I.V.A.F., Suárez-Ruiz, I., Sýkorová, I., Valentine, B.J., 2015. Standardization of reflectance measurements in dispersed organic matter: results of an exercise to improve interlaboratory agreement. Marine and Petroleum Geology 59, 22–34.
Hagemann, H., Hollerbach, A., 1981. Spectral fluorometric analysis of extracts a new method for the determination of the degree of maturity of organic matter in sedimentary rocks. Bulletin Des Centres De Recherchese Exploration-Production Elf-Aquitaine 5 (2), 635–650.
Harris, A.G., Harris, L.D., Epstein, J.B., 1978. Oil and gas data from Paleozoic rocks in the Appalachian basin; maps for assessing hydrocarbon potential and thermal maturity (conodont color alteration isograds and overburden isopachs). U.S. Geological Survey Miscellaneous Investigations Series Map I–917–E, 4 sheets, scale 1:2,500,000.
Harwood, R.J., 1977. Oil and gas generation by laboratory pyrolysis of kerogen. American Association of Petroleum Geologists Bulletin 61, 2082–2102.
Herron, S.L., 1991. Situ evaluation of potential source rocks by wireline logs: Chapter 13, Source and Migration Processes and Techniques for Evaluation. In: Merrill, R.K. (Ed.), Treatise of Petroleum Geology, Handbook of Petroleum Geology. American Association of Petroleum Geologists, pp. 127–134.

Hood, A., Gutjahr, C.C.M., Heacock, R.L., 1975. Organic metamorphism and the generation of petroleum. American Association of Petroleum Geologists Bulletin 59, 986–996.

Horsfield, B., 1989. Practical criteria for classifying kerogens: some observations from pyrolysis-gas chromatography. Geochimica et Cosmochimica Acta 53, 891–901.

Houseknecht, D.W., Weesner, C.M.B., 1997. Rotational reflectance of dispersed vitrinite from the Arkoma basin. Organic Geochemistry 26, 191–206.

Hunt, J.M., 1996. Petroleum Geochemistry and Geology, second ed. W. H. Freeman, New York. 743 pp.

Jacob, H., 1985. Disperse solid bitumen as an indicator for migration and maturity in prospecting for oil and gas. Erdol Kohle 38, 365–374.

Jacob, H., 1989. Classification, structure, genesis and practical importance of natural solid bitumen (migrabitumen). International Journal of Coal Geology 11, 65–79.

Jarvie, D.M., 1991. Total organic carbon (TOC) analysis. In: American Association of Petroleum Geologists Treatise of Petroleum Geology: Source and Migration Processes and Evaluation Techniques, pp. 113–118.

Jones, R.W., Edison, T.A., 1978. Microscopic observations of kerogen related to geochemical parameters with emphasis on thermal maturation. In: Oltz, D.F. (Ed.), Low Temperature Metamorphism of Kerogen and Clay Minerals. S.E.P.M. Pacific Section, Los Angeles, pp. 1–12.

Katz, B.J., Elrod, L.W. 1983. Organic geochemistry of DSDP Site 467, offshore California, Middle Miocene to Lower Pliocene strata: Geochimica et Cosmochimica Acta 47, 389–396.

Katz, B.J., Pheifer, R.N., Schunk, D.J., 1988. Interpretation of discontinuous vitrinite reflectance profiles. American Association of Petroleum Geologists Bulletin 72, 926–931.

Kuncheva, L.I., Charles, J.J., Miles, N., Collins, A., Wells, B., Lim, I.S., 2008. Automated kerogen classification in microscope images of dispersed kerogen preparation. Mathematical Geosciences 40, 639–652.

Landis, C.R., Castaño, J.R., 1995. Maturation and bulk chemical properties of a suite of solid hydrocarbons. Organic Geochemistry 22, 137–149.

Lang, W.H., 1994. The determination of thermal maturity in potential source rocks using interval transit time/interval velocity. The Log Analyst 35, 47–59.

Larter, S.R., Douglas, A.G., 1980. A pyrolysis-gas chromatographic method for kerogen typing. In: Douglas, A.G., Maxwell, J.R. (Eds.), Advances in Organic Geochemistry, 1979. Pergamon Press, New York, pp. 579–584.

Law, B.E., Nuccio, V.F., Barker, C.E., 1989. Kinky vitrinite reflectance well profiles: evidence of paleopore pressure in low-permeability, gas-bearing sequences in Rocky Mountain foreland basins. American Association of Petroleum Geologists Bulletin 73, 999–1010.

Lo, H.B., 1993. Correction criteria for the suppression of vitrinite reflectance in hydrogen-rich kerogens: preliminary guidelines. Organic Geochemistry 20, 653–657.

Marshall, J.E.A., 1991. Quantitative spore colour. Journal of the Geological Society of London 148, 223–233.

Massoud, M.S., Kinghorn, R.R.F., 1985. A new classification for organic components of kerogen. Journal of Petroleum Geology 8, 85–100.

McTavish, R.A., 1998. Applying wireline logs to estimate source rock maturity. Oil and Gas Journal 96, 76–79.

Meissner, F.F., 1978. Petroleum geology of the Bakken formation, Williston Basin, North Dakota and Montana. In: Estelle, D., Miller, R. (Eds.), The Economic Geology of the Williston Basin, 1978 Williston Basin Symposium. Montana Geological Society, pp. 207–230.

Meyer, B.L., Nederlof, M.H., 1984. Identification of source rocks on wireline logs by density/resistivity and sonic transit time/resistivity crossplots. American Association of Petroleum Geologists Bulletin 68, 121–129.

Mukhopadhyay, P.K., Hagermann, H.W., Gormly, J.R., 1986. Characterization of kerogen as seen under the aspect of maturation and hydrocarbon generation. Erdol Kohle Erdgas Petrochem 38, 7–18.

Noble, R.A., 1991. Geochemical techniques in relation to organic. In: Merrill, R.K. (Ed.), Source and Migration Processes and Evaluation Techniques. American Association of Petroleum Geologists, Tulsa, pp. 97–102.

Palciauskas, V.V., 1991. Primary migration of petroleum. In: American Association of Petroleum Geologists Treatise of Petroleum Geology: Source and Migration Processes and Evaluation Techniques, pp. 13–22.

Passey, Q.R., Bohacs, K., Esch, W.L., Klimentidis, R., Sinha, S., 2010. From oil-prone source rock to gas-producing shale reservoir - geologic and petrophysical characterization of unconventional shale gas reservoirs. In: Society of Petroleum Engineers Conference Paper131350, Presented at the CPS/SPE International Oil & Gas Conference and Exhibition, Beijing, China, June 8–10, 2010 29 p.

Passey, Q.R., Creaney, S., Kulla, J.B., Moretti, F.J., Stroud, J.D., 1990. A practical model for organic richness from porosity and resistivity logs. American Association of Petroleum Geologists Bulletin 74, 1777–1794.

Pearson, D.L., 1984. Pollen/Spore Color Standard, Version #2. Phillips Petroleum Company, Exploration Project Section. Bartelsville, Oklahoma.

Peters, K.E., 1986. Guidelines for evaluating petroleum source rock using programmed pyrolysis. American Association of Petroleum Geologists Bulletin 70, 318–329.

Peters, K.E., Ishiwatari, R., Kaplan, I.R., 1977. Color of kerogen as index of organic maturity. American Association of Petroleum Geologists, Bulletin 61, 504–510.

Petersen, N.F., Hickey, P.J., 1987. California Plio-Miocene oils: evidence of early generation. In: Meyer, R.F. (Ed.), Exploration for Heavy Crude Oil and Natural BitumenAmerican Association of Petroleum Geologists Studies in Geology, vol. 25, pp. 351–359.

Powell, T.G., Greaney, S., Snowdon, L.R., 1982. Limitations of use of organic petrographic techniques for identification of petroleum source rocks. American Association of Petroleum Geologists Bulletin 66, 430–435.

Robert, P., 1973. Analyse microscopique des charbons et des bitumens disperses dans les roches et mesure de leur pouvoir reflecteur, Application a l'etude de la paleogeothermie des bassins sedimentaires et de la genese des hydrocarbures. In: Advances in Organic Geochemistry, International Meeting of Organic Geochemistry, Program and Abstracts, 6, pp. 549–569.

Ronov, A.B., 1958. Organic carbon in sedimentary rocks (in relation to the presence of petroleum). Geochemistry 5, 497–509.

Saxby, J.D., 1970. Isolation of kerogen in sediments by chemical methods. Chemical Geology 6, 173–184.

Schmoker, J., 1979. Determination of organic content of Appalachian Devonian shales from formation-density logs. American Association of Petroleum Geologists Bulletin 63, 1504–1537.

Schmoker, J.W., 1981. Determination of organic-matter content of Appalachian Devonian shales from Gamma-ray logs. American Association of Petroleum Geologists Bulletin 65, 1285–1298.

Schmoker, J.W., Hester, T.C., 1983. Organic carbon in Bakken formation, United States portion of Williston Basin. American Association of Petroleum Geologists Bulletin 67, 2165–2174.

Schmoker, J.W., Hester, T.C., 1990. formation resistivity as an indicator of oil generation – Bakken formation of North Dakota and Woodford shale of Oklahoma. The Log Analyst 31, 1–9.

Senftle, J.T., Landis, C.R., 1991. Vitrinite reflectance as a tool to assess thermal maturity. In: American Association of Petroleum Geologists Treatise of Petroleum Geology: Source and Migration Processes and Evaluation Techniques, pp. 119–125.

Smagala, T.M., Brown, C.A., Nydegger, G.L., 1984. Log-derived indicator of thermal maturity, Niobrara Formation, Denver Basin, Colorado, Nebraska, Wyoming. In: Woodward, J., Meissner, F.F., Clayton, J.L. (Eds.), Hydrocarbon Source Rocks of the Greater Rocky Mountain Region. Rocky Mountain Association of Geologists, pp. 355–363.

Sondergeld, C.H., Newsham, K.E., Comisky, J.T., Rice, M.C., Rai, C.S., 2010. Petrophysical considerations in evaluating and producing shale gas. In: SPE Conference Paper Number 131768 Presented at SPE Unconventional Gas Conference, Pittsburgh, Pennsylvania, February 23–25, 2010 34 p.

Stach, E., Mackowsky, M-Th., Teichmüller, M., Taylor, G.H., Chandra, D., Teichmüller, R., 1982. Stach's Textbook of Coal Petrology, second ed. Gebrüder Borntraeger, Berlin. 535 p.

Staplin, F.L., 1969. Sedimentary organic matter, organic metamorphism and oil and gas occurrence. Bull. Canadian Petroleum Geology 17, 47–66.

Stasiuk, L.D., Osadetz, K.G., Potter, J., 1990. Fluorescence spectral analysis and hydrocarbon exploration: examples from Paleozoic source rocks. In: Beck, L.S., Harper, C.T. (Eds.), Innovative Exploration Techniques. Saskatchewan Geological Society, Special Publication No. 12, pp. 242–251.

Sweeney, J.J., Burnham, A.K., Braun, R.L., 1987. A model of hydrocarbon generation from Type I kerogen: application to Uinta Basin, Utah. American Association of Petroleum Geologists Bulletin 71, 967–985.

Teichmuller, M., 1982a. Application of coal petrological methods in geology including oil and natural gas prospecting. In: Stach's Textbook of Coal Petrology. Gebrubder Borntraeger, Berlin, pp. 381–413.

Teichmüller, M., 1982b. Fluorescence Microscopical Changes of Liptinites and Vitrinites during Coalification and Their Relationship to Bitumen Generation and Coking Behaviour: The Society for Organic Petrology. Special Publication No. 1, 74 p.

Teichmuller, M., 1989. The genesis of coal from the viewpoint of coal petrology. International Journal of Coal Geology 12, 1–87.

Teichmüller, M., Durand, B., 1983. Fluorescence microscopical rank studies on liptinites and vitrinites in peat and coals, and comparison with results of the Rock-Eval pyrolysis. International Journal of Coal Geology 2, 197–230.

Thompson-Rizer, C.L., 1987. Some optical characteristics of solid bitumen in visual kerogen preparations. Organic Geochemistry 11, 385–392.

Thompson, C.L., Dembicki Jr., H., 1986. Optical characteristics of amorphous kerogen and the hydrocarbon generating potential of source rocks. International Journal of Coal Geology 6, 229–249.

Thompson-Rizer, C.L., Woods, R.A., 1987. Microspectrofluorescence measurements of coals and petroleum source rocks. International Journal of Coal Geology 7, 85–104.

Tissot, B., 1984. Recent advances in petroleum geochemistry applied to hydrocarbon exploration. American Association of Petroleum Geologists Bulletin 68, 546–563.

Tissot, B., Durand, B., Espitalie, J., Combaz, A., 1974. Influence of the nature and diagenesis of organic matter in the formation of petroleum. American Association of Petroleum Geologists Bulletin 58, 499–506.

Tissot, B.P., Welte, D.H., 1984. Petroleum Formation and Occurrence. Springer-Verlag, New York. 699 p.

Tissot, B., Pelet, R., Ungerer, P., 1987. Thermal history of sedimentary basins, maturation indices, and kinetic of oil and gas generation. American Association of Petroleum Geologists Bulletin 71, 1445–1466.

van Gijzel, P., 1979. Manual of the Techniques and Some Geological Applications of Fluorescence Microscopy: American Association of Stratigraphic Palynologists, 12 Annual Meeting Workshop. Dallas, 55 p.

Vandenbroucke, M., Largeau, C., 2007. Kerogen origin, evolution and structure. Organic Geochemistry 38, 719–833.

Wenger, L.M., Pottorf, R.J., Macleod, G., Otten, G., Dreyfus, S., Justwan, H., Sekula-Wood, E., 2009. Drillbit Metamorphism: Recognition and Impact on Show Evaluation. SPE paper 125218, 9 p.

Williams, L.A., 1984. Subtidal stromatolites in Monterey formation and other organic-rich rocks as suggested source contributors to petroleum formation. American Association of Petroleum Geologists Bulletin 68, 1879–1893.

CHAPTER

4

Interpreting Crude Oil and Natural Gas Data

OUTLINE

Introduction 136

Bulk Properties of Crude Oil and Natural Gas 136
- *API Gravity* 136
- *Gas Specific Gravity* 137
- *Viscosity* 137
- *Pour Point* 137
- *Cloud Point* 137
- *Flash Point* 138
- *Weight Percent Sulfur* 138
- *Nickel (Ni) and Vanadium (V) Contents* 138
- *S-A-R-A* 138
- *Solution Gas–Oil Ratio and Solution Oil–Gas Ratio* 138
- *Heating Value of Natural Gas* 139

Phase Behavior 139

Crude Oil and Natural Gas Alteration 141
- *Thermal Alteration* 141
- *Biodegradation* 142
- *Water-Washing* 147
- *Gas-Washing* 147
- *Deasphalting* 147
- *Devolatilization* 148
- *Thermochemical Sulfate Reduction* 149
- *Contamination* 149

Oil-to-Oil and Oil-to-Source Rock Correlations 150
- *Biomarkers* 151
- *Biomarker Analysis* 152
- *Comparing Data for Oil-to-Oil and Oil-to-Source Rock Correlations* 154

Crude Oil Inversion 160
- *Presence of Individual Biomarker Indicators* 160
- *Relative Abundance of Groups of Compounds* 163
- *Biomarker Ratios* 167

Strategies and Obstacles in Oil Correlation and Oil Inversion Studies 170

Natural Gas Data 173

The Source of Natural Gas: Biogenic Versus Thermogenic 173

The Maturity of Thermogenic Natural Gas 176

Gas-to-Gas and Gas-to-Source Rock Correlations 178

Strategies and Obstacles in Interpreting Gas Data 183

References 184

Practical Petroleum Geochemistry for Exploration and Production
http://dx.doi.org/10.1016/B978-0-12-803350-0.00004-0

INTRODUCTION

Once wells are drilled and hydrocarbons are encountered, the scope of exploration geochemistry expands to include the analysis and interpretation of the oil and gas found. These hydrocarbons indicate that a working petroleum system is present and can give clues to help decipher the elements and processes comprising the petroleum system. This process usually begins by establishing the characteristics of the hydrocarbons. Was the oil waxy or naphthenic? Was the crude sweet or sour? Is the gas dry or wet? Is the oil and gas in the subsurface a single phase or is there a separate gas cap in the reservoirs? The answers to these questions and others can help define the nature of the hydrocarbons, better understand how the hydrocarbons might behave in the reservoir, and begin to give some indications of how the oil and gas may have formed.

With additional careful detailed analyses, the oil and gas found may be related, or correlated, to other oils and gases previously found in the vicinity. These correlations can establish a genetic link between the hydrocarbons suggesting they were derived from the same source. If source rock samples are available, these correlations may be extended to linking a particular source rock interval to families of related oils or gases in oil-to-source rock or gas-to-source rock correlation. This type of information is vital to understanding where the generative "kitchens" are in a basin and suggesting migration pathways from source to trap. This knowledge when combined with information about the basin's development can lead to additional discoveries.

The focus of this chapter is to the describe analyses used in the characterization and correlation of discovered hydrocarbons and methods for interpreting these data. There will also be discussions of processes that may alter the composition of the oil and gas in the reservoir and how these alterations can interfere with making valid interpretations.

BULK PROPERTIES OF CRUDE OIL AND NATURAL GAS

The bulk properties of crude oils described here are part of a larger set of analyses called a crude oil assay that help to assess the market value of the oil, as well as indicate its refining and transportation characteristics. The measurements discussed below are common properties available to exploration geoscientists and have some geochemical significance. Prior to detailed geochemical analysis of crude oils, many of these properties were also used as rudimentary oil correlation parameters. These brief definitions will provide a basis for understanding how these properties are used in deciphering crude oil alteration processes, as well as how they can provide some insight in oil correlation work.

API Gravity

This is the American Petroleum Institute's (API) expression of the specific gravity of crude oils and condensates measured at 60°F (16°C). Specific gravity is the ratio of the mass of a body to the mass of an equal volume of water at a specified temperature. It can be thought of as a proxy for density. The API gravity is defined as (141.5/specific gravity)-131.5 and is reported in degrees (°). Oils are classified by their API gravities: heavy oils are less than 25°; medium gravity oils are 25°–35°; light oils are 35°–45°; and condensates are greater than 45°.

Gas Specific Gravity

Gas specific gravity is defined as the ratio of the density of the gas to the density of air at 1 atm pressure at 60°F (standard conditions). If ideal gas law behavior is assumed, gas specific gravity is the molecular weight of the gas divided by the molecular weight of air. Because natural gas is a mixture, the molecular weight used is a weighted average molecular weight based on the components of the gas.

Gas specific gravity is easy to measure at the wellhead and is used as an indication of the composition of the gas. The specific gravity of natural gas ranges from about 0.55 to about 0.87, due to variation in natural gas composition, with richer gases (higher wet gas content) typically having higher specific gravity. However, high nitrogen and/or carbon dioxide may also influence specific gravity.

Viscosity

Viscosity is the internal friction due to molecular cohesion in fluids which results in resistance to flow. It is measured in poise (P), 0.100 $kg\,m^{-1}\,s^{-1}$, but usually expressed as centipoise (cP), 0.001 $kg\,m^{-1}\,s^{-1}$. For a frame of reference, the viscosity of some common materials is as follows: water ~1 cP, olive oil ~80 cP, honey 2000–10,000 cP, peanut butter ~250,000 cP, and chocolate ~1,000,000 cP.

Viscosity is influenced by the composition of the crude oil, the temperature, dissolved gas content, and the pressure. As temperature increases, the viscosity will decrease. As a result, viscosity measurements are always reported with the temperature at which the measurement is made. By measuring viscosity at several temperatures, a relationship can be determined to help predict the viscosity of the fluid at other temperatures. Dissolved gas will influence the viscosity of petroleum. As dissolved gas content increases, the fluid viscosity decreases.

Gas viscosities are important for determining flow characteristic and production rates. They usually range from 0.01 to 0.03 cp. It is very difficult to directly measure these low viscosity values accurately. Instead, gas viscosities are typically approximated based on the composition of the gas and the temperature–pressure conditions.

Pour Point

The pour point is the temperature at which a crude oil becomes semisolid and will no longer flow. Pour point is important to recovery and transportation and is always determined. Pour points can range from about 32°C to −57°C. High pour points usually occur in crude oils that have significant paraffin content. Paraffins, especially the n-paraffins, will start to precipitate as temperature decreases causing wax deposition as well as flow issues. This phenomenon is not restricted to heavy oils, but can occur in light oils as well.

Cloud Point

The temperature at which turbidity is noticed in a crude oil due to the settling out of solid paraffin waxes as the oil is chilled is called the cloud point. It is sometimes referred to as the wax precipitation temperature or the wax appearance temperature. The cloud point, like pour point, is an indicator of potential problems with the recovery and transportation of the crude oil.

Flash Point

The temperature at which the vapors rising off the surface of a heated oil will ignite with a flash when a flame is passed over the surface of the oil is called the flash point. Although this measurement has little geochemical significance, it does have safety implications as an indicator of fire hazard. It is also required for shipping crude oil samples from the field to the lab for analysis.

Weight Percent Sulfur

The weight percent sulfur is a measure of both the free and bound sulfur in a crude oil. The free sulfur is elemental sulfur dissolved in the crude oil, while bound sulfur is the sulfur in the organic compounds that makes up some of the crude oil. Weight percent sulfur is usually inversely proportional to the API gravity. Large amounts of sulfur in an oil can be an indicator of the kerogen type that generated it, e.g., type II-S, or it may be an indicator of alteration processes, such as biodegradation (discussed below). Sour crude oil will have greater than 0.5% sulfur.

Nickel (Ni) and Vanadium (V) Contents

As chlorophyll degrades in sediments, it loses the magnesium ion from the center of the porphyrin structure and picks up either nickel or vanadium. The porphyrins carry the nickel and vanadium in the same proportions as established in the depositional environment. During generation and migration, these porphyrin structures are incorporated into the crude oil and carried along to the reservoir, preserving this information about the proportions of nickel and vanadium in the source rock. The nickel and vanadium content is also important is establishing a value for the oil. Nickel and vanadium will poison catalysts during the refining process. As such, high concentrations of nickel and vanadium in a crude oil will reduce the market value of the oil.

S-A-R-A

S-A-R-A refers to the *s*aturate, *a*romatic, *r*esin, and *a*sphaltenes fractions derived from separation analysis by liquid chromatography (discussed in Chapter 3) of the crude oil. Saturates are saturated hydrocarbon compounds. Aromatics are aromatic compounds, mostly hydrocarbons, but also including aromatic sulfur compounds. Resins are nitrogen, sulfur, and oxygen (NSO) bearing compounds. And the asphaltenes are the high molecular weight complex molecules held in colloidal solution within crude oils (see definition of asphaltenes in Chapter 1). S-A-R-A data are useful in classifying crude oils and helping decipher alteration processes that may have occurred.

Solution Gas–Oil Ratio and Solution Oil–Gas Ratio

For crude oils, the solution gas–oil ratio, usually referred to as GOR, is the volume of gas that can be exsolved out of a unit volume of oil under standard temperature and pressure conditions. It is expressed in standard cubic feet of gas/stock tank barrel of oil (SCF/STB). In contrast, the solution oil–gas ratio, usually referred to as OGR, is the amount of condensate that can be condensed out of a gas under standard pressure and temperature conditions. OGR is sometimes referred to as the liquid content of the gas. It is expressed in stock tank barrels of oil/million standard cubic feet of gas, STB/MMSCF. The GOR and OGR are useful in understanding the behavior of the fluid in the reservoir and during production, as well as being used to determining the volume of the fluids in the reservoir.

Heating Value of Natural Gas

The heating value of natural gas is the energy released per unit mass or per unit volume of the gas when the gas is completely burned, usually expressed in BTU/SCF. It is calculated by summing the product of the heating value and the mole or volume fraction for each gas component.

While the heating value of natural gas is used mainly for determining market value, it can also provide insight into the composition of the gas. The heating value for methane is ~900 BTU/SCF, while the heating values for butanes ~3000 BTU/SCF. As a gas becomes richer (higher wet gas content), the heating value will increase. Heating values can be lower than 900 (100% methane) due to the presence of noncombustible gases such as nitrogen or carbon dioxide.

PHASE BEHAVIOR

Determining the phase behavior of petroleum is a means of describing and predicting how the reservoir fluid will respond to changes in *p*ressure, *v*olume, or *t*emperature, PVT, in both the subsurface and during production. Because phase behavior is dependent on the fluid composition as well as PVT conditions, each reservoir fluid will be different. In the reservoir, it is normally assumed that the volume is initially fixed and any changes in fluid behavior will be a result of changes in pressure or temperature. To build a pressure–temperature, or P-T, phase diagram, as shown in Fig. 4.1, samples of the reservoir fluid must be collected

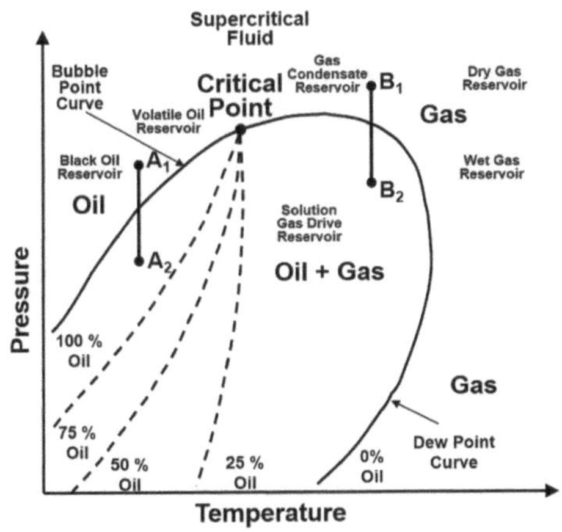

FIGURE 4.1 A temperature–pressure diagram used to explain phase behavior of fluid in the subsurface during production, where p is pressure, T_c is the critical temperature, p_c is the critical pressure, p_b is the bubble point pressure, and p_d is the dew point pressure. Also included is a classification of petroleum reservoirs in pressure–temperature space following the criteria of Calhoun (1953).

under pressure. In the laboratory, the reservoir fluid is transferred to a high pressure–high temperature cell and brought to reservoir conditions. By varying the temperature and pressure in the cell, the behavior of the fluid can be observed and the collected data are used to construct a phase diagram in P-T space.

In the phase diagram in Fig. 4.1, the curve to the left of the critical point is the bubble point curve. The bubble point curve is the boundary between the single-phase oil field above the curve and the two phase oil + gas field below the curve. An oil at point A_1 is a single-phase oil with all gas in solution. If it experiences a reduction in pressure to point A_2, gas will begin to bubble out of the oil as it crosses the bubble point curve and a separate gas phase will form. The farther the oil goes into the two phase field, the more gas is exsolved. To the right of the critical point, in Fig. 4.1, is the dew point curve. The dew point curve is the boundary between the single-phase gas field above and to the right of the curve and the two phase oil + gas field below and to the left of the curve. If a gas at point B_1 experiences a reduction in pressure to point B_2, it begins as a single-phase gas with all oil in solution. As it crosses the dew point curve, the oil begins to condense out of the gas and forms a separate liquid phase. The critical point is where the bubble point curve meets the dew point curve. In the area above the critical point, liquid and gas phases are indistinguishable and behave as a supercritical fluid exhibiting properties of both.

In addition to helping understand phase behavior of a subsurface fluid, P-T diagrams are used to classifying petroleum reservoirs into six categories based on fluid type as shown in Fig. 4.1. Black oils reservoirs are undersaturated with dissolved gas, have initial gas–oil ratios of less than 2000 scf/stb, API gravities of less than 45°, and the reservoir pressure is always greater than the bubble point pressure (McCain, 1990). As pressure drops in the reservoir and the fluid goes through the bubble point, gas will come out of solution forming a separate phase and the reservoir becomes a solution drive reservoir. The formation of this separate gas phase provides most of the reservoir's drive energy for production.

Volatile oil reservoirs typically have initial gas–oil ratios around 1000 to 8000 scf/stb with the oil having an API gravity of 45° or higher with the reservoirs near the critical temperature and pressure of the fluid, but always below the critical temperature (McCain, 1990). As pressure drops in the reservoir and the fluid goes through the bubble point, gas will come out of solution forming a separate phase and the reservoir also becomes a solution drive reservoir.

Gas condensate reservoirs have gas–oil ratios between 70,000 and 1000,000 scf/stb and API gravities greater than 60° (McCain, 1990). They are similar to volatile oil reservoirs in that they are near the critical temperature and pressure of the fluid. But these gas reservoirs are always at temperatures higher than the critical temperature. As pressure is reduced in a gas condensate gas reservoir, the fluid will pass through the dew point and light liquids will condense out of the gas in the reservoir.

Wet gas reservoirs have gas–oil ratios greater than 1000,000 scf/stb and exist solely as a gas phase reservoir throughout the reduction in reservoir pressure due to production (McCain, 1990). The wet gas contains significant amounts of heavy hydrocarbon gas components such as propane, butanes, pentanes, hexanes, and some liquid hydrocarbons. Unlike gas condensates, no liquid is formed inside the reservoir and liquids condensate out only when the gas is produced at the surface.

Dry gas reservoirs are void of condensate or liquid hydrocarbons. The gas is primarily methane with small amounts of ethane and propane. The fluid exists solely as gas in the reservoir and there is no condensable liquids formed either in the reservoir or at surface. The pressure path line does not enter into the phase envelope in the phase diagram, thus there is only dry gas in the reservoir.

Understanding the phase behavior of a reservoired fluid is key to efficient production and obtaining the most value from the asset discovered. Additional information about the sample collection and laboratory measurements going into PVT analysis as well as some discussion of data interpretation can be found in overviews by Freyss et al. (1989) and Whitson (1992).

CRUDE OIL AND NATURAL GAS ALTERATION

When petroleum reaches a reservoir, it may or may not remain in the same condition as it arrived. The oil and gas may be subject to alteration by processes active in the reservoir. These processes include thermal alteration, biodegradation, water-washing, gas-washing, deasphalting, devolatization, thermochemical sulfate reduction (TSR), and contamination. As a result of these processes, diagnostic physical properties and chemical characteristics of the crude oil and natural gas can become obscured making it difficult to make interpretations about their origin and their relationship to other hydrocarbons found in the vicinity. It is therefore important to recognize these alteration processes and understand how they impact crude oil and natural gas so that any interpretation made can be compensated for any changes.

Thermal Alteration

Often referred to as crude oil maturation or oil cracking, thermal alteration is the result of increased heating or prolonged exposure to heat in the reservoir. This is usually in response to increased depth of burial of the reservoir and the subsequent increase in reservoir temperature. Thermal alteration of oil can begin as low as 60°C (Hunt, 1996) and is characterized by the cracking of the larger molecules to form smaller molecules, as shown in the C_7+ portions of the gas chromatograms in Fig. 4.2. While these chromatograms show the overall pattern of a shift toward smaller, lower molecular weight compounds with increasing thermal alterations, it should be noted not all compounds groups react at the same rate. Some compound types are more thermally stable and persist longer, while others are more thermally labile and are altered more quickly or are completely eliminated. As thermal alteration progresses, asphaltenes and N-S-O compounds concentrations decrease, gasoline range (C_6–C_{12}) hydrocarbons and the paraffins and naphthenes in the saturate fraction increase (Milner et al., 1977). As a result of the cracking of larger compounds into smaller compounds, the API gravity will increase and the gas range (C_1–C_4) compounds formed will cause the gas-to-oil ratio to increase. Continued or extensive thermal alteration will eventually eliminate all liquids from the reservoir, leaving only gas and a carbon residuum. This main phase of cracking of oil to gas usually begins at about 150–160°C and is usually complete by about 250°C (Hunt, 1996).

Because thermal alteration can modify and eventually destroy compounds found in crude oil, this process can have a severe impact on interpretations of crude oil characteristics and any oil correlation work. Prior to making these interpretations, it is necessary to assess the extent of the thermal alteration and decide if the compounds being used have been adversely

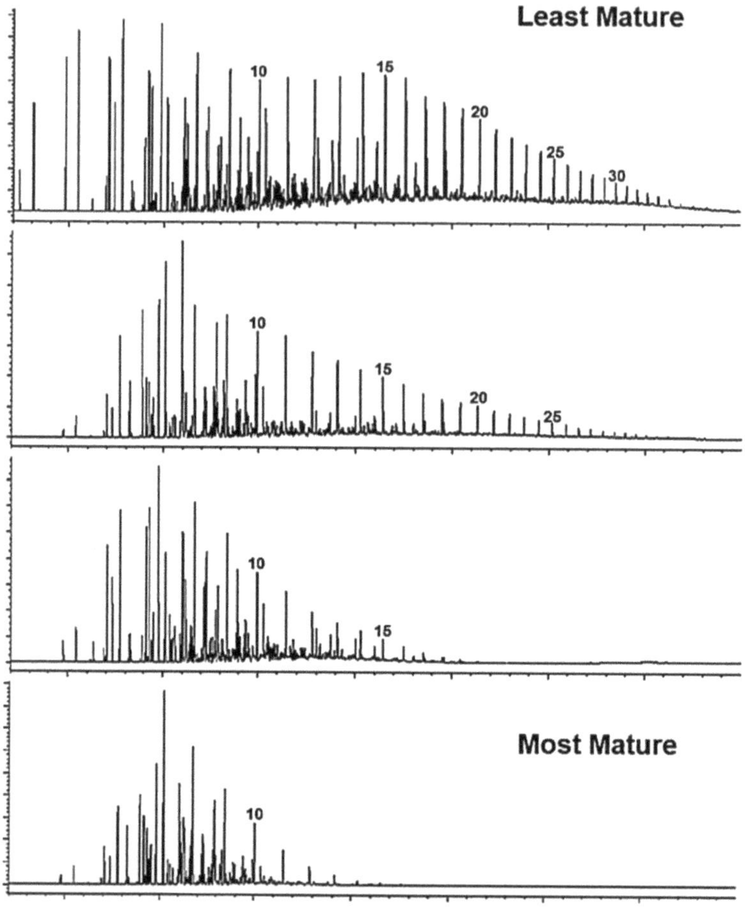

FIGURE 4.2 Whole oil gas chromatograms showing the thermal evolution of crude oils from the same source. The carbon numbers of selected n-paraffin peaks are shown for reference. Evaporative loss of some material in the less than C_7 range is evident in the lower three gas chromatograms.

affected. The biomarker compounds are especially susceptible to high thermal stress which can cause isomerization as well as destruction. These assessments must be made for each class of compounds being used because the effects of thermal alteration will vary depending on compound type and structure.

Once all liquids have been cracked to gas, thermal alteration may continue with the cracking of wet gas components. The thermodynamic stability of the wet gas components decreases with increasing carbon number. All the pentanes will eventually be lost, followed by the butanes, propane, and finally ethane. The end point is a reservoir containing methane only.

Biodegradation

Microbial alteration of crude oil in the reservoir is referred to as biodegradation. This alteration is carried out by microbes that are indigenous to the reservoir sediments at time of deposition and remain dormant until hydrocarbons are present (Head et al., 2003). Due to the lack of oxygen in the subsurface, these microbes must be anaerobic (Jones et al., 2008) and because of their need for water-soluble nutrients, the biological activity of biodegradation is focused along the oil–water contact in the reservoir (Head et al., 2003). For biodegradation to occur, the reservoir conditions must be conducive for the microbes. These conditions include a suitable reservoir temperature, formation water salinity, sufficient nutrients in the formation water, and an adequate hydrocarbon substrate to promote the microbial growth.

Biodegradation is temperature sensitive. At reservoir temperatures about 35–45°C, maximum microbial activity can be expected in the reservoir (Larter et al., 2006). As the temperature increases from 45 to 65°C, microbial activity begins to diminish. In the 65–80°C range, microbial activity is limited and biodegradation rates are greatly reduced. At temperatures greater than 80°C, the majority of the microbes are eliminated, or "pasteurized," ending biodegradation (Head et al., 2003).

Reservoirs can also be "paleo-pasteurization" prior to the introduction of hydrocarbons. If a reservoir is buried to depths where temperatures exceed 80°C, the microbes are eliminated. Even if the reservoir is then uplifted and the basin inverted such that the reservoir temperature now falls below 80°C, any hydrocarbons that find their way into the reservoir will not experience biodegradation due to the "paleo-pasteurization" of the sediments.

Although not as clearly understood as reservoir temperature, formation water salinity is also a control on biodegradation. As the salinity of the formation water in the water leg increases, the rate of biodegradation decreases (Mille et al., 1991; Later et al., 2006), with high water-leg salinities near saturation severely curtailing microbial alteration. More detailed studies of the impact of salinity are underway to better define the control it exerts on biodegradation.

There must be a sufficient supply of nutrients, such as nitrate, phosphate, iron, and potassium to support the microbial communities responsible for biodegradation. To deliver sufficient nutrients to the oil–water contact, active water column mixing must occur (Head et al., 2003). Similarly, an adequate supply of fresh hydrocarbons capable of being degraded must also be delivered to the oil–water contact, which requires active oil column mixing (Head et al., 2003). This can result in compositional gradients in the oil column showing progressively less biodegradation with increasing distance from the oil–water contact in actively biodegrading reservoirs (Later et al., 2006).

Trap shape and water-leg height can also influence the amount and rate of biodegradation experienced by a reservoired oil. Trap shape controls the oil–water contact area, the effective area where microbial activity can take place. The larger the area, the more potential for biodegradation. The water-leg height is an indicator of the volume of water in the reservoir that can supply nutrients for biodegradation. Larger water-leg heights can supply more nutrients and influence the rate of the microbial activity. The combination of these factors can control whether a reservoir experiences extensive alteration or merely low levels of biodegradation (Later et al., 2006).

What does biodegradation do to a reservoired oil? The microbial activity will change the composition of the oil as well as reduced its overall volume. With respect to bulk properties of the oil, the weight percent S, viscosity, asphaltenes content, and metal concentration will

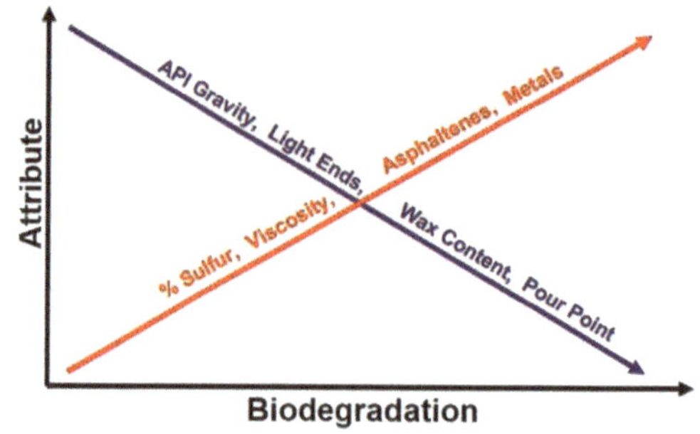

FIGURE 4.3 Response of crude oil bulk properties to biodegradation.

increase, while the API gravity, light ends content, wax content, and pour point will decrease, as shown in Fig. 4.3. The changes in these properties mainly reflect changes in the composition of the oil.

How biodegradation affects crude oil composition is well documented. The microbes consume the components of crude oil according to the ease which the compounds can be utilized. Small hydrocarbons are consumed first. This includes the C_1–C_5 gases (James and Burns, 1984). As the biodegradation progresses, the size of compounds consumed increases, first to the intermediate range C_6–C_{15} compounds and then the C_{15}+ compounds. The apparent order in which components of the crude oil are consumed is also influenced by the compound's structure. In the saturated hydrocarbons, the straight-chained n-paraffins are consumed preferentially followed by branched-chain compounds and then the side chains of ringed compounds. Aromatic compounds can also be altered, primarily through the alteration and removal of side chains off the main aromatic ring structures.

An overview of the geochemically significant changes during crude oil and natural gas biodegradation is given in Fig. 4.4. These changes include both the removal and formation of compound types. This summary diagram clearly shows which compounds are more easily altered and which compounds are more resistance. The biomarker compounds are of particular interest because of their use in making interpretations about crude oils. As shown in Fig. 4.4, these compounds are preserved retaining their geochemical information in crude oils up to the later stages of biodegradation.

Compositional changes in the crude oil are also how biodegradation is recognized and the extent of the alteration is gauged. This is accomplished by examining the gas chromatograms of whole crudes, as shown in the examples in Fig. 4.5. These chromatograms are from oils that have the same source and maturity, but are from different wells and exhibit different amounts of biodegradation. The progressive changes in the chromatograms show the loss of lower molecular weight compounds first, as well as the reduction in the n-paraffins as discussed above. Less resistant compounds are removed, while more resistant compounds are concentrated. The chromatograms also show the development of the "hump" of unresolved complex mixture, or UCM, below the peaks that is diagnostic of biodegraded crude oils. As the oil experiences heavy biodegradation, few recognizable compounds remain and the chromatogram is dominated by the UCM.

As the hydrocarbons are consumed and the volume of oil is reduced in the reservoir, where does the carbon consumed go? To answer this, it is necessary to look at the microbial

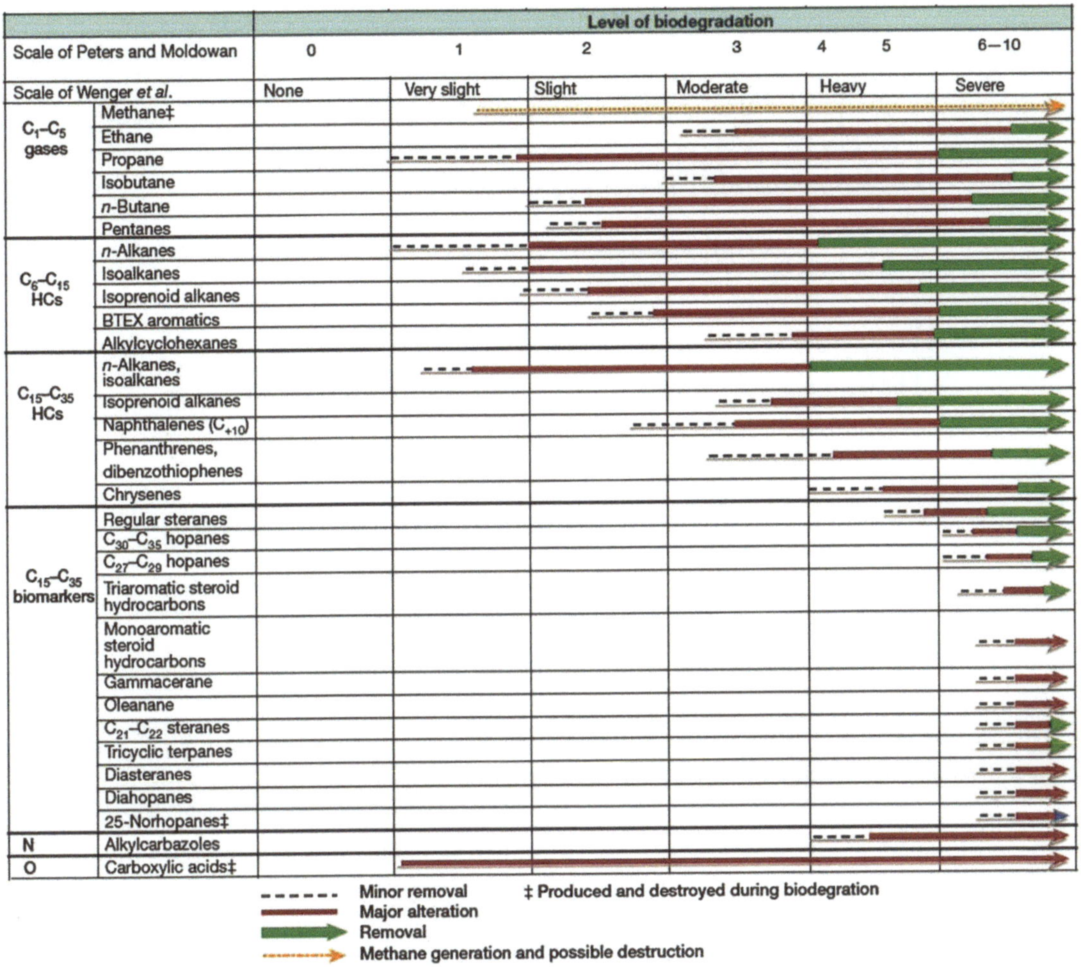

FIGURE 4.4 Scale of biodegradation intensity. *Modified from Head, I.M., Jones, D.M., Larter, S.R., 2003. Biological activity in the deep subsurface and the origin of heavy oil. Nature 426, 344–352; Based on Peters, K.E., Moldowan, J.M., 1993. The Biomarker Guide, first ed. Prentice Hall, Englewood Cliffs, New Jersey, 363 p.; Wenger, L.M., Davis, C.L., Isaksen, G.H., 2002. Multiple controls on petroleum biodegradation and impact on oil quality. Society of Petroleum Engineers Reservoir Evaluation and Engineering 5, 375–383.*

community responsible for the bulk of biodegradation of oil in the subsurface. Beginning in the late 1990s, studies such as Zengler et al. (1999) documented the conversion of model compounds found in crude oil, such as n-paraffins, to methane and carbon dioxide by methanogenic microorganisms in laboratory experiments. Field observation then lead Pallasser (2000) to suggest that biodegradation of crude oil resulted in the formation of secondary biogenic gas frequently found in reservoir with biodegraded oil. Laboratory experiments using naturally occurring microbial communities from geologic environments, such as Jones et al. (2008), demonstrated crude oil degradation with generation of biogenic gas. And finally investigations of the microbial communities at and below the oil–water contact in biodegraded crude

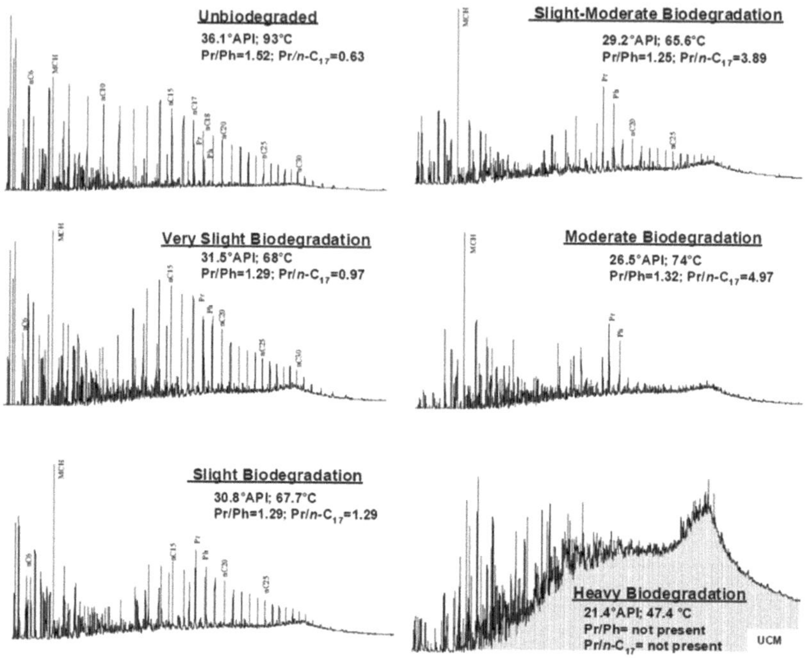

FIGURE 4.5 Whole oil gas chromatograms showing the progress of biodegradation. Peak labels include Pr for pristane; Ph for phytane; MCH for methyl cyclohexane; n-Cx for the n-alkane series where x is the carbon number. *From Wenger, L.M., Davis, C.L., Isaksen, G.H., 2002. Multiple controls on petroleum biodegradation and impact on oil quality. Society of Petroleum Engineers Reservoir Evaluation and Engineering 5, 375–383.*

oil reservoirs, such as Grabowski et al. (2005) and Bennett et al. (2013), confirmed the presence of methanogenic organisms at the site of the subsurface biodegradation. All these lines of evidence taken together show that biodegradation of crude oil results in the generation of biogenic methane and carbon dioxide. The carbon dioxide can be free gas or dissolved in formation water as HCO_3^- and may contribute to the formation of isotopically light calcite cements in the reservoir (Pallasser, 2000). The biogenic methane gas may be dissolved into the remaining crude oil or, if the oil is saturated with respect to gas, form a separate phase as a gas cap.

The amount of biogenic methane produced is significant. Zengler et al. (1999) show that biodegradation of 1 mol of n-hexadecane would produce more than 12 mol of methane and nearly 4 mol of carbon dioxide. Putting it in more relevant units of measure for petroleum exploration, Dessort et al. (2003) estimated that one barrel of oil biodegrades to ~15,100 cu. ft. of biogenic methane gas. As a result, this secondary biogenic gas may form substantial accumulations in the subsurface. These accumulations may be associated with the biodegraded oil or the gas may have migrated away from the oil-bearing reservoir to form a separate accumulation.

The exploration significance of encountering substantial amounts of secondary biogenic gas is counter intuitive. Instead of thinking that biogenic gas is simply an accumulation of primary biogenic gas formed early in the sediments history, this secondary biogenic gas in an indicator of a functioning oil-prone petroleum system. More details about the exploration significance of secondary biogenic gas from biodegradation of oil will be given in the section on interpreting the source of natural gas later in this chapter.

Water-Washing

Water-washing is the removal of mostly low molecular weight compounds from a reservoired oil by flowing formation waters (Bailey et al., 1973). The susceptibility of the components of crude oils to water-washing is related to their solubility in water under reservoir conditions. The amount of hydrocarbon removed from the oil also depends on the amount of water flowing past the oil–water contact. For each type of hydrocarbon, solubility decreases with increasing carbon number, for a given carbon number, aromatic compounds are typically more soluble than saturated compounds, and within the saturated compounds ringed compounds are usually more soluble than the branched-chain compounds which are more soluble than the straight-chained compounds (Lafargue and Barker, 1988). The net effect is usually a decrease or removal of gasoline range hydrocarbons ($<C_{10}$) and in the C_{15}+ fraction a decrease in some aromatic compounds and naphthenes (Lafargue and Barker, 1988). Little or no changes have been observed in the saturated biomarker of water-washed oils (Palmer, 1984; Lafargue and Barker, 1988). These compositional changes can result in changes to some of the bulk properties, such as a decrease in API gravity and increases in the weight percent sulfur and asphaltenes content (Palmer, 1984). Because water-washing often occurs in conjunction with biodegradation, the effects of biodegradation will generally result in more alteration of the crude oil than the water-washing.

Gas-Washing

Gas-washing, also referred to as evaporative fractionation (Thompson, 2010) and gas stripping, occurs when gas migrating through an oil column dissolves some of the oil and carry it along in a gas solution as a gas condensate (Losh et al., 2009). When this gas condensate reaches the final accumulation, the dissolved oil may remain in solution until it is condensed out during production. Gas-washing can occur in either stratigraphic or structural trapping situations where seals are permeable to gas but not oil. The example shown in Fig. 4.6 is a fault trap. The composition of the gas-washed oil will change by showing a depletion of the lighter end material (Losh et al., 2009), as compared to the original oil. Often, the gas and "condensate" will have different maturities, the gas usually being more mature than the "condensate" indicating different origins (Losh et al., 2002). This difference in maturity in the migrated gas condensate is very diagnostic of gas-washing.

The loss of lighter hydrocarbons as a result of gas-washing can lower the API gravity and increase the viscosity of the original oil, as well as substantially diminish its value (Losh et al., 2009). However, because gas-washing has its greatest impact on the less than C_{15} fraction, little or no changes should be observed in the biomarkers used in making interpretations of crude oils.

Deasphalting

The asphaltenes are held in colloidal solution in crude oils. These solutions are not always inherently stable and can be disturbed by several processes. Deasphalting occurs when the colloid solution are destabilized initiating the precipitation of asphaltenes from the oil. Asphaltene precipitation can be triggered by a number of mechanisms including loss of light ends, water-washing, gas-washing, leaky seals, biodegradation, pressure drops, mixing of oils, and

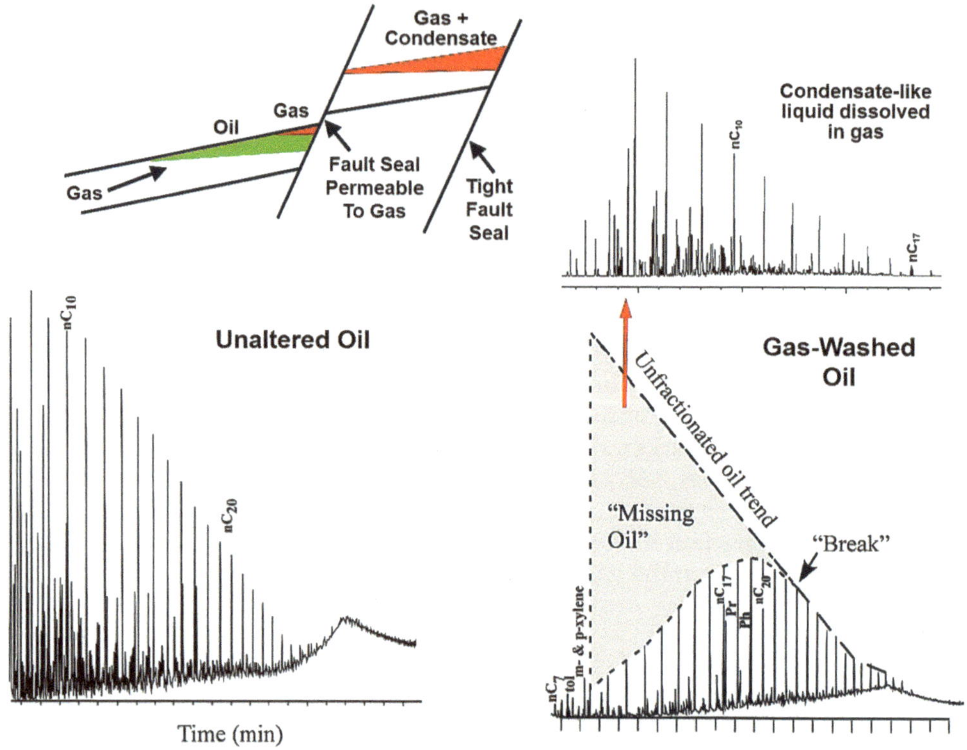

FIGURE 4.6 Examples of potential gas-washing scenarios, one in a stratigraphic trap with a seal permeable to gas but not oil and the other in a fault trap with a gas permeable seal. *Modified after Losh, S., Swart, D., Dickinson, A., 2009. Gas washing pattern and economics in an area of continental shelf, offshore Louisiana. Gulf Coast Association of Geological Societies Transactions 59, 477–484.*

gas mixing with an oil (Akbarzadeh et al., 2007). The precipitated asphaltenes may be present as a mat at the oil–water contact or in the form of solid bitumen within the reservoir porosity. While the precipitation of asphaltenes in itself does little to hinder making geochemical interpretation about crude oils, some of the triggering mechanisms, such as biodegradation, may pose problems. Deasphalting is a greater problem as it relates to hydrocarbon production and will be discussed in more detail in the Chapter 5.

Devolatilization

When reservoirs are exhumed or when a seal is breached near the surface, low molecular weight compounds in the crude oil can evaporate into the atmosphere because of their high vapor pressure. This process is usually referred to as devolatilization, or inspissation, and is associated most often with tar sand deposits. This selective loss of components up to about C_{15} (Barbat, 1967) causes high viscosities that interfere with conventional recover. Devolatilization is nearly always concurrent with biodegradation and oxidation of the oil. Frequently when the reservoir is exposed at the surface, the devolatilization helps in the formation of a tar seal which can reduce further loss of hydrocarbons further below the surface.

Thermochemical Sulfate Reduction

TSR was originally described as a process by which hydrocarbon gases at high temperature would react with anhydrite present in the reservoir. The reactions would consume the hydrocarbon gases producing hydrogen sulfide, H_2S, and carbon dioxide (Orr, 1974). The carbon dioxide will often react with the calcium from the anhydrite to form calcium carbonate usually in the form of calcite cements. The process is autocatalytic as H_2S is both a product and a catalyst (Hunt, 1996).

In gas reservoirs, these reactions are considered to initiate at temperature of about 120°C (248°F) (Heydari, 1997) with significant hydrocarbon destruction occurring at temperatures greater than 140°C (284°F) (Worden et al., 1995). However, Orr (1977) was able to carry out these reactions experimentally at temperatures of 77°C to 121°C (170°F to 250°F) suggesting that liquid hydrocarbons were also likely to be susceptible to TSR. This is supported by experimental evidence showing oil reacting with sulfate to produce organic sulfur compounds such as mercaptans, thiophenes, and benzothiophenes in addition to H_2S (Yue et al., 2014) and field observations, e.g., the Alabama portion of the Smackover trend (Sassen and Wade, 1994; Wade and Sassen, 1994).

Reservoirs with H_2S contents higher than 10% are often attributed to TSR. TSR can be very destructive to a hydrocarbon resource. Sassen and Moore (1988) reported the Black Creek field in the Smackover trend had a gas composition of 2% methane, 20% carbon dioxide, and 78% H_2S as a result of TSR.

Contamination

Up to this point, only natural processes that could alter crude oils and natural gases have been considered. The one artificial process that needs to be included is contamination. Contamination is the introduction of foreign organic matter into a crude oil or natural gas sample. In the case of crude oils, this usually occurs as a consequence of exposure to or mixing with drilling fluid in the reservoir or borehole or during sampling. Drilling muds can be formulated with diesel, crude oils, and synthetic oils. They can also contain numerous organic compounds used to control or alter mud properties such as defoaming agents, viscosity controls additives, lubricants, and tracers. Crude oils can also be contaminated by storage in plastic bottles or in improperly cleaned glass or metal containers.

As described in the last chapter, it is therefore highly recommended to periodically collect drilling mud samples and when possible samples of drilling mud additives for reference. Most crude oil contamination can be recognized by distinctive features displayed in whole oil gas chromatograms. These features can then be matched to gas chromatograms from the drilling mud or mud additives. Depending on the composition of the contamination, the contaminant may or may not interfere with making interpretations from the crude oil data. The base oils for most synthetic oil-based muds do not contain biomarker compounds that will compromise interpretations. However, if the drilling mud has been used for more than one well, the synthetic base oil may have extracted biomarker compounds from the cuttings circulation in the mud during drilling. Careful analysis of the drilling mud reference samples is needed to ascertain if any problems may exist.

Gas samples may also be compromised. If the sample container is not completely purged prior to capturing a sample, air may be introduced into the sample. Air contamination can often be recognized by the presence of oxygen and the ratio of oxygen-to-nitrogen being similar to air. Leaking sample containers may also be a problem. Even minor leaks can alter both the composition and isotopic signature of a gas sample. To aid in recognizing leakage, the pressure at which the gas samples were collected should be recorded and compared to the pressure in the sample container at the time of analysis.

OIL-TO-OIL AND OIL-TO-SOURCE ROCK CORRELATIONS

Oil-to-oil and oil-to-source rock correlations are used to determine if there are any geochemical relationships between oils and between oils and source rocks based on source-controlled geochemical parameters. Oil-to-oil correlations identify how many source rocks are active in a basin by determining how many oil families are present. The ability to correlate one oil to another also provides a means of detecting the migration of oil within or between reservoirs. Oil-to-source rock correlations define petroleum systems by establishing the genetic relationship between the generating source rock and resulting reservoired oil. When used in conjunction with the structural analysis of a basin, oil-to-source rock correlations can help to determine the timing and pathway of the migration.

Oil correlation parameters are the geochemical characteristics used to demonstrate the relationships between oils and between oils and their source rocks. The parameters should have characteristic distributions or profiles that allow individual source units or crude oils to be differentiated. At the same time, they should not be greatly affected by any process acting on the crude oil or source rock after their physical separation. And, they should be based on ubiquitous components of oils and source rock extracts. Unfortunately, there are no chemical characteristics that meet all these requirements in all situations. Instead, it is necessary to use geochemical characteristics that best fit these criteria for the sample set under study and to use multiple parameters for making interpretations to avoid arriving at ambiguous or erroneous conclusions.

When looking for geochemical compounds that can be used for oil correlation work, it is also essential to consider the potential geochemical information a compound may carry as well as how difficult it is to obtain that information. As the size and structural complexity of organic molecules increase, the potential information they carry increases. However, with increasing size and structural complexity, our ability to easily analyze the compounds decreases, leaving much of this information unattainable. It is, therefore, necessary to find a compromise between the amount of potential geochemical information and our analytical abilities.

Based on this consideration, methods for correlate oils to oils and oils to their source rocks have gone through an evolution linked to developments in analytical chemistry. Early attempts at oil correlation were confined to simple comparisons of bulk properties of crude oils. As analytical chemistry began to develop methods for characterizing hydrocarbon in petroleum, more detailed comparisons were made. However, these were initially confined to characterizing the gasoline fraction by distillation methods (Barr et al., 1943). This approach was eventually expanded to the use of correlation indexes which measure the relative amounts of paraffins, naphthenes, and aromatics in each distillation fraction of the whole crude oil and by the sulfur and nitrogen contents of the oils (Barat, 1967). During the early 1970s, the use of gas chromatography in petroleum geochemistry was a major step forward to improve oil correlation methods. Crude oil correlations were then based on the distributions of C_4–C_7 hydrocarbon and C_{15}+ n-paraffin from gas chromatograms along with determination of the carbon-isotopic compositions of the topped crude and the saturate, aromatic, N–S–O, and asphaltic fractions (Erdman and Morris, 1974; Williams, 1974). Shortly thereafter, improvements in mass spectrometer technology allowed the use of biomarker distributions in oil-to-oil and oil-to-source rock correlations (Welte et al., 1975; Leythaeuser et al., 1977; Seifert, 1977). Currently, biomarker distributions and carbon isotope ratios of the whole oil or extract and their liquid chromatography fractions are the basis for most oil correlation studies.

There are, however, occasions where biodegradation has severely altered or completely destroyed all biomarkers in the oils under study. This may occur when trying to correlate heavy oils, tar sands, or solid bitumens to each other or to more typical crude oils. In these instances, trace elements (Hitchon and Filby, 1984) and heavy metals (Curiale, 1987) have been found to be useful oil correlation parameters.

Biomarkers

In the context of petroleum geochemistry, biomarkers are compounds found in sediments and crude oils that have the same basic structure as compounds found in living organisms and could only have been derived from the diagenetic alteration of biological molecules. Many early works also called these compounds chemical or molecular fossils. Biomarker was originally an abbreviation for biological marker hydrocarbon. However, as classes of compounds other than hydrocarbons, such as porphyrins, were eventually included, a broadening of the definition was adopted. A classic example of a biomarker and its relationship to the biological precursor is cholestane and cholesterol, shown in Fig. 4.7. By simply removing the OH group and saturating a double bond, the biological compound cholesterol is converted to the geological compound cholestane, which is commonly found in source rock extracts and crude oils. One of the earliest applications of biomarkers was by Treibs (1934) when he recognized chlorophyll-like structures in crude oil, thereby establishing the basis for the science of petroleum geochemistry.

Not all the geochemical compounds considered to be biomarkers are actually used in oil correlation work. Over the years, many compounds have been suggested as correlation tools, but have not gained widespread acceptance usually due to limited occurrence in sediments and crude oils or difficult analytical procedures to obtain the data. The primary oil correlation tools used today are the distributions of isoprenoid hydrocarbons, tricyclic terpanes, tetracyclic terpane, pentacyclic terpanes, steranes, and aromatized steranes. Examples of the basic structure of these compounds are shown in Fig. 4.8.

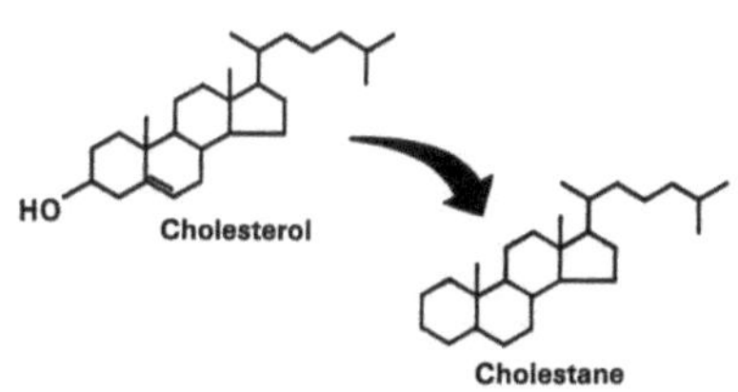

FIGURE 4.7 Conversion of the biochemical cholesterol into the geochemical biomarker cholestane.

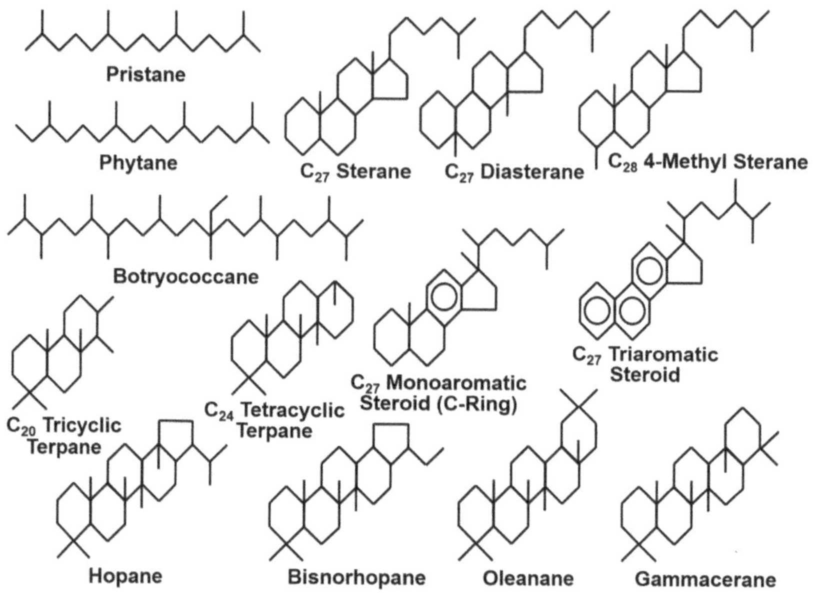

FIGURE 4.8 Representatives of the major groups of hydrocarbon biomarker compounds. Details of the stereo stereochemistry are not included, see Waples and Machihara (1991) for illustrations of the stereochemistry.

Biomarker Analysis

To obtain biomarker data, analysis of the hydrocarbons is done on a gas chromatograph in tandem with a mass spectrometer, GC–MS. For the terpanes and steranes, the saturated hydrocarbon fractions from liquid chromatography of a crude oil or source rock extract are used. The saturated hydrocarbon fraction is first separated in the gas chromatograph. The column effluent is then sent to the sample inlet of a mass spectrometer. A mass spectrometer, shown schematically in Fig. 4.9, can be thought of as a very sensitive detector for the gas chromatograph that can be set to measure specific compound groups. As the hydrocarbons enter the ion source, they are bombarded with electrons that cause the hydrocarbons to ionize and, if enough energy is applied, to break apart into fragments. The biomarker compounds will break at specific places in their structure and form fragments with a characteristic mass, as shown for the hopanes and steranes in Fig. 4.9. These mass fragments are described in units of m/z, which is mass divided by charge. The hopane fragment has a mass of 191 atomic mass units (amu), while the sterane fragment has a mass of 217 amu. Both the hopane and sterane fragments have a single charge so they are described as being m/z 191 and m/z 217, respectively.

The ions are then formed into a beam using electrical fields and sent through the mass filter. The mass filter is used to focus molecular fragments of a single mass onto the ion detector. The mass filter can be changed very rapidly (milliseconds) so that multiple ions can be measured in rapid succession. In this way, several mass fragments can be detected and measured nearly simultaneously. The recording of each mass fragment is displayed in a mass chromatogram. Using the mass fragment of m/z 191, the distribution of the terpanes is recorded, as shown in Fig. 4.10. The terpanes consist of tricyclic and tetracyclic compounds as well as the pentacyclic terpanes, which includes the predominant hopanes and compounds like oleanane

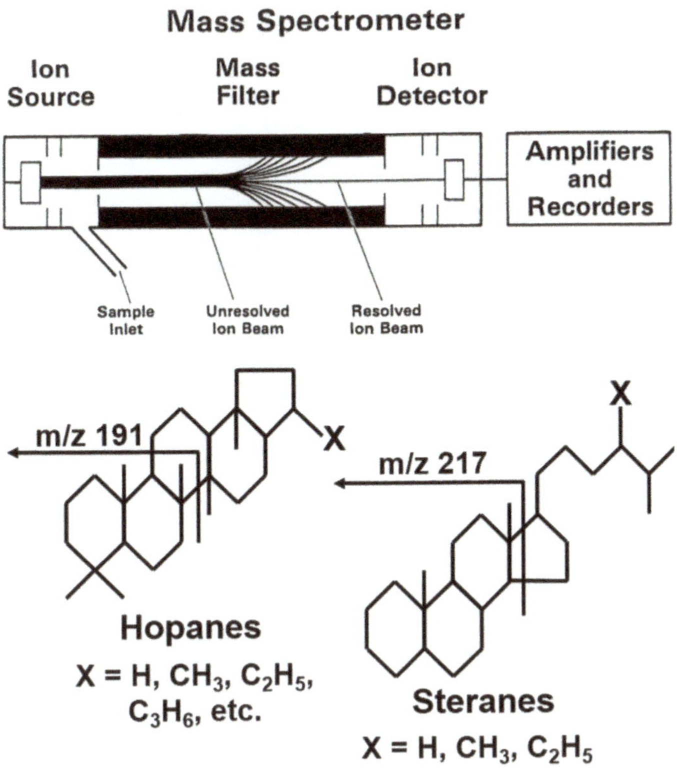

FIGURE 4.9 A schematic of a mass spectrometer and the fragmentation patterns for hopanes and steranes.

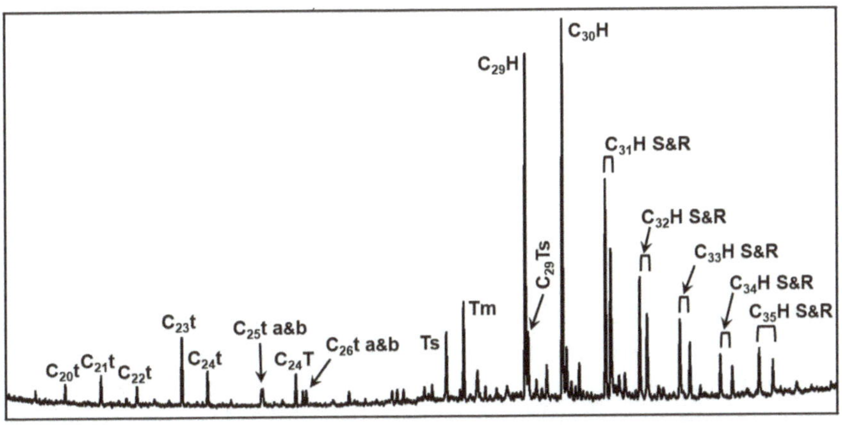

FIGURE 4.10 Typical m/z 191 terpanes data for a crude oil. Tri- and tetracyclic terpane peak labels: $C_{20}t$–C_{20} tricyclic terpane; $C_{21}t$–C_{21} tricyclic terpane; $C_{22}t$–C_{22} tricyclic terpane; $C_{23}t$–C_{23} tricyclic terpane; $C_{23}t$–C_{23} tricyclic terpane; $C_{25}t$–C_{25} tricyclic terpane a and b; $C_{24}T$–C_{24} tetracyclic terpane, des-E-hopane; $C_{26}t$–C_{26} tricyclic terpane. Hopane labels: Ts–Ts, 18α(H), 22, 29, 30-trisnorhopane; Tm–Tm, 17α(H), 22, 29, 30-trisnorhopane; $C_{29}H$–C_{29} Tm 17α(H), 21β(H)-norhopane; $C_{29}Ts$–C_{29} Ts 18α(H), 30-norneohopane; $C_{30}H$–C_{30} 17α(H) 21β(H)-hopane; C_{31}HS&R–C_{31} 22S and 22r 17α(H), 21β(H)-homohopanes; C_{32}HS&R–C_{32} 22S and 22r 17α(H), 21β(H)-homohopanes; C_{33}HS&R–C_{33} 22S and 22r 17α(H), 21β(H)-homohopanes; C_{34}HS&R–C_{34} 22S and 22r 17α(H), 21β(H)-homohopanes; C_{35}HS&R–C_{35} 22S and 22r 17α(H), 21β(H)-homohopanes.

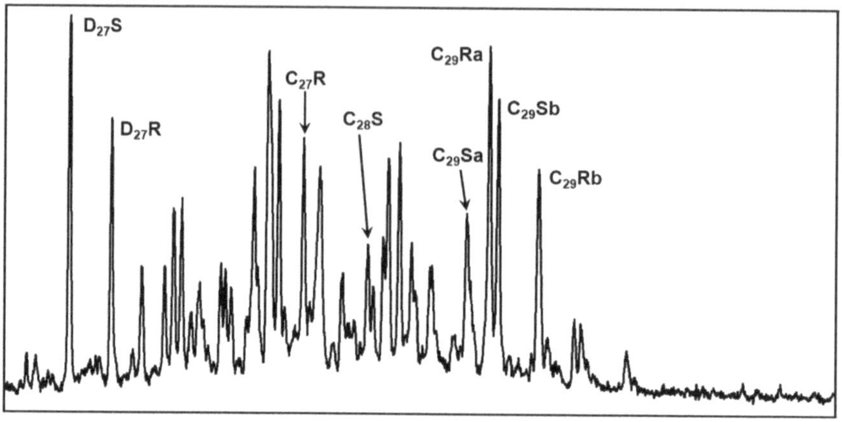

FIGURE 4.11 Typical m/z 217 steranes data for a crude oil. Due to co-elution of the C_{28}–C_{29} diasteranes and the C_{27}–C_{28} regular steranes, positive peak assignments are more difficult. Sterane peak labels: $D_{27}S$–C_{27} βα 20S diasterane; $D_{27}R$–C_{27} βα 20R diasterane; $C_{27}R$–C_{27} αα 20R sterane; $C_{28}S$–C_{28} αα 20S sterane; $C_{29}Sa$–C_{29} ααα 20S sterane; $C_{29}Ra$–C_{29} αββ 20R sterane; $C_{29}Sb$–C_{29} αββ 20S sterane; $C_{29}Rb$–C_{29} ααα 20R sterane.

and gammacerane. The mass chromatogram for m/z 217 captures the sterane distribution, as shown in Fig. 4.11, with the main compounds of interest being the C_{27}–C_{30} steranes. Because of the number of isomers of each carbon number and the overlap in the retention time ranges for the C_{28}–C_{29} diasteranes and the C_{27}–C_{28} steranes, it is more difficult to identify individual compounds in the m/z 217 mass chromatogram. Attention is typically focused on those steranes that can be clearly identified.

The monoaromatic and triaromatic steroids are analyzed in a similar fashion to the terpanes and steranes, only using the aromatic fraction from liquid chromatography. Diagnostic ions monitored for the monoaromatic and triaromatic steroids are m/z 253 and m/z 231, respectively.

The isoprenoid hydrocarbon data are usually obtained from either whole oil/extract gas chromatograms or from saturate fraction gas chromatograms. They can also be analyzed by GC–MS if the chromatograms do not clearly show the isoprenoid compounds of interest.

Comparing Data for Oil-to-Oil and Oil-to-Source Rock Correlations

Oil correlation studies are based on comparisons of geochemical data to determine the degree of similarity and/or dissimilarity between oils and between oils and their source rocks. The process of making oil-to-oil and oil-to-source rock correlations utilizes the same multiple parameter approach followed in source rock evaluation. That entails that any observation of similarities with one data type should be corroborated with at least one additional data type. Moreover, the extent of similarity or dissimilarity observed within a specific data type needs to be evaluated with respect to any of the potential alteration processes discussed above. Thermal alteration and biodegradation are the two alteration processes most likely to have an impact on making oil correlations and the data should be examined closely looking for their effects. Different data types may also require different methods and criteria for comparison. Specific interpretation methods have been developed for specific data, such as carbon isotopes. And finally, the number of samples to be compared has a direct bearing on how the comparisons are made. Let us begin by reviewing some of the methods used with biomarker data.

There are several approaches for comparing biomarker data while attempting to make oil-to-oil and oil-to-source rock correlations. The first and foremost method is using the mass chromatograms as "fingerprints" of the distribution of different compound groups and comparing them visually. The overall similarity of the distributions of compounds in the mass chromatogram, the presence/absence of individual biomarker compounds, and the relative abundance of individual or groups of biomarker compounds are the main criteria for comparison. An example of this type of comparison is shown in Fig. 4.12. Of the three oils shown, the sterane distributions for oils A and B show a high degree of similarity to each other, while there are substantial differences between them and oil C.

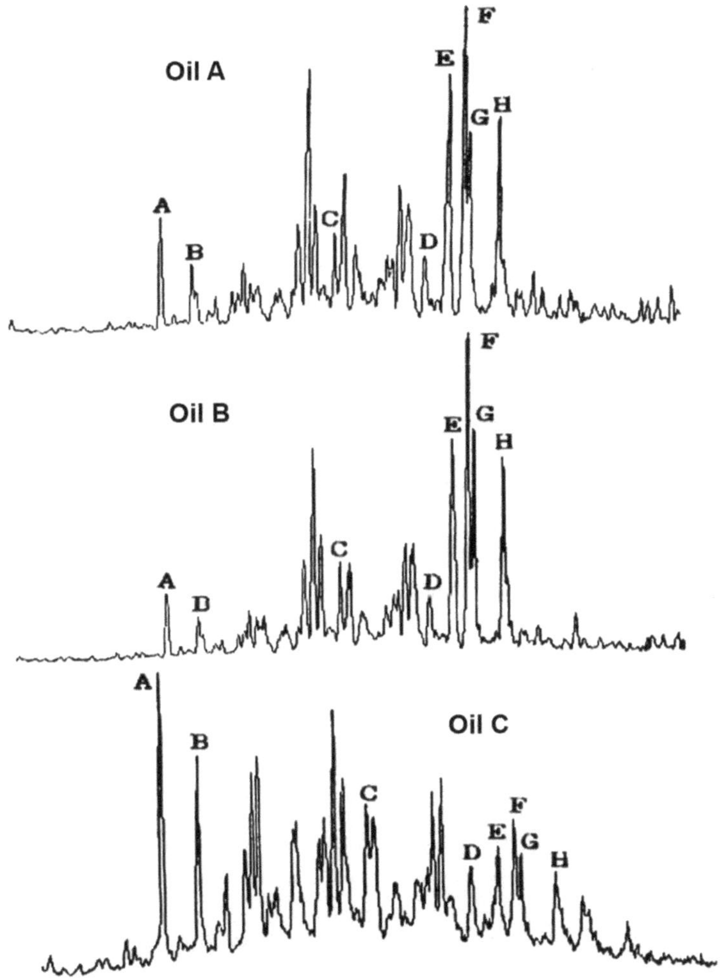

FIGURE 4.12 Comparison of the m/z 217 mass fragmentograms of three oils showing similarities between A and B, while C is dissimilar.

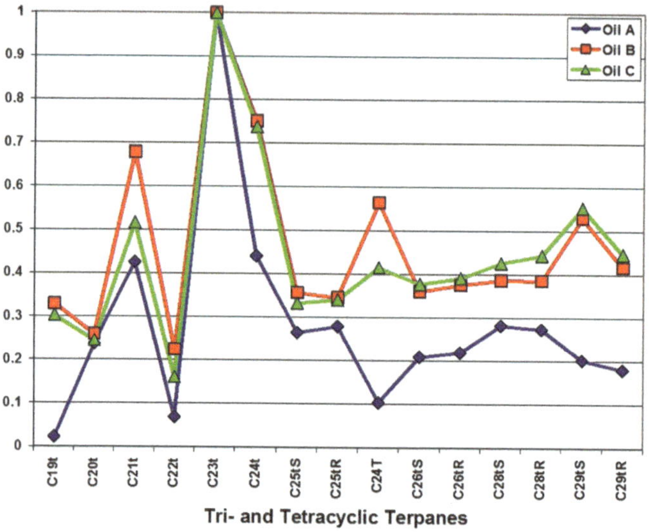

FIGURE 4.13 Normalized distributions of the tricyclic and tetracyclic terpanes of three oils showing similarities between B and C, while A is dissimilar.

Another means of comparison uses either peak heights or peak areas measured from the mass chromatograms. These measurements are made by the software used to capture the data from the mass spectrometer and should be provided by the laboratory that has done the biomarker analysis. By taking either the peak heights or peak areas for one type of biomarker compound and normalizing the data with respect to the largest peak, the data can be plotted on the same set of axes where a more direct comparison of the distributions can be made. The example shown in Fig. 4.13 shows the normalized distributions of the tri- and tetracyclic terpanes of three oils. Oils B and C can be seem to be very similar in most of the data; however, there are a few points of deviation such as the C_{24} tetracyclic terpane-labeled C24T. In contrast, oil A is very different at a number of compounds in the distribution.

Another way of utilizing the sterane data is shown in Fig. 4.14. The C_{27}, C_{28}, and C_{29} steranes are individually summed up and percentages calculated for each carbon number. The percentages are then plotted on a ternary diagram (Moldwan et al., 1985). This comparison is based more on a bulk property of the steranes and does not utilize the more obvious as well as subtle difference that can be observed in looking at mass chromatograms. Oils and source rock samples with similar sterane distributions often cluster together, but clustering does not necessarily indicate genetic relationships, rather it indicates that similar organic matter generated the oils. Corroboration by other parameters needs to establish a genetic link. This ternary plot does have other interpretations associated with it that will be discussed in the following section on crude oil inversion.

The ratios of two or more biomarkers can also be used for comparison in oil correlation studies. Ratios are calculated using either peak heights or peak areas measured from the mass chromatograms, as described above. These ratios often have implications about the nature of the oil and/or the source rock that generated and these interpretations will be discussed in more detail in the description of crude oil inversion to follow. Comparisons are made using two biomarker ratios plot against each other (cross-plots), as shown in Fig. 4.15, with related

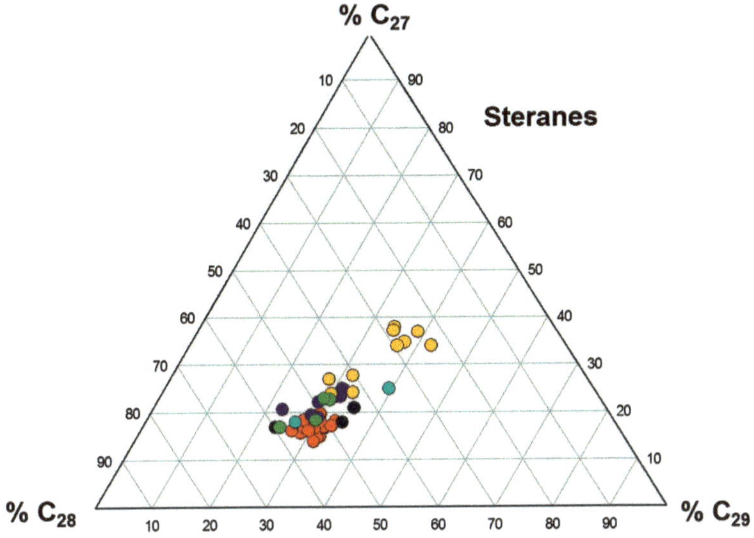

FIGURE 4.14 A ternary plot of the percentage of C_{27}, C_{28}, and C_{29} steranes of a large group of crude oils. Clustering of data points indicates similarity.

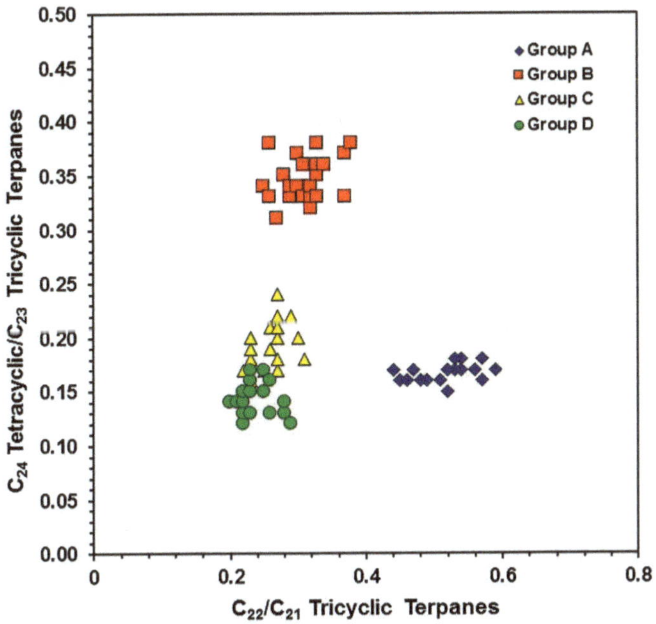

FIGURE 4.15 A cross-plot of two biomarker ratios, the C_{22}/C_{21} tricyclic terpanes and C_{24} tetracyclic/C_{23} tricyclic terpanes. Clustering of data points indicates similarity.

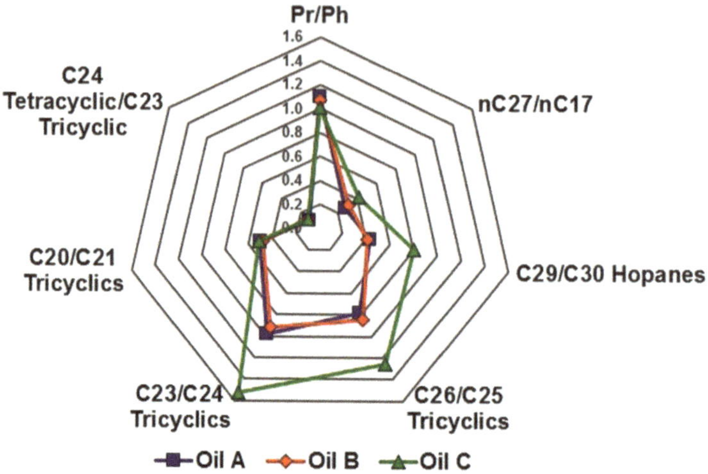

FIGURE 4.16 A radar plot of seven biomarker ratio for oil-to-oil correlation. Oils A and B show similarity, while oil C shows dissimilarity.

oils or source rock extracts clustering together. A partial list of biomarker ratios that might be used in correlation includes pristane/phytane, C_{22}/C_{21} tricyclics terpanes, C_{26}/C_{25} tricyclics terpanes, C_{24} tetracylic/C_{23} tricyclic terpanes, C_{27} and C_{29} Ts/Tm hopanes, C_{29}/C_{30} hopanes, and $C_{35}S/C_{34}S$ hopanes.

Perhaps a more effective way of utilizing biomarker ratios for oil-to-oil or oil-to-source rock correlations is to use radar plots, also called star diagrams. Instead of utilizing just two ratios, a radar plot can use numerous biomarker ratios to compare oils and/or source rocks for similarities and differences as shown in Fig. 4.16. These simultaneous comparisons of multiple biomarker ratios can streamline correlations by providing easily recognizable information about variations among the samples for each ratio. This type of information is also useful in preparing for crude oil inversion studies, as discussed below.

In addition to biomarkers, the carbon isotope ratios of the whole oil or source rock extract in addition to their liquid chromatography fractions can be used in oil correlation studies. One approach is to use isotope curves by plotting the individual carbon isotope ratios for the saturate, aromatic, NSO, and asphaltenes fractions on a single axis (Stahl, 1978), as shown in Fig. 4.17. Occasionally, the carbon isotope ratio of the whole oil or whole source rock extract is also included in the isotope curve. A more common method of comparison is cross-plotting the carbon isotope ratio of the saturate fraction versus aromatic fraction (Sofer, 1984), shown in Fig. 4.18. Agreement of less than ~1‰ indicates a positive correlation, while similarity is indicated by less than ~2‰ variations among the data (Sofer, 1984; Peters et al., 2005a). Keeps in mind that maturity difference among the samples may show isotopic variation of up to ~2–3‰ (Peters et al., 2005a). Differences greater than ~2–3‰ usually indicates a lack of correlation (Peters et al., 2005a). Oils and source rock samples with carbon isotope ratios that track closely on the isotope curves or that cluster together do not always indicate genetic relationships. Similarity in isotopic signature may only be indicating that similar organic matter generated the oils or rock extracts. As with the other parameters, corroboration by other parameters is needed to establish a genetic link.

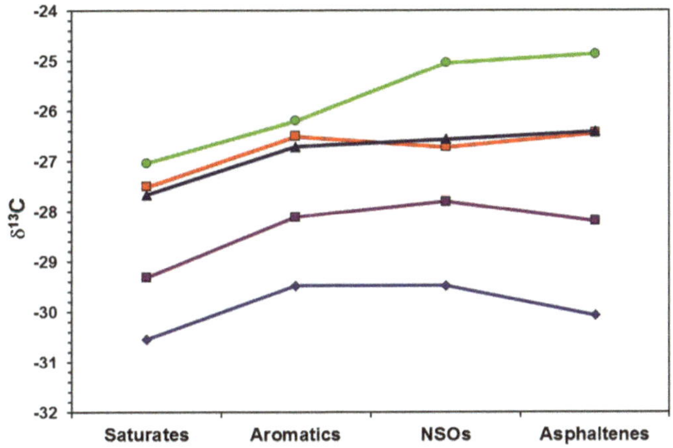

FIGURE 4.17 A plot of the carbon isotope ratios, in $\delta^{13}C$ notation, of the saturate, aromatic, resin, and asphaltenes chromatographic fractions (S-A-R-A) of four crude oils as use in oil correlation studies, following the method of Stahl (1978).

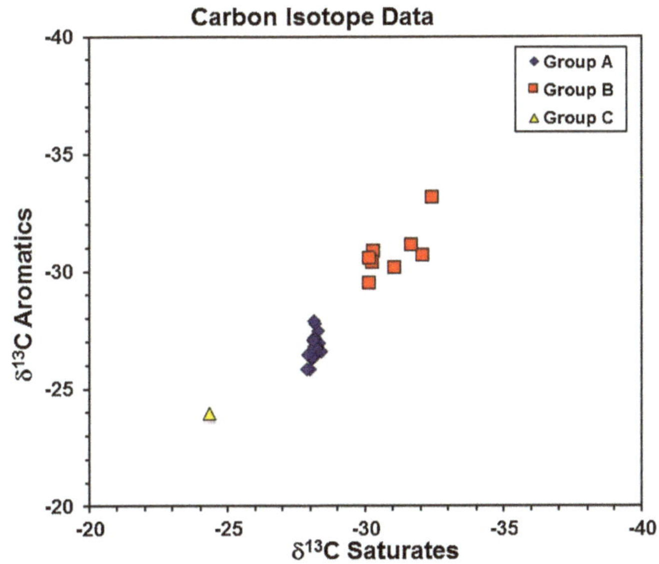

FIGURE 4.18 A cross-plot of the carbon isotope ratios, in $\delta^{13}C$ notation, of the saturate and aromatic chromatographic fractions of three groups of crude oils as use in oil correlation studies.

When oil correlation studies involve a large numbers of samples, the sheer volume of data makes the simple visual comparisons methods described above difficult to employ. In these instances, multivariate statistical techniques, such as hierarchical cluster analysis and principal components analysis, have been found to be effective (e.g., Sofer et al., 1986; Telnaes and Dahl, 1986; Pasadakisa et al., 2004; Peters et al., 2008a,b). These statistical techniques can handle biomarker distribution, biomarker ratios, and isotopic data together or individually.

If biomarker distributions are used, it is necessary to normalize the data so all peak values fall between 0.0 and 1.0. It is highly recommended that at least some visual comparisons be used to corroborate the statistical results to verify their validity.

CRUDE OIL INVERSION

The key to defining a petroleum system is making an oil-to-source rock correlation. These correlations are also important elements in delineating migration pathways that help exploration geologists find additional accumulations. As important as oil-to-source correlations are to petroleum exploration, they are not always successful. So, what do you do if an oil-to-source rock correlation fails?

First, consider if all possible source rock units in the area have been sampled. Is it possible that there is a source rock deeper than the sediment column drilled? Or perhaps the oil has migrated from a source area not covered by the samples. If so, there may have been an organic facies change in the source interval not represented in the samples. It is also possible that the oil is a product of the mixing of hydrocarbons from two or more sources such that any indications of a contribution from a single source are likely obscured.

While these are important and useful steps to take in trying to define the source rock for a particular oil or family of oils, the most useful strategy is to use the geochemical characteristics of the crude oil to provide clues about the nature of the source rocks that generated it. This is called crude oil inversion. Crude oil inversion uses certain aspects of a crude oil's geochemistry to provide indications of some of the characteristics of its source rock such as kerogen type, depositional setting, lithology, maturity, and, in some case, the geologic age (Bissada et al., 1992). Because the information provided by crude oil inversion may eventually point toward a likely source rock, it is considered a proxy for oil-to-source rock correlations. And these interpretations are not confined only to reservoired crude oils. They are also applicable to oil stains and seeps, making them a potentially valuable source of information even before an oil accumulation is found.

There are several method employed in crude oil inversion to gain information about the source rock that generated the oil. Some rely on the presence of individual biomarker compounds, while others rely on the relative abundance of groups of compounds. However, the most frequently employed utilize biomarker ratios to discern characteristics of crude oils. The following discussion will provide examples of these three interpretation methods used in crude oil inversion.

Presence of Individual Biomarker Indicators

The presence of an uncommon biomarker or an unusual amount of an individual biomarker compound in an oil can be an indicator of the nature of the source rock that they were generated from. To illustrate this, let us review the interpretation of the occurrence of the pentacyclic triterpanes gammacerane, bisnorhopane, and oleanane and the acyclic isoprenoid botryococcane.

Gammacerane is commonly found in low concentrations in source rock extracts and crude oils. However, moderate to high concentrations of gammacerane, as shown in Fig. 4.19, are

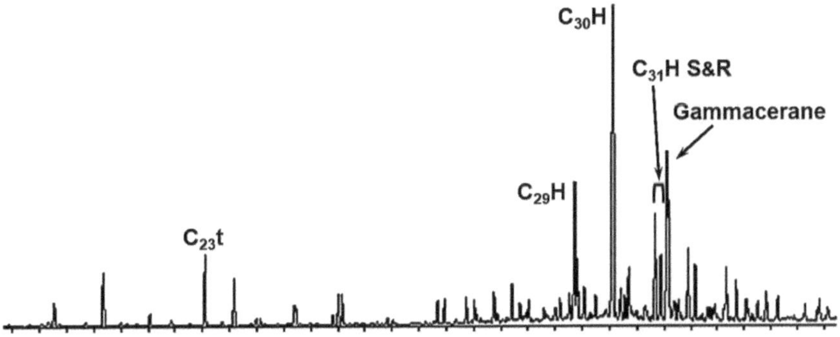

FIGURE 4.19 An examples of using an individual biomarker indicator, gammacerane, in a crude oil inversion study.

unusual and significant. Abundant gammacerane was initially thought to be an indicator of source rock deposition in lacustrine settings due to the high concentration of gammacerane found in the lacustrine Eocene Green River shale (Anders and Robinson, 1971). However, further investigation revealed that gammacerane was actually an indicator of source rocks deposited in restricted depositional systems with stratified water columns and highly saline to hypersaline waters (ten Haven et al., 1988; de Leeuw and Sinninghe Damste, 1990). High gammacerane content associated with these depositional conditions is often found in high salinity lacustrine deposits (Brassell et al., 1988) and in some marine carbonates and evaporates from lagoonal settings (Mello et al., 1988).

Although bisnorhopane is relatively rare in sediment bitumens and crude oils, it is often a very prominent component in sediments from euxinic environments and their derived oils, as shown in Fig. 4.20. Examples of such depositional settings are the Miocene Monterey Formation of the California coastal basins (Williams, 1984) and parts of the Jurassic Kimmeridge Clay in and around the North Sea (Cornford et al., 1983; Schou et al., 1985). It is most likely derived from microbial input to sediments when deep anoxia has developed in a sulfur-rich depositional environment (Grantham et al., 1980; Katz and Elrod, 1983; Schoell et al., 1992). Because of its association with deep anoxia and sulfur-rich depositional conditions,

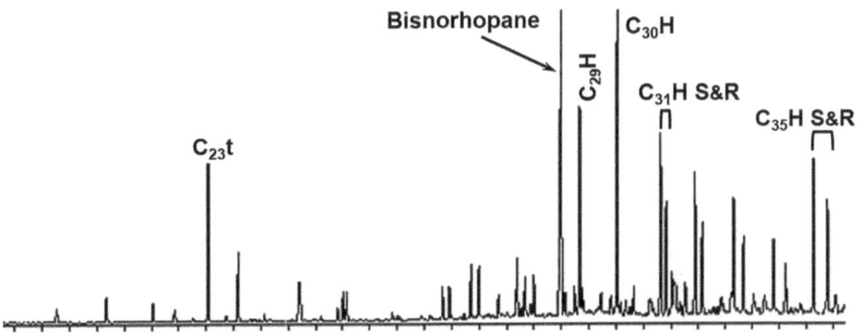

FIGURE 4.20 An examples of using an individual biomarker indicator, bisnorhopane, in a crude oil inversion study.

bisnorhopane is often assumed to reflect generation from source rocks containing Type II-S kerogen. However, there are many Type II-S source rocks, such as the Permian Phosphoria, that do not contain any recognizable bisnorhopane. It is likely that a specific microbial community is responsible for the contribution of bisnorhopane to the sediment (Williams, 1984) and not just deeply anoxic sulfur-rich depositional settings.

Oleanane is thought to be derived from organic matter contributions from angiosperms (Ekweozor and Udo, 1988). Because angiosperms only evolved in the very late Jurassic and did not flourishing until the Cretaceous, the presence of oleanane in a crude oil, such as shown in Fig. 4.21, is used as an indicator of a source rock deposited in the Cretaceous or Tertiary (Ekweozor and Udo, 1988; Riva et al., 1988; Moldowan et al., 1994). As one of only a few geologic age indicators used in crude oil inversion, oleanane contributes important information about the source rock. However, very few source rocks deposited in the Cretaceous or Tertiary, or their related oils, actually contain oleanane. This suggests the presence of oleanane is likely related to contributions of terrestrial organic matter inputs from specific plant sources and not just angiosperms in general.

The presence of the isoprenoidal hydrocarbon botryococcane in an oil indicates contributions from the green algae *Botryococcus braunii*, which is known to live predominantly in fresh to brackish lacustrine depositional environments (Moldowan and Seifert, 1980; McKirdy et al., 1986). As a result, the presence of botryococcane is used as an indicator of generation from a source rock deposited under fresh to brackish lacustrine conditions. However, the lack of botryococcane does not indicate that deposition of the source rock was not under fresh to brackish water lacustrine conditions (Derenne et al., 1988).

Although very useful when found, these four individual biomarker indicators of source rock characteristics, and others like them, have limited utility in crude oil inversion because of their infrequent occurrence. While their presence has significance, their absence is ambiguous. Crude oil alteration can also impact these compounds. Thermal alteration can eliminate these compounds, thereby erasing their information about the source rock. Biodegradation can have the opposite effect on some of these compounds. Gammacerane, oleanane, and bisnorhopane are more resistant to biodegradation than the pentacyclic triterpanes of the regular hopane series (Wenger et al., 2002; Head et al., 2003). Because of their resistance, gammacerane, oleanane, and bisnorhopane can appear to increase in concentration as biodegradation

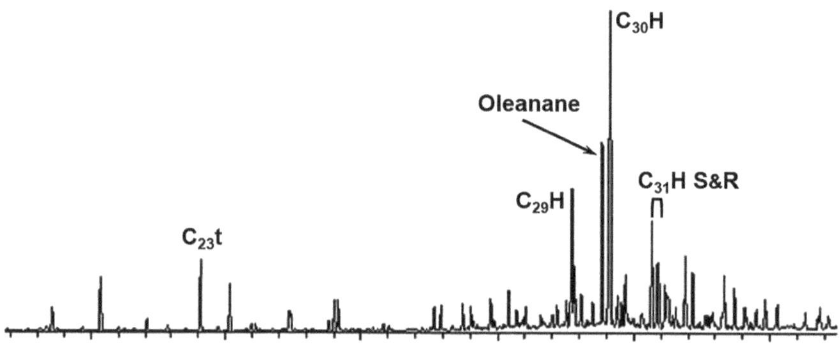

FIGURE 4.21 An examples of using an individual biomarker indicator, oleanane, in a crude oil inversion study.

progresses. This can cause an oil with an initial low concentration of one of these compounds to appear to have a significantly higher concentration. Consequently, caution must be used while interpreting these biomarkers.

Relative Abundance of Groups of Compounds

Another source of information for crude oil inversions is the relative abundance of all or part of a group of compounds. To illustrate this, let us review the interpretation of the relative abundance of the regular steranes, diasteranes, thiophenes, and diamondoids to provide insight into a crude oil's source.

Steranes are ubiquitous components of crude oils and source rock bitumens. The amounts and types of steranes found in these geological materials are controlled mainly by the type of sterols in photosynthetic organisms contributing to the sediment. Whereas the contributions of marine photosynthetic organisms are dominated by the C_{27} sterols, nonmarine higher plant contributions are dominated by C_{29} sterols (Huang and Meinschein, 1976). Huang and Meinschein (1979) later proposed this relationship could be extended to steranes in older sediments and used a ternary plot of the C_{27}, C_{28}, and C_{29} steranes to illustrate this relationship, as shown in Fig. 4.22. While this approach is generally useful in making depositional facies interpretations, it is not always straightforward. Not all marine sediments show a predominance of C_{27} steranes (Volkman, 1986). This may be reflecting in part the transportation of terrestrial organic matter to offshore environments. C_{29} steranes have also been found to be abundant in lower Paleozoic (Moldowan et al., 1985) and Precambrian (Fowler and Douglas, 1987) sediments, prior to the appearance of land plants. Fowler and Douglas (1987) have suggested that some cyanobacteria could have produced C_{29} steranes in these older sediments.

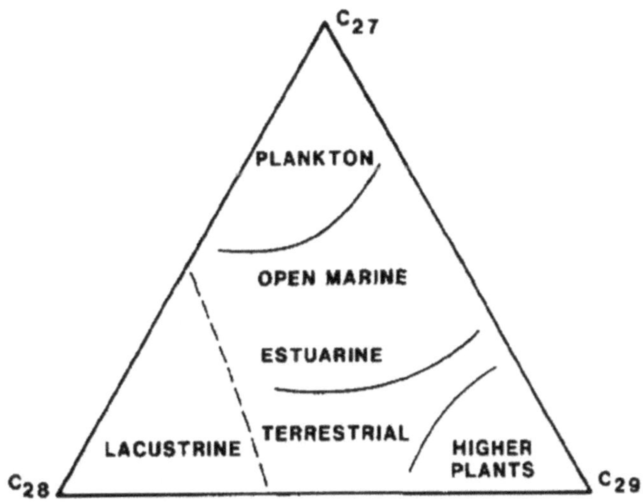

FIGURE 4.22 A ternary plot of the percentage of C_{27}, C_{28}, and C_{29} steranes used for interpreting depositional environments for use in crude oil inversion studies. *From Huang, W-Y., and Meinschein, W.G., 1979. Sterols as ecological indicators: Geochimica et Cosmochimica Acta 43, 739–745.*

While the C_{27}, C_{28}, and C_{29} steranes are ubiquitous components of crude oils and source rock bitumens, the C_{30} steranes are not always encountered. Moldowan et al. (1985) have proposed that the presence of C_{30} steranes indicates contributions of marine organic matter to the source sediments. Whereas this observation has been confirmed by others (e.g., Mello et al., 1988; McCaffrey et al., 1994), not all marine sediments may contain recognizable concentrations of C_{30} steranes (Waples and Machihara, 1990). This may occur due to the presence of steranes from large amounts of terrestrial organic matter diluting the marine-derived C_{30} steranes. The absence of C_{30} steranes in marine sediments of Cambrian and Precambriann age has also been noted by Moldowan et al. (1985). They attribute this to an evolutionary lag in the biochemical pathway producing C_{30} sterols in early marine organisms.

In contrast to regular steranes, diasterane, or rearranged sterane, concentrations are controlled mainly by maturity and lithology. C_{27}–C_{29} diasteranes are present in most samples that are at least moderately mature, with the C_{27} diasteranes being the most easily recognized as they are well separated from other peaks in the m/z 217 mass chromatograms. The relative amount of diasteranes compared to regular steranes seems to depend on both sediment type and maturity. The rearrangement of regular steranes to diasteranes is a normal response to increasing maturity in sediments. However, the rearrangement can also be catalyzed by the presence of clays in the host sediment (Rubinstein et al., 1975). Because of this, high diasterane contents are often used as an indicator of clastic, or more correctly clay-rich, facies, while clay-poor sediments such as carbonate and siliceous rocks have low diasteranes contents (Zumberge, 1984; Mello et al., 1988). The example in Fig. 4.23 shows Oil A displaying high diasteranes likely from a clay-rich source, while Oil B has low diasteranes from a clay-poor source. It must be remembered that diasteranes will increase with increasing maturity in sediments, thereby becoming more dominant. Diasteranes are also more resistant to biodegradation than regular steranes (Wenger et al., 2002). As biodegradation intensity increases, diasteranes will become more concentrated as the regular steranes are consumed.

Another feature of the sterane mass chromatograms that can be used in crude oil inversion is the isomerization of the C_{29} regular sterane compounds. As can be seen in the chromatograms in Fig. 4.23, there are four isomers: the C_{29} $\alpha\alpha\alpha$ 20S and 20R steranes and the C_{29} $\alpha\beta\beta$ 20S and 20R steranes. The $\alpha\alpha\alpha$ isomer peaks, especially the 20R, are initially large compared to the $\alpha\beta\beta$ isomer peaks. With increasing maturity, the $\alpha\beta\beta$ isomer peaks increase as the $\alpha\alpha\alpha$ isomer peaks decrease, as a result of the higher intrinsic thermal stability of the $\alpha\beta\beta$ isomers (Seifert and Moldowan, 1981). In the example in Fig. 4.23, Oil B has slightly more abundant $\alpha\alpha\alpha$ isomer, especially the 20R, suggesting it is slightly less mature than Oil A. This relationship between the C_{29} sterane isomers has also been the source of at least two biomarker ratios sometimes used as maturity indicators (see below).

Although not true biomarkers, thiophenes can also provide information about the lithology of a source rock. Thiophenes are organosulfur compounds formed by the interaction of elemental sulfur in the sediments and kerogen bound sulfur with organic matter early in the maturation of the sediments (Gransch and Posthuma, 1974). Hughes (1984) showed that the concentration and diversity of thiophene compounds increased in oils derived from carbonate (nonsiliciclastic) sediments as compared to siliciclastic sediments. The phenomenon was attributed to the higher concentration of iron in the siliciclastic sediments reacting with sulfur in the sediments to inhibit thiophene formation. The example shown in Fig. 4.24 shows the oil from the clay-poor source having higher relative concentrations of the dimethyldibenzothiophenes and trimethyldibenzothiophenes, as well as a greater variety of these compounds in the distribution.

Another group of nonbiomarker compounds that can be useful in crude oil inversions are the diamondoids. Diamondoid molecules are cage-like, ultra-stable, ringed saturated hydrocarbons which have a diamond-like structure consisting of a number of six-member carbon rings fused together (Mansoori, 2007). As shown in Fig. 4.25, they consist of repeating units of 10 carbon atoms called adamantine that form tetracyclic cages systems (Marchand, 2003). They are called "diamondoid" because their carbon–carbon framework constitutes the fundamental repeating unit in the diamond lattice structure (Mansoori, 2007).

Diamondoids were first identified as components of petroleum with the discovery of the simplest diamondoid, adamantine, in some European crude oils by Landa et al. (1933). Diamondoids are found in mature high-temperature petroleum fluids including volatile oils, condensates, and wet gases. It is believed that they form from rearrangements of cyclic hydrocarbons catalyzed on clay mineral during the cracking of oil to gas (Dahl et al., 1999).

Because of this high-temperature origin, diamondoids have been used as indicators of crude oil cracking and high thermal maturity. As shown in Fig. 4.26, the concentration of diamondoids, in this case, the methyldiamantanes, starts out very low, but increases rapidly when the oil enters into the high thermal maturity cracking zone (Dahl et al., 1999). The diamondoid concentration is also useful as an indicator of mixed source oils. If an oil exhibits

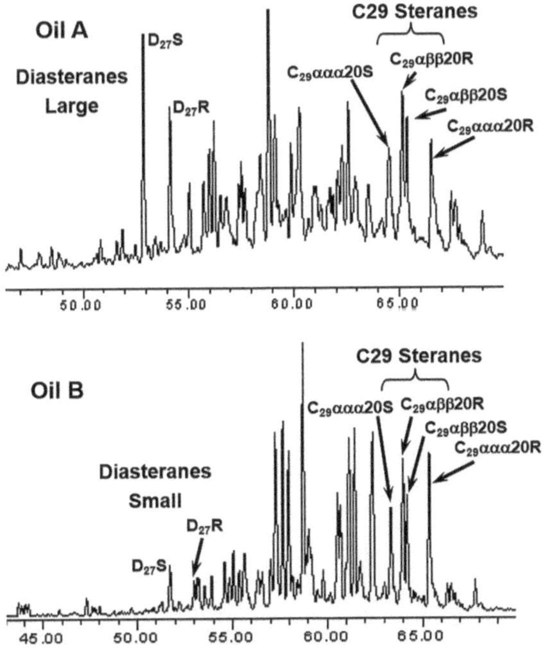

FIGURE 4.23 m/z 217 mass chromatograms of two oils. Oil A exhibits large diasterane peaks suggesting a clay-rich source rock and a pattern of C_29 steranes isomers indicating a maturity near peak oil generation, while Oil B has small diasterane peaks suggesting a clay-poor source rock and a pattern of C_{29} sterane isomers indicating a maturity less than peak oil generation.

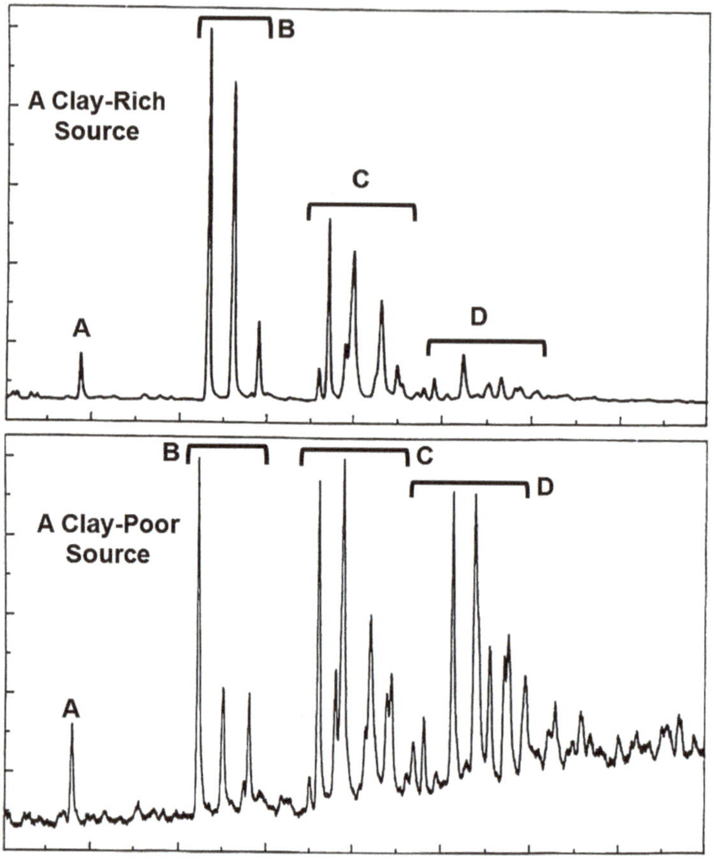

FIGURE 4.24 Thiophene distributions for clay-rich versus clay-poor source rock. Data is from a flame photometric detector gas chromatography. Compounds labels: A–dibenzothiophene, B–methyldibenzothiophenes; C–dimethyldibenzothiophenes; and D–trimethyldibenzothiophenes.

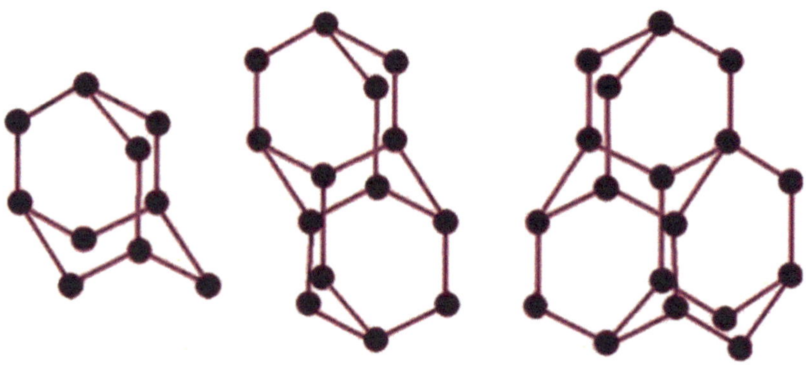

FIGURE 4.25 From left to right, the molecular structures (carbon skeletons) of adamantane, diamantane, and trimantane, the smaller diamondoids, with chemical formulas $C_{10}H_{16}$, $C_{14}H_{20}$, and $C_{18}H_{24}$, respectively. *From Mansoori, G.A., 2007. Diamondoid molecules. Advances in Chemical Physics 136, 207–258.*

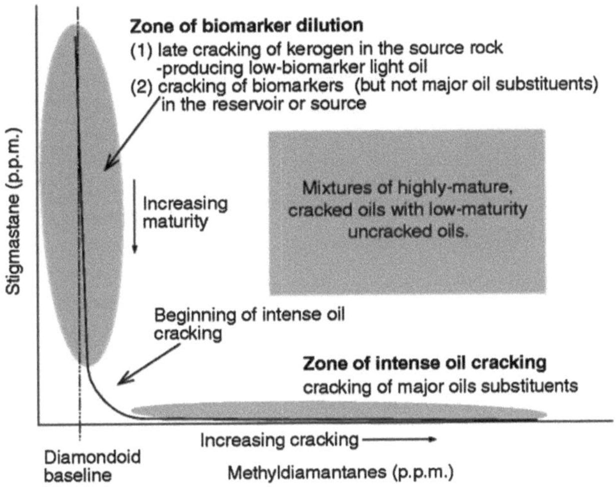

FIGURE 4.26 The correlation between the concentration of diamondoids (in this case the methyldiamantanes) and the C_{29} sterane biomarker stigmastane (in this case the ααα 20R isomer) in oils of different thermal maturities. Note that the diamondoid concentration remains low until the oil cracking zone is reached. The area where both the stigmastane and diamondoid concnetrations are high indicates a mixture of lower maturity uncracked oil with a higher maturity cracked oil. *From Dahl, J.E., Moldowan, J.M., Peters, K.E., Claypool, G.E., Rooney, M.A., Michael, G.E., Mello, M.R., Kohnen, M.L., 1999. Diamondoid hydrocarbons as indicators of natural oil cracking. Nature 399, 54–57.*

both high diamondoid concentrations and low biomarker thermal maturity indicators, a mixture of lower maturity uncracked oil with a higher maturity cracked oil is suggested.

Biomarker Ratios

One of the main tools used in crude oil inversion studies are biomarker ratios, usually presented in cross-plots. There are a substantial number of biomarker ratios that have been put forth as indicators of source rock characteristics such as lithology (clastic versus carbonate, clay-poor versus clay-rich), depositional setting (marine versus lacustrine, high anoxia, high salinity), and maturity. A short list of some of the more commonly used biomarker ratios used in crude oil inversion studies is given in Fig. 4.27. For a comprehensive list of these biomarker indicators, please refer to Peters et al. (2005a,b).

The basic method employed uses a pair of biomarker ratios that are indicators of the same source rock characteristic, such as lithology. When plotted against each other on a set of axes, a cross-plot, the data can reveal either corroboration or contradiction between the two ratios. If they corroborate each other, the interpretation made is reinforced, while contradiction would indicate an investigation into the reason for the lack of agreement is needed. An example of how this is applied is shown in Fig. 4.28. Four cross-plots are used to represent relative maturity, lithology (clastic versus carbonate), anoxia, and depositional setting (marine versus lacustrine).

The maturity indicators used in Fig. 4.28 are the ratios of the thermally more stable Ts and thermally less stable Tm forms of the C_{27} and C_{29} hopanes. The ratio used here is Ts/Tm, however, it can also be expressed as Ts/(Ts + Tm). As the value of the Ts/Tm ratio increases,

	Biomarker Ratio	Interpretation
Lacustrine vs. Marine	Hopanes/Steranes	> 1.6 indicates lacustrine < 1.6 indicates marine Moldowan et al. (1985)
	C_{26}/C_{25} Tricyclic Terpanes	> 1.0 indicates lacustrine <1.0 indicates marine Zumberge (1987)
Carbonate vs. Clastic	C_{29}/C_{30} Hopanes	> 0.8 indicates carbonates < 0.8 indicates clastics Connan et al. (1986)
	C_{21}/C_{22} Tricyclic Terpanes	> 0.6 indicates carbonates < 0.6 indicates clastics Peters et al. (2005b)
Anoxia	C_{35}Hopane (S)/C_{34}Hopane (S)	> 1.0 indicate strong anoxia Peters et al. (2005b)
Maturity	C_{27} Ts/Tm Hopanes	Increase with increasing maturity Seifert and Moldowan (1978)
	C_{29} Ts/Tm Hopanes	Increase with increasing maturity Fowler and Brooks (1990)

FIGURE 4.27 A table of some biomarker ratios and their interpretation that are used in crude oil inversion studies.

it indicates higher thermal maturity for the oil or source rock (Seifert and Moldowan, 1978). The Ts/Tm ratio is only a relative maturity indicator and no reliable conversion of the ratio to equivalent vitrinite reflectance values exists. This is mostly likely due to influences of the source organic material on the ratio (Moldowan et al., 1985).

Other biomarker ratios used as maturity indicators include the ratios of the αββ/(αββ + ααα) 20R C_{29} steranes and the ααα 20S/(ααα 20S + ααα 20R) C_{29} steranes (Seifert and Moldowan, 1986). Both show increasing maturity with the increasing value of the ratio. Like the Ts/Tm ratio, these two C_{29} steranes ratios are only relative indicators of maturity (Moldowan et al., 1985). Additional maturity indicators such as the methylphenanthrene index (Radke and Welte, 1983) and ratios using the monoaromatic steroid (Seifert and Moldowan, 1978) and the triaromatic steroids (Beach et al., 1989; Peters et al., 2005b) may also be helpful deciphering maturity.

The C_{22}/C_{21} tricyclic terpane ratio (Peters et al., 2005b) and C_{29}/C_{30} hopane ratio (Connan et al., 1986) are used in the example found in Fig. 4.28 as indicators of clastic versus carbonate lithology for the source sediments. Connan et al. (1986) original set the cutoff value for the C_{29}/C_{30} hopane ratio at 1.0, but a predominant carbonate lithology may be indicated as low as 0.8. Another biomarker ratio that is useful for indicating carbonate source rocks is the C_{24}/C_{23} tricyclic terpane ratio (Peters et al., 2005b). Ratio values <0.60 suggest a carbonate tendency for an oil's source (Peters et al., 2005b).

Anoxia is indicated by the $C_{35}S/C_{34}S$ hopane ratio (Peters et al., 2005b) in the example found in Fig. 4.28. The other ratio in this cross-plot, the C_{24} tetracyclic terpane/C_{23} tricyclic terpane does not indicate any known specific characteristics about the source rock. However, it does exhibit significant variation in geologic materials and is often used for segregating families of oils and their source rocks. Another anoxia indicating biomarker ratio is the bisnorhopane/hopane ratio (Curialie et al., 1985). However, the sparse occurrence of bisnorhopane limits its usefulness.

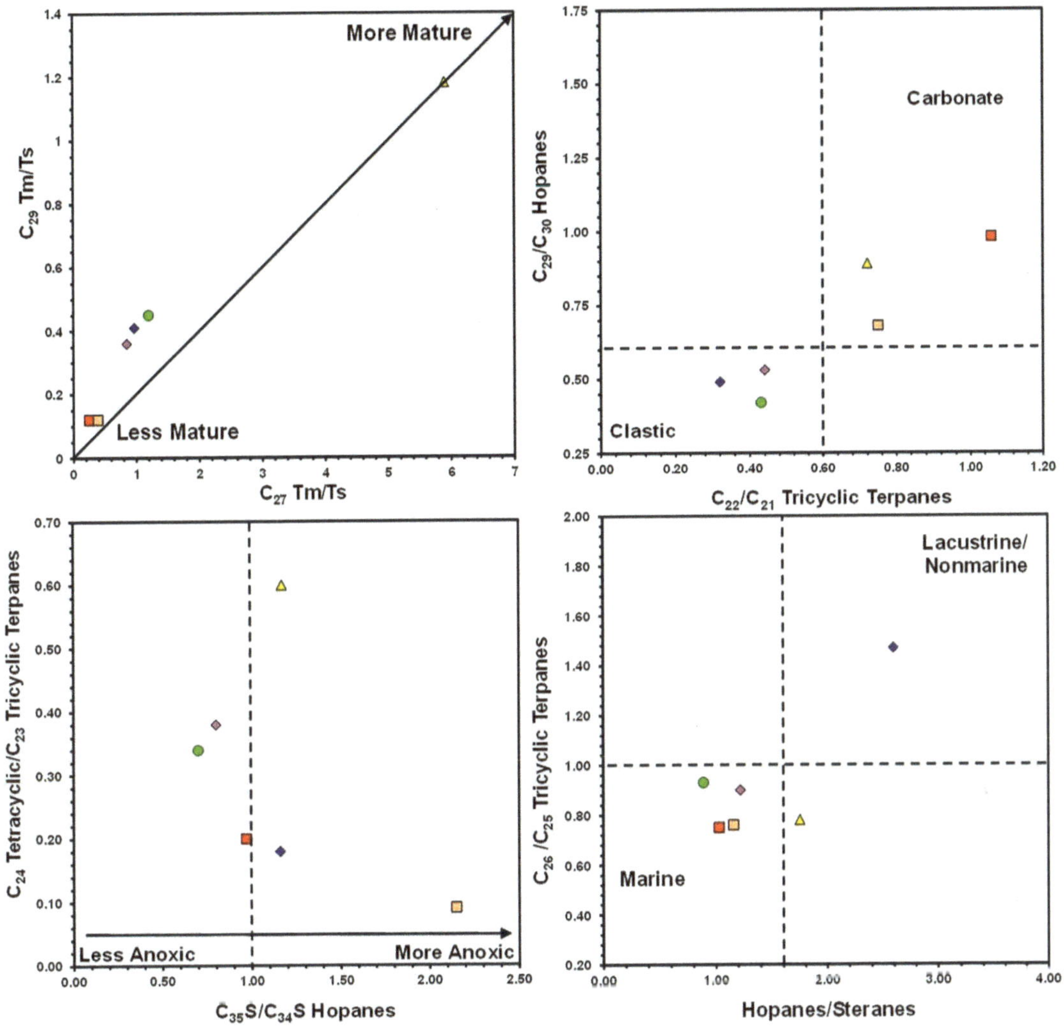

FIGURE 4.28 Four cross-plots of some biomarker ratios and their interpretation that are used in crude oil inversion studies. A typical suite of cross-plots as shown would address relative maturity, carbonate versus clastic source rock lithology, depositional environment anoxia, and marine versus clastic depositional setting.

The lacustrine versus marine indicators used in the example found in Fig. 4.28 are the C_{26}/C_{25} tricyclic terpane ratio (Peters et al., 2005b) and the hopanes/steranes ratio (Moldowan et al., 1985). The hopane/sterane is used instead of the usual sterane/hopane ratio to make the comparison of the two ratios more straightforward with both ratios increasing to show lacustrine tendencies. While a variety of other biomarker ratios have been suggested as possible indicators of lacustrine versus marine source rock contributions, there has not been enough evidence gathered to validate many of these observations at this time.

There are several advantages to using biomarker ratios in crude oil inversion. First, they are typically based on commonly occurring biomarker compounds found during routine analysis. The interpretations using biomarker ratios are semiquantitative and thereby less prone to bias. And with the cross-plotting method, two indicators can often be used to confirm the interpretations made conforming to the multiple parameter approach.

There are also some disadvantages to using biomarker ratios. There is a need for consistent analytical results and data from more than one lab may result in problems. Interlaboratory variations can stem from differences in the instruments and analytical conditions used. Potential problem areas including instrument tuning, variations in gas chromatographic column resolution, integration or measurement differences, and approaches used in the handling of co-eluted peaks. Individual biomarker ratio interpretations may not always be applied consistently and basin-specific adjustments may be needed. Hopane- and sterane-based ratios can be influenced by biodegradation and thermal alteration, while the tricyclic terpanes used in many of the ratios are more resistant to both microbial degradation and thermal processes than other biomarkers. And finally, all biomarkers ratios may work some of the time, but not all biomarker ratios work all of the time. As a result, some inconsistencies may result from applying biomarker ratios. These conflicts will need to be resolved to make a coherent interpretation.

STRATEGIES AND OBSTACLES IN OIL CORRELATION AND OIL INVERSION STUDIES

Prior to beginning any oil correlation or crude oil inversion study, take time to assess each oil sample and source rock extract to be used with respect to data quality and potential alteration. Look for biodegradation, maturation, contamination, and other processes that may influence interpretations. If any alteration is recognized, determine what effect, if any, the alteration may have on the ability to make correlations. In addition, check that source rock candidates have adequate source potential and are mature enough, with respect to both the samples being used and the overall source interval in the basin, to have generated commercial quantities of hydrocarbons.

For the biomarker data, determine if all the analytical work has been down by one laboratory. If not, consider the potential for interlaboratory inconsistencies in both analytical procedures and measurement methods. They can stem from differences in the instruments themselves, instrument conditions used, and the analytical protocols. Determine if the data from one lab can be compared to the results from another.

If numerical data, such as peaks height or peak areas, are being used, inspect the data for measurement errors. This is done by comparing the measurements to the mass chromatograms to see if they are consistent. Match the peak labels on the chromatogram to the values in the data table. Occasionally, there could be a shift in retention times during an analysis and some peaks may be miss-identified. It is also important that if numerical data are being used, the measurements are consistently based on either peak height or peak area and not a mix. To minimize these potential problems, use the same lab with a consistent analytical scheme when possible.

Use multiple correlation or inversion parameters to determine relationships between oils and between oils and their source rocks or to decipher source rock characteristics. Single parameter correlation may be erroneous and misleading. Recognize that the visual comparison techniques using compound distributions and cross-plot of biomarker ratios or isotope data are subjective methods. While statistical methods in oil correlation may be more objective, their results still require verification with some visual comparisons to insure their validity.

For oil correlation studies, after correlations are made based on the similarity of geochemical data, it should be determined if these correlations are geologically reasonable. For oil-to-oil correlations, it should be determined if there could be communication between the reservoirs and if the reservoirs could be sourced by a common rock unit. For oil-to-source rock correlation, it should be determined if the source rock is in a good stratigraphic position to source the reservoir(s) and if there is a clear migration path between the source rock and the reservoirs.

For crude oil inversion studies, after potential source rock characteristics are determined, it should be determined if these characteristics are geologically reasonable. For example, do source rocks with these properties exist in the vicinity? Could these source rocks be mature enough to have generated and expelled commercial quantities of hydrocarbons? Is there a clear migration path(s) between the source rock and the reservoir(s)?

Just as in source rock evaluation studies, the oil-to-oil or oil-to-source rock geochemical correlation and crude oil inversion interpretations cannot stand alone, they need to make good geologic sense or they have little value.

Oil correlations and crude oil inversions are key elements in defining a petroleum systems and finding additional accumulations. But, there are many possible complications that can hinder or prevent a successful completion of these studies. It is important to understand why correlations and/or inversions could fail in order to assess what might be done next in the exploration program.

Crude oil alteration processes are an obvious place to start. Thermal alteration, if severe enough, can remove some or all of the source-controlled geochemical characteristics of an oil, as well as the source rock's bitumen. The result is a composition that is controlled by thermodynamic conditions rather than the source organic matter. Biodegradation is also a major source of interference. If biodegradation is severe enough, many of the biomarker distributions used in correlations and inversion can be altered or completely eliminated. In some instance, the biomarkers distributions are altered, but not enough to be obvious, possibly resulting in misleading interpretations.

Contamination from oil-based drilling muds is also a concern. Drilling muds formulated with crude oils pose obvious problems for both source rocks and crude oils samples. However, diesel and synthetic oil-based mud can be just as problematic, especially when oil samples are collected with downhole sampling tools such an MDT (modular formation dynamics tester) tool. The base oils used to formulate oil-based drilling muds can extract source rocks during drilling and obtain a biomarker signature as mentioned earlier. Biomarkers from crude oil can also be introduced into the drilling mud once an oil-filling reservoir has been encountered. Even at low concentrations, these biomarkers can alter the indigenous biomarker distributions and cause interpretation problems. And, if the oil-based drilling muds are used more than once without reconditioning, these contaminant biomarkers can be carried over to the next well.

The samples collected for these studies can also be problems. With respect to the source rock, we are limited in where we can sample to wherever wells have been drilled or where outcrops exist. These locations are not always where we envision the actual hydrocarbon generation and expulsion to have occurred that are responsible for forming the crude oil accumulation. As a result, it is not possible to ascertain if the source rock samples available represent the same organic facies or are at a similar maturity as the sediments in the generative area. Significant difference on the molecular level may exist between these locations, potentially leading to ambiguous or erroneous conclusions.

Even when we can sample the source rock in the "generative kitchen" for a particular oil accumulation, it is uncertain if the bitumen will be representative of the hydrocarbons in the reservoir. This uncertainty stems from three possible sources. First, a basic assumption in oil correlation and crude oil inversion is that the biomarker compound distributions in the source rock and resulting oil are not significantly altered during expulsion and migration. As discussed in Chapter 2, there are observed compositional differences that exist at least on a gross basis between the oil expelled and the bitumen that remains behind in the source rock (Deroo, 1976). But it is not clearly understood whether these compositional differences extend down to the molecular level and impact the biomarker distributions.

Second, while the source rock samples used in these analyses represent the maturity level of the sediments at the time of sample collection, the hydrocarbons delivered to the reservoir do not. As a reservoir fills over time, it is receiving hydrocarbons from the source rock at progressively increasing maturity levels. The final product is a mixture of all these contributions over time. Additional heating in the reservoir may continue the maturation process for the oil (thermal alteration) which may also alter biomarker distributions. Under these circumstances, comparisons of source rock extracts and crude oils may not show the level of similarity demonstrating a clear correlation.

The third potential source of uncertainty lies in the physical sampling of the source rock. The generation and expulsion in a source rock that results in an oil accumulation may occur over an area of at least tens of square kilometers and involve sediment thickness of tens of meters or more. However, the sampling of a source rock for comparison to an oil represents a single location, the well, and may represent a thickness of only a few centimeters, if a sidewall core is used, to a few meters, if a cuttings sample is used. Katz et al. (1993) and Curiale (2008) observed that there can be significantly different biomarker distributions in cuttings and sidewall core samples taken in the same source rock only a few meters apart in the same well. Considering there are likely both vertically and laterally variations in organic matter richness and type, this is not surprising. If we further consider the oil reservoir is receiving contributions from the entire thickness of the source rock and over the full area of the "generative kitchen," it is not difficult to envision how our sampling may be nonrepresentative.

The crude oil used in these studies can also be a source of concern. As mentions before, a reservoir can receive contributions from more than one source rock. Depending on the proportion of each oil in the mix, the resulting mixture can make correlating one oil to another or correlating the mixed oil to a source rock difficult if not impossible. But even when an accumulation is derived from a single source, difficulties may arise. A migrating crude oil itself can act as a solvent and essentially extract organic material from source rocks, bitumens, and coals encountered along its migration pathway or in the reservoir (Philp and Gilbert, 1982; Hughes and Dzou, 1995; Curiale, 2002). This is termed migration contamination (Curiale, 2002). The amount of material extracted as well as the distributions of the biomarkers added to the oil will determine if this is a major barrier to making a successful oil correlation or crude oil inversion.

While there is little to be done to overcome some of these obstacles to oil correlation and crude oil inversion studies, being aware of the problems is an important aspect of these studies. This information can provide the petroleum geologist with reasons for the failure of an oil correlation or crude oil inversion or clues to why the results of a study appears to be erroneous or contrary to the geology of the proposed petroleum system. It may also help determine what additional samples or analytical work may be needed to try to resolve these difficulties and add value to the study.

NATURAL GAS DATA

Natural gases are the least compositionally complex form of petroleum. Unlike crude oils that have a large number of geochemical characteristics for making interpretations, natural gases because of their simplicity are limited in the type and amount of information that can be obtained from them. Basically, only the composition of the gas and its isotopic signatures are available for use in making interpretations of the origins of the gas, its maturity, and making gas-to-gas and gas-to-source rock correlations.

The composition of natural gases is determined by measuring the concentrations of the individual components through a series of analyses using gas chromatography. Both the hydrocarbon and nonhydrocarbon gas components of the gas are measured. The hydrocarbons typically encountered are methane, ethane, propane, iso-butane, and n-butane. Iso-pentane, n-pentane, and hexanes may also be reported, but they are often considered part of the gasoline range hydrocarbons, as well as usually being in low concentration. In addition to the hydrocarbon gases, nonhydrocarbons including nitrogen, carbon dioxide, hydrogen sulfide, helium, and hydrogen may also be present. The overall composition of the gas will be a list each component, both hydrocarbon and nonhydrocarbon, usually expressed in mole percent.

The other diagnostic feature of the components of natural gas is their stable isotope composition. Stable isotopes are measured using mass spectrometers. For the carbon isotopes, each individual hydrocarbon component of the gas is separated by gas chromatography. These individual gases are then burned in a close system to form carbon dioxide. Carbon isotope measurements are then made on the carbon dioxide. For the deuterium/hydrogen isotopic analysis, the water formed during the burning of the individual gases are collected for measurement. Isotope data for carbon and hydrogen in the hydrocarbons, usually expressed in δ notation (see Chapter 1), are important tools for deciphering the origin and maturity of the gas. Typical isotopic measurements reported for natural gas include $\delta^{13}C$ methane, $\delta D/H$ methane, $\delta^{13}C$ ethane, and $\delta^{13}C$ propane. $\delta^{13}C$ on the butanes may also be attempted, but often the concentration of the butanes in the gas is too low for accurate determination. Similarly, $\delta D/H$ may also be attempted on ethane and propane, but low concentrations of these gases are frequently an obstacle to obtaining good data. For the nonhydrocarbon gases, the $\delta^{13}C$ for carbon dioxide can be a useful tool if the concentration of the gas is adequate for analysis.

THE SOURCE OF NATURAL GAS: BIOGENIC VERSUS THERMOGENIC

As previously discussed in Chapter 2, natural gas can be derived from either biogenic or thermogenic processes. Biogenic gas is the product of microbial action on organic matter, while thermogenic gas is the product of thermal decomposition of complex organic matter during hydrocarbon generation and oil cracking. Both biogenic and thermogenic gases can be recognized by using their combined composition and isotopic signatures.

With respect to the hydrocarbon portion of natural gas, biogenic gas is composed almost entirely of methane, usually >99.95%, and is isotopically light, with the $\delta^{13}C$ for the methane ranging from −110‰ to −60‰. The amount of biogenic gas in the subsurface is significant and accounts for about 20% of the world's natural gas resources (Rice, 1992). Thermogenic gas is very different from biogenic gas. Except for very thermally mature natural gases, the C_2+ hydrocarbon compounds are typically >5% of the total hydrocarbons. Isotopically, the methane is heavier in thermogenic gas, exhibiting $\delta^{13}C$ values from about −60 to −20.

While the composition of biogenic gas is essentially constant over its history, thermogenic gas can experience significant changes. As discussed in Chapter 2, thermogenic gas in source rocks exhibits a trend with depth as a result of continued maturation due to increasing time and temperature. With increasing maturity, the methane $\delta^{13}C$ becomes progressively heavier (less negative). There is also a corresponding change in gas composition in the source rock. The C_2+ compounds begin with low concentrations that increase to a maximum as the source rock nears peak oil generation. After peak generation, there is a steadily decrease in C_2+ compounds until they are nearly absent in highly mature gases. These compositional and isotopic changes are reflected in the gas migrated from the source rock and accumulation in reservoirs. These trends in the evolution of thermogenic gas and the composition of biogenic gas are summarized in the cross-plot in Fig. 4.29 of the $\delta^{13}C$ of methane versus the percentage of

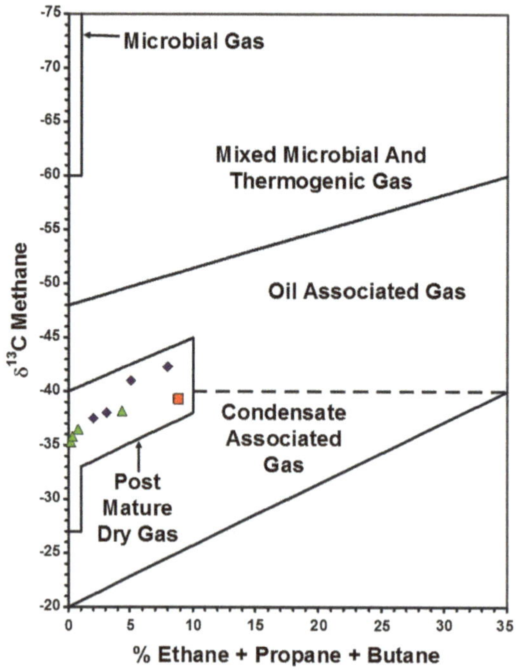

FIGURE 4.29 A cross-plot of $\delta^{13}C$ of methane versus the summed concentrations of ethane, propane, and butanes used in classifying natural gases and assessing their relative maturity. *Modified from Schoell, M., 1983. Genetic characterization of natural gases. American Association of Petroleum Geologists Bulletin 67, 2225–2238.*

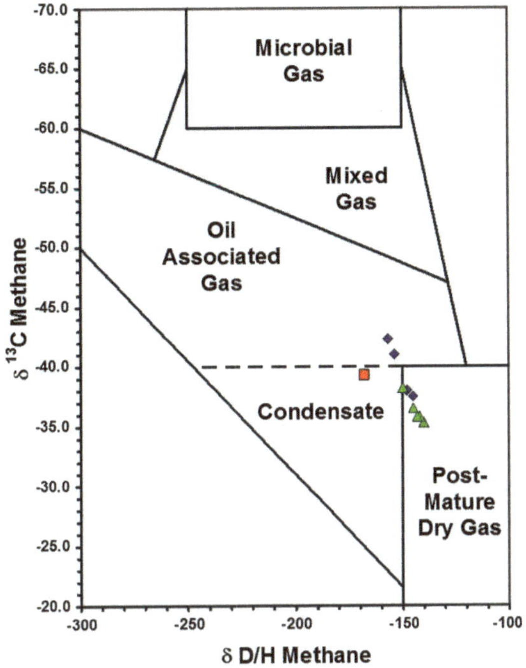

FIGURE 4.30 A cross-plot of $\delta^{13}C$ of methane versus $\delta D/H$ of methane used in classifying natural gases and assessing their relative maturity. *Modified from Schoell, M., 1983. Genetic characterization of natural gases. American Association of Petroleum Geologists Bulletin 67, 2225–2238.*

the C_2+ hydrocarbon gases based on the work of Schoell (1983). This cross-plot is a convenient way to display gas data to illustrate the origin of the gas along with the relative maturity. Schoell (1983) also developed a similar diagram cross-plotting the $\delta^{13}C$ versus the $\delta D/H$ of methane, shown in Fig. 4.30. It is used to make the same type of observations about the origin of the gas and relative maturity using purely stable isotope data. These two diagrams are often used together to corroborate interpretations.

Because of the ease by which natural gas can move about in the subsurface, it is always a possibility that a natural gas accumulation may have more than one source or receive gas from its source during more than one expulsion episode. It may be difficult to recognize the mixing of natural gases because of their compositional simplicity, but it is something that needs to be kept in mind while trying to fit together the different components of a petroleum system.

The exploration significance of encountering thermogenic gas is clear: hydrocarbon generation and migration from a source rock has occurred indicating a petroleum system is likely present. However, the exploration significance of encountering biogenic gas is a more complex issue and depends on how the biogenic gas was formed. Initially biogenic gas was thought to form only by microbial decomposition of contemporary organic matter in recently deposited sediments primarily in the near surface (Rice and Claypool, 1981). This primary biogenic gas would accumulate in the interstitial spaces of the sediment, usually dissolve in the pore water and it was believed this biogenic gas could migrate to form commercial accumulations (Rice and Claypool, 1981). However, the mechanisms by which this biogenic gas would exsolve from the pore water, migrated, and accumulated has been unclear. Without traps or seals forming early in the evolution of these sediments, it is probable that a large portion of this early biogenic gas is lost during sediment dewatering due to compaction, making it suspect as a major source of commercial biogenic gas.

The exploration significance of encountering primary biogenic gas is that no generation has occurred, this methane has accumulated from microbial production early in the sediments history, and no petroleum system is operating. Only additional primary biogenic gas accumulations may be found in the vicinity. Typically, this is not looked upon favorably in an exploration program.

However, a secondary source of biogenic gas is the biodegradation of crude oil (Head et al., 2003) as discussed above. Because of the mobility of gas in the subsurface, secondary biogenic gas may be associated with the biodegraded oil or it may have migrated away from the oil-bearing reservoir where it was formed and now resides in a separate reservoir/trap system. The amount of this secondary biogenic methane in oil and gas reservoirs is significant. Milkov (2010) estimated that as much as 66,500 trillion cubic feet of secondary microbial methane could have been generated in existing accumulations of biodegraded oils and bitumen through their geological history. And many studies have also demonstrated that secondary biogenic gas is an important component in oil, gas, and condensate reservoirs around the world (e.g., Jeffrey et al., 1991; Pallasser, 2000; Lillis et al., 2007; Milkov and Dzou, 2007). The conclusion drawn from these data is that large biogenic gas accumulations are likely to be formed from secondary biogenic gas.

The presence of a secondary biogenic gas accumulation is a very positive indicator in an exploration program. Because secondary biogenic gas forms from the biodegradation of crude oil, a petroleum system must exist that generated, migrated, and accumulated an oil deposit. If the biogenic gas is not directly associated with a biodegraded oil, a deeper or down dip biodegraded oil accumulation must be present. It also suggests that there may be other additional deeper or down dip oil accumulations that may be altered or unaltered in the area.

If evidence of biodegraded oil is not present in the reservoir, distinguishing between primary and secondary biogenic gas is not always straightforward, but there are a few clues. Secondary biogenic gas frequently has associated CO_2 that is isotopically depleted with respect to ^{12}C with $\delta^{13}C > +2‰$ (Pallasser, 2000; Milkov, 2011). Another clue may be found in any liquids that are entrained in the gas. Secondary biogenic gas often contains a small amount of liquids dissolved in the gas (concentrations of about 1 barrel or less of liquids per million cubic feet of gas). On close examination, these liquids are nearly always found to be biodegraded oil, most likely from the reservoir where the secondary biogenic gas was originally formed.

THE MATURITY OF THERMOGENIC NATURAL GAS

Determining a more precise thermal maturity for thermogenic natural gas is important to understanding its origin as well as a being a clue to its source. Thermal maturity estimates for natural gas are based on the carbon isotope data of the individual hydrocarbon gases, specifically the methane, ethane, and propane. Because of low concentration, it is sometimes

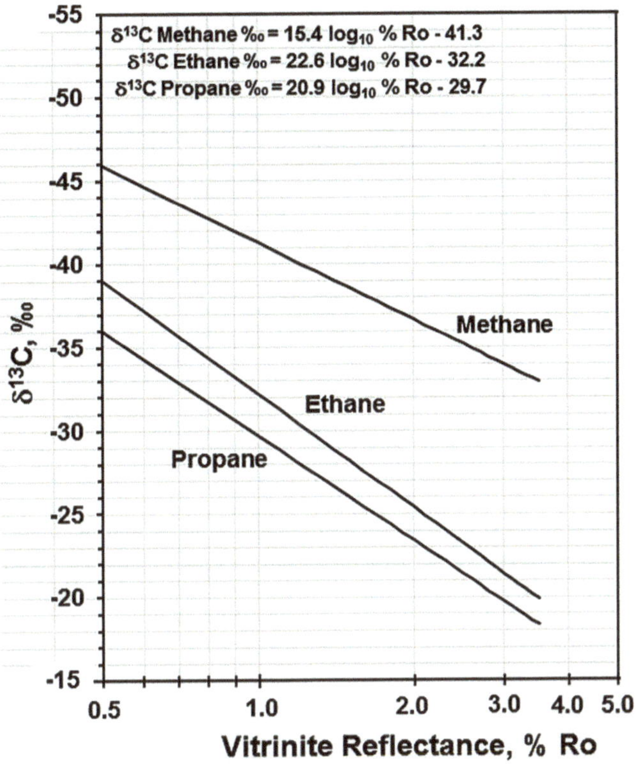

FIGURE 4.31 The observed relationship between the $\delta^{13}C$ of methane, ethane, and propane versus vitrinite reflectance, based on the equations of Stahl (1977).

difficult to obtain accurate isotopic analysis of propane. In these instances, maturity assessments rely solely on the methane and ethane and are more tentative without corroboration from the propane.

The relationship between the carbon isotope ratios of methane, ethane, or propane in a natural gas and the vitrinite reflectance of the source was established initially by Stahl and Koch (1974) with later refinement by Stahl (1977) and Faber (1987). These relationships are shown in Fig. 4.31. A common way of using these relationships to estimate the maturity of a natural gas is by cross-plotting two trends, typically methane versus ethane and propane versus ethane as shown in Fig. 4.32. Natural gas samples that have not been altered and are not the product of the mixing of two or more gases should plot along these trends within a zone of about ± 1.5‰ (Berner and Faber, 1988). There should be a general agreement between the maturity estimates of the two trends to confidently assign a maturity to the gas. Because of possible variations in the carbon-isotopic signature of the original source organic material, the maturity assessments of natural gases based on these trends should only be considered approximate.

Some carbon isotope data for natural gases are observed to plot off these trends. These deviations from the trends are usually due to the mixing of two or more gases of different

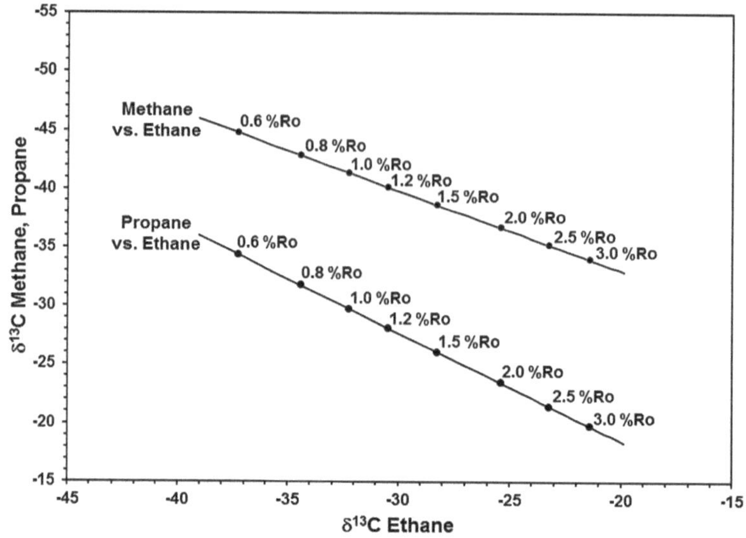

FIGURE 4.32 Cross-plots of the $\delta^{13}C$ of methane versus ethane and propane versus ethane form estimating the thermal maturity of natural gases in equivalent vitrinite reflectance, % Ro, based on the equations of Stahl (1977). These cross-plots can also be used to indicate biodegradation and mixing of gases.

origin or different maturity. Samples deviating by plotting above the methane–ethane trend suggest mixing of a biogenic gas with a thermogenic gas, while deviating by plotting below the trend indicates the mixing of two thermogenic gases (Berner and Faber, 1988). Deviation below the propane–ethane trend is also indicative of the mixing of two thermogenic gases (Berner and Faber, 1988).

Deviations from these trends may also be signaling alteration of the gas. Biodegradation of an associated crude oil would be expected to contribute isotopically lighter (more negative) biogenic methane (Head et al., 2003) to the gas also resulting in data points plotting above the methane–ethane trend. In contrast, microbial alteration of the gas itself would initially focus on the wet gas components and the likely outcome would be isotopically heavier (more positive) ethane and propane (James and Burns, 1984), resulting in data points plotting below the trends.

GAS-TO-GAS AND GAS-TO-SOURCE ROCK CORRELATIONS

Gas-to-gas correlations attempt to find geochemical relationships between gases in two or more reservoirs or the same reservoir in two or more traps. These relationships suggest the gases were generated from the same source rock at approximately the same time or maturity. Gas-to-source rock correlations extend this concept to attempt, when possible, to identify the actual source rock that generated the gas. This information, when used in conjunction with reconstructions of the structural development of the basin, can be applied to delineating migration pathways between reservoirs, traps, and sources that can point toward finding additional gas accumulations.

As mention early, the compositional simplicity of natural gas limits the geochemical characteristics that can be used for these endeavors. For both isolated gases and gases associated with oil, there are two avenues for comparison: one is based on composition of the gas and the other utilizes carbon-isotopic signatures of the individual C_1 through C_4 components (Erdman and Morris, 1974).

Proper sampling is a key component to successful gas-to-gas correlations. The most desirable gas samples are from pressurized downhole sampling tools, with the next best samples being from long-term production tests or actual production. Gas sample associated with drilling mud from Isotubes (covered in the Chapter 5), Isojars (discussed in the Chapter 3), or canned headspace gas samples (discussed in the Chapter 3) will not have the same hydrocarbon and nonhydrocarbon composition as the reservoired gas due to partitioning of the gas components between drilling mud (Isotubes) or water (Isojars or canned headspace gas samples) and the gas volume above it. However, the isotopic signature of the gases in these types of samples has been found to be essentially identical to equivalent reservoir gases. If these are the only samples that are available, the gas-to-gas correlation study must rely only on the isotopic signatures of the gases.

A good place to start the correlation process is by using the standard gas classification plots in Figs. 4.29 and 4.30. These snapshots of the origin and relative maturity of the gases can provide a preliminary classification and grouping of the gas samples that can be supplemented by more detailed comparisons of the gas composition and isotopic signature.

Comparing the composition of two or more gases for correlation can be accomplished in several ways. The most obvious is to simply plot the relative concentration of the hydrocarbons gases as shown in Fig. 4.33. These types of direct comparisons are useful, but not

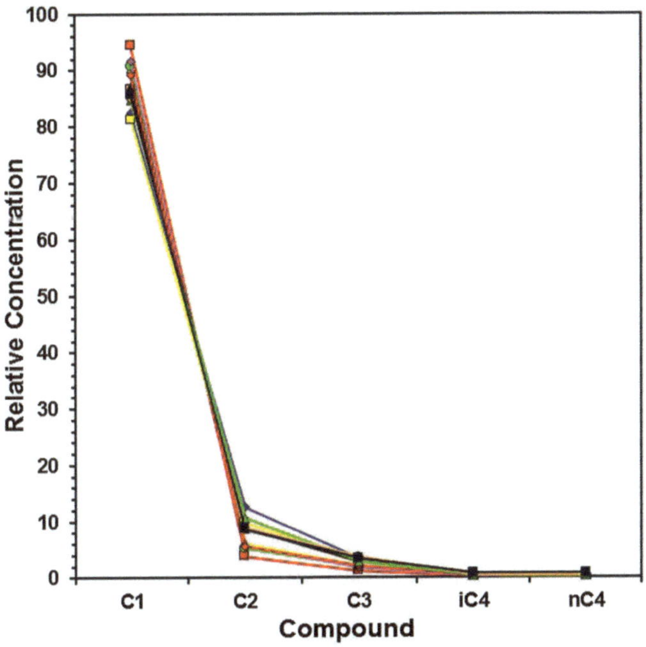

FIGURE 4.33 An example of the comparison of C_1–C_4 gas composition for use in gas-to-gas correlation.

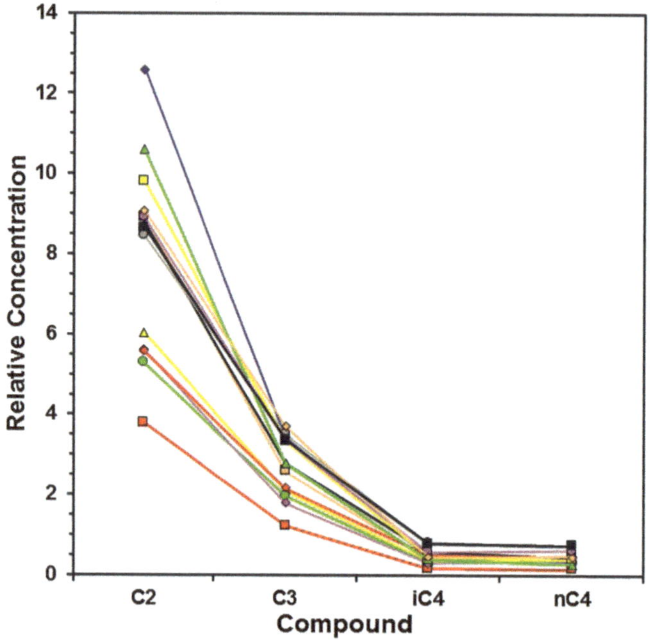

FIGURE 4.34 An example of the comparison of C_2–C_4 gas composition for use in gas-to-gas correlation.

always very informative. This is usually because most natural gases are dominated by methane, which makes it difficult to clearly see difference in the ethane, propane, and butanes. In these instances, a second look at the relative composition of the hydrocarbon gases excluding methane, as shown in Fig. 4.34, can provide more insights into the difference and similarities between the gases.

It is often beneficial to cross-plot some of the hydrocarbon gas ratios to illustrate similarities and difference between the gases, as shown in Fig. 4.35. Some of the hydrocarbon ratios that can be used in these comparisons are the % wet gas $((\Sigma\ \%C_{2-4})/(\Sigma\ \%C_{1-4}))$, C_1/C_2, C_2/C_3, and iC_4/nC_4. These ratios can also be used in cross-plotted with carbon isotope data of individual gases to assist in grouping gases of similar origin.

The nonhydrocarbon portion of natural gases is also useful in gas-to-gas correlations. Plotting the nitrogen, carbon dioxide, and hydrogen sulfide concentrations can demonstrate similarities and difference between the gases, as shown in Fig. 4.36. Care must be taken when using the carbon dioxide concentration due to the multiple sources of this gas in the subsurface (see Chapter 2). Similarly, helium is not likely to be definitive because it is derived from radioactive decay and not a byproduct of hydrocarbon generation.

Comparing carbon isotope data from the individual gas components is an important tool in gas-to-gas correlations. When a complete set of carbon isotope data is available for the C_1 to C_4 components, James (1983) suggested plotting the data on a single set of axis as shown in Fig. 4.37, providing direct comparison of all components in a single image. Frequently, not all the C_1 to C_4 components can be analyzed for their carbon isotope signature, usually due to low concentration of the propane and butanes. In these cases, simple cross-plots of the

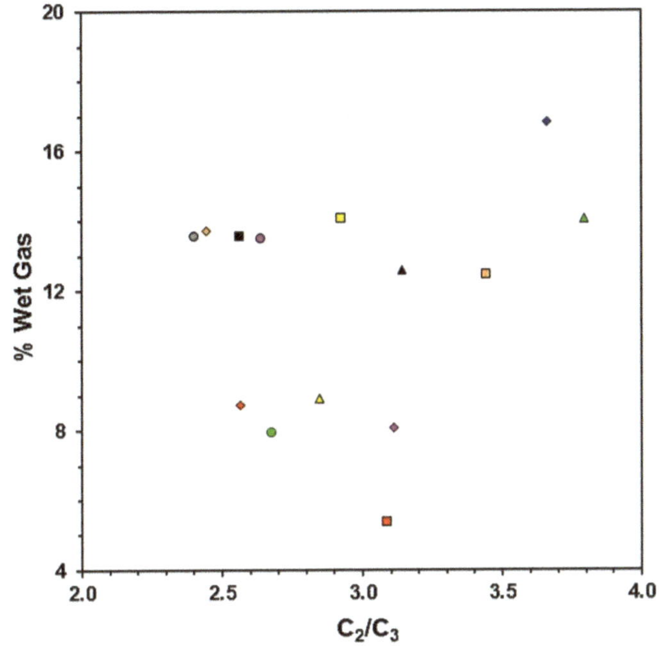

FIGURE 4.35 An example of using a cross-plot of natural gas ratios in gas-to-gas correlation.

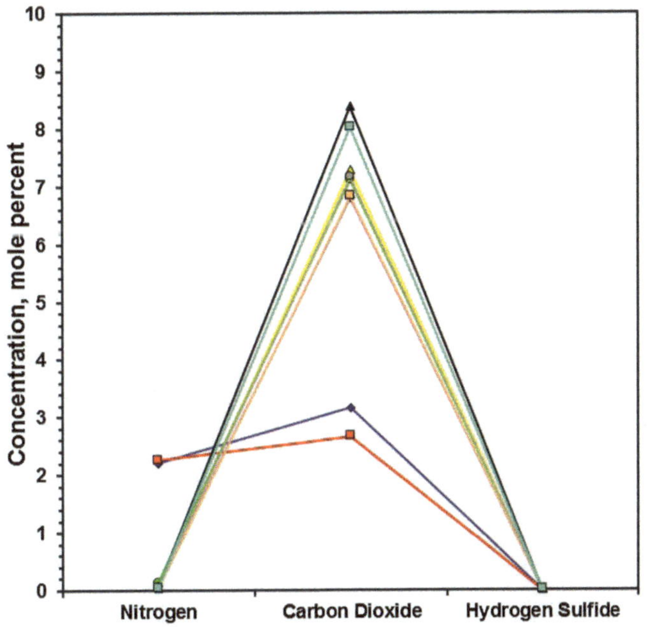

FIGURE 4.36 An example of the comparison of nonhydrocarbon gas composition for use in gas-to-gas correlation.

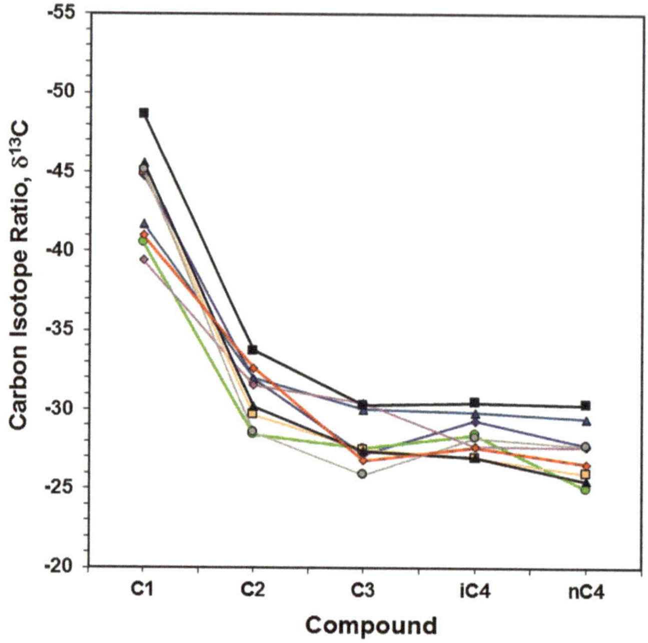

FIGURE 4.37 An example of the comparison of C_1–C_4 carbon isotope ratios, $\delta^{13}C$, for use in gas-to-gas correlation.

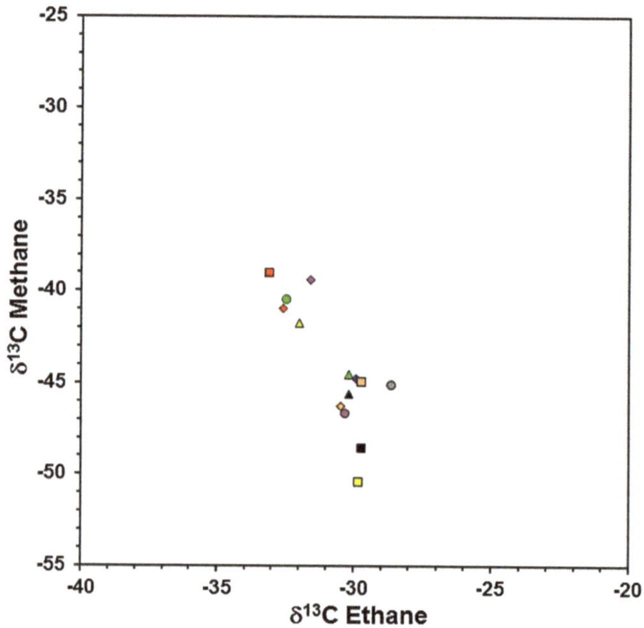

FIGURE 4.38 An example of a cross-plot of the $\delta^{13}C$ of methane versus ethane in gas-to-gas correlation.

methane versus ethane, as shown in Fig. 4.38, can be used to compare the gases. These cross-plots are also helpful for illustrating differences and similarities between gas samples even when isotopic data are available for all the gas components.

The process used in correlating gases to their source rocks is identical to correlating gases to gases. However, the ability to make these correlations is usually limited by the availability of good quality gas samples from source rock intervals. In most instances, the gas samples from source rock intervals are limited to Isotubes or a headspace gas sample derived from cuttings. While these samples may provide representative isotopic data, compositional information will be skewed due to partitioning, as mentioned above. As a result, most gas-to-source rock correlations (e.g., Stahl and Cary, 1975) are based on isotope data alone (James, 1983; Whiticar, 1994).

STRATEGIES AND OBSTACLES IN INTERPRETING GAS DATA

As with source rock and crude oil studies, gas interpretations require background information about how the well was drilled and the geological setting of the stratigraphic sequence encountered. A complete understanding of how samples were collected and their location in the wellbore is also needed.

Because of the limited amount of information available for interpreting natural gas data, sample quality is of prime importance. As mentioned above, compositional alteration due to sampling error can be large (Erdman and Morris, 1974) and samples like Isotubes or a headspace gas can only provide reliable isotopic data.

Natural gases are also very mobile in the subsurface and mix easily. This can obscure the origin and maturity of the individual contributing gases, as well as prevent gas correlations from being made. Similarly, alteration in the reservoir is possible. Cracking of oil to gas can provide a secondary source for the gas. Even the cracking of wet gas components to methane can occur. Biodegradation can also change both the composition and isotopic signature for the reservoired gas. It is important to look for indications of these processes, such as contradictory maturity indicators that might suggest mixing and/or alteration.

Remember, maturity assessments of natural gases are only approximate. Variations in the isotopic makeup of the source organic material may cause deviations from the isotopic maturity trends of a generated gas. You need to look for corroboration among the carbon isotope maturity indicators and the composition of the gas to increase the confidence in these interpretations.

Gas-to-gas and gas-to-source rock correlations should be based on both composition and isotopic signature. However, because of sampling issues these correlations may only be based on isotopic data. In these instances, the correlations may need to be considered tentative until more data become available.

As with oil-to-source rock correlations, gas-to-source rock correlations may be hampered by the lack of source rock gas samples deep enough to have encountered the actual source rock. In these instances, it may be necessary to make a tentative correlation based on which source rocks are closest to the maturity of the generated gas. In some cases, the source rock maturity may have been extrapolated from a vitrinite reflectance trend or estimated from basin modeling studies, adding additional uncertainty to the interpretation.

And just as in source rock evaluation, oil correlation, and crude oil inversion studies, the geochemical interpretations of gas data cannot stand alone. They need to make geologic sense or they have little value.

References

Anders, D.E., Robinson, W.E., 1971. Cycloalkane constituents of the bitumen from Green River shale. Geochimica et Cosmochimica Acta 35, 661–678.

Akbarzadeh, K., Hammami, A., Kharrat, A., Zhang, D., Allenson, S., Creek, J., Kabir, S., Jamaluddin, A., Marshall, A.G., Rodgers, R.P., Mullins, O.C., Solbakken, T., 2007. Asphaltenes – problematic but rich in potential. Oilfield Review 19 (2), 22–43.

Bailey, N.J.L., Krouse, H.R., Evans, C.R., Rogers, M.A., 1973. Alteration of crude oil by waters and bacteria–evidence from geochemical and isotope studies. American Association of Petroleum Geologists Bulletin 57, 1276–1290.

Barbat, W.N., 1967. Crude-oil correlations and their role in exploration. American Association of Petroleum Geologists Bulletin 51, 1255–1292.

Barr, K.W., Morton, V., Richards, A.R., 1943. Application of chemical analysis of crude oils to problems of petroleum geology study of crude oils of Forest Sands of Bernstein Field, Trinidad, B. W. I. American Association of Petroleum Geologists Bulletin 27, 1595–1617.

Beach, F., Peakman, T.M., Abbott, G.D., Sleeman, R., Maxwell, J.R., 1989. Laboratory thermal alteration of triaromatic steroid hydrocarbons. Organic Geochemistry 14, 109–111.

Bennett, B., Adams, J.J., Gray, N.D., Sherry, A., Oldenburg, T.B.P., Huang, H., Larter, S.R., Head, I.M., 2013. The controls on the composition of biodegraded oils in the deep subsurface – Part 3. The impact of microorganism distribution on petroleum geochemical gradients in biodegraded petroleum reservoirs. Organic Geochemistry 56, 94–105.

Berner, U., Faber, E., 1988. Maturity related mixing model for methane, ethane and propane, based on carbon isotopes. Organic Geochemistry 13, 67–72.

Bissada, K.K., Elrod, L.W., Darnell, L.M., Szymczyk, H.M., Trostle, J.L., 1992. Geochemical inversion – a modern approach to inferring source-rock identity from characteristics of accumulated oil and gas. In: Proceedings of the 21st. Annual Convention of the Indonesian Petroleum Association, vol. 1, pp. 165–199.

Brassell, S.C., Guoying, S., Jiamo, F., Eglinton, G., 1988. Biological markers in lacustrine Chinese oil shales. In: Fleet, A.J., Kelts, K., Talbot, M.R. (Eds.), Lacustrine Petroleum Source Rocks. Blackwell, Oxford, pp. 299–308.

Calhoun, J., 1953. Fundamentals of Reservoir Engineering. University of Oklahoma Press, Norman. 417 p.

Connan, J., Bouroullec, J., Dessort, D., Albrecht, P., 1986. The microbial input in carbonate-anhydrite facies of a sabkha palaeoenvironment from Guatemala: a molecular approach. In: Leythaeuser, D., Rullkrtter, J. (Eds.), Advances in Organic Geochemistry 1985. Pergamon Press, Oxford, pp. 29–50.

Cornford, C., Morrow, J.A., Turlington, A., Miles, J.A., Brooks, J., 1983. Some geological controls on oil composition in the U.K. North Sea. In: Brooks, J. (Ed.), Petroleum, Geochemistry and Exploration of Europe. The Geological Society of London Special Publication 12. Blackwell Scientific, Oxford, pp. 35–50.

Curiale, J.A., Cameron, D., Davis, D.V., 1985. Biological marker distribution and significance in oils and rocks of the Monterey Formation, California. Geochimica et Cosmochimica Acta 49, 271–288.

Curiale, J.A., 1987. Distribution and occurrence of metals in heavy crude oils and solid bitumens–implications for petroleum exploration. In: Meyer, R.F. (Ed.), Exploration for Heavy Crude Oil and Natural BitumensAmerican Association of Petroleum Geologists Studies in Geology, vol. 25, pp. 207–219.

Curiale, J.A., 2002. A review of the occurrences and causes of migration–contamination in crude oil. Organic Geochemistry 33, 1389–1400.

Curiale, J.A., 2008. Oil-source rock correlations – limitations and recommendations. Organic Geochemistry 39, 1150–1161.

Dahl, J.E., Moldowan, J.M., Peters, K.E., Claypool, G.E., Rooney, M.A., Michael, G.E., Mello, M.R., Kohnen, M.L., 1999. Diamondoid hydrocarbons as indicators of natural oil cracking. Nature 399, 54–57.

de Leeuw, J.W., Sinninghe Damste, J.S., 1990. Organic sulphur compounds and other biomarkers as indicators of palaeosalinity. In: Orr, W.L., White, C.M. (Eds.), Geochemistry of Sulfur in Fossil FuelsAmerican Chemical Society Symposium Series, vol. 429, pp. 417–443.

Derenne, S., Largeau, C., Casadevall, E., Connan, J., 1988. Comparison of torbanites of various origins and evolutionary stages. Bacterial contribution to their formation. Causes of the lack of botryococcane in bitumens. Organic Geochemistry 12, 43–59.

Deroo, G., 1976. Correlations between crude oil and source rocks on the scale of sedimentary basins: research Centre Pau Societe Nationale des Petrole d'Aquitaine. Bulletin 10, 317–335.

Dessort, D., Poirier, Y., Sermondadez, G., Levache, D., 2003. Methane generation during biodegradation of crude oil. In: Abstracts of the 21st International Meeting on Organic Geochemistry. Krakow, Poland, September 2003.

Ekweozor, C.M., Udo, T.O., 1988. The oleananes: origin, maturation and limits of occurrence in southern Nigeria's sedimentary basins. Organic Geochemistry 13, 131–140.

Erdman, J.G., Morris, D.A., 1974. Geochemical correlation of petroleum. American Association of Petroleum Geologists Bulletin 58, 2326–2337.

Faber, E., 1987. Zur Isotopengeochemie gasformiger Kohlenwasserstoffe. Erdol, Erdgas & Kohle 103, 210–218.

Fowler, M.G., Brooks, P.W., 1990. Organic geochemistry as an aid in the interpretation of the history of oil migration into different reservoirs at the Hibernia K-18 and Ben Nevis I-45 wells, Jeanne d'Arc Basin, offshore eastern Canada. Organic Geochemistry 16, 461–475.

Fowler, M.G., Douglas, A.G., 1987. Saturated hydrocarbon biomarkers in oils of late Precambrian age from eastern Siberia. Organic Geochemistry 11, 201–213.

Freyss, H., Guieze, P., Varotsis, N., Khakoo, A., Lestelle, K., Simper, D., 1989. PVT analysis for oil reservoirs. Oilfield Review 37 (1), 4–15.

Grabowski, A., Nercessian, O., Fayolle, F., Blanchet, D., Jeanthon, C., 2005. Microbial diversity in production waters of a low-temperature biodegraded oil reservoir. FEMS Microbiology Ecology 54, 427–443.

Gransch, J.A., Posthuma, J., 1974. On the origin of sulphur in crudes. In: Tissot, B., Bienner, F. (Eds.), Advances in Organic Geochemistry 1973. Editions Technip, Paris, pp. 727–739.

Grantham, P.J., Posthuma, J., DeGroot, K., 1980. Variation and significance of the C27 and C28 triterpane content of a North Sea core and various North Sea crude oils. In: Douglas, A.G., Maxwell, J.R. (Eds.), Advances in Organic Geochemistry 1979. Pergamon, Oxford, pp. 29–38.

Head, I.M., Jones, D.M., Larter, S.R., 2003. Biological activity in the deep subsurface and the origin of heavy oil. Nature 426, 344–352.

Heydari, E., 1997. The role of burial diagenesis in hydrocarbon destruction and H_2S accumulation, Upper Jurassic Smackover Formation, Black Creek field, Mississippi. American Association of Petroleum Geologists Bulletin 81, 26–45.

Hitchon, B., Filby, R.H., 1984. Use of trace elements for classification of crude oils into families - example from Alberta, Canada. American Association of Petroleum Geologists Bulletin 68, 838–849.

Huang, W.-Y., Meinschein, W.G., 1976. Sterols as source indicators of organic materials in sediment. Geochimica et Cosmochimica Acta 40, 323–330.

Huang, W.-Y., Meinschein, W.G., 1979. Sterols as ecological indicators. Geochimica et Cosmochimica Acta 43, 739–745.

Hughes, W.B., 1984. Use of thiophenic organosulfur compounds in characterizing crude oils derived from carbonate versus siliciclastic sources. In: American Association of Petroleum Geologists Studies in Geology 18: Petroleum Geochemistry and Source Rock Potential of Carbonate Rocks, pp. 181–196.

Hughes, W.B., Dzou, L.I.P., 1995. Reservoir overprinting of crude oils. Organic Geochemistry 23, 905–914.

Hunt, J.M., 1996. Petroleum Geochemistry and Geology, second ed. W. H. Freeman, New York. 743 pp.

James, A.T., 1983. Correlation of natural gas by use of carbon isotopic distribution between hydrocarbon components. American Association of Petroleum Geologists Bulletin 67, 1176–1191.

James, A.T., Burns, B.J., 1984. Microbial alteration of subsurface natural gas accumulations. American Association of Petroleum Geologists Bulletin 68, 957–960.

Jeffrey, A.W., Alimi, H.M., Jenden, P.D., 1991. Geochemistry of Los Angeles basin oil and gas systems. In: Biddle, K.T. (Ed.), Active Margin Basins, Tulsa, OklaAmerican Association of Petroleum Geologists Memoir, vol. 52, pp. 197–219.

Jones, D.M., Head, I.M., Gray, N.D., Adams, J.J., Rowan, A.K., Aitken, C.M., Bennett, B., Huang, H., Brown, A., Bowler, B.F.J., Oldenburg, T., Erdmann, M., Larter, S.R., 2008. Crude-oil biodegradation via methanogenesis in subsurface petroleum reservoirs. Nature 451, 176–181.

Katz, B.J., Breaux, T.M., Colling, E.L., Darnell, L.M., Elrod, L.W., Jorjorian, T., Royle, R.A., Robison, V.D., Szymczyk, H.M., Trostle, J.L., Wicks, J.P., 1993. Implications of stratigraphic variability of source rocks. In: Katz, B.J., Pratt, L.M. (Eds.), Source Rocks in a Sequence Stratigraphic FrameworkAmerican Association of Petroleum Geologists Studies in Geology, vol. 37, pp. 5–16.

Katz, B.J., Elrod, L.W., 1983. Organic geochemistry of DSDP 467, offshore California, Middle Miocene to lower Pliocene strata. Geochimica et Cosmochimica Acta 47, 267–275.

Lafargue, E., Barker, C., 1988. Effect of water washing on crude oil compositions. American Association of Petroleum Geologists Bulletin 72, 263–276.

Landa, S., Machacek, V., Mzourek, M., Landa, M., 1933. "Title unknown", Chimica e lIndustria Special Publication. In: (Abstracts of the 12th Conference of Industrial Chemistry, Prague, Sept. 1932), vol. 506, p. 5949 Reprinted in Chemical Abstracts, 1933, vol. 27.

Larter, S., Huang, H., Adams, J., Bennett, B., Jokanola, O., Oldenburg, T., Jones, M., Head, I., Riediger, C., Fowler, M., 2006. The controls on the composition of biodegraded oils in the deep subsurface: Part II-geological controls on subsurface biodegradation fluxes and constraints on reservoir-fluid property prediction. American Association of Petroleum Geologists Bulletin 90, 921–938.

Leythaeuser, D., Hollerbach, A., Hagemann, H.W., 1977. Source rock/crude oil correlation based on distribution of C_{27+}-cyclic hydrocarbons. In: Campos, R., Goni, J. (Eds.), Advances in Organic Geochemistry, 1975: Madrid, Enadimsa, pp. 3–20.

Lillis, P.G., Warden, A., Claypool, G.E., Magoon, L.B., 2007. Petroleum systems of the San Joaquin basin province – geochemical characteristics of gas types. In: Hosford Scheirer, A. (Ed.), Petroleum Systems and Geologic Assessment of Oil and Gas in the San Joaquin Basin Province, California, US Geological Survey Professional Paper 1713 (Chapter 10), 30 p.

Losh, S., Swart, D., Dickinson, A., 2009. Gas washing pattern and economics in an area of continental shelf, offshore Louisiana. Gulf Coast Association of Geological Societies Transactions 59, 477–484.

Losh, S., Walter, L., Meulbroek, P., Martini, A., Cathles, L., Whelan, J., 2002. Reservoir fluids and their migration into the South Eugene Island Block 330 reservoirs, offshore Louisiana. American Association of Petroleum Geologists Bulletin 86, 1463–1488.

Mansoori, G.A., 2007. Diamondoid molecules. Advances in Chemical Physics 136, 207–258.

Marchand, A.P., 2003. Diamondoid hydrocarbons–delving into nature's bounty. Science 299, 52–53.

McCaffrey, M.A., Dahl, J.E., Sundararaman, P., Moldowan, J.M., Schoell, M., 1994. Source rock quality determination from oil biomarkers II–A case study using Tertiary-reservoired Beaufort Sea Oils. American Association of Petroleum Geologists Bulletin 78, 1527–1540.

McCain Jr., W.D., 1990. The Properties of Petroleum Fluids, second ed. PennWell, Tulsa, OK. 548 p.

McKirdy, D.M., Cox, R.E., Volkman, J.K., Howell, V.J., 1986. Botryococcane in a new class of Australian non-marine crude oils. Nature 320, 57–59.

Milkov, A.V., 2010. Methanogenic biodegradation of petroleum in the West Siberian basin (Russia): significance for formation of giant Cenomanian gas pools. American Association of Petroleum Geologists Bulletin 94, 1485–1541.

Milkov, A.V., 2011. Worldwide distribution and significance of secondary microbial methane formed during petroleum biodegradation in conventional reservoirs. Organic Geochemistry 42, 184–207.

Milkov, A.V., Dzou, L., 2007. Geochemical evidence of secondary microbial methane from very slight biodegradation of undersaturated oil in a deep hot reservoir. Geology 35, 455–458.

Mille, G., Almallah, M., Bianchi, M., van Wambeke, F., Bertrand, J.C., 1991. Effect of salinity on petroleum biodegradation. Fresenius' Journal of Analytical Chemistry 339, 788–791.

Milner, C.W.D., Rogers, M.A., Evans, C.R., 1977. Petroleum transformations in reservoirs. Journal of Geochemical Exploration 7, 101–153.

Mello, M.R., Telnaes, N., Gaglianone, E.C., Chicarelli, M.I., Brassell, S.C., Maxwell, J.R., 1988. Organic geochemical characterisation of depositional palaeoenvironments of source rocks and oils in Brazilian marginal basins. In: Mattavelli, L., Novelli, L. (Eds.), Advances in Organic Geochemistry 1987. Pergamon Press, Oxford, pp. 31–45.

Moldowan, J.M., Huizinga, B.J., Dahl, J.E., Fago, F.J., Taylor, D.W., Hickey, L.J., 1994. The molecular fossil record of oleanane and its relationship to angiosperms. Science 265, 768–771.

Moldowan, J.M., Seifert, W.K., 1980. First discovery of botryococcane in petroleum. Journal of the Chemical Society, Chemical Communications 912–914.

Moldowan, J.M., Seifert, W.K., Gallegos, E.J., 1985. Relationship between petroleum composition and depositional environment of petroleum source rocks. American Association of Petroleum Geologists Bulletin 69, 1255–1268.

Orr, W.L., 1974. Changes in sulfur content and isotopic ratios of sulfur during petroleum maturation—study of Big Horn Paleozoic oils. American Association of Petroleum Geologists Bulletin 58, 2295–2318.

Orr, W.L., 1977. Geologic and geochemical controls on the distribution of hydrogen sulfide in natural gas. In: Campos, R., Goñi, J. (Eds.), Advances in Organic Geochemistry 1975, Empresa Nacional Adaro De Investigaciones Mineras, Madrid, pp. 571–597.

Pallasser, R.J., 2000. Recognising biodegradation in gas/oil accumulations through the $\delta^{13}C$ compositions of gas components. Organic Geochemistry 31, 1363–1373.

Palmer, S.E., 1984. Effect of water washing on C_{15}+ hydrocarbon fraction of crude oils from Northwest Palawan, Philippines. The American Association of Petroleum Geologists Bulletin 68, 137–149.

Pasadakisa, N., Obermajerb, M., Osadetz, K.G., 2004. Definition and characterization of petroleum compositional families in Williston Basin, North America using principal component analysis. Organic Geochemistry 35, 453–468.

Peters, K.E., Moldowan, J.M., 1993. The Biomarker Guide, first ed. Prentice Hall, Englewood Cliffs, New Jersey. 363 p.

Peters, K.E., Walters, C.C., Moldowan, J.M., 2005a. The Biomarker Guide, second ed. Biomarkers and Isotopes in the Environment and Human History, vol. 1. Cambridge University Press, Cambridge. 471 p.

Peters, K.E., Walters, C.C., Moldowan, J.M., 2005b. The Biomarker Guide, second ed. Biomarkers and Isotopes in Petroleum Exploration and Earth History, vol. 2. Cambridge University Press, Cambridge. 683 p.

Peters, K.E., Hostettler, F.D., Lorenson, T.D., Rosenbauer, R.J., 2008a. Families of Miocene Monterey crude oil, seep, and tarball samples, coastal California. American Association of Petroleum Geologists Bulletin 92, 1131–1152.

Peters, K.E., Ramos, L.S., Zumberge, J.E., Valin, Z.C., Bird, K.J., 2008b. De-convoluting mixed crude oil in Prudhoe Bay field, North Slope, Alaska. Organic Geochemistry 39, 623–645.

Philp, R.P., Gilbert, T.D., 1982. Unusual distribution of biological markers in Australian crude oil. Nature 299, 245–247.

Radke, M., Welte, D.H., 1983. The methylphenanthrene index (MPI): a maturity parameter based on aromatic hydrocarbons. In: Bjorøy, M., Albrecht, P., Cornford, C., de Groot, K., Eglinton, G., Galimov, E., Leythaeuser, D., Pelet, R., Rullkötter, J., Speers, G. (Eds.), Advances in Organic Geochemistry 1981. John Wiley and Sons, New York, pp. 504–512.

Rice, D.D., 1992. Controls, habitat, resource potential of ancient bacterial gas. In: Vitaly, R. (Ed.), Bacterial Gas, Editions Technip, pp. 91–118.

Rice, D.D., Claypool, G.E., 1981. Generation, accumulation, and resource potential of biogenic gas. American Association of Petroleum Geologists Bulletin 65, 5–25.

Riva, A., Caccialanza, P.G., Quagliaroli, F., 1988. Recognition of 18β(H)-oleanane in several crudes and Cainozoic-Upper Cretaceous sediments. Definition of a New Maturity Parameter: Organic Geochemistry 13, 671–675.

Rubinstein, I., Sieskind, O., Albrecht, P., 1975. Rearranged sterenes in a shale: occurrence and simulated formation. Journal of the Chemical Society, Perkin Transactions 1, 1833–1836.

Sassen, R., Moore, C.H., 1988. Framework of hydrocarbon generation and destruction in eastern Smackover trend. American Association of Petroleum Geologists Bulletin 72, 649–663.

Sassen, R., Wade, W.J., 1994. Alteration of Liquid Hydrocarbons Associated with Thermochemical Sulfate Reduction. Western Canadian and International Expertise Exploration Update '94 Program book with Expanded Abstracts, Calgary, May 9–12. , p. 197.

Schoell, M., 1983. Genetic characterization of natural gases. American Association of Petroleum Geologists Bulletin 67, 2225–2238.

Schoell, M., McCaffrey, M.A., Fago, F.J., Moldowan, J.M., 1992. Carbon isotopic compositions of 28,30- bisnorhopanes and other biological markers in a Monterey crude oil. Geochimica et Cosmochmica Acta 56, 1391–1399.

Schou, L., Eggen, S., Schoell, M., 1985. Oil-oil and oil-source rock correlation, Northern North Sea. In: Thomas, B.M. (Ed.), Petroleum Geochemistry in Exploration of the Norwegian Shelf, pp. 101–117.

Seifert, W.K., 1977. Source rock/oil correlations by C27-C30 biological marker hydrocarbons. In: Campos, R., Goni, J. (Eds.), Advances in Organic Geochemistry, 1975: Madrid, Enadimsa, pp. 21–44.

Seifert, W.K., Moldowan, J.M., 1978. Application of steranes, terpanes and monoaromatics to the maturation, migration and source of crude oils. Geochimica et Cosmochimica Acta 42, 77–79.

Seifert, W.K., Moldowan, J.M., 1981. Paleoreconstruction by biological markers. Geochimica et Cosmochimica Acta 45, 783–794.

Seifert, W.K., Moldowan, J.M., 1986. Use of biological markers in petroleum exploration. In: Johns, R.B. (Ed.), Methods in Geochemistry and Geophysics 24. Elsevier, Amsterdam, pp. 261–290.

Sofer, Z., 1984. Stable carbon isotope compositions of crude oils: application to source depositional environments and petroleum alteration. American Association of Petroleum Geologists Bulletin 68, 31–49.

Sofer, Z., Zumberge, J.E., Lay, V., 1986. Stable carbon isotopes and biomarkers as tools in understanding genetic relationship, maturation, biodegradation and migration of crude oils in the northern Peruvian Oriente (Maranon) basin. Organic Geochemistry 10, 377–389.

Stahl, W.J., 1977. Carbon and nitrogen isotopes in hydrocarbon research and exploration. Chemical Geology 20, 121–149.

Stahl, W.J., 1978. Source rock-crude oil correlation by isotopic type-curves. Geochimica et Cosmochimica Acta 42, 1573–1577.

Stahl, W.J., Cary Jr., B.D., 1975. Source-rock identification by isotope analysis of natural gases from fields in the Val Verde and Delaware basins, west Texas. Chemical Geology 16, 257–267.

Stahl, W., Koch, J., 1974. $^{13}C/^{12}C$-Verhalltnis nordeutscher Erdgase - Reifemerkmal ihrer Muttersubstanzen. Erdoel Kohle, Erdgas, Petrochemie 27, 10–36.

Telnaes, N., Dahl, B., 1986. Oil-oil correlation using multivariate techniques. Organic Geochemistry 425–432.

ten Haven, H.L., de Leeuw, J.W., Sinninghe Damste, J.S., Schenck, P.A., Palmer, S.E., Zumberge, J.E., 1988. Application of biological markers in the recognition of paleo-hypersaline environments. In: Fleet, A.J., Kelts, K., Talbot, M.R. (Eds.), Lacustrine Petroleum Source Rocks. The Geological Society, London, Blackwell, Oxford, pp. 123–130.

Thompson, K.F.M., 2010. Aspects of petroleum basin evolution due to gas advection and evaporative fractionation. Organic Geochemistry 41, 370–385.

Treibs, A., 1934. The occurrence of chlorophyll derivatives in an oil shale of the upper Triassic. Annalen 517, 103–114.

Volkman, J.K., 1986. A review of sterol biomarkers for marine and terrigenous organic matter. Organic Geochemistry 9, 83–99.

Wade, W.J., Sassen, R., 1994. Controls on H_2S Content and Thermochemical Sulphate Reduction in Smackover Reservoirs, Alabama and Arkansas. Western Canadian and International Expertise Exploration Update '94 Program book with expanded abstracts, Calgary, May 9–12, p. 198.

Waples, D.W., Machihara, T., 1990. Application of sterane and triterpane biomarkers in petroleum exploration. Bulletin of Canadian Petroleum Geology 38, 357–380.

Waples, D.W., Machihara, T., 1991. Biomarkers for geologists–a practical guide to the application of steranes and triterpanes. In: Petroleum GeologyAmerican Association of Petroleum Geologists Methods in Exploration, 9, 91 pp.

Welte, D.H., Hagemann, H.W., Hollerbach, A., Leythaeuser, D., 1975. Correlation between petroleum and source rock. In: 9th World Petroleum Congress, vol. 2, pp. 179–191.

Wenger, L.M., Davis, C.L., Isaksen, G.H., 2002. Multiple controls on petroleum biodegradation and impact on oil quality. Society of Petroleum Engineers Reservoir Evaluation and Engineering 5, 375–383.

Whiticar, M., 1994. Correlation of natural gases with their sources. In: American Association of Petroleum Geologists Memoir 60: The Petroleum System–from Source to Trap, pp. 261–283.

Whitson, C.H., 1992. Petroleum reservoir fluid properties: Part 10. Reservoir engineering methods. In: Morton-Thompson, D., Woods, A.M. (Eds.), American Association of Petroleum Geologists Methods in Exploration Series No. 10: Development Geology Reference Manual, pp. 504–507.

Williams, J.A., 1974. Characterization of oil types in Williston basin. American Association of Petroleum Geologists Bulletin 58, 1243–1252.

Williams, L.A., 1984. Subtidal stromatolites in Monterey formation and other organic-rich rocks as suggested source contributors to petroleum formation. American Association of Petroleum Geologists Bulletin 1879–1893.

Worden, R.H., Smalley, P.C., Oxtoby, N.H., 1995. Gas souring by thermochemical sulfate reduction at 140°C. American Association of Petroleum Geologists Bulletin 79, 854–863.

Yue, C., Li, S., Song, H., 2014. Thermochemical sulfate reduction simulation experiments on the formation and distribution of organic sulfur compounds in the Tuha crude oil. Bulletin of the Korean Chemistry Society 35, 2057–2064.

Zengler, K., Richnow, H.H., Rossello-Mora, R., Michaelis, W., Widdel, F., 1999. Methane formation from long-chain alkanes by anaerobic microorganisms. Nature 401, 266–269.

Zumberge, J.E., 1984. Source rocks of the La Luna formation (upper Cretaceous) in the Middle Magdalena valley, Colombia. In: Palacas, J.G. (Ed.), Geochemistry and Source Rock Potential of Carbonate Rocks American Association of Petroleum Geologists Studies in Geology No. 18, pp. 127–133.

CHAPTER

5

Reservoir Geochemistry

OUTLINE

Introduction 189

Pay Zone Detection 190

Mud Gas Analysis 191

Mud Gas Data Interpretation 192

Fluorescence and Cut 195

Isotubes 196

Rock-Eval Pyrolysis 197

Solvent Extraction/Gas Chromatography 197

Thermal Extraction–Gas Chromatography 198

High-Molecular Weight Waxes 199

Asphaltenes 200

Reservoir Continuity 202

Reservoir Continuity Using Mud Gas Data 203

Reservoir Continuity Using Oil Samples 203

Reservoir Continuity Using Gas Samples 206

Production Allocation 207

Production Problems and Periodic Sampling 208

Monitoring Enhanced Oil Recovery 210

Reservoir Souring 212

Strategies in Reservoir Geochemistry 212

References 213

INTRODUCTION

The role of petroleum geochemistry was once thought to be solely to support exploration efforts. However, the concepts, tools, and methods that have been developed for evaluating source rocks, crude oils, and natural gases are applicable to many problems in petroleum production and field development. These applications broadly fall into two main groups: reservoir appraisal and production applications. While reservoir appraisal is mainly concerned with the identification of shows and pay zones, the preliminary assessment of oil column quality, and determining fluid contacts, production applications deal more with the production problems and field development issues. These include organic deposition issues, reservoir continuity questions, production monitoring, commingling problems, production allocation, monitoring enhanced oil recovery programs, and reservoir souring.

Practical Petroleum Geochemistry for Exploration and Production
http://dx.doi.org/10.1016/B978-0-12-803350-0.00005-2

When dealing with production problems, time is of the essence, with any loss or slow down in production resulting in loss of revenue. Although many of these problems can be addressed with conventional petroleum engineering techniques, these techniques typically require interruptions in production. The geochemical approach to these production problems can often be much less expensive and require less time to execute.

The tools for reservoir geochemistry include familiar geochemical methods such as Rock-Eval pyrolysis, whole oil gas chromatography, biomarker and isotopic analysis of oils, and compositional and isotopic analysis of natural gases. Sometimes these methods are used in ways already described, while other times, they are applied in a unique way to solve a particular problem. In addition, information from mud gas logging, Isotubes, and thermal extraction–gas chromatography (TEGC) is used to supplement the more conventional geochemical techniques to solve problems.

The following discussion will cover the main issues in reservoir geochemistry and provide some examples of how these problems can be solved. It will show the value of using petroleum geochemistry in reservoir management and field development to find quick solutions to these problems and help restore lost production.

PAY ZONE DETECTION

Obviously, finding producible petroleum is the key to a successful commercial exploration and production program. But it is not always easy to recognize intervals of subsurface hydrocarbons and assess if they are producible or not. While techniques such as petrophysical analysis of wireline log data are often the standard for this task, some difficult interpretation problems can be encountered, such as low-resistivity pay or a fractured reservoir, that make finding producible hydrocarbons more challenging. Geochemical techniques for pay zone detection are often better in these situations, as well as being good supplemental tools to help find or corroborate the presence of more typical reservoirs.

To start this discussion, it is useful to define what we are looking for. A pay zone is an interval of reservoir rock which contains oil and gas in exploitable quantities. Anything less is called a show, a noncommercial quantity of oil or gas encountered while drilling. The difference between exploitable, or commercial, quantities and noncommercial quantities of oil and gas is dictated by local economic conditions. For example, a well that can produce 200 barrels of oil a day may be a success onshore, but in a remote offshore area a well capable of 2000 barrels/day may be considered only a show. So it is important to understand the local economic thresholds when evaluating well results.

While not as economically important, shows are significant sources of information in an exploration program. They can point toward a potential petroleum system and can provide clues to the nature of active source rocks in the area. However, there is some ambiguity in the definition of a show. By defining a show as any noncommercial quantity of oil or gas encountered while drilling, it could range in size from something just below the commercial/economic limit to mere traces of petroleum. Schowalter and Hess (1982) addressed this problem by proposing a classification for shows. They recommended four types of shows be recognized. The first is the *continuous phase oil or gas* show consisting of a filament of oil or gas with a continuous connection through the pore network of a water-saturated porous rock. This is the easily observable oil staining or saturation that most would readily recognize as a show. The second is the *isolated droplets of oil or gas* that likely represents water displaced residual hydrocarbons remaining in the pores of a rock, such as grain coatings or bitumen-filled pores. It is visually observable but not widespread in the rock. The third is *dissolved hydrocarbons*, molecular scale dissolved or dispersed hydrocarbons occurring in solution in pore fluids or sorbed on the rock framework. And the fourth is hydrocarbons *associated with kerogen* consisting of any soluble organic material, bitumen, associated with kerogen in a potential source rock. This bitumen may already exist in the source rock or may be created by the heating effect of the drill bit or heating during sample preparation (sample heating in ovens or retorts). In terms of petroleum exploration and production, all four types of shows have some value, but only the *continuous phase oil or gas* show is a significant indicator of a potential working petroleum system and nearby producible hydrocarbons.

To have an impact of a drilling program, the detection of pay zones and significant shows needs to be real time. This is accomplished in two ways, one using logging while drilling or measurement while drilling petrophysical tools and the other using geochemical data from mud gas logging. In terms of geochemical applications to petroleum exploration and production, mud gas logging is a frontline technique. It detects gaseous hydrocarbons from C_1–C_5 (sometimes up to C_8) released from sediments by the drill bit and carried to the surface with the returning drilling mud. This information can then be used to detect the presence of hydrocarbons in the subsurface, distinguish between gas and oil, indicate fluid contacts, and sometimes assess producibility.

There are also some ancillary techniques that can assist in detecting pay zones and significant shows. A few, such as fluorescence and cut, are routinely performed at the well site and provide real-time data. Others like Isotube analysis, Rock-Eval pyrolysis, solvent extraction/gas chromatography, and TEGC are laboratory-based analyses and can provide supplemental information to confirm petrophysical and mud gas observations. These methods can also be used to insure that potential pay intervals have not been overlooked using the real-time data.

Mud Gas Analysis

Hydrocarbon mud logging began commercially in 1939 with the simple detection of gas released into the mud, drilling mud by the drill bit (Hunt, 1996). These early systems used thermal conductivity, or "hot wire," detectors that were bulk gas detectors and were not able to distinguish between hydrocarbon and nonhydrocarbon gases. Mud gas logging did not begin to develop as a real geochemical tool until it gained the ability to separate and identify the individual component gases with the advent of gas chromatography in the 1970s. It was still limited by the use of thermal conductivity detectors, and there were difficulties separating some of the gases adequately to identify hydrocarbon from nonhydrocarbon. Eventually, flame ionization detectors were used that detected only hydrocarbon gases and with greater sensitivity. These systems provided reliable analysis for the C_1–C_4 hydrocarbons and sometimes extended up to C_5.

While these analytical improvements help provide better data on the gas sampled, problems still existed in the gas traps used to extract the gas from the drilling mud. Most of the traps were placed in a fixed position in the possum belly, as shown in Fig. 5.1. Fluctuating mud levels during drilling would change the volume of the headspace gas extracted by the

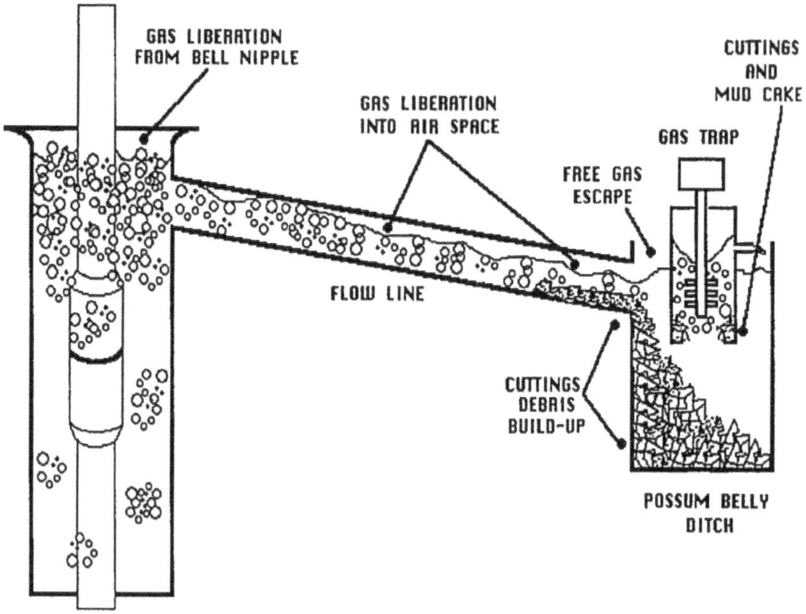

FIGURE 5.1 A schematic of the return mud flow on a drilling rig showing the location of the gas trap. *Whittaker, A., 1991. Mud Logging Handbook. Prentice-Hall, Englewood Cliffs, NJ. 531 p.*

gas trap resulting in inconsistency in the samples. The temperature and pressure also influences the amount of gas released from the drilling mud into the headspace. With the drilling mud temperature in the borehole increasing with increasing depth and the variable influence of the weather on both temperature and pressure, reproducible sampling of the mud gas was very challenging. As a result, these problems made it difficult to obtain consistently comparable mud gas data over the entire borehole.

Recent improvements in gas trap design take samples of drilling mud from the return flow line and pump it to a constant volume and constant temperature mud gas extractor (e.g., Blanc et al., 2003) providing consistent reproducible sampling. This technology has also been coupled in some configurations with a mass spectrometer as the detector for the gas chromatograph to increase the analytical range of the method. With these new systems, it is possible to detect both hydrocarbon and nonhydrocarbon gases in the C_1–C_8 range faster and with more sensitivity. There has also been an effort by several groups to bring real-time carbon isotope analysis of mud gas to the well site that may provide important additional information in the near future.

Mud Gas Data Interpretation

While there have been significant changes in gas trap design and well site analytical tools, the interpretation of mud gas data typically relies on two basic schemes. These two main methods for interpreting mud gas data utilize mud gas ratios on Pixler plots (Pixler, 1969) and/or apply the so-called Haworth mud gas parameters (Haworth et al., 1985).

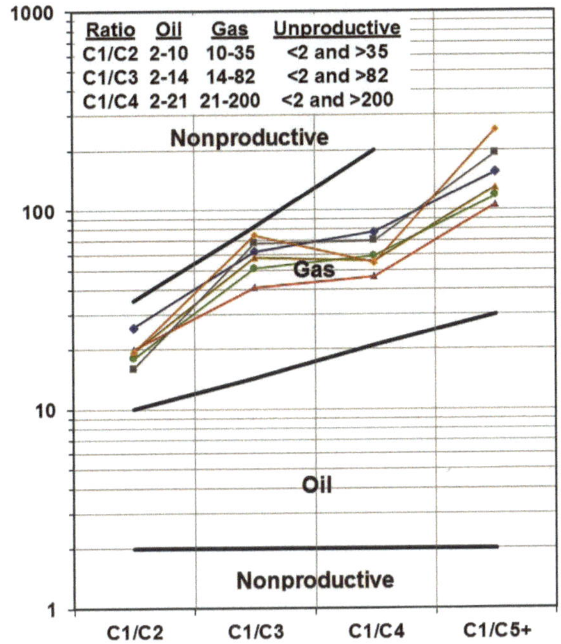

FIGURE 5.2 An example of mud gas interpretation using a standard Pixler plot with ratios calculated and interpretation cutoff values. *Following Pixler, B.O., 1969. Formation evaluation by analysis of hydrocarbon ratios. Journal of Petroleum Technology 24, 665–670.*

Pixler (1969) uses the C_1/C_2, $C1/C_3$, C_1/C_4, and C_1/C_5 ratios calculated from the gas chromatographic analysis of mud gas for interpreting reservoir contents. In the C_1/C_4 ratios, the C_4 includes both n-butane and isobutane, while in the C_1/C_5 ratio, the C_5 includes the n-pentane, isopentane, and neopentane. When the pentanes are not present in the mud gas data or the well site analytical equipment does not provide pentane data, the ratio $(10 \times C_2)/C_3$ can be substituted for C_1/C_5 ratio (Whittaker, 1991).

These ratios are plotted on diagrams, as shown in Fig. 5.2, that have come to be known as Pixler plots. The individual ratios are plotted for each mud gas sampling point in a reservoir interval and are connected by a line. The nonproductive area at the bottom of the diagram usually represents residual oil in the reservoir, while the nonproductive area at the top of the diagram is noncommercial gas (Whittaker, 1991). Pixler (1969) cautions that if the C1/C2 ratio is low in the oil and the C1/C4 ratio is high in the gas section for the zone or if any ratio is lower than a preceding ratio, the zone is probably nonproductive. Productive dry gas zones may show only C_1, but abnormally high shows of C_1 only are usually indicative of salt water. This approach is also not definitive for low-permeability zones.

A more complex approach to mud gas data interpretation is the use of the Haworth mud gas parameters (Haworth et al., 1985). This approach utilizes C_1–C_5 hydrocarbon mud gas data to calculate three parameters: wetness, balance, and character, as shown in Fig. 5.3. The wetness is a standard % wet gas calculation extended out to C_5, while the balance is a ratio comparing the lighter gases, C_1 and C_2, to the heavier gases, C_3–C_5. The character uses the

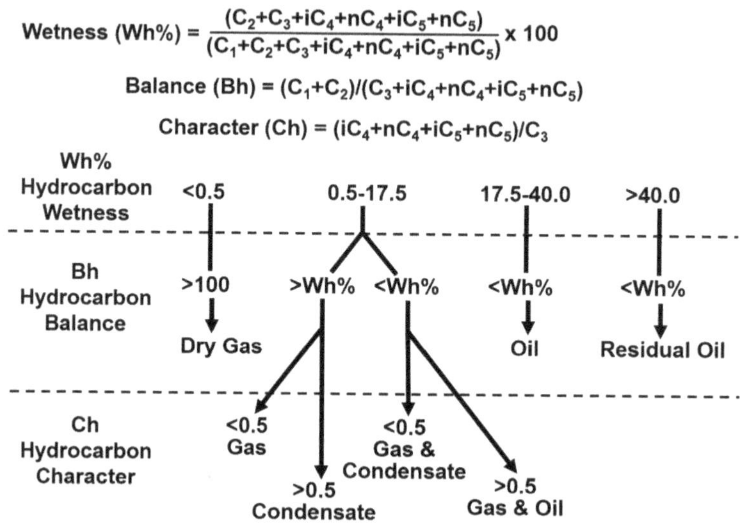

FIGURE 5.3 The equations for the Haworth mud gas parameters and a flow diagram for their interpretation. *Based on Haworth, J.H., Sellens, M., Whittaker, A., 1985. Interpretation of hydrocarbon shows using light C_1–C_5 hydrocarbon gases from mud-log data. American Association of Petroleum Geologists Bulletin 69, 1305–1310.*

C_3–C_5 components to interpret shows associated with gas-caps, oil/gas contacts, and water-wet zones. The interpretation rules for using these three parameters are summarized in the flow chart in Fig. 5.3. An example of using the Haworth parameters is shown in Fig. 5.4. In this data set, the upper gas zone can be seen to transition through the lower gas zone into the oil zone at the base of the section.

Very often, the Haworth parameters are used in conjunction with the Pixler plots. This utilizes the strengths of both methods to provide a cross-check of the interpretations which should result in a better description of the fluid's character.

When using either the Pixler plots or the Haworth parameters, all the data must be corrected for lag time to place it in a proper depth context. In addition, the gas data need to be corrected for the effects of oil-based muds, background/recycled gas, and connection/trip gas. The total gas in the drilling mud is made up of more than just gas release into the drilling mud from reservoirs. It also contains contributions from the sediments penetrated on the way to the reservoir. This is usually called the background gas. In theory, the gas circulated to the surface should be purged from the drilling mud prior to the mud being pumped back into the borehole. In reality, some of this gas is recycled in the drilling mud and can contribute to the signal. This residual gas is called recycled gas and is most prevalent after drilling through a reservoir interval. Both background gas and recycled gas are more problematic when an oil-based drilling mud is used. The oil in the mud increases the amount of dissolved gas the drilling mud can hold. The oil in the mud can also interfere with the purging of the gas from the mud system.

In addition to background and recycled gas, there are periodic sharp increases in the mud gas signal due to drilling operations. Connection gas is the extra gas that accumulated in the borehole when mud circulation is stopped while adding a new section of drill pipe. The extra

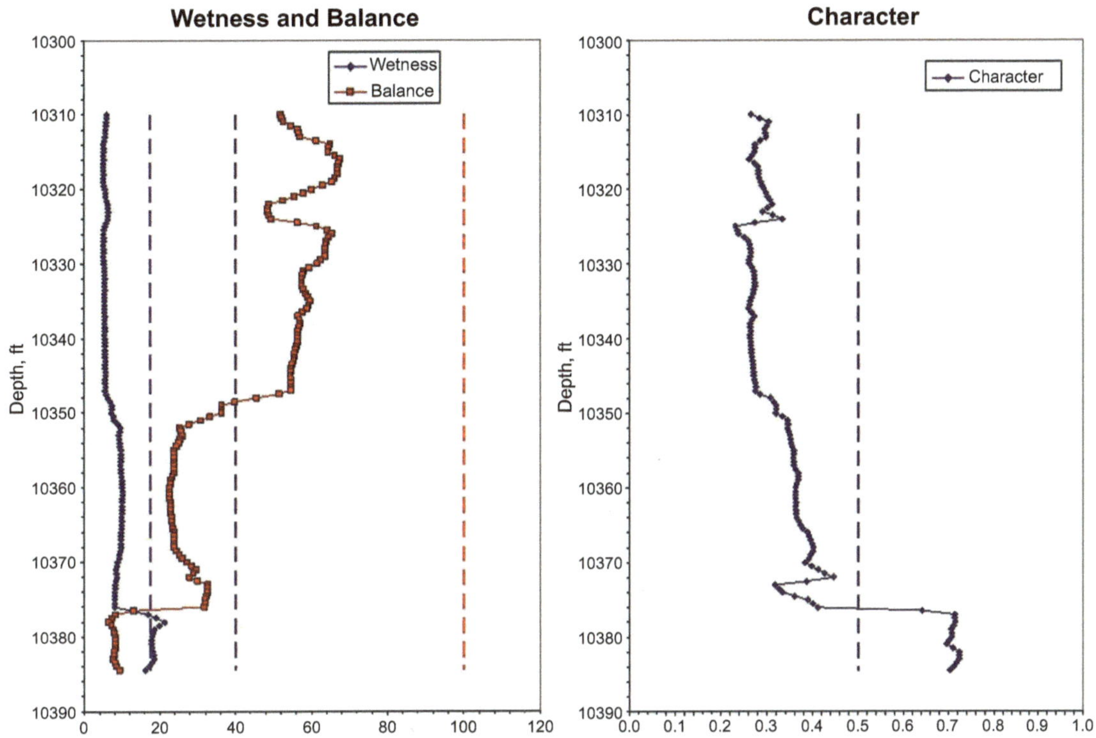

FIGURE 5.4 An example of applying the Haworth mud gas parameters. Gas is indicated above 10,348 ft, between 10,348 and 10,376 ft the parameters indicate gas in proximity to oil, and below 10,376 ft, oil is indicated.

gas can cause a sharp increase in gas above the background when mud circulation resumes. Trip gas is similar to connection gas in that it is the extra gas that accumulates when mud circulation is stopped during a trip. It will also cause a sharp increase in gas above background when mud circulation begins and may also cause a temporary increase in the overall background gas.

When interpreting mud gas data, it is imperative that more than one reading per reservoir intervals be used to make an interpretation. Multiple samplings of the mud gas over the reservoir interval provide corroboration of the signal and can show trends in the data which may indicate changes in the reservoir contents. At the well site for real-time evaluation, this information should be integrated with the lithologic descriptions of the cuttings from the interval and the fluorescence and cut observations before drawing any conclusions about the contents of the reservoir.

Fluorescence and Cut

Fluorescence and cut are part of the visual inspection of the cuttings recovered from the potential reservoir interval. After the cuttings' lithology and grain size have been described, a visual inspection of the cuttings is done for any visible indications of oil staining. At the same time, a check for any odor of hydrocarbons that may be present is usually made. Staining may not be obvious in the cuttings. To look for the more subtle forms of staining, the cuttings should be placed under an ultraviolet light (both long and short wavelengths) to determine if any fluorescence is present. Fluorescence present may be from traces of hydrocarbons in the sample, or it could be from minerals or some contaminant. To help distinguish between fluorescence from hydrocarbons versus minerals or contaminants, the color, intensity, and distribution of the fluorescing material is noted. Oil fluorescence ranges from blue-white to white to yellow to red-brown. Fluorescing minerals are usually intense and related to only specific grains in the cuttings sample. To help distinguish fluorescence of oil from minerals, a few drops of solvent are added to the cuttings while still under UV light to induce a cut or mobilization of the oil from the cuttings. The appearance of the cut, (uniform or streaming), color, intensity, and the rate at which it forms can provide information about the oil's mobility and reservoir permeability. A more detailed discussion of fluorescence and cut can be found in Whittaker (1991).

While fluorescence and cut can provide very useful information about potential shows, there are a few obstacles to universally applying it. First, if the drilling mud has been formulated with diesel or crude oil, the technique cannot be used. In addition, pipe dope and some drilling mud additives may also fluoresce giving false positive indications. And finally cavings or recycled oil in the drilling mud can be observed and be misleading.

Isotubes

Isotubes provide a means of capturing a sample of the mud gas for later isotope analysis. This is done by fitting a sampling manifold in the mud gas flow line between the gas trap and the well site analytical system. This manifold usually resides in the mud logging trailer for ease of operation. The laboratory analysis of the gas in the Isotube provides a composition as well as the carbon and hydrogen/deuterium isotopic signature of the gases, given the individual gases occur in high enough concentration. Although the composition of mud gas is not directly comparable to the composition of reservoir gas due to partitioning between the gas phase and the mud (dissolved gas), the isotopic signature of the mud gas has been found to be comparable to the reservoir gas.

Applying Isotube data to confirm the presence of a hydrocarbon-bearing reservoir consists of using both the composition and the isotopic data plotted versus depth. The amount of gas in each Isotube, as well as its composition, when plotted versus depth provides indications of shows and pay zones. Rapid increases in the amount of gas with depth, especially over a short depth interval, suggest a hydrocarbon-bearing reservoir. Depending on the drilling mud weight and the formation pressure, this increase in gas content may or may not decrease once the drill bits have moved beyond the reservoir. This increase in the amount of gas in the Isotubes is often accompanied by a change in composition in the gas reflecting the differences in the background gas versus the reservoired gas.

When the carbon isotope ratio of methane from the Isotube data is plotted versus depth, there should be a trend of heavier (less negative) methane with increasing depth. This trend reflects the normal evolution of the interstitial gases in the sediment as local organic matter matures with increasing depth and temperature. This trend can be used to estimate maturity using some of the isotope data trends versus vitrinite reflectance discussed in Chapter 4.

Reservoir rocks containing potential shows or pay typically deviate from this trend, usually indicating more mature gas has migrated up and into the reservoir.

The Isotube data are often used in conjunction with wireline logs and mud gas data to define the limits of potential shows and pay zones. It is not used as a standalone tool for finding hydrocarbon-bearing reservoir intervals.

Rock-Eval Pyrolysis

The S1 peak from Rock-Eval pyrolysis can be used to detect potential pay zones through the analysis of cuttings collected over the suspected reservoir internal (Dow and Talukdar, 1991; Jarvie et al., 2001). The ratio of S1/TOC is used to normalize the S1 signal for the amount of organic matter present in the sediment. In hydrocarbon-bearing reservoir rocks, the volatile material that makes up the S1 peak should be substantially larger than the amounts present in even rich source rocks, as shown in the example in Fig. 5.5.

Baskin and Jones (1993) have also shown that Rock-Eval data from cuttings or sidewall cores from the reservoir interval can be used to predict some physical properties of oils, such as API gravity and viscosity, if the oils are derived from the same source. These predictions require calibration with measurements of crude oil physical properties and are best applied when large-scale development of a field is under way.

Solvent Extraction/Gas Chromatography

This technique utilizes a simple solvent extraction and gas chromatographic analysis of cuttings or sidewall cores from suspected reservoir intervals (Baskin et al., 1995; Jarvie et al., 2001) to indicate the presence of reservoired hydrocarbons. A few grams of rock is placed in a vial with a few milliliters of solvent, usually dichloromethane, to extract

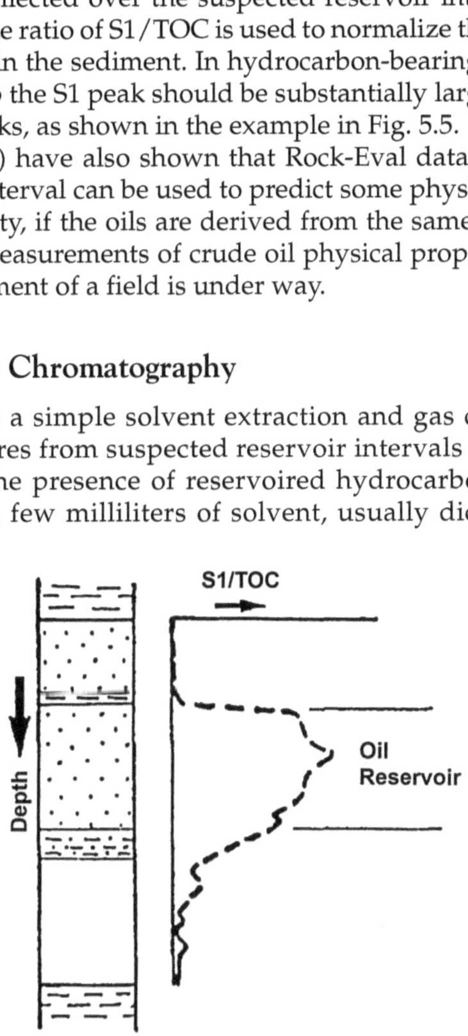

FIGURE 5.5 An example of using Rock-Eval S1 data to indicate the presence of an oil-bearing reservoir. *Modified after Dow, W.G., Talukdar, S.C., January 1991. Petroleum geochemistry in oil production. Houston Geological Society Bulletin 33, 44–52.*

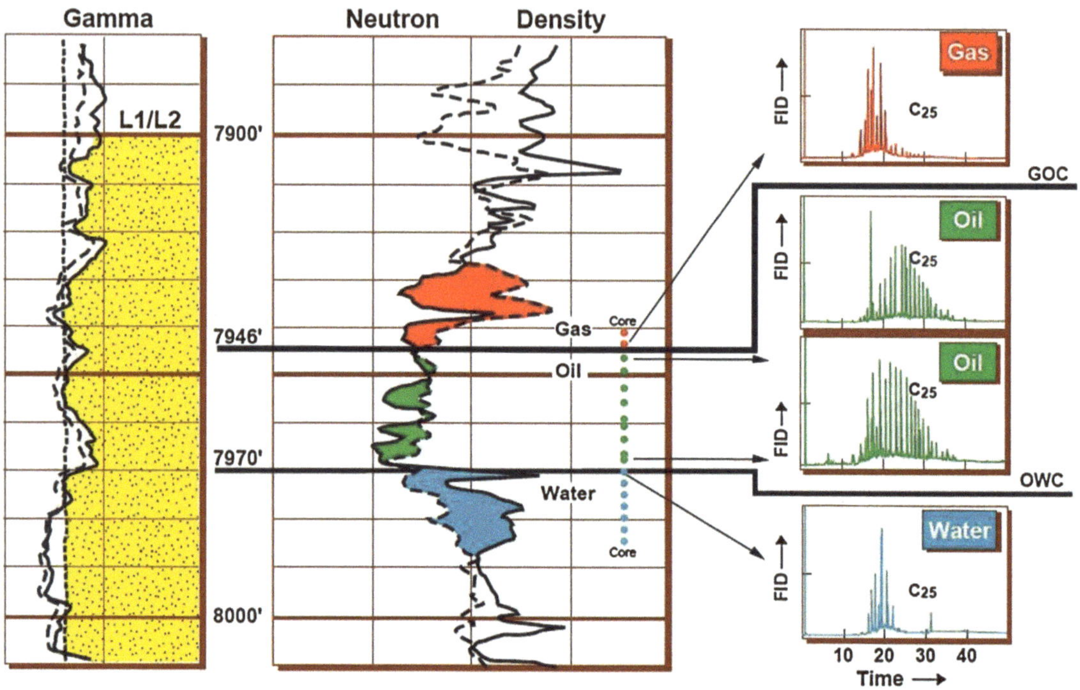

FIGURE 5.6 An example of using whole oil gas chromatogram "fingerprinting" to confirm wireline log interpretations of the contents of selected reservoir and the locations of fluid contacts. *From Baskin, D.K., Hwang, R.J., Purdy, R.K., 1995. Predicting gas, oil and water intervals in Niger Delta reservoirs using gas chromatography. American Association of Petroleum Geologists Bulletin 79, 337–350.*

any hydrocarbons in the rock. The extract is used in a simple whole extract gas chromatographic analysis (described in Chapters 3 and 4) to get a "fingerprint" distribution of the hydrocarbons. As shown in Fig. 5.6, the chromatograms can be used to compare the composition of multiple zones to look for diagnostic characteristics that indicate gas, oil, and/or water in the reservoir rock. This type of data can also be used to recognize compartmentalization, reservoir alteration processes, fluid contacts, bypassed pay, and tar mats. Oil-based muds will interfere with this method by obscuring the signal from the formation. Synthetic oils used in drilling muds may not totally conceal the in situ hydrocarbons.

Thermal Extraction–Gas Chromatography

This technique can be used to obtain data similar to the solvent extraction/gas chromatography method described earlier. Instead of solvent extraction, the cuttings are thermally extracted in a specially designed inlet of a gas chromatograph (see pyrolysis-gas chromatography in Chapter 3). The subsequent gas chromatographic analysis provides a "fingerprint" of the hydrocarbon distribution in the rock similar to the solvent extract, and it can be used in the same fashion.

In addition to rock samples, TEGC has been used with drilling mud samples to detect oil-bearing reservoirs (Dembicki, 1986). C_5+ hydrocarbons from oil-bearing reservoirs are also released into the drilling mud along with C_1–C_4 gases. TEGC allows these heavier hydrocarbons to be detected and analyzed. Application of this technique is limited to wells drilled with water-based drilling mud.

HIGH-MOLECULAR WEIGHT WAXES

High-molecular weight waxes (>C_{40}) in crude oils can cause problems during production. These waxes are held in solution in the crude oil in the subsurface. As the oil is produced and leaves the reservoir, the temperature of the oil can decrease causing the high-molecular weight waxes to drop out of solution. This can result in wax deposition in the near borehole environment, the production lines to the surface, the flow lines from the wellhead, surface storage, and pipelines.

The role of petroleum geochemistry in high-molecular weight wax issues is to provide a means of early recognition of the potential for these problems. This can be accomplished using high-temperature gas chromatography, which allows routine analysis of >C_{40} hydrocarbons present in petroleum (del Rio and Philp, 1992; Carlson et al., 1993). An example of a high-temperature gas chromatogram showing high-molecular weight waxes is shown in Fig. 5.7. Typically, the analysis is done on either the saturate fraction from an oil or the deasphalted whole oil. However, Thanh et al. (1999) demonstrated that the deasphalting of an oil or rock extract could induce the precipitation of both the asphaltenes and the high-molecular weight waxes. If only the saturate fraction or the deasphalted whole oil was analyzed, the presence of these high-molecular weight waxes could be missed. A special separation scheme described by Thanh et al. (1999) is needed to isolate the waxes and get an accurate assessment of potential wax deposition problems.

It may be thought that wax deposition problems are only relevant to waxy oil generated from Type I kerogen in lacustrine depositional settings. However, Carlson et al. (1993) showed

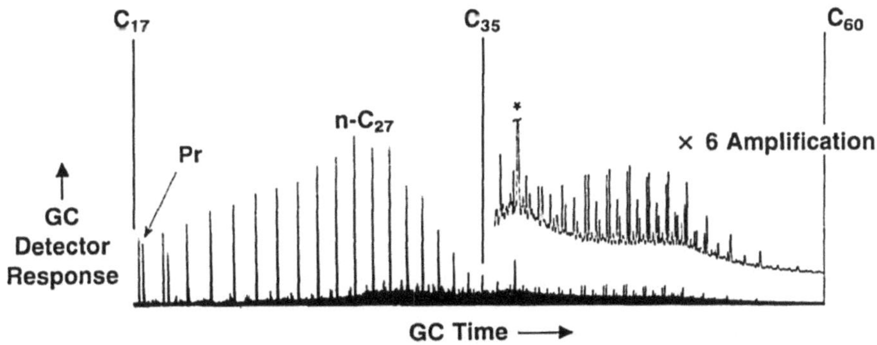

FIGURE 5.7 An example of a high-temperature gas chromatographic analysis for >C_{40} waxes in a crude oil. *GC*, Gas chromatography. *From Carlson, R.M.K., Teerman, S.C., Moldowan, J.M., Jacobson, S.R., Chan, E.I., Dorrough, K.S., Seetoo, W.C., Mertani, B., 1993. High temperature gas chromatography or high wax oils. In: Indonesian Petroleum Association, 22nd Annual Convention Proceeding, pp. 483–507.*

that these high-molecular weight waxes can occur in both marine Type II sourced oils as well as the lacustrine Type I sourced oils. And high-molecular weight wax deposition can even be a problem for condensates (Leontaritis, 1998). It is therefore recommended that all crude oils should be routinely examined for the presence of high-molecular weight waxes to avoid potential wax deposition problems. The cost of the analysis is by far less than the cost of lost production and treatment. If the potential for wax deposition appears to exist, additional laboratory studies can be conducted to predict under what conditions wax deposition may occur (Leontaritis, 1996).

When it happens, wax deposition can be mitigated by treatment with solvents, such as xylene (Fan and Llave, 1996). It can also be prevented if the crude oil is maintained at temperatures above the pour point of the oil or wax inhibitors are employed (Gloczynski and Kempton, 2006).

ASPHALTENES

In addition to high-molecular weight waxes, organic deposition in the form of asphaltenes can also occur. Asphaltenes are sometimes referred to as solid hydrocarbon, reservoir bitumen, or tar. Asphaltene deposits can occur in the subsurface, in production lines to the surface, in flow lines from the wellhead, in surface storage, and in pipelines. Subsurface asphaltene deposition can pose significant barriers to flow in petroleum reservoirs. But unlike high-molecular weight waxes, these deposits may not be completely soluble in organic solvents and as such are more difficult to deal with. The problems that asphaltenes pose include reduction or loss of effective porosity and permeability, which in turn reduces production and can interfere with enhanced oil recovery (Lomando, 1992). Because asphaltene deposition is not readily identifiable on wireline logs, it can have a negative impact on estimating recovery factors and making reserve calculations (Lomando, 1992). The presence of these asphaltenes may even change the reservoir's wettability from water wet to oil wet (Buckley et al., 1997; Amroun and Tiab, 2001).

Asphaltene deposition may occur as a result of natural processes, and it can be induced by production. In naturally occurring asphaltene deposition, it is most frequently encountered in the subsurface as the filling of interstitial spaces, mainly pores but also fractures and faults. These interstitial asphaltenes occur mainly in reservoirs impregnating the sediments as either grain coatings or as partial to complete pore fillings. It may be pervasive in the reservoir or occur in discrete zones. Occasionally, it is confined to the vicinity of the oil-water contact.

In core, the interstitial asphaltenes appear as dark brown to black, highly viscous material coating grains and filling pore space. The asphaltene frequently comprises up to 10–15% of the rock with the remaining being 80–85% mineral matter and around 4–6% water. Asphaltenes can be differentiated from oil stain by the nonfluorescent nature of asphaltenes.

Naturally occurring interstitial asphaltenes are formed by precipitation from an unstable oil due to pressure/temperature drop, mixing with a lighter oil, late gas migration, biodegradation, in situ oil cracking to gas, or a combination of processes. Pressure or temperature drop in a reservoir could be the result of uplift and erosion or a structural repositioning of the reservoir. Concentration of asphaltenes due to biodegradation is the result of removal of saturate and aromatic hydrocarbons. The addition of biogenic gas as well as loss of light ends during biodegradation can also contribute to the process.

While naturally deposited interstitial asphalt can be a detriment to production, there is little that can be done to mitigate its occurrence. Treatments with solvents, such as xylene, may provide some short-term improvements, but these treatments will likely need to be repeated to maintain rates (Mansoori, 2010).

Naturally occurring subsurface asphaltenes (tar) may also occur in massive apparently void-filling bodies with little or no sediment incorporated (Han et al., 2010). These asphaltene bodies may occur either parallel to bedding in sill-like bodies, or cross-cutting bedding, filling along faults/fractures in dike-like bodies (Romo et al., 2007). These tar bodies are nearly always associated with salt, usually just below or adjacent to a salt body. This type of asphaltene deposit appears in cuttings as a shiny black substance, often with conchoidal fracture, usually brittle, can be easily crushed into a dark brown powder, and is virtually sediment free. These asphalt bodies are not detectable in seismic data.

Their exact mode of formation is unknown. However, it appears oil may have migrated up faults or along salt to the paleoseafloor filling large fault zones and/or forming mounds or bedded tar. This can be observed today in many places in the Gulf of Mexico (e.g., MacDonald et al., 2004; Hewitt et al., 2008; Williamson et al., 2008), the Santa Barbara Channel (Valentine et al., 2010), and in the deepwater off Angola (Jones et al., 2014). Subsequently buried by sediments or entrainment by the salt canopy may have incorporated these sediment-free asphalt bodies into the stratigraphic record. When encountered during drilling, these asphalt bodies can become ductile and flow into the borehole if the overburden pressure and subsurface temperature have reached critical levels (Han et al., 2010). If these asphalt bodies become mobilized, the lower portion of the wellbore often needs to be abandoned, and the well is side tracked (Weatherl, 2007). At this time, these mobile asphaltene appear to be confined to the Gulf of Mexico and may be related to the dynamic salt tectonics experienced in this region.

During production, asphaltene precipitation may also occur as a result of decreases in pressure or temperature, gas injection, water injection, or mixing of two oils in the wellbore (multizone production) or in surface facilities (Leontaritis and Mansoori, 1988). These precipitated asphaltenes may be present as solid bitumen within the reservoir porosity near the wellbore, deposits in the wellbore, deposits in flow lines, and/or deposits in surface facilities. Mitigation of asphaltene deposition problems in the reservoir near the wellbore and the wellbore itself is possible, but it is usually expensive, time consuming, and results in lost production. It is better to be proactive and test the crude oil about to be produced to determine if asphaltene precipitation will be a problem and under what circumstances precipitation may occur. Most frequently, the cause of precipitation is related to pressure–temperature changes. PVT studies can be undertaken that will define the asphaltene precipitation envelope in P–T space (Leontaritis, 1996), such as the generic one shown in Fig. 5.8. By knowing the reservoir conditions that may trigger precipitation of asphaltenes, steps can be taken to maintain the temperature and pressure changes during production to remain within the stability limits of the asphaltenes. This can be applied to the reservoir, borehole, or topside production facilities. This same approach can be employed to natural gas reinjection or nitrogen or carbon dioxide injection to predict when asphaltene precipitation may occur (Burke et al., 1990).

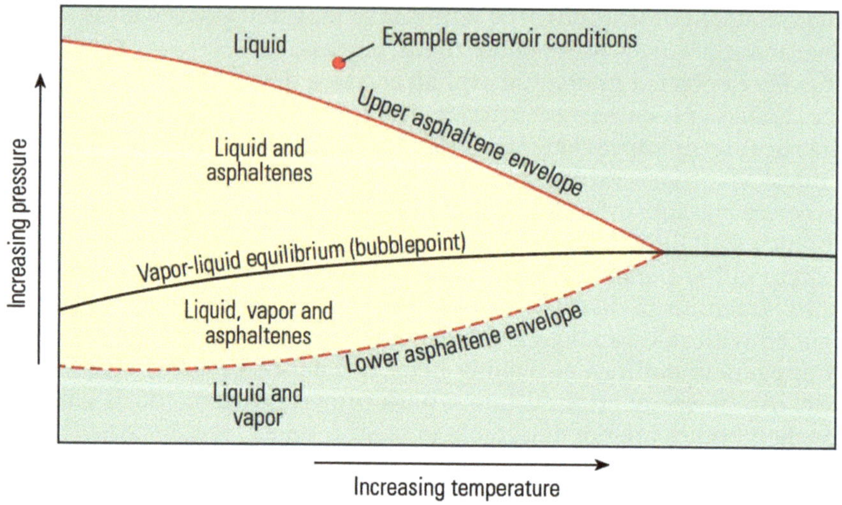

FIGURE 5.8 A generic asphaltene precipitation envelope in pressure–temperature space used to predict asphaltene deposition. *From Akbarzadeh T.S.K., Hammami, A., Kharrat, A., Zhang, D., Allenson, S., Creek, J., Kabir, S., Jamaluddin, A., Marshall, A.G., Rodgers, R., Mullins, O.C., 2007. Asphaltenes – problematic but rich in potential. Oil Field Review, Schlumberger 19 (2), 22–43. (Summer 2007).*

If asphaltene precipitation cannot be avoided or has already occurred, treatment options are available for the wellbore and near wellbore reservoir including the use of solvents, such as xylene, to reduce deposits and improve production (Mansoori, 2010). Inhibitor may also be used to prevent future deposition from occurring.

RESERVOIR CONTINUITY

Reservoir continuity is defined as the absence of vertical flow barriers in a hydrocarbon column within a single well and/or lateral flow barriers within a hydrocarbon-bearing interval between wells. When a reservoir has been compartmentalized, flow barriers exist that divide the reservoir interval into a series of discrete containers. To efficiently develop a field and maximize the amount of hydrocarbons recovered, reservoir compartments need to be identified.

Reservoir compartments are often identified by repeat formation tester pressures, pressure decline curves, oil–water contact depths, and/or fault juxtaposition. They can also be recognized by compositional difference of the fluids sampled from different locations within the reservoir interval. As a reservoir fills over time, it is receiving hydrocarbons from the source rock at progressively increasing maturity levels. The final product is a mixture of all these contributions over time. In a reservoir with more than one compartment, the filling history of each compartment is likely to be different, resulting in subtle differences in the hydrocarbons found in each compartment. These subtle differences can be observed in the mud gas composition while drilling, as well as in oil and gas samples recovered from the reservoir during testing.

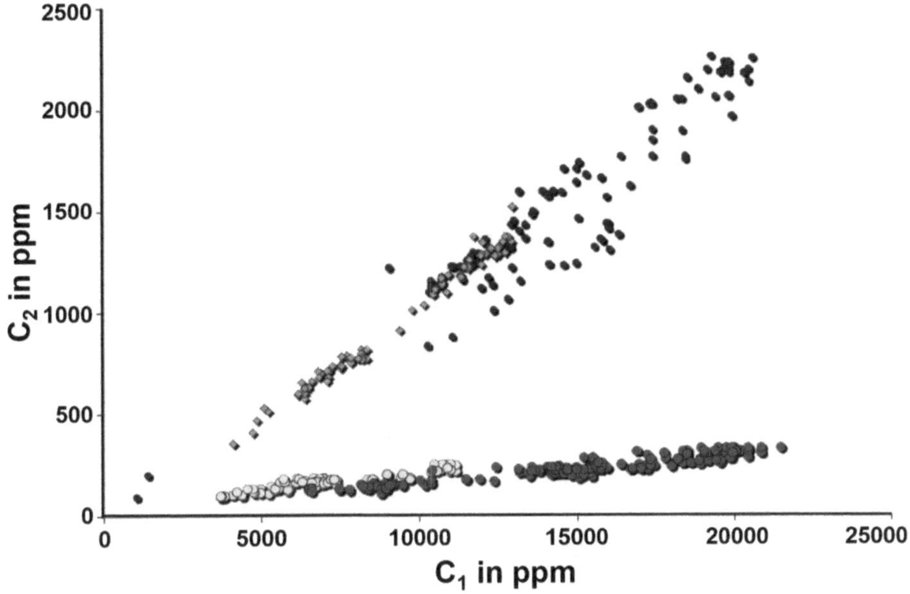

FIGURE 5.9 A cross-plot of the methane versus ethane from the mud gas samples collected from a series of small sands. The two trends show potential separation of these sands into two vertical compartments.

Reservoir Continuity Using Mud Gas Data

The first indications of potential vertical compartmentalization in a reservoir may be found in the mud gas data. Reservoir intervals that are connected will exhibit a consistent mud gas composition, while reservoir intervals that are not connected will exhibit different compositions (Blanc et al., 2003; McKinney et al., 2007). Differences in composition can be observed in simple cross-plots of gas composition, such as the methane–ethane cross-plot shown in Fig. 5.9. Data from the mud gas samples from four small sands are plotted in the diagram. The two trends show potential vertical separation between the two upper sands and the two lower sands suggesting two vertical compartments. Another method using mud gas data is to plot the Pixler ratios on a radar plot, as shown in Fig. 5.10. This is similar to the approach used in gas-to-gas correlation discussed in Chapter 4. Similar patterns on the radar plot suggest communication between adjacent sands, while different patterns suggest potential vertical separation of the sands. And finally, mud gas isotopes measured from Isotubes can be used to show similarities and differences in adjacent sands.

While these indications of vertical continuity and/or compartmentalization from mud gas data are useful, they should be considered tentative and require corroboration with other data from actual reservoir samples.

Reservoir Continuity Using Oil Samples

When testing the continuity of an oil-bearing reservoir, several methods can be employed. Occasionally, simple whole oil chromatograms can be used to demonstrate easily recognizable

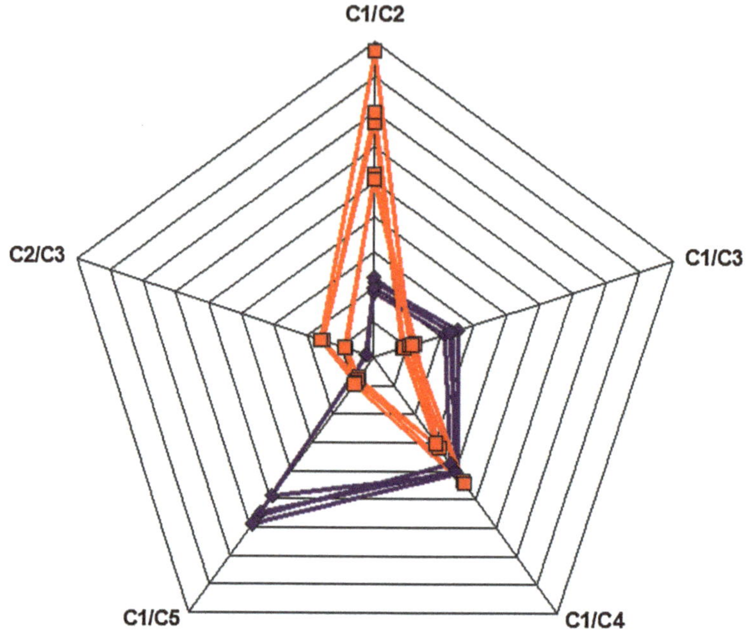

FIGURE 5.10 An example of using mud gas Pixler ratios plotted on a radar plot to show differences in composition suggesting vertical compartmentalization in a well.

difference between compartments, such as differential biodegradation (Edman and Burk, 1999). In other instances, the biomarker data used in oil-to-oil correlation studies, as described in Chapter 4, can be applied to these problems (Peters and Fowler, 2002; Pomerantz et al., 2010). However, the common method of comparing oils in reservoir continuity studies is the use of high-resolution whole oil gas chromatograms (Kaufman et al., 1990; Dow and Talukdar, 1991). Because reservoir continuity studies look for subtle differences in the hydrocarbon composition of the oils being compared, the whole oil gas chromatography used must be done in a distinct way. The temperature program used in this analysis is usually slowed, often doubling the analysis time, to provide higher resolution. In addition, analytical blanks, standards, and duplicate samples are run more frequently to insure that the retention times are reproducible, and there is no carry over of hydrocarbons from previous analyses. In the event that the oils being analyzed are heavy and carry over may be a problem, blank may have to run between samples to make sure the column has cleaned up.

Very often, the whole oil chromatograms being compared will look almost identical on a gross scale. However, the details of the lower concentration compounds can provide a means of demonstrating similarities and differences (Kaufman et al., 1990). An example of this is shown in the high-resolution gas chromatograms from whole oil analyses used in a reservoir continuity study in Fig. 5.11. The three oils come from the same reservoir interval in three different wells, and the wells are separated from each other by faults. The question is whether the faults are open and there is communication between the wells or are the faults closed isolating the wells from each other. The chromatogram in the upper left shows the full whole oil chromatogram from one of the oils, which on a gross scale is identical to the other two oils. On the chromatogram, a window from 10 to 15 min retention time is indicated as

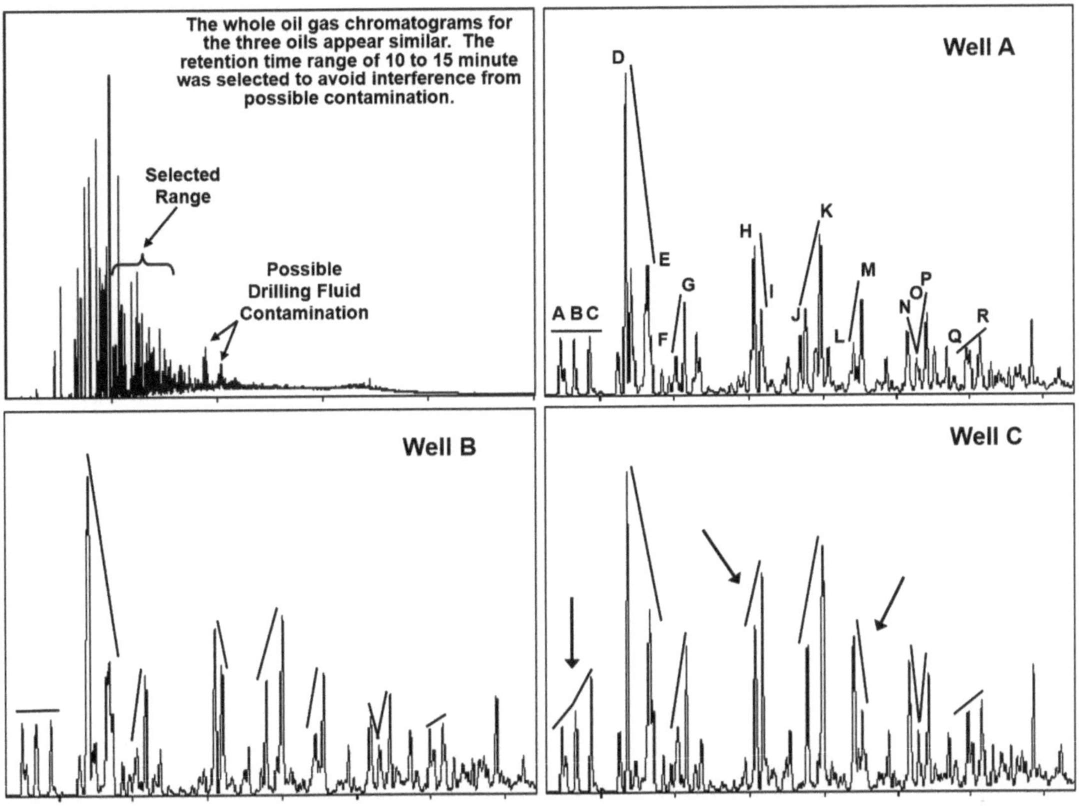

FIGURE 5.11 An example of using high-resolution gas chromatograms from whole oil analyses used in reservoir continuity studies. The upper left chromatogram shows the full analytical range with the 10–15 min retention time window selected for comparison. The three remaining chromatograms show the 10–15 min retention time windows for the oils from the same reservoir interval in Wells A, B, and C, as well as the peaks selected for comparison. The arrows in the Well C chromatogram indicate significant difference between it and the oils in Wells A and B.

the selected time range for comparison. Also indicated is the presence of possible synthetic oil-based drilling mud contamination outside the window for comparison. The three remaining chromatograms in Fig. 5.11 show the 10–15 min retention time windows for the oils in the same reservoir interval in Wells A, B, and C. The peaks selected for comparison are shown in the chromatogram for Well A. The bars drawn over the selected peaks provide a preview of the ratios to be calculated and give a preliminary assessment of similarities and difference between the oils. The arrows in the Well C chromatogram indicate significant difference between it and the oils in Wells A and B.

Peak ratios are then calculated, usually based on peak heights, and compared graphically using radar plots. The results of the calculated peak ratios from the three whole oil chromatograms in Fig. 5.11 plotted on a radar plot are shown in Fig. 5.12. While the oils in Wells A and B show a high degree of similarity, the oil in Well C is significantly different. These similarities and differences along with the geologic setting were then used to make the interpretation of connectivity between the wells shown in the cross-section in Fig. 5.12. From the data, there appears to be communication of reservoir fluids across the fault between Wells

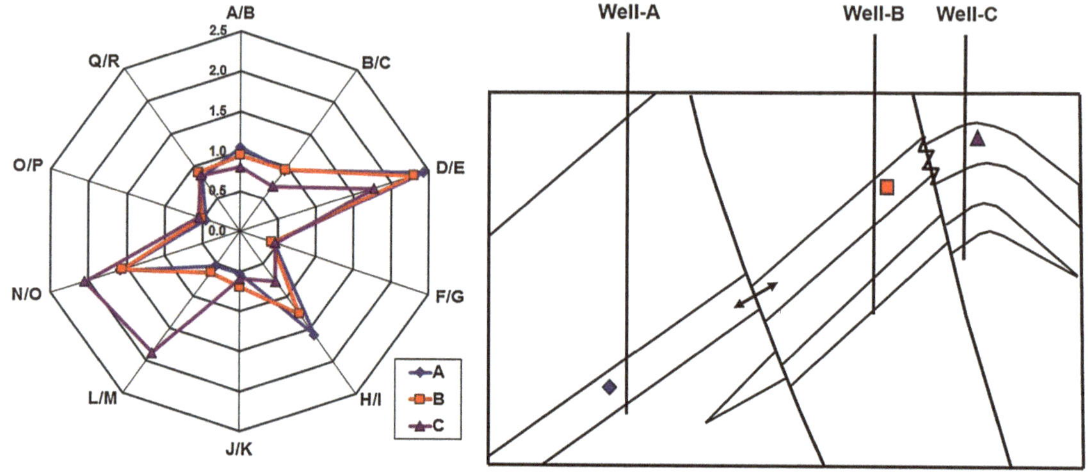

FIGURE 5.12 The results of the calculated peak ratios from the three whole oil chromatograms in Fig. 5.11 plotted on a radar plot showing the basis for the interpretation in the cross-section on the right. The data indicate communication across the fault between Wells A and B, while the fault between Wells B and C appears to isolate the two compartments.

A and B, while the fault between Wells B and C appears to isolate the two compartments. In this simple example, the graphic display of data on radar plots is an adequate way to discern between oils in different compartments. However, if large data sets are being examined, the use of multivariate statistical analysis may be employed in a fashion similar to the method described for oil correlation in Chapter 4.

In addition to the high-resolution whole oil gas chromatograms, distributions of alkyl benzene in crude oils have been used in reservoir continuity studies (Fox and Bowman, 2010). These compounds can be identified in gas chromatograms of the aromatic hydrocarbon fractions; however, the typical way to obtain alkyl benzene data is by gas chromatography–mass spectrometry analysis monitoring the characteristic m/z 92 fragment. These data are usually normalized to the highest peak in the distribution and are employed in the same fashion as the selected peak ratios from the high-resolution whole oil gas chromatograms.

Reservoir Continuity Using Gas Samples

Reservoir continuity can also be tested in gas reservoirs applying the same general method using a combination of both the compositional information and the carbon isotope ratios to establish similarities and differences between gases from the same producing horizon in different wells (Weissenburger and Borbas, 2004; Milkov et al., 2007). The method of comparison is essentially identical to the gas-to-gas correlation described in Chapter 4, with the exception that the geochemical differences from the differences in filling history for gases may be more subtle than observed for oils.

Because gases migrate more easily in the subsurface, mixing of gases from different sources or different maturities is more likely to be encountered. Coupled with the potentially smaller compositional and isotopic differences between compartments, this may obscure dissimilarities in the gases that might indicate reservoir compartmentalization.

PRODUCTION ALLOCATION

When more than one producing horizon is encountered in a field, the oil or gas from these separate horizons may be commingled either during production, in the surface facilities, or in a pipeline. It is sometimes useful and necessary to be able to determine contribution of individual oils or gases to the overall production. This may be for assessing the performance of individual producing zones for reservoir management or for determining revenue allocation from the production when different lease holders have rights to the different producing horizons.

The basic method used for estimating production allocation uses the high-resolution whole oil gas chromatographic data and peak ratios similar to that used in reservoir continuity studies (Kaufman et al., 1990). The method is based on exploiting the compositional differences between individual oils that went into to making the mixture. To demonstrate this, a simple mixing model is employed to allocate production from two separate producing horizons, shown in the example in Fig. 5.13. The two horizons have been established to be isolated from one another by geochemical comparison. In this example, three peak ratios that are significantly different in each of the two oils have been selected for estimating the amount of each oil's contribution to the overall mixture. Using simple linear mixing models, the changes in

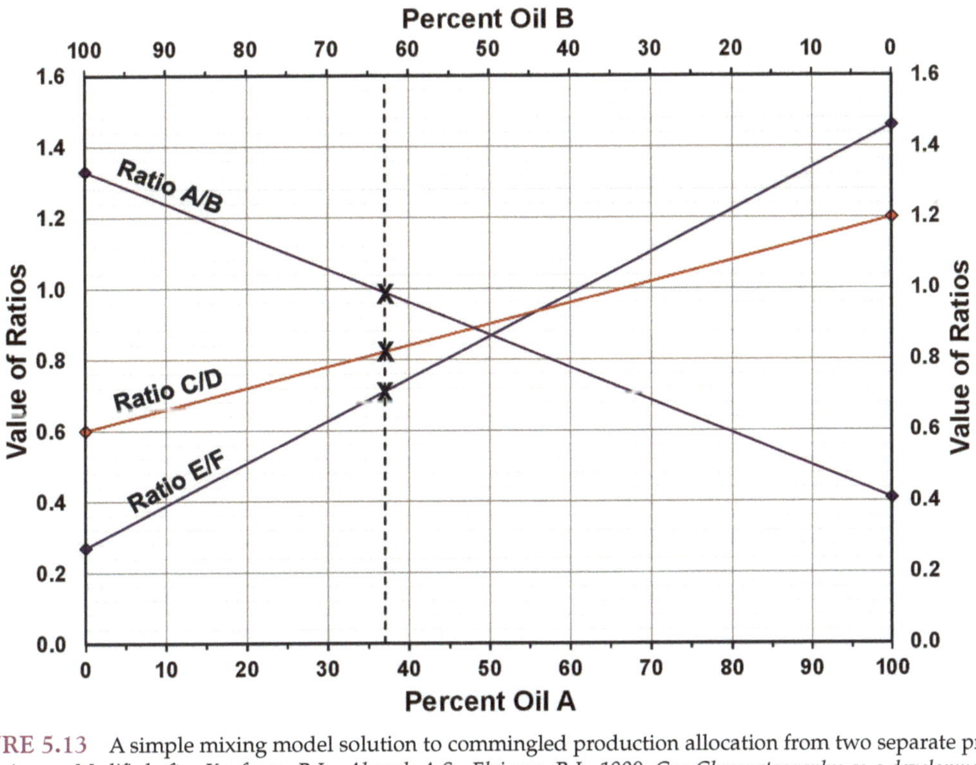

FIGURE 5.13 A simple mixing model solution to commingled production allocation from two separate producing horizons. *Modified after Kaufman, R.L., Ahmed, A.S., Elsinger, R.J., 1990. Gas Chromatography as a development and production tool for fingerprinting oils from individual reservoirs: applications in the Gulf of Mexico. In: Schumaker, D., Perkins, B.F. (Eds.), Proceedings of the 9th Annual Research Conference of the Society of Economic Paleontologists and Mineralogists, October 1, 1990: New Orleans, pp. 263–282.*

the ratios can go from 100% Oil A on the right to 100% Oil B on the left along the designated trends. By plotting the ratios observed in the mixed oil, as shown by the Xs along the trends, a composition of the mixed oil will be indicated, in this case, approximately 37% Oil A and 63% Oil B. In practice, there are often some variations observed in the results from each ratio. If these variations are small, an average of the three results is used for the final composition. While this approach is useful for binary mixtures, when more than two oils are commingled, it is necessary to employ multivariate statistical analysis to ascertain the contribution of each oil (Hwang et al., 2000; McCaffrey et al., 2011).

In gas reservoirs, the approach to deconvolving mixtures is more complex. While mixing using compositional data may follow simple linear models, the composition of the gases alone may not provide enough differences to adequately distinguish the contribution of each gas to the mixture. Using isotopic data in combination with the compositional data, as discussed in the section on reservoir continuity studies, can provide additional characteristics to distinguishing each gas going into the mixture. However, because of the way stable isotope ratios are calculated using the δ notation, they cannot be used in simple linear mixing models. Instead, a combined compositional and isotopic approach has been proposed by McCaffrey et al. (2011) which uses a series of mixing equations. While these equations do not supply a unique solution, the results can be used to approximate the most likely mixtures of gases.

PRODUCTION PROBLEMS AND PERIODIC SAMPLING

After production begins, there are a number of problems that might arise that can reduce or stop the flow of hydrocarbons and thereby impact the profitability of a field. These problems can be related to cement or packer failures, tubing or casing leaks, or organic deposits to name just a few possible causes. Many times, these production problems can be recognized in changes in the composition of the hydrocarbons being produced. Ideally, the composition of the produced hydrocarbons should be monitored on a periodic basis, perhaps every 6 months, to look for changes that may signal potential difficulties. This type of information can also help insure production in the field is being done efficiently, and the recovery of hydrocarbons is maximized. However, a surveillance program for producing wells is difficult to sell to production engineers unless specific production problems have been encountered or are anticipated.

An alternative to periodic analysis of the produced fluids is periodic sampling for archival purposes. For oils, a 25–50 ml sample collected every six months, sealed and stored in a cool dark place, is an insurance policy against potential production problems. These samples are inexpensive to collect and store but are on hand and invaluable for geochemical analysis if complications do occur. Not only do these samples provide data to help diagnose what the production problem might be, they can also provide a temporal framework that may be important to understanding the underlying cause for the problem.

A classic published example of this concept is the case study presented by Kaufman et al. (1990) documenting a tubing string failure in a well from in the Gulf of Mexico. A schematic of the development of this problem is shown in Fig. 5.14. Initially, two zones with distinct fluid compositions were brought on production about 5 years apart. Archival samples were collected from each producing zone on a periodic basis. When it was suspected that there

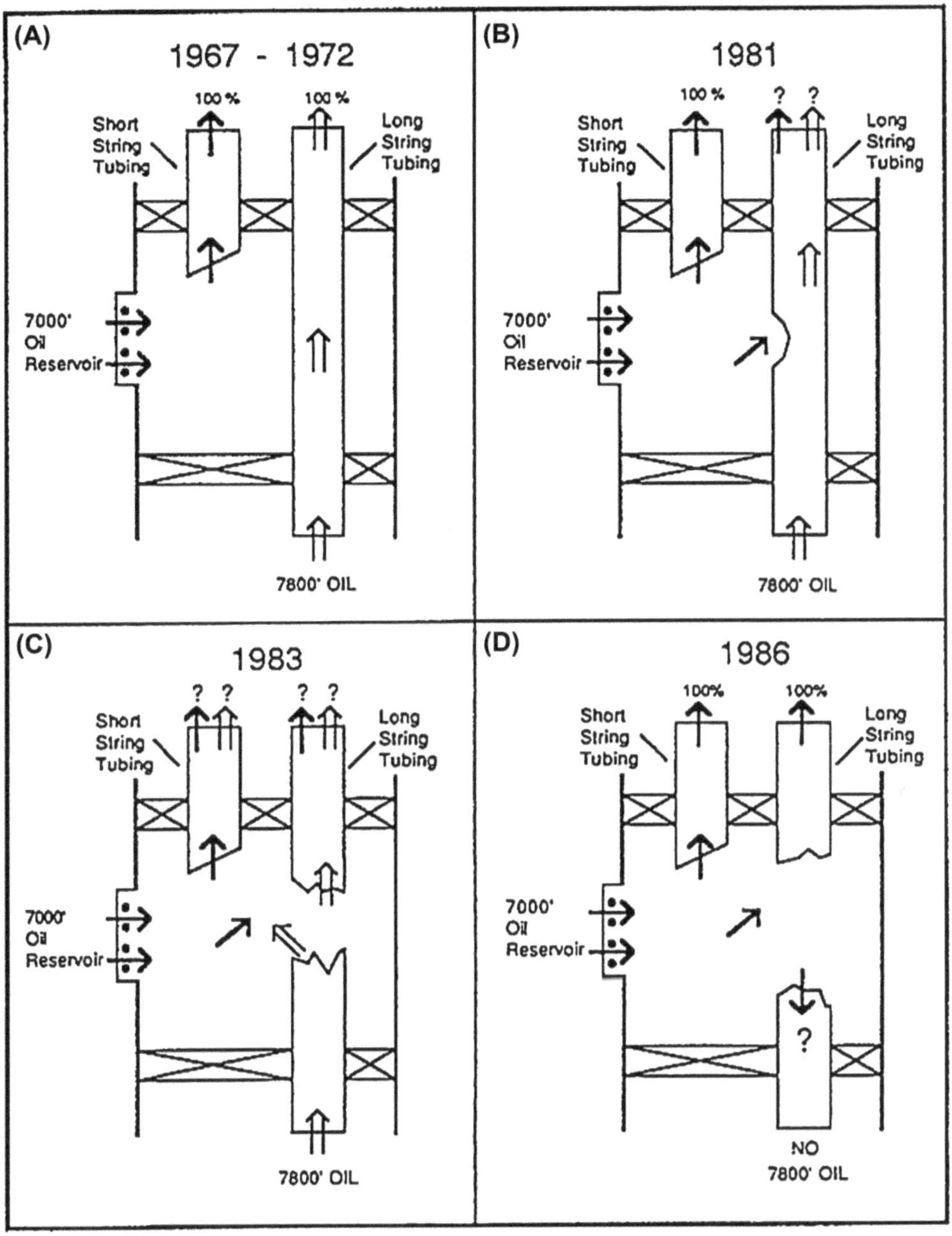

FIGURE 5.14 The steps involved in tubing leakage problem in a well decipher from reservoir geochemistry data. *From Kaufman, R.L., Ahmed, A.S., Elsinger, R.J., 1990. Gas Chromatography as a development and production tool for fingerprinting oils from individual reservoirs: applications in the Gulf of Mexico. In: Schumaker, D., Perkins, B.F. (Eds.), Proceedings of the 9th Annual Research Conference of the Society of Economic Paleontologists and Mineralogists, October 1, 1990: New Orleans, pp. 263–282.*

was communication between the two tubing strings in the well, the geochemists involved were able to go to the archived samples. By applying the high-resolution whole oil fingerprinting technique used in reservoir continuity studies and the method for production allocation, both described earlier, they were able to establish there was indeed communication between the tubing strings and built the scenario defining the progressive steps in its development shown in Fig. 5.14. After completing this study, production from the lower zone was reestablished increasing the overall production of the well, and steps were taken to prevent future tubing failures.

Contrary to oil samples, the collecting and archiving of gas samples from producing wells is not as practical. The need for expensive pressure cylinders for gas samples and the amount of space needed to store them make archiving impractical. There is also the potential for the gas cylinders to leak, altering the composition and isotopic signature of the gas. Analysis of composition and stable isotopes for natural gas is relatively inexpensive, compared to crude oil analysis. Because of these considerations, it is more practical and cost effective to analyze produced gas samples on a periodic basic and store the data. In this way, the data will be readily available for investigation if problems arise.

MONITORING ENHANCED OIL RECOVERY

With only 10–20% of the original oil in place being produced by primary recovery techniques, enhanced oil recovery has become a common practice to yield more oil from already discovered fields to maximize the investment in exploration. Although there has been very little published on the application of petroleum geochemistry to the monitoring of enhanced oil recovery, there is great potential for this application. Petroleum geochemistry can be applied to both the monitoring and assessing of the efficiency of enhanced oil recovery. These techniques are applicable to water, steam, CO_2, and surfactant floods, as well as natural gas reinjection. The basic concept entails developing a set of baseline data for the reservoir prior to beginning any enhanced oil recovery. This could be accomplished by developing a vertical geochemical profile through the reservoir at a given location, typically between an injection well and a production well. The vertical geochemical profile would be built using rotary sidewall cores to obtain samples of the fluids in the reservoir at fixed positions in the borehole that can be resampled after the flood front has passed. The data collected would consist of the quantities and distributions of hydrocarbons present at each sampled depth point. The distributions could be obtained by either TEGC or conventional solvent extraction followed by whole oil gas chromatography. The best results have been obtained when using sidewall cores that are frozen at the well site to help preserve the loss of light hydrocarbon material. After the flood front has passed, the same coring points are used to collect a second set of sidewall cores. The second set of samples should be collected and treated in exactly the same fashion as those collected to establish the baseline profile. Changes in the quantity and composition of the hydrocarbons can be used to confirm the flood front's passage as well as to estimate the efficiency of the process to extract additional oil.

The example in Fig. 5.15 shows the results from two sidewall cores collected from the same depth in a monitoring well for a steam flood pilot project. The data shown are from TEGC of frozen sidewall cores collected before and after the flood front passed the monitor well

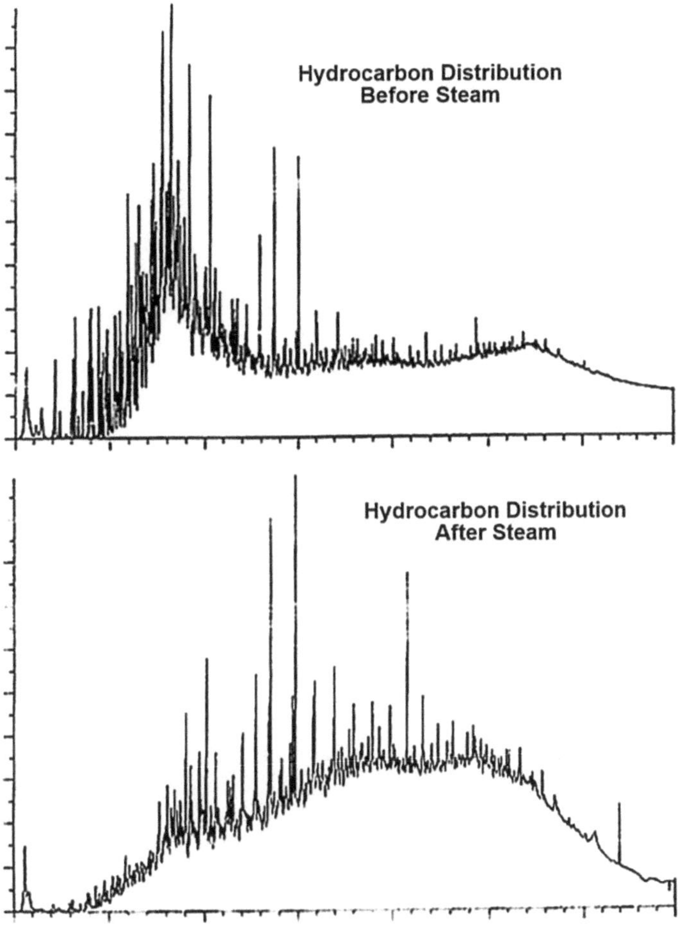

FIGURE 5.15 An example of using thermal extraction-gas chromatography data to monitor enhanced oil recovery. The sidewall cores analyzed were collected at the same depth before and after a steam flood front has passed the well location.

location. The original oil in the before chromatogram shows the oil to be extensively biodegraded. The after chromatogram clearly shows the shift in the hydrocarbon distribution even in the unresolved complex mixture represented by the "hump" below the peaks. Additional quantitative data were obtained from the sidewall cores to indicate the amount of the oil that was stripped out of the sediments by the steam.

These analyses are fast and relatively inexpensive, in addition to providing information about the effectiveness of enhanced oil recovery. It is expected that more use of this technique will be seen in the future as enhanced oil recovery becomes an even more important aspect of production operations.

RESERVOIR SOURING

Reservoir souring is an observed increasing in the hydrogen sulfide (H_2S) content of reservoir fluids over time. It most often occurs in reservoirs undergoing water flooding usually within a few months to a few years after injection begins. Increased H_2S can lead to a reduction in the quality of the produced hydrocarbons, reduced well productivity, and added safety and health risks and liabilities. It can also increase the potential for sulfide stress cracking and corrosive failure of downhole equipment, flow lines, and surface facilities (Iverson, 1987).

The increase in H_2S during water flooding is the result of sulfate-reducing bacteria that are introduced with the injected water biodegrading the oil in the reservoir (Cord-Ruwisch et al., 1987). It occurs more commonly in offshore fields where sulfate-rich seawater is being injected, replenishing the supply of sulfate for the microbes. The induced biodegradation can be recognized by changes in the composition of the produced oil that is consistent with natural biodegradation. It also has a temperature sensitivity similar to natural biodegradation, with an upper limit of 80°C for this bacterial activity. Reservoir souring from sulfate-reducing bacteria can be mitigated by either adding biocide to the injection water and/or removing sulfate from the injection water (Vance and Thrasher, 2005).

While occurring more frequently in water-flooded reservoir, reservoir souring may also occur due to thermochemical sulfate reduction (TSR) induced by steam flooding the reservoir (Hoffmann and Steinfatt, 1993; Kowalewski et al., 2008). During steam floods, reservoir rock may reach temperatures greater than 150°C. If sulfate is present, the TSR reactions can take place consuming part of the hydrocarbons and producing hydrogen sulfide.

The role of petroleum geochemistry in reservoir souring is that of monitoring the production. Periodic sampling of produced fluids should be done to help recognize geochemical changes in the hydrocarbons, both liquids and gases, that can signaling souring is occurring in order that remedial actions can be taken. Production monitoring is cost effective, considering the lost production, diminished value of the petroleum, and expense of repairing corrosion in the production facilities that can result from reservoir souring.

STRATEGIES IN RESERVOIR GEOCHEMISTRY

Although petroleum geochemistry has been applied to reservoir and production problems since the late 1970s, the advancement of this aspect of the science has not progressed as much as source rock evaluation or oil and gas correlation. There are still opportunities to discover new ways to apply petroleum geochemistry to reservoir problems in ways not previously recognized using the tools already available. We need to be looking for innovative applications for reservoir geochemistry to increase its value in field development and production practices.

There are a few particular recommendations that can be made for applying petroleum geochemistry to reservoir and production problems. To begin, it is essential to be specific about what the problem is, when it occurred, and what you want to know from the study. These specifications set down in the definition of the problem will determine what samples will be required and what analytical program needs to be followed. They will also define the expectations of the study and how it will help resolve the issues.

Next, it cannot be emphasized enough that for reservoir geochemistry reservoir fluid samples must be collected routinely and archived. Begin collecting early in the life of the reservoir, and periodically resample at least every 6 months. If a problem develops, you will need these samples to establish a baseline and help determine the timeline for when the problem started.

It is always best to try to resolve production problems with geochemistry early. It is often quicker and less expensive than the more conventional engineering procedures. The sooner the problem is solved, the faster the production can be restored. If needed, the engineering option is still available.

And finally, as with all petroleum geochemistry applications, place the geochemical results in a geologic context. The interpretations cannot stand alone. They need to make geologic sense, or they have little value.

References

Akbarzadeh, T.S.K., Hammami, A., Kharrat, A., Zhang, D., Allenson, S., Creek, J., Kabir, S., Jamaluddin, A., Marshall, A.G., Rodgers, R., Mullins, O.C., 2007. Asphaltenes – problematic but rich in potential. Oil Field Review, Schlumberger 19 (2), 22–43 (Summer 2007).

Amroun, H., Tiab, D., 2001. Alteration of Reservoir Wettability due to Asphaltene Deposition in Rhourd-Nouss Sud Est Field, Algeria. Society of Petroleum Engineers Paper 71060. 11 p.

Baskin, D.K., Jones, R.W., 1993. Prediction of oil gravity prior to drill-stem testing in Monterey Formation reservoirs, offshore California. American Association of Petroleum Geologists Bulletin 77, p1479–1487.

Baskin, D.K., Hwang, R.J., Purdy, R.K., 1995. Predicting gas, oil and water intervals in Niger Delta reservoirs using gas chromatography. American Association of Petroleum Geologists Bulletin 79, 337–350.

Blanc, P., Brevière, J., Laran, F., Chauvin, H., Boehm, C., Fréchin, N., Capot, M., Benayoun, A., 2003. Reducing Uncertainties in Formation Evaluation Through Innovative Mud Logging Techniques. Society of Petroleum Engineers Paper 84383. 19 p.

Buckley, J.S., Liu, Y., Xie, X., Morrow, N.R., 1997. Asphaltenes and crude oil wetting – the effect of oil composition. Society of Petroleum Engineers Journal 2, 107–119.

Burke, N.E., Hobbs, R.E., Kashou, S.F., 1990. Measurement and modeling of asphaltene precipitation. Journal of Petroleum Tecnology 42, 1440–1446.

Carlson, R.M.K., Teerman, S.C., Moldowan, J.M., Jacobson, S.R., Chan, E.I., Dorrough, K.S., Seetoo, W.C., Mertani, B., 1993. High temperature gas chromatography or high wax oils. In: Indonesian Petroleum Association, 22nd Annual Convention Proceeding, pp. 483–507.

Cord-Ruwisch, R., Kleinitz, W., Widdel, F., 1987. Sulfate-reducing bacteria and their activities in oil production. Journal of Petroleum Technology 39, 97–106.

del Rio, J.C., Philp, R.P., 1992. High-molecular-weight hydrocarbons: a new frontier in organic geochemistry. Trends in Analytical Chemistry 11, 187–193.

Dembicki Jr., H., 1986. C_5+ hydrocarbon mud logging by thermal extraction. Society of Petroleum Engineers Formation Evaluation 1, 331–334.

Dow, W.G., Talukdar, S.C., January 1991. Petroleum geochemistry in oil production. Houston Geological Society Bulletin 33, 44–52.

Edman, J.D., Burk, M.K., 1999. Geochemistry in an integrated study of reservoir compartmentalization at Ewing Bank 873, offshore Gulf of Mexico. Society of Petroleum Engineers Reservoir Evaluation & Engineering 2, 520–526.

Fan, Y., Llave, F.M., 1996. Chemical Removal of Formation Damage from Paraffin Deposition Part I - Solubility and Dissolution Rate. SPE Paper 31128. 10 p.

Fox, R.J., Bowman, M.B.J., 2010. The challenges and impact of compartmentalization in reservoir appraisal and development. In: Reservoir Compartmentalization, vol. 347. Geological Society of London Special Publication, pp. 9–23.

Gloczynski, T.S., Kempton, E.C., March 2006. Understanding wax problems leads to deepwater flow assurance solutions. World Oil 7–10.

Han, G., Osmond, J., Zambonini, M., 2010. A USD 100 million "rock": bitumen in the deepwater Gulf of Mexico. Society of Petroleum Engineers Drilling & Completion 25, 290–299.
Haworth, J.H., Sellens, M., Whittaker, A., 1985. Interpretation of hydrocarbon shows using light C_1-C_5 hydrocarbon gases from mud-log data. American Association of Petroleum Geologists Bulletin 69, 1305–1310.
Hewitt, A.T., Smith, J.L., Weiland, R.J., 2008. AUV and ROV data integration to predict environmentally sensitive biological communities in deep-water. In: Offshore Technology Conference Proceedings Paper 19358 14 p.
Hoffmann, G.G., Steinfatt, I., 1993. Thermochemical sulfate reduction at steam flooding processes – a chemical approach. American Chemical Society Petroleum Chemistry Division Preprint 38, 181–184.
Hunt, J.M., 1996. Petroleum Geochemistry and Geology, second ed. W. H. Freeman, New York. 743 pp.
Hwang, R.J., Baskin, D.K., Teerman, S.C., 2000. Allocation of commingled pipeline oils to field production. Organic Geochemistry 31, 1463–1474.
Iverson, W.P., 1987. Microbial corrosion of metals. Advances in Applied Microbiology 32, 1–36.
Jarvie, D.M., Morelos, A., Han, Z., 2001. Detection of pay zones and pay quality, Gulf of Mexico: application of geochemical techniques. Gulf Coast Association of Geological Societies Transactions 51, 151–160.
Jones, D.O.B., Walls, A., Clare, M., Fiske, M.S., Weiland, R.J., O'Brien, R., Touzel, D.F., 2014. Asphalt mounds and associated biota on the Angolan margin. Deep Sea Research Part I: Oceanographic Research Papers 94, 124–136.
Kaufman, R.L., Ahmed, A.S., Elsinger, R.J., 1990. Gas Chromatography as a development and production tool for fingerprinting oils from individual reservoirs: applications in the Gulf of Mexico. In: Schumaker, D., Perkins, B.F. (Eds.), Proceedings of the 9th Annual Research Conference of the Society of Economic Paleontologists and Mineralogists, October 1, 1990: New Orleans, pp. 263–282.
Kowalewski, I., Fiedler, C., Parra, T., Adam, P., Albrecht, P., 2008. Preliminary results on the formation of organosulfur compounds in sulfate-rich petroleum reservoirs submitted to steam injection. Organic Geochemistry 39, 1130–1136.
Leontaritis, K.J., 1996. The asphaltene and wax deposition envelopes. Fuel Science and Technology International 14, 13–39.
Leontaritis, K.J., 1998. The wax deposition envelope of gas condensates. In: Proceeding of the Offshore Technology Conference, Paper 8776 8 p.
Leontaritis, K.J., Mansoori, G.A., 1988. Asphaltene deposition: a survey of field experiences and research approaches. Journal of Petroleum Science and Engineering 1, 229–239.
Lomando, A.J., 1992. The influence of solid reservoir bitumen on reservoir quality. American Association of Petroleum Geologist Bulletin 76, 1137–1152.
MacDonald, I.R., Bohrmann, G., Escobar, E., Abegg, F., Blanchon, P., Blinova, V., Bruckmann, W., Drews, M., Eisenhauer, A., Han, X., Heeschen, K., Meier, F., Mortera, C., Naehr, T., Orcutt, B., Bernard, B., Brooks, J., de Farago, M., 2004. Asphalt volcanism and chemosynthetic life in the Campeche Knolls, Gulf of Mexico. Science 304, 999–1002.
Mansoori, G.A., 2010. Remediation of asphaltene and other heavy organic deposits in oil wells and in pipelines. State Oil Company of Azerbaijan Republic Proceeding 12–23 Issue 4.
McCaffrey, M.A., Ohms, D.S., Werner, M., Stone, C.L., Baskin, D.K., Patterson, B.A., 2011. Geochemical Allocation of Commingled Oil Production or Commingled Gas Production. Society of Petroleum Engineers Paper 144618. 19 p.
McKinney, D., Flannery, M., Elshahawi, H., Stankiewicz, A., Clarke, E., Breviere, J., Sharma, S., 2007. Advanced Mud Gas Logging in Combination with Wireline Formation Testing and Geochemical Fingerprinting for an Improved Understanding of Reservoir Architecture. Society of Petroleum Engineers Paper 10986. 15 p.
Milkov, A.V., Goebel, E., Dzou, L., Fisher, D.A., Kutch, A., McCaslin, N., Bergman, D.F., 2007. Compartmentalization and time-lapse geochemical reservoir surveillance of the Horn Mountain oil field, deep-water Gulf of Mexico. American Association of Petroleum Geologists Bulletin 91, 847–876.
Peters, K.E., Fowler, M.G., 2002. Applications of petroleum geochemistry to exploration and reservoir management. Organic Geochemistry 33, 5–36.
Pixler, B.O., 1969. Formation evaluation by analysis of hydrocarbon ratios. Journal of Petroleum Technology 24, 665–670.
Pomerantz, A.E., Ventura, G.T., McKenna, A.M., Cañas, J.A., Auman, J., Koerner, K., Curry, D., Nelson, R.K., Reddy, C.M., Rodgers, R.P., Marshall, A.G., Peters, K.E., Mullins, O.C., 2010. Combining biomarker and bulk compositional gradient analysis to assess reservoir connectivity. Organic Geochemistry 41, 812–821.

Romo, L.A., Prewett, H., Shaughnessy, J., Lisle, E., Banerjee, S., Willson, S., 2007. Challenges Associated with Subsalt Tar in the Mad Dog Field. Society of Petroleum Engineers Paper 110493. 15 p.

Schowalter, T.T., Hess, P.D., 1982. Interpretation of subsurface hydrocarbon shows. American Association of Petroleum Geologists Bulletin 66, 1302–1327.

Thanh, N.X., Hsieh, M., Philp, R.P., 1999. Waxes and asphaltenes in crude oils. Organic Geochemistry 30, 119–132.

Valentine, D.L., Reddy, C.M., Farwell, C., Hill, T.M., Pizarro, O., Yoerger, D.R., Camilli, R., Nelson, R.K., Peacock, E.E., Bagby, S.C., Clarke, B.A., Roman, C.N., Soloway, M., 2010. Asphalt volcanoes as a potential source of methane to late Pleistocene coastal waters. Nature Geoscience 3, 345–348.

Vance, I., Thrasher, D., 2005. Reservoir souring: mechanisms and prevention. In: Ollivier, B., Magot, M. (Eds.), Petroleum Microbiology. ASM Press, Washington, D.C, pp. 123–142.

Weissenburger, K.S., Borbas, T., 2004. Fluid properties, phase and compartmentalization: Magnolia field case study, deepwater Gulf of Mexico, U.S.A. In: Cubitt, J.M., England, W.A., Larter, S.R. (Eds.), Understanding Petroleum Reservoirs: Towards an Integrated Reservoir Engineering, vol. 237. Geological Society of London Special Publication, pp. 231–256.

Weatherl, M.H., 2007. Encountering an Unexpected Tar Formation in a Deepwater Gulf of Mexico Exploration Well. Society of Petroleum Engineers/International Association of Drilling Contractors Paper 105619. 10 p.

Whittaker, A., 1991. Mud Logging Handbook. Prentice-Hall, Englewood Cliffs, NJ. 531 p.

Williamson, S.C., Zois, N., Hewitt, A.T., 2008. Integrated site investigation of seafloor features and associated fauna, Shenzi Field, deepwater. In: Gulf of Mexico Offshore Technology Conference Proceedings Paper 19356 8 p.

CHAPTER

6

Surface Geochemistry

OUTLINE

Introduction 217

Microseepage 219

Direct Indicators of Hydrocarbon Microseepage 220

Soil Gas 220

Adsorbers 221

Indirect Indicators of Hydrocarbon Microseepage 222

Microbial 222

Gamma-Ray Spectrometry 223

Iodine 223

Induced Polarization and Resistivity 223

Magnetic Contrasts 224

Microseepage Survey Design and Interpretation 224

Onshore Macroseepage 229

Offshore Macroseepage 230

Locating Potential Seafloor Seep Sites 231

Sea Surface Slicks 232

Water Column "Sniffers" 234

Seafloor Features Associated With Seepage 234

Locating Potential Seafloor Seep Features 235

Sampling Potential Seafloor Seep Sites 240

Analyzing Seafloor Sediments for Thermogenic Hydrocarbons 241

Sea Surface Slicks 244

Oil Slick Sampling and Analysis 246

Offshore Gas Seep Sampling and Analysis 247

Beach Tar Sampling and Analysis 247

References 248

INTRODUCTION

Any review of the history of petroleum exploration would show that oil and gas seeps were important clues for early discoveries. Many of the initial giant oil fields discoveries were made as the direct result of drilling on or near seepage. By studying seeps and the reasons for their occurrence, valuable insight into petroleum systems and fluid flow regimes in an exploration area can be gained that can help reduce exploration risk.

Practical Petroleum Geochemistry for Exploration and Production
http://dx.doi.org/10.1016/B978-0-12-803350-0.00006-4

Surface geochemistry is the search for direct or indirect near-surface expressions of hydrocarbons seepage from deeper subsurface hydrocarbon accumulations. It is based on the premise that all seals over petroleum reservoirs are imperfect and leak to some extent and that these leaked hydrocarbons are driven toward the surface by natural buoyancy (Klusman and Saeed, 1996). This seepage is then detected at or near the surface as elevated hydrocarbon concentrations, changes in sediment mineralogy, and/or a biological response to the presence of the hydrocarbons.

The hydrocarbon seepage sought after in surface geochemistry is usually divided into main two categories, microseepage and macroseepage, as illustrated in Fig. 6.1. MicroSeepage is low-concentration seepage where only small amounts of hydrocarbons are leaking by vertical migration to the surface. This type of seepage is not obvious to an observer at the surface and must be pursued usually through the geochemical analysis of near-surface sediments. In contrast, macroseepage is high-concentration seepage where large amounts of hydrocarbons are leaking and resulting in visible or easily detectable oil and gas at the surface. This type of seepage is usually closely associated with faults or fracture zones that extend down to the vicinity of the reservoir.

Surface geochemistry is a controversial topic. Nearly every explorationist and petroleum geochemist can see the benefit of utilizing information gleaned from finding and analyzing

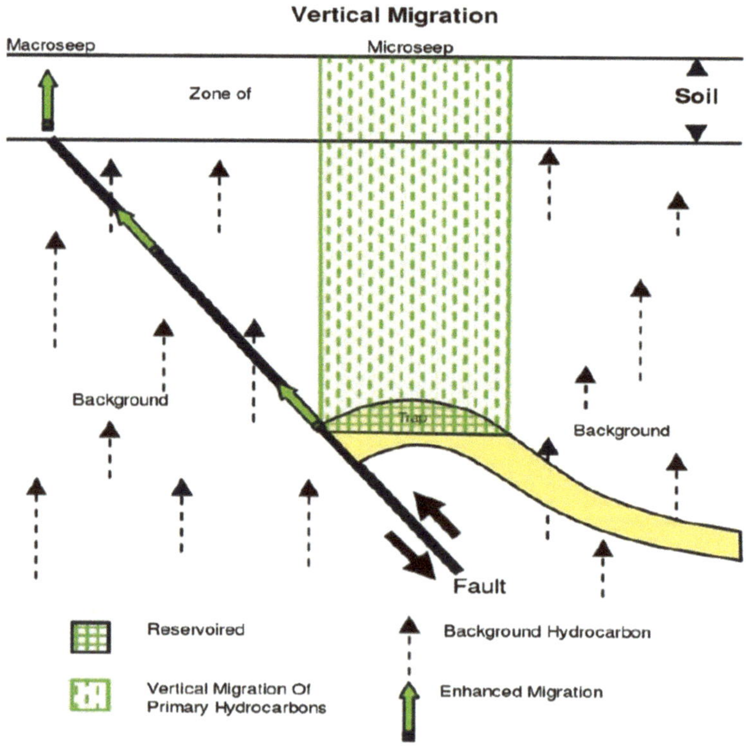

FIGURE 6.1 Schematic of seepage styles. *From Rasheed, M.A., Srinivasa Rao, P.L., Annapurna, B., Zaheer Hasan, S., 2015. Implication of soil gas method for prospecting of hydrocarbon microseepage. International Journal of Petroleum and Petrochemical Engineering 1, 31–41.*

hydrocarbons in macroseepage, but microseepage is quite another story. While some believe whole heartedly in its use, others will deny that microseepage even exists. And of those who concede the existence of microseepage, some do not believe it can actually be used to find subsurface petroleum accumulations. In addition, published reports of the successful application of the detection of microseepage exist for some of the techniques, while other methods are poorly documented and are held suspect by many. There seems to be no middle ground on this topic.

The goal of this discussion of surface geochemistry is to provide the background information about these methods that will enable the reader to begin to formulate their own opinions. It will start with microseepage, focusing on the direct detection of leaking hydrocarbons, followed by a review of how to utilize information from macroseepage both onshore and offshore.

MICROSEEPAGE

The search for microseepage is based on three main assumptions (Schumacher, 1996). The first assumption, as stated earlier, is that petroleum accumulations are dynamic, their seals are imperfect, and therefore all reservoirs leak to some extent. For microseepage, the amount of leakage needed is extremely small in comparison to the volume of hydrocarbons in an accumulation. As a result, for all practical considerations the seal is still effective for containing the accumulation.

The second assumption is that the leaking hydrocarbons can move vertically through thousands of meters of sediments without observable faults or fractures in relatively short periods of time (Schumacher, 1996). While vertical migration is not a process by which large volumes of high-molecular weight hydrocarbons can move easily through the subsurface, it may be adequate for small quantities of low-molecular weight hydrocarbons. Early researchers thought that diffusion was the force behind this vertical migration until Klusman and Saeed (1996) showed it was more likely buoyancy driving the process. Though rates for this vertical movement are still unclear, one study by Rice et al. (2002) calculated hydrocarbon vertical migration rates of 0.6–2 m/day based on changes observed in soil–gas hydrocarbon concentrations over a producing field in response to a waterflood.

The third assumption is that seepage produces changes in the near-surface environment that can be detected, mapped, and related back to the hydrocarbons at depth (Schumacher, 1996). Detectable manifestations of microseepage can take many forms. The most obvious is an increase in thermogenic hydrocarbons in near-surface sediments. However, the presence of these hydrocarbons can also cause changes in the host sediments that can be detected at the surface, such as alteration of the mineralogy that can in turn modify the magnetic and/or electrical properties of the sediments. The presence of seeped hydrocarbons may also have an influence on the microbial community in the near-surface sediment or on the surface vegetation that can be observed and mapped.

The use of microseepage is an onshore exploration tool. Attempts have been made to extend the technique offshore, and there are some reported successes (e.g., Hitzman et al., 2009). However, the conditions observed in the marine environment and the logistical challenges involved in conducting an offshore survey, especially in deep water, are not conducive for the detection of microseepage.

The discussions that follow will review both the direct and indirect indicators of microseepage, as well as some of the conceptions and practices used for survey design and data interpretations.

DIRECT INDICATORS OF HYDROCARBON MICROSEEPAGE

Using direct indicators of hydrocarbon microseepage concentrates on analyzing for the presence of thermogenic hydrocarbons in soils or near-surface sediments. While the majority of the direct methods focus on detecting microseepage using the C_1–C_5 hydrocarbon gases in soils (the soil gas methods), there are a few others that look at some of the higher molecular weight components by employing an adsorber buried in the sediments (passive collection or adsorber method). The manner in which these two classes of methods are applied is very similar even if the background theory and method of investigation are not.

Soil Gas

Soil gas methods look for C_1–C_5 hydrocarbon gases in the surface soil profile and near-surface sediments. These light hydrocarbons are thought to be the most likely manifestation of direct vertical migration characterized by microseepage. There are two basic methods for sampling soil gas for analysis. The first uses a gas probe inserted at least a meter into the soil (Jones and Drozd, 1983; von der Dick et al., 2002). The probe is connected to a pumping system to pull the gas out of the soil. Often, this method incorporates a portable gas chromatograph for field analysis of the gas, however, some versions capture the gas for later analysis in the laboratory. The second method is to dig down at least a meter to sample the soil and quickly place the sample in a sealed can for later laboratory analysis. Because of the logistics of the gas probe, pumping system, and portable gas chromatograph (if used), most soil gas surveys are conducted using canned soil samples.

Once the canned soil samples are back in the laboratory, there are several methods that can be used to obtain gas data. The most common approach is to simply sample the headspace gas of the canned soil samples and analysis for light hydrocarbons by gas chromatography (GC) (Jones et al., 2000). It relies on only free gas from the interstitial spaces of the soil to supply the microseepage signal, and it is very similar to the headspace cuttings gas analysis described in Chapter 3. This calibrated analysis gives both the total amount of hydrocarbon gases present in the headspace and a quantitative distribution of individual gas components. Efforts are usually made to warm the canned soil samples to the same temperature and adjust the results for the amount of soil in each individual sample to help normalize the data.

In addition to looking at the headspace gas, some workers believe a significant portion of the seepage may be adsorbed on the mineral grain and needs to be liberated for proper analysis. To do this, analytical schemes have been developed that will first do a headspace gas analysis followed by opening the can and placing the soil into a sealed blender (Abrams, 1992) or ball mill (Bjoroy and Ferriday, 2002) to liberate adsorbed gases for analysis. Unfortunately, both of these methods have fallen under suspicion because frictional heating with these devices can potentially increase temperatures to the point where hydrocarbons may be generated from organic matter native to the soil. Another approach to looking at adsorbed hydrocarbons was developed by Horvitz (1939, 1969). The soil sample is placed in a sealed chamber and put under vacuum to remove any residual interstitial hydrocarbon gases. The soil's mineral grains are then acid etched to release any adsorbed hydrocarbons, which can then be isolated and analyzed by GC. Some workers in this field use both the interstitial and adsorbed gas to make interpretations, while others use only the adsorbed gas. Adsorbed microseepage has the advantage that the accumulated hydrocarbons should represent an integrated signal over time which can overcome some of the shortcomings of soil gas sampling discussed later.

In addition to the gas chromatographic analysis for the C_1–C_5 hydrocarbons used in all of these methods, carbon isotope analysis should also be attempted for the methane if its concentration is high enough. As stated in Chapters 2 and 4, methane can be either biogenic or thermogenic, and the carbon isotope ratio is a key piece of evidence for making this determination. The carbon isotope ratios of ethane and propane can also be useful for making interpretations if these gases are abundant enough for analysis.

This essentially instantaneous sampling of soil gas, either by probe or by headspace gas analysis of canned soil sample, can be problematic. The flux rate of hydrocarbon microseepage can be impacted by a large variety of phenomena which in turn can have an influence on the concentration of the microseepage in the soil at any given time. Atmospheric conditions such as changing temperature and pressure, humidity, wind velocity, air turbulence, and rainfall (Klusman and Webster, 1981), as well as diurnal temperature changes of the soil (Jones et al., 2000), can influence the short-term concentration of seeping hydrocarbons. In addition, earth tides can cause both the compression and dilation of fractures and faults that influence vertical migration of microseepage (Calhoun and Hawkins, 2002). These influences make it sometimes difficult to get comparable samples over the several hours or days needed to complete a soil gas survey.

Contamination can also be a problem in soil gas sampling. Hydrocarbons from internal combustion engines and spilled petroleum products can find their way into soil samples. When collecting soils for analysis, the sampling area should be inspected for any possible pollution and be clear of any motor vehicle exhaust.

Adsorbers

Microseepage sampling with adsorbers, sometimes referred to as passive soil gas sampling, is a well-established surface geochemistry technique. The use of adsorbers to sample for microseepage has two distinct advantages over soil gas methods. First, adsorbers collect potential microseepage over a period of days or weeks to eliminate the atmospheric and other effects that make it difficult to collect comparable samples with soil gas. Second, they collect a larger molecular weight range that provides more opportunities to recognize thermogenic hydrocarbons. And while the surveys using adsorbers require more field time and expense, these advantages more than compensate for these efforts.

The first practical use of adsorbers for microseepage detection was the Petrex system (Klusman and Voorhees, 1983; Voorhees and Klusman, 1986). It consisted of wire filaments coated with activated charcoal. These filaments were placed under a small glass dome and buried in the soil for at least a week. After retrieval, the wire filaments were "desorbed" in the inlet of a mass spectrometer for analysis of the collected organic materials with molecular weights of up to about 240, equivalent to about a C_{16} n-alkane.

In 1996, W.L. Gore & Associates, Inc., purchased Petrex and developed a more advanced adsorber system. The samplers consist of dual filaments coated with proprietary adsorbents that are sealed into a Gore-Tex sheath. These adsorbers are buried at least a meter deep and left for a period of 7–10 days. The advantage of using the Gore-Tex sheath is that it acts as a membrane to prevent water vapor from saturating the adsorption sites on the filaments while still allowing volatile organic compounds to pass through and be collected (Schrynemeeckers and Silliman, 2014). Back in the laboratory after retrieval, the filaments are "desorbed" into the inlet of a gas chromatograph–mass spectrometer system for analysis. Up to 82 volatile and semivolatile organic compounds from C_2 to C_{20} are identified (Hellwig, 2011). To interpret the data, it is necessary to look at the signal intensity and composition of the compound distribution. The compounds found are separated into those typically found in petroleum, compounds that are typically found in the soil background (biogenic input), and probable alteration products (Schrynemeeckers and Silliman, 2014). From the data, the signal received is classified as background, subsurface oil signature, or subsurface gas signatures and assigned an intensity. When possible, an exemplar compound distribution from a local crude oil is used for comparison.

The service is currently marketed by Amplified Geochemical Imaging LLC (AGI). There are several other adsorber-based technologies available for the collection and analysis of volatile and semivolatile organic compounds in soils, and they are all similar in concept to the AGI system. However, these other techniques are currently configured only for environmental work but could be adapted for use in petroleum exploration.

INDIRECT INDICATORS OF HYDROCARBON MICROSEEPAGE

Not all microseepage exploration methods look directly for low concentrations of thermogenic hydrocarbons in near-surface sediments. Many are focused on searching for evidence of hydrocarbon-induced alteration of sediments that may have taken place, while others look for microbial or botanical changes that could be attributed to hydrocarbon seepage (Schumacher, 1996). In some instances, these methods are indirectly sensing something that may be indirectly indicating the possible presence of seeped hydrocarbons. While these methods may have plausible theoretical bases, there is often more than one possible explanation for the phenomenon they describe. These methods for indirectly indicating hydrocarbon seepage may be useful in some situations, but they generally lack sufficient documented studies to fully support their claims. Considerable research will be needed to understand what is being detected and why, before determining the full value of these indirect methods for hydrocarbon exploration. Brief descriptions of some of the more prevalent methods in use follow. See Klusman (1993) or Schumacher (1996) for more information on indirect detection methods.

Microbial

Microbial surveys entail collecting soil sample and culturing the microbial communities present (Rasheed et al., 2013). The objective is to identify microorganisms capable of metabolizing light hydrocarbons, mainly the ethane through butane-oxidizing microbes (Tucker and Hitzman, 1996). Elevated populations of these organisms are then used as an indicator of the potential presence of elevated light hydrocarbon concentrations due to hydrocarbon seepage. Methane oxidizers are usually not considered due to the potential for biogenic methane production in the soil. While many of the organisms looked for are capable of metabolizing hydrocarbons, they do not exclusively consume hydrocarbons and may be in high concentration for other reasons.

Gamma-Ray Spectrometry

Spectral gamma ray surveys utilize the same principles used for detection source rocks with the spectral gamma ray logging tool. The gamma ray signal detected in near-surface sediments is derived primarily from the naturally occurring radioactive elements uranium, potassium-40, and thorium. Uranium compounds are water soluble when the sediment's Eh is oxidative, but under reducing conditions, uranium compounds precipitate and become fixed in the sediments. Organic matter in sediments, in this case seeping hydrocarbons, can cause reducing conditions to develop which in turn precipitates uranium from natural waters (Fertl and Chilingar, 1988). Gamma-ray spectrometry allows the uranium signal to be separated from the thorium and potassium signals. These instruments can be deployed using aircraft or surface vehicles to conduct surveys for surface geochemistry (LeSchack and Van Alstine, 2002).

Iodine

Using soil iodine as an indicator of hydrocarbon seepage is based on the affinity for light hydrocarbons seeping to the surface to interact with iodine to form low-volatility iodorganic compounds in surface soils (Leaver and Thomasson, 2002). Although other chemical characteristics of soils may affect the amount of iodine present, it is thought to be primarily dependent on the concentration of one or more of the light alkanes ethane, propane, and the butanes (Gallagher, 1984). The source of the iodine was initially thought to be atmospheric (Gallagher, 1984), however, Klusman (1993) suggested a biogenic origin. These iodorganic compounds are metastable and will decompose in a matter of months if the hydrocarbons source is shut off (Leaver and Thomasson, 2002). Because of this, analysis for iodine must be done as soon as possible after sample collection.

Induced Polarization and Resistivity

Induced polarization (IP) and resistivity are two electrical properties measured in near-surface sediments as indirect hydrocarbon indicators. They are usually measured at the same time by inserting two electrodes into the earth surface and passing a current through them. After the resistivity measurement is made, the current is shut off and the IP is measured. The IP is caused by the current placed into the earth "charging" specific mineral phases similar to a capacitor. The polarization measures the slow decay of voltage from this stored charge after the flow of electric current ceases (Sumner, 1978).

Both the increases in resistivity and IP are thought to be in response to the presence of diagenetic minerals that could be related to hydrocarbon seepage. The increase in resistivity can be from precipitation of carbonate minerals occluding pore space. The carbonate can be induced to precipitate when seeping hydrocarbons are oxidized to carbon dioxide by microbes, and the carbon dioxide reacts with excess calcium and magnesium in the pore water (Oehler and

Sternberg, 1984). The IP is thought to be from shallow sulfide minerals, such as pyrite and marcasite (Oehler and Sternberg, 1984). Seeping hydrocarbon can be consumed by sulfate-reducing bacteria in the near-surface sediments producing hydrogen sulfide which in turn reacts with iron in the sediment to form sulfide mineral.

Magnetic Contrasts

The presence of seeping hydrocarbon may either induce the formation of some magnetic minerals, such as magnetite and pyrrhotite, or the destruction of others, such as hematite (Machel and Burton, 1991). Magnetotactic bacteria can produce magnetite, magnetic pyrrhotite, or greigite under anaerobic conditions (Machel and Burton, 1991). The presence of sulfate-reducing bacteria may also contribute to the formation of magnetic sulfide minerals. Under the same conditions, hematite is unstable and is subject to dissolution or reduction. The presence of seeping hydrocarbons helps to establish a reducing environment in these near-surface sediments as well as acting as a food source for the microbial community. Because the presence of seeping hydrocarbons may either increase or reduce concentrations of magnetic minerals, magnetic contrasts and not necessarily high localized magnetism are used as the indicator of hydrocarbon seepage (Machel, 1996). Both airborne and surface magnetic surveys can be used to obtain the data.

MICROSEEPAGE SURVEY DESIGN AND INTERPRETATION

Microseepage surveys can range in size from regional investigations screening for areas of interest to reconnaissance studies of leased acreage down to a focused prospect size. While large-scale regional surveys may be desirable, the sampling density required to provide a meaningful appraisal of potential seepage for the area may not be logistically feasible or economical. More concentrated efforts over small areas are likely to provide more insight into the potential for subsurface oil and gas, especially if the subsurface geology is integrated with the surface geochemical data.

When looking at a focused survey dealing with one or more prospects, the size and shape of the survey must be large enough to accommodate the suspected subsurface targets as well as extend beyond expected anomalies into background areas. Efforts must be made to sample in background areas to establish a baseline for comparison for recognition of anomalous signal. This includes estimating the signal-to-noise ratio for the surface geochemistry technique used to help establish the potential variation that is tolerable in the background signal. Matthews (1996) suggests that background will not be found unless sampling is done beyond a distance equal to the target width from the edge of the target.

Surveys are done either in a grid pattern or as a series of lines. Lines might be used if there is limited access to the area under investigation or limited resources that can be applied to the survey. Another reason for a line survey may be to follow 2-D seismic lines to assist in the integration with subsurface geology. Sample density should be great enough to have at least four samples within the target width (Matthews, 1996), and lines should extend well beyond the target into the background zone. Grid sampling should follow similar guidelines with multiple samples over the target and adequate sampling out into the background areas. Remember that surveys need to include field blanks, laboratory blanks, replicate samples, and replicate analyses of some samples to provide adequate data quality assurance and quality control for the study.

For interpretation, soil gas data are typically plotted on a series of maps at the grid points or at the sample points along transect lines. These data normally include the total C_1–C_5 gas concentration as well as the concentrations of individual gases and any ratios calculated. Some interpretation schemes will eliminate methane from consideration, because it can be either biogenic or thermogenic and concentrate solely on the C_2–C_5 gases. If a grid pattern was used, contours of the data (either as contour lines or color) can be drawn as part of the process of identifying potential anomalies, as shown in the example in Fig. 6.2. For either grid patterns or transects, dot distribution maps (see example in Fig. 6.3) are also useful. This type of map uses a circle (dot) at the sample location with the size of the circle scaled in proportion to the intensity of the variable. The utility of both of the contour or dot distribution maps of surface geochemistry data can be greatly enhanced if they are superimposed on a subsurface structure map at the prospect level.

In addition to maps, X-Y plots, where X is the distance along the transect and Y is the magnitude of the variable under investigation, are very useful interpretation tools. Multiple variables can be plotted on the Y axis to increase the amount of data being compared. These X-Y plots can be paired with geologic cross-sections plotted at the same scale to assist in the integration of the geochemistry with the geology, as shown in Fig. 6.4.

When microseepage is detected at the surface or in near-surface sediments, there are two types of geochemical anomalies that are found (Schumacher, 1996), as shown in Fig. 6.5. An apical anomaly, shown on the left side of Fig. 6.5, is the result of a direct path to the surface for seeping hydrocarbon with the anomaly directly over the leaking accumulation. A halo anomaly, shown on the right side of Fig. 6.5, is the result of seepage-induced mineral precipitation directly over the leaking accumulation. These changes in the sediments divert the vertical hydrocarbon migration around the precipitation zone placing the anomaly over the edges of the accumulation.

Recognizing anomalies is not always easy. Increases in soil gas hydrocarbon concentrations above the background levels are the main indicators for the presence of seepage. Background concentrations need to be estimated from areas on the maps where no seepage is anticipated. Typical background concentrations for soil gases are usually only a few parts per million. Problems may arise when signal levels outside the background areas are also small, perhaps only double or triple the background. Taking into consideration potential sampling and analytical error in making these measurement as well as some variations in the background levels, it is easy to see that recognizing an anomaly in low-concentration data may be difficult in some instances. To help recognize anomalies in the hydrocarbon signal, ratios of the gases may be employed. In this case, the changes in the ratios may be used to indicate differences in the source of the hydrocarbon signal. Some of the ratios that may be used include the percent wet gas, the Pixler ratios, C_2/C_3, and iC_4/nC_4. Again, the ratios should be compared back to the designated background signal. Often the data are also plotted on a Pixler plot or simply cross-plotted to get an additional perspective. And if isotope data on the gases are available, the standard gas isotope interpretation schemes to separate biogenic from thermogenic gas, as discussed in Chapter 4, can be applied. Some additional guidelines to assist in recognizing anomalies from background can be found in Klusman (1993) and Jones et al. (2000).

While the assumption that all seals are imperfect and leak to some extent is not difficult to accept, the ability for the leaked hydrocarbons to migrate vertically and reach the surface

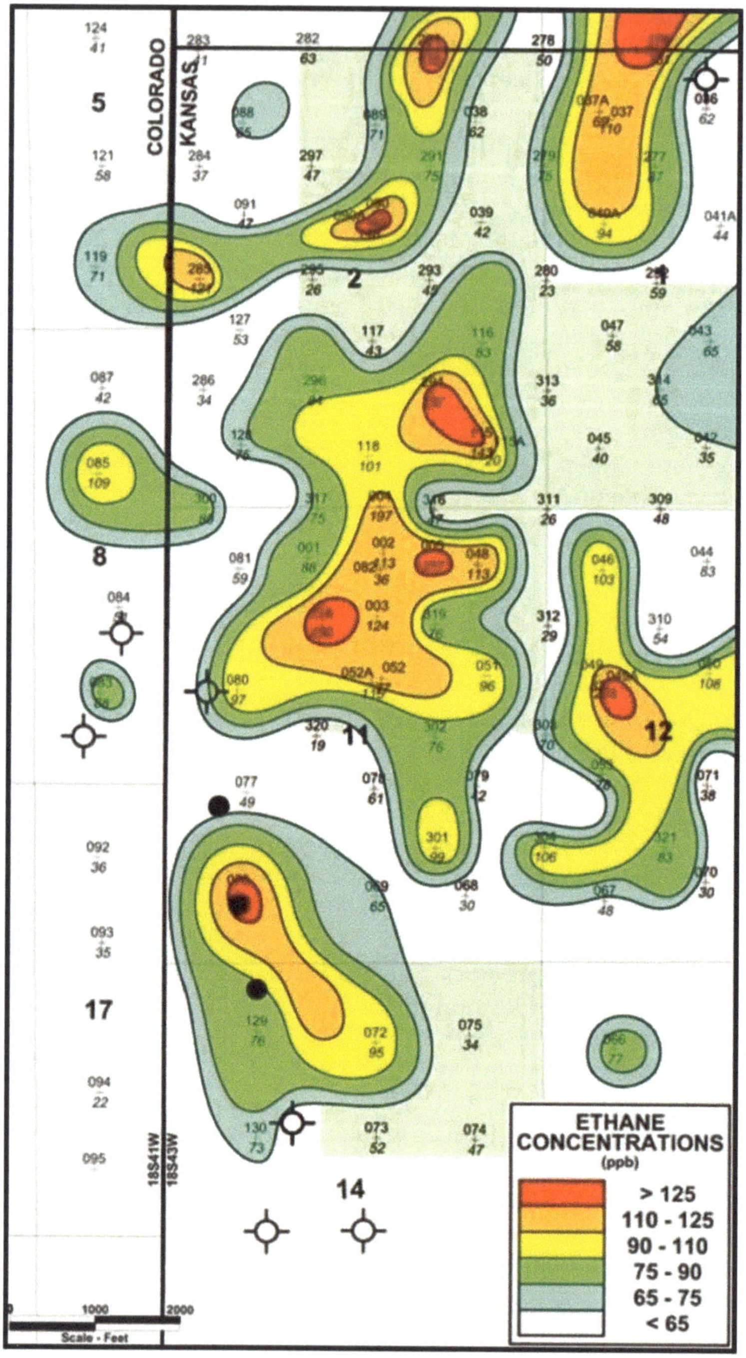

FIGURE 6.2 Contoured ethane concentration map from soil gas. *Data modified from Jones, V.T., LeBlanc, Jr., R.J., 2004. Moore-Johnson (Morrow) Field, Greeley County Kansas: a successful integration of surface soil gas geochemistry with subsurface geology and geophysics. American Association Petroleum Geologists Search and Discovery Article #20022 (2004), 30 p.*

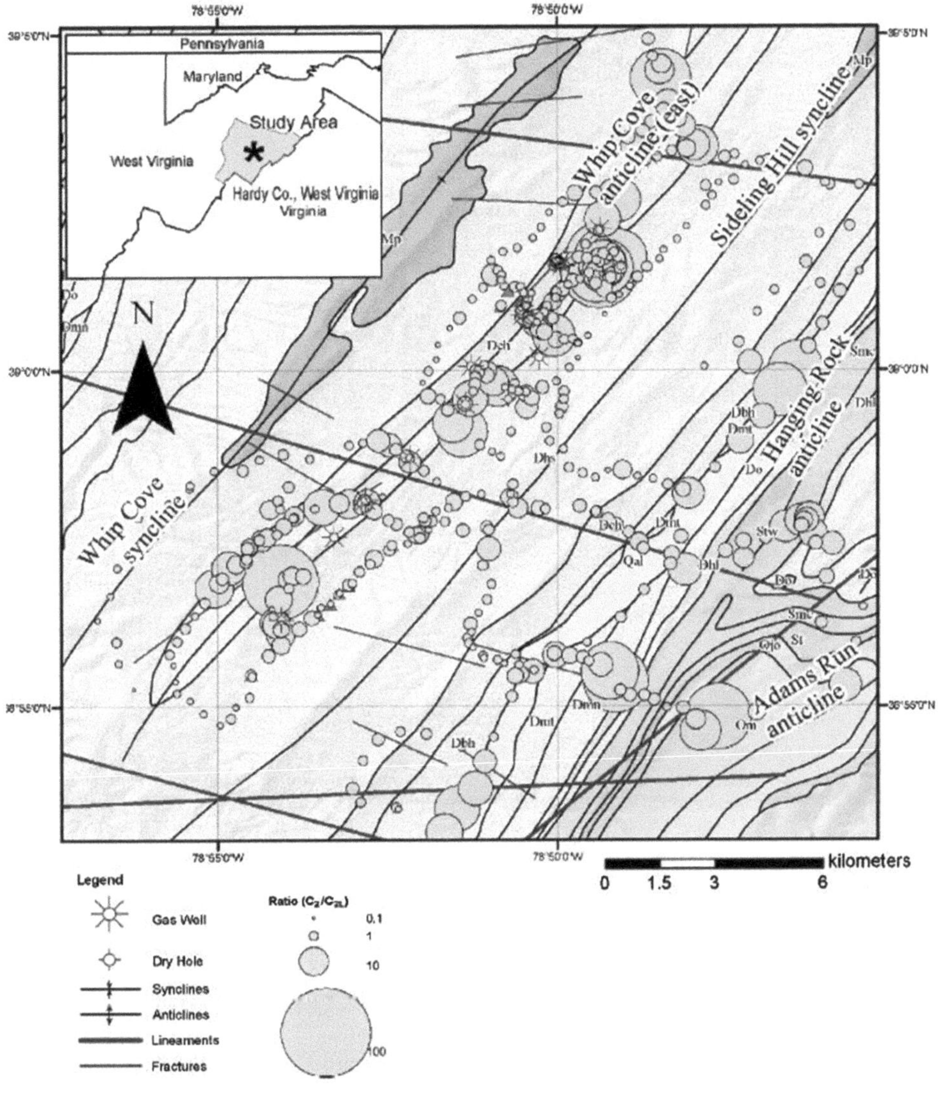

FIGURE 6.3 Dot map of the ethane/ethane ratio from soil gas. Note the geochemical data are overlain on a structure map. *Data from Harbert, W., Jones, V.T., Izzo, J., Anderson, T.H., 2006. Analysis of light hydrocarbons in soil gases, Lost River region, West Virginia: relation to stratigraphy and geological structures. American Association of Petroleum Geologists Bulletin 90, 715–734.*

in detectable quantities is often more difficult to believe. Whether the seepage will reach the surface and be detectable should be dependent on the amount of hydrocarbon leakage and the intervening geology. Faults and fracture zones, as well as porous and permeable sediments, can provide pathways of less resistance for the migrating hydrocarbons and divert them from their vertical migration (Thrasher et al., 1996), and very low permeability sediments can form barriers. Groundwater flow between the leaking accumulation and the

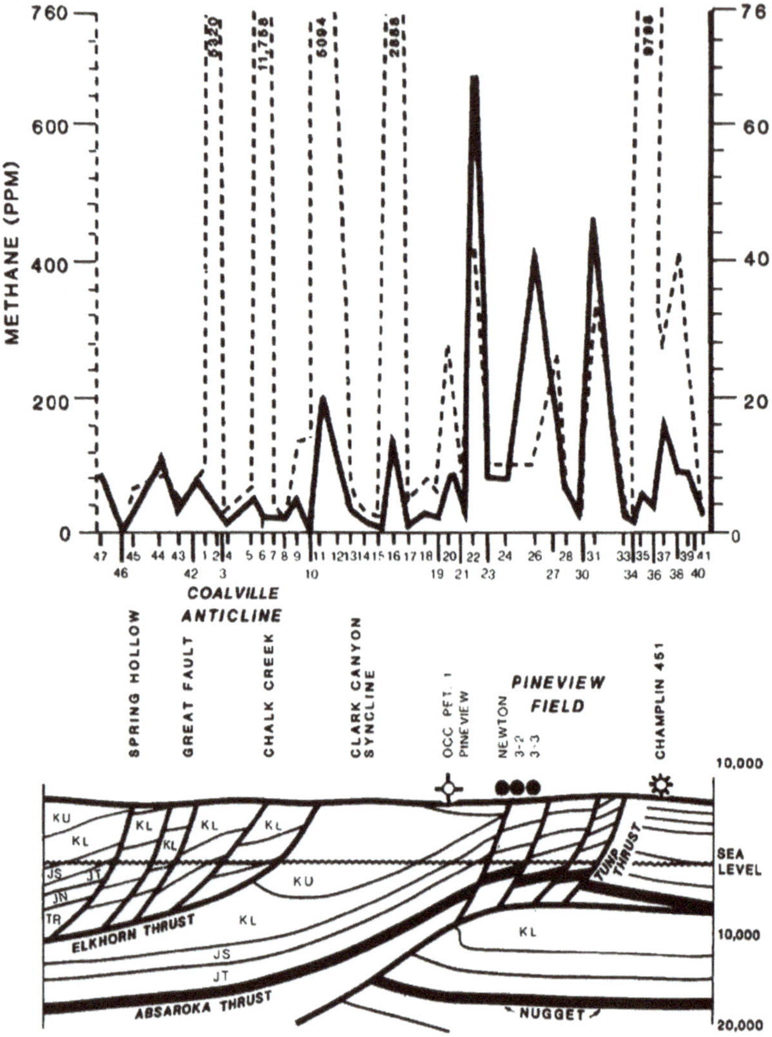

FIGURE 6.4 Transect surface geochemistry data plotted on an X-Y plot, where X is the distance along the transect and Y is the amplitude of the signal, in this case methane as the *dashed line* and propane + butanes as the *solid line*. It is paired with a geologic cross-section to aid in the interpretation of the data. *From Jones, V.T., Drozd, R.J., 1983. Predictions of oil or gas potential by near-surface geochemistry. American Association Petroleum Geologists Bulletin 67, 932–952.*

surface can also diminish and divert this vertical migration preventing the development of clearly defined near-surface anomalies (Holysh and Toth, 1996). Toth (1996) has suggested that a hydrogeologic component needs to be factored into the vertical migration model to better understand where anomalies may occur and the impact of groundwater flow on the intensity.

As a result of these factors, the interpretations of surface geochemistry indications of microseepage cannot stand alone. It is imperative that these interpretations be integrated

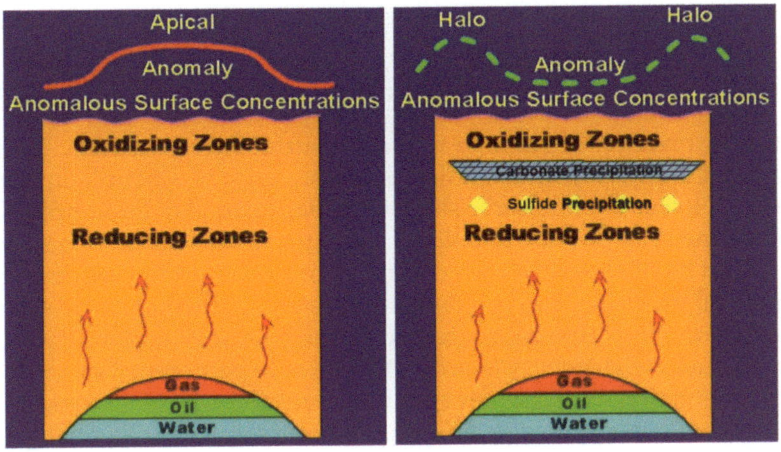

FIGURE 6.5 The two types of microseepage geochemical anomalies. *Modified from Schumacher, D., 1996, Hydrocarbon-induced alteration of soils and sediments. In: Schumacher, D., Abrams, M.A. (Eds.), Hydrocarbon Migration and its Near-surface Expression: American Association Petroleum Geologists Memoir, vol. 66, pp. 71–89.*

with other exploration data, and the interpretations make good geologic sense. If not, they have little value in an exploration program or a petroleum systems analysis.

ONSHORE MACROSEEPAGE

While microseepage may be controversial, macroseepage, especially onshore, is a proven exploration tool that goes back to the beginning of the modern era of petroleum exploration (for a history of seep exploration, see Link, 1952). By recognizing that oil and gas seeps were the product of leakage from petroleum accumulations and that understanding their geology could point back to the location of that accumulation, the petroleum industry got its start. Drake's first well in 1859 was drilled near an oil seep in Titusville, Pennsylvania which set off an early exploration boom focused on oil seeps in Ohio, Kentucky, Indiana, and Illinois. By the mid-1860s, oil seep exploration had spread to East Texas and Wyoming and by 1875 it had reached the Los Angeles basin in California. Initial exploration success in the early 1900s in Iran and Iraq has been attributed to exploiting the presence of oil seeps. And the first exploration efforts in Canada, Europe, Trinidad, Venezuela, Columbia, Ecuador, Argentina, Indonesia, and Burma (now Myanmar) were all driven by seeps.

As discussed earlier, macroseepage resulting in oil and gas seeps is often closely associated with faults or fracture zones that extend down to the vicinity of the reservoir. These faults and fractures provide a migration pathway for significant volumes of petroleum to move from the reservoir toward the surface. In addition to faults and fracture zones, macroseepage may be the result of a structural readjustment that exposes a carrier bed at the surface, exhumation of a reservoir, migration up the sides of salt structures, and migration along unconformities. This macroseepage can manifest as surface pools (tar pools), gas vents including natural "eternal flames," gas bubbling up in springs, streams, ponds, or lakes, oil staining on or oil flowing from outcrops, and mud volcanoes.

While domestic exploration no long relies heavily on seeps, onshore macroseepage in frontier areas can still be useful. As stated earlier, the key is understanding the local geology to relate the surface expression of seepage to the subsurface accumulation. In terms of what can be learned from macroseepage, it indicates the presence of a petroleum system, helping to identify areas of lower exploration risk. The seepage can also be sampled, and the geochemical data obtained can provide detailed information about the oil or gas, its source rock, thermal history, and its relationship with other accumulations in the area using the crude oil correlation, crude oil inversion, and gas data interpretation schemes discussed in Chapter 4.

With respect to sampling oil seeps, try to sample the "freshest" looking material in the seep (e.g., the least viscous) in the vicinity of any active flow. Expect seeped oil to be biodegraded, to have lost volatile material through evaporation, and to be partially oxidized. The extent of these alterations will determine the amount of geochemical insight you may receive from the sample.

Gas seeps in springs, streams, ponds, or lakes can be sampled by placing an Isojar or 1 quart paint can in the water, filling the container with water, inverting the can or jar over the seep, and allowing the bubbling gas to displace at least half the water before placing the lid on the can or jar. Both gas composition and carbon isotopes ratios may be obtained from the sample.

OFFSHORE MACROSEEPAGE

In contrast to onshore, offshore macroseeps are still being searched for as part of comprehensive offshore exploration programs. If found and sampled, offshore macroseeps can provide valuable insight into the petroleum systems and fluid flow regimes operating in the basin prior to any exploration drilling. Seafloor seeps can help identify areas with high potential and help to risk prospects, and they can provide detailed information about the oil, its source rock, and thermal history. Taken together with the subsurface geology, seafloor seeps can be used to identify and map exploration play fairways ahead of the drill bit.

Unfortunately, seafloor macroseeps do not occur in every sedimentary basin offshore. They tend to be found in basins where there are rich source rocks capable of charging large traps, where oil generation and/or migration is ongoing or has occurred in the geologically recent past, and where vertical conduits, such as faults or salt diapirs, come near or reach the seafloor (Bolchert et al., 2000; Abrams et al., 2001). Seafloor seeps do not usually occur where source rocks are absent, lean, or immature or if generation and/or migration has occurred in the distant geologic past. And they will likely be absent where thick, unfaulted, and undisturbed Tertiary cover limits or prevents the migration of accumulated hydrocarbons to the seafloor.

There are a few things to keep in mind while using seafloor seep data. While seafloor seeps indicate the presence of oil/gas in the subsurface, they do not indicate how much petroleum there is in the accumulation. Only the drill bit can answer that question. Although seeps document leakage from a hydrocarbon accumulation, they do not indicate blown seals. And, the lack of seeps does not indicate the lack of a petroleum system. It only indicates a lack of seepage.

The workflow for using seafloor seeps in an exploration program is not as simple as for onshore macroseeps. Potential seep sites must first be located on the seafloor by various means. Then these sites must be sampled and the recovered sediments analyzed for thermogenic hydrocarbons. After analysis, the data can be interpreted as to whether thermogenic hydrocarbons are present or absent and if present what can be deciphered from the data. All this information needs to then be integrated with the subsurface geology to see what impact it has on the exploration program and petroleum system analysis. The following discussions will provide more details of each of these workflow steps and some of the difficulties that might be encountered along the way.

LOCATING POTENTIAL SEAFLOOR SEEP SITES

To find seafloor seeps, it helps to understand their anatomy. A good example of seep structure is in the model of the Coal Oil Point Seeps (offshore California) shown in Fig. 6.6. Macroseepage from an oil reservoir leaks to the seafloor. A portion of the petroleum remains at or just below the sediment–water interface while some of the hydrocarbons are released

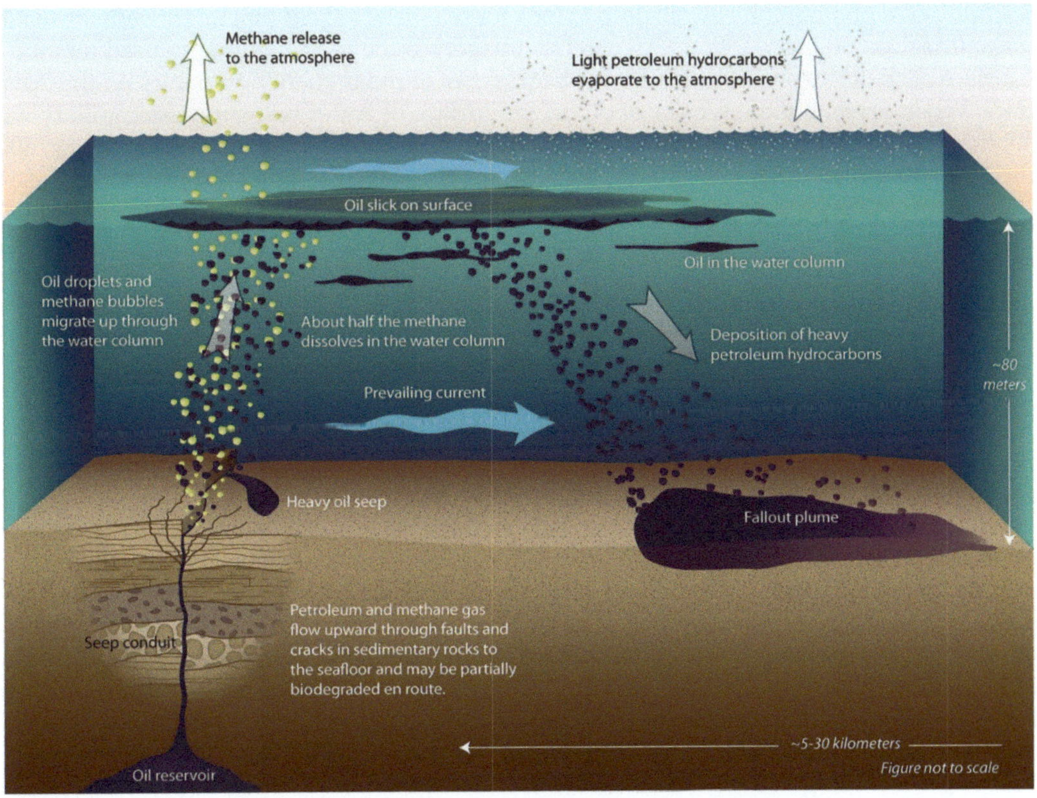

FIGURE 6.6 Anatomy of a seafloor seep and associated surface slick. *Image from Woods Hole Oceanographic Institution.*

to the water column, with gas bubbles and oil droplets heading toward the surface. Some of this oil and gas will go into solution in the seawater, and if the seep is deep enough, all the rising gas bubbles may go into solution in the seawater before they reach the surface (MacDonald et al., 2002). Oil droplets that do reach the sea's surface will "pancake" out forming a very thin layer of oil as a slick. The slick is acted upon by a variety of processes including evaporation, biodegradation, oxidation, and emulsification (ITOPF, 2011). Eventually, some of the oil may congeal into dense masses and sink to the ocean floor. This model indicates that clues to the locating seafloor seep can be found on the sea's surface, in the water column, and on the seafloor.

Locating seafloor seeps is not always an easy task, especially in deep water. The search usually begins with an evaluation of a basin's geology to determine if seepage is probable. After the geologic assessment, the actual search often begins with a search for sea surface slicks as initial indicators of possible macroseepage. Focus will then shift to the water column and seafloor to provide additional location information.

Sea Surface Slicks

Observations using synthetic aperture radar (SAR) satellite imagery, airborne laser fluorescence, and aerial photography are all used to look for sea surface slicks that might indicate seafloor seepage (MacDonald et al., 1996). SAR satellite imagery is the most commonly used method for sea surface slick detection. SAR detects the backscatter of radar energy off waves on the ocean's surface. When present, the viscosity of oil on the surface of water will suppress capillary wave formation, causing what appears to be a "calm spot" on the ocean surface (Lennon et al., 2005). A slick appears in the SAR satellite image as dark spots against a lighter background because the slick reflects back less energy to the satellite due to this suppression of capillary waves (Williams and Lawrence, 2002). An example of some SAR images of oil slicks is shown in Fig. 6.7.

Hydrocarbon seepage is not the only cause for sea surface slicks. Pollution from offshore drilling and production operations, as well as shipping discharges and accidents can form slicks. Seabed bathymetry and currents or tidal flow can combine to dampen waves and mimic slicks on SAR images (Jones et al., 2005). And, biological events such as algal blooms, coral spawning, and schooling fish feeding can produce biological slicks that can mimic natural seepage slicks in their appearance (Jones et al., 2005).

To help distinguish slicks from natural seepage versus slicks formed from pollution or other causes, the slick's location, size, shape, and direction of flow are recorded along with its relationship to subsurface and near-surface geological features (Kanaa et al., 2005). Repeat data also help to identify slicks from natural seepage, with persistent natural seeps being observed on repeat dates (Williams and Lawrence, 2002). However, in areas of intermittent seepage, it may be more difficult to document the repeatability of natural seepage.

Extreme ocean conditions can hinder slick detection by SAR. Rough seas create more backscatter, so the slick blends into the background (Williams and Lawrence, 2002). In calm seas, little backscatter bounces back from the ocean surface, and the slick blends in with the background (Williams and Lawrence, 2002).

Although not being actively used today, airborne laser fluorescence is an effective means of locating hydrocarbon slicks (e.g., O'Brien et al., 2002). Aromatic hydrocarbons from oil will fluoresce when excited by ultraviolet light. A laser light source mounted on an aircraft pointing

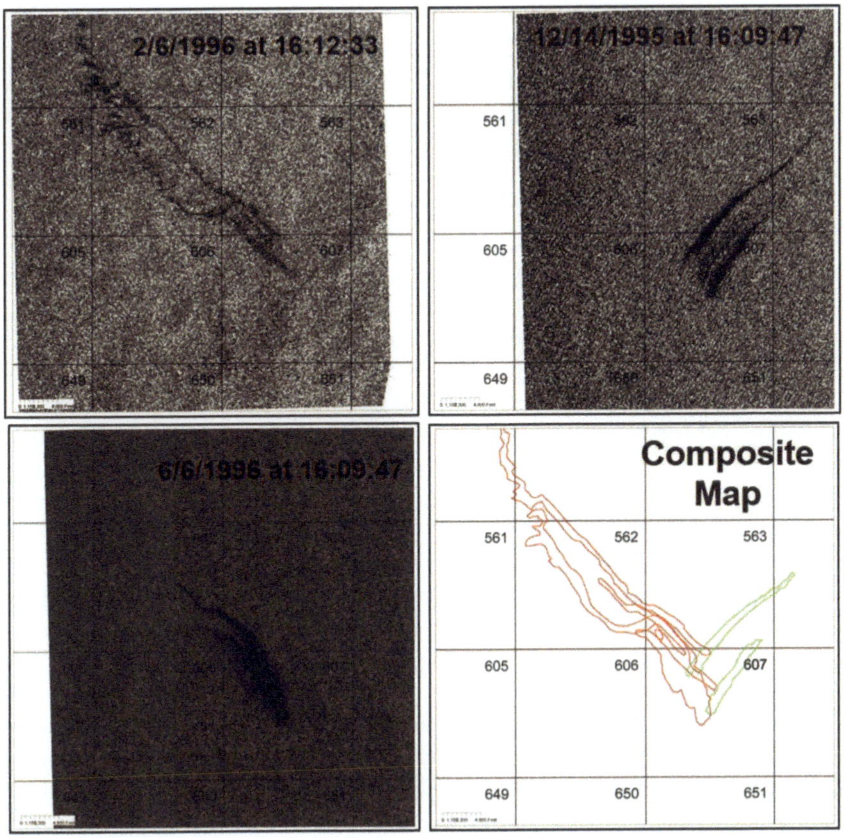

FIGURE 6.7 Three synthetic aperture radar images over the same area of the Gulf of Mexico on different dates. Dembicki Jr., H., Samuel, B.M., 2008. Improving the detection and analysis of seafloor macro-seeps: an example from the Marco Polo field, Gulf of Mexico, USA. In: International Petroleum Technology Conference Proceedings Paper 12124 18 p. The composite map shows the sea surface slicks are repeatedly forming with their focal point located in the north central part of Block 607.

down toward the sea's surface is used as the excitation light source to produce a fluorescence signal (Clarke et al., 1988). The spectra and other characteristics of the fluorescence allows oil to be distinguished from naturally occurring biological fluorescence (Williams, 1996). The major shortcoming of this method is that a low sea state is important in the successful application of the technique, limiting when and where it can be used. Improvements in the instrumentation used in airborne laser fluorescence have made it less sensitive to sea state, and the technique is currently used for oil spill monitoring (e.g., Lennon et al., 2006). A revival of the petroleum exploration application may be in the offing.

While used more often in oil spill and pollution studies, aerial photography of coastal and offshore regions can also be used to locate sea surface slicks from natural seepage (MacDonald et al., 1996; Garcia-Pineda et al., 2016). The higher resolution of aerial photography when compared to satellite imagery can often provide additional insight about the origin of the slick, the spreading of the oil, and eventual dispersion.

Finally, observation from ships, especially seismic vessels, provides good information about sea surface slick. In addition to locating slicks, ship observation can add information about the color of the sheen on the water, any odor of hydrocarbons ("fuel-like smell"), and the presence of oil drops reaching the sea surface, as well as providing an opportunity to sample the slick (see section on slick sample).

Water Column "Sniffers"

In addition to sea surface slicks, there are clues to the location of seafloor seeps that can be found in the water column. In the 1960s and 1970s, research groups began to notice light hydrocarbon gases could be detected in the water column above natural hydrocarbon seeps (e.g., Bernard et al., 1976). Most of this early work dealt with water samples taken at depth and analyzed in shore-based laboratories. One of these research groups at Gulf Oil Company was able to develop a shipboard system (Jeffery and Zarella, 1970; Mosseau and Glezen, 1980) for "sniffing" hydrocarbon gases in the seawater. It consisted of a towed "fish" at some depth below the surface with an inlet for a pumping system that would bring water samples to a surface ship. The water would be stripped of its dissolved gases which were then analyzed by GC. The position of the ship at the time of sampling was also recorded.

Water column gas data are most valuable when there is a good understanding of the subsurface geology, and the seepage detected can be related to the potential migration pathway from reservoir to the surface and potential seep features on the seafloor (e.g., O'Brien et al., 2002). The water column gas data are typically collected at the same time as 2-D seismic acquisition, putting the gas data into context with subsurface geological data (Mosseau, 1980). In addition to the subsurface geology, the seismic data provides information about the seafloor bathymetry and water column features such as gas bubble plumes (Geyer and Sweets, 1973; Sweets, 1974). Because the pathway from the seafloor toward the sea surface may not be direct due to subsea currents, this acoustic imaging of the water column and seafloor is essential for proper interpretation of this data.

Seafloor Features Associated With Seepage

Identifying the location of potential seafloor seeps ultimately comes down to recognizing seafloor features that may be associated with seeps. Bathymetric features such as seafloor mounds and hills, fault scarps, and seafloor depressions and craters, as well as high impedance contrast at the seafloor and near-subsurface chaotic, amorphous, and wipe-out zones, may indicate the presence of seafloor seeps and possibly associated chemosynthetic communities and/or authigenic crust or pavements (Roberts, 1995).

Mounds may be bulging sediments from hydrocarbon charging (Roberts, 1995), bioherm-like buildups from chemosynthetic communities (Sassen et al., 1993), or hydrates extruding from the seafloor (Sassen et al., 1999), while larger hill-like structures may be mud volcanoes (Roberts, 1995). These features may be related to salt diapirs or other salt-related seabed features, which provide migration pathways for seeped hydrocarbons to reach the seafloor. Chemosynthetic communities are colonies of chemoautotrophic organisms using seeping H_2S, methane and heavier hydrocarbons as an energy source. They usually occur in water depths greater than 500 m. Mud volcanoes are cone-like structures that form when mud, a slurry of sediment with water and gas, are extruded at the seafloor. And hydrates are ice-like solids made of methane and water where the methane molecules are enclosed in molecular cages of water molecules. Methane hydrates are only stable at the seafloor and in seabed sediments at temperatures below about 4°C and at water depths greater than about 350 m. Fault scraps may be the intersection of the migration conduit for the seeping hydrocarbons with the seafloor, and seafloor depressions or craters could represent fluid expulsion features, often called pockmarks (Roberts, 1995). Fluid expulsion features indicate focused flow discharging water, oil, and/or gas at sediment–water interface. They are often associated with faults and fracture zones.

High impedance contrast at the seafloor may be indicative of oil or asphalt saturation of bottom sediments, chemosynthetic communities, and/or hardgrounds of authigenic carbonate crusts or pavements (Roberts et al., 2006). Authigenic carbonate buildups are formed from HCO_3^- from microbial oxidation of hydrocarbons at or near the seafloor reacting with calcium in seawater to form carbonate minerals (Fang, 1991). Other diagenetic minerals such as barite and pyrite may also be included depending on the microbial community presence and the chemistry of the sediments and pore waters. Occasionally, the high impedance contrast is slightly below the seafloor. This may be the result of a buried chemosynthetic community or a paleohardground.

Near subsurface acoustic data exhibiting a chaotic or amorphous character are often called wipe-out zones. These wipe-out zones may be indicative of shallow gas charging of the sediments or hydrates disrupting the sediments bedding (Roberts, 1996). If the gas charging is associated with a fault and extends below the near-surface sediments, a gas chimney could be indicated (Ligtenberg, 2005; Connolly et al., 2008).

Occasionally, near-vertical bands of reflections that extending up from the seafloor toward the sea surface are observed. These could represent columns of gas bubbles escaping from the seafloor (Sweets, 1974). These gas bubble plumes do not always reach the sea surface, especially in deep water. In these instances, the escaping gas goes into solution in the seawater.

In deep-looking seismic data, areas with bottom simulating reflectors (BSRs) may be indicating the presence of gas hydrates. BSRs roughly parallel the seafloor reflection. If caused by a hydrate, the BSR marks the contrast between gas hydrate and the underlying gas-saturated sediments (Kvenvolden and Barnard, 1982). This will occur in deep water where subseafloor sediments are within the hydrate stability field.

Locating Potential Seafloor Seep Features

The most effective way of finding potential seep features offshore is acoustic mapping of the seafloor. Acoustic mapping can be accomplished using conventional seismic, side scan sonar, and/or multibeam sonar. To further demonstrate why these observed seafloor features should be considered seeps, the seafloor maps are then integrated with deeper looking seismic data to confirm that the seafloor features are tied via migration conduits to potential subsurface hydrocarbon sources. This process usually starts with seismic data.

Seismic data are almost always available in a seep search area, or it is in the process of being acquired. While 2-D seismic data can see bubble plumes in the water column above seep features (Sweets, 1974) and can be an effective means of locating seeps (Roberts, 1995), it is limited to imaging the seafloor only along the seismic grid lines. This is often too restricting for a successful seep search. 3-D seismic data, however, provide greater areal coverage but needs to be processed properly for the extraction of seafloor data including amplitude information (Roberts et al., 1996, 2000). This seafloor extraction will give both bathymetry and some indications of acoustic impedance to help recognize potential seafloor seep features, as shown in the example in Fig. 6.8. Depending on acquisition parameters and water depth, problems sometimes arise with this approach because of the relatively low resolution of the 3-D seismic data. Small to moderately sized seep features (a few meters or tens of meters in diameter) may be easily overlooked due to the lower resolution of the 3-D data (Dembicki and Samuel, 2007).

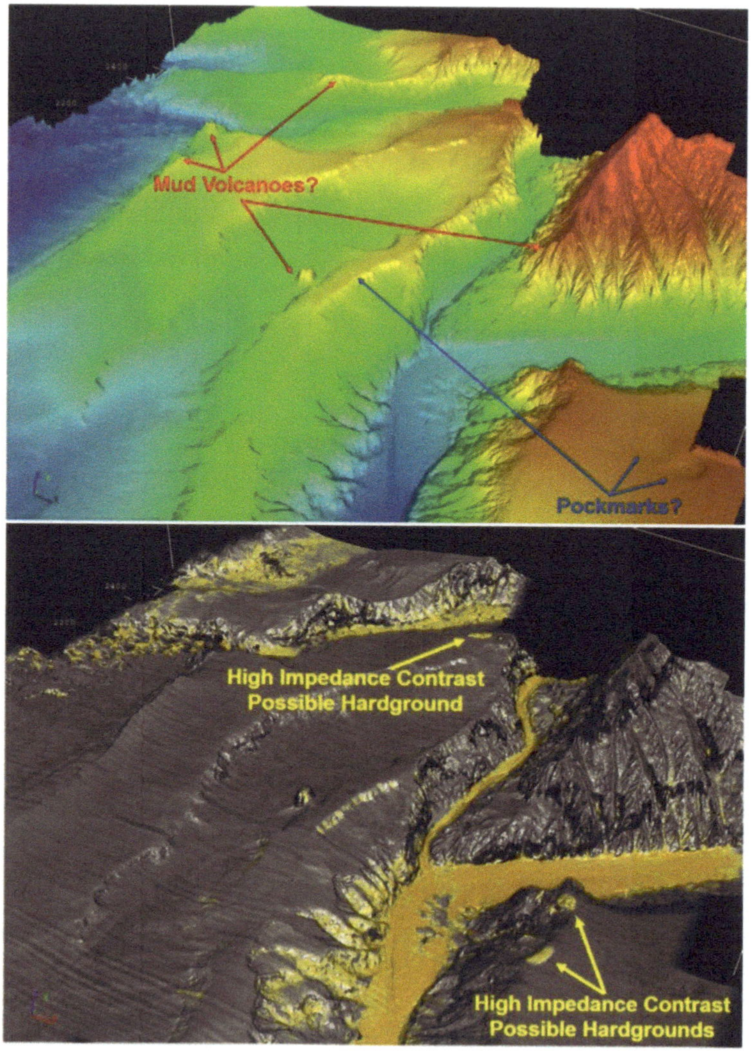

FIGURE 6.8 Potential seeps features recognized in the bathymetry (upper image) and near-surface amplitudes (lower image) extracted from 3-D seismic. *From Dembicki, Jr., H., 2014. Confirming the presence of a working petroleum system in the eastern Black Sea Basin, offshore Georgia using SAR imaging, sea surface slick sampling, and geophysical seafloor characterization. American Association of Petroleum Geologists Search and Discovery Article #10610 (2014), 21 p.*

When only 2-D seismic is available or better seafloor resolution is needed, side scan and/or multibeam sonars can be used effectively. Side scan sonar is usually acquired using a towed source/receiver vehicle at some depth. The side scan sonar images of the seafloor are obtained by emitting a continuous fan-shaped beam of sonar pulses while moving. The intensity of the acoustic reflections from the seafloor (backscatter) creates a series of image slices, which are then put together to form a mosaic image of the seafloor, as demonstrated in Fig. 6.9.

In contrast, multibeam sonar is usually attached to a surface vessel's hull. It also uses a fan-shaped beam and measures the time difference between sound emission and reception to find depth, as well as record backscatter energy (Orange et al., 1999). While the depth data provide a means of making a bathymetric map of the seafloor, the backscatter gives a measure of how strongly an acoustic pulse reflects from the seafloor and the amount of scattering created by the seafloor (Orange et al., 2010). This reflectivity depends on the composition of the seafloor (mud, sand, gravel, bedrock, or a mixture of these) and how rough or lumpy it is (Zhang and McConnell, 2010). An example of combined multibeam bathymetry and backscatter data is given in Fig. 6.10. A recent development in multibeam sonar now allows bubble plumes to be detected in the water column in addition to the standard data (e.g., Barnard et al., 2015). Because it gives both bathymetry and backscatter information, multibeam is the preferred seafloor mapping method for potential seep feature locations.

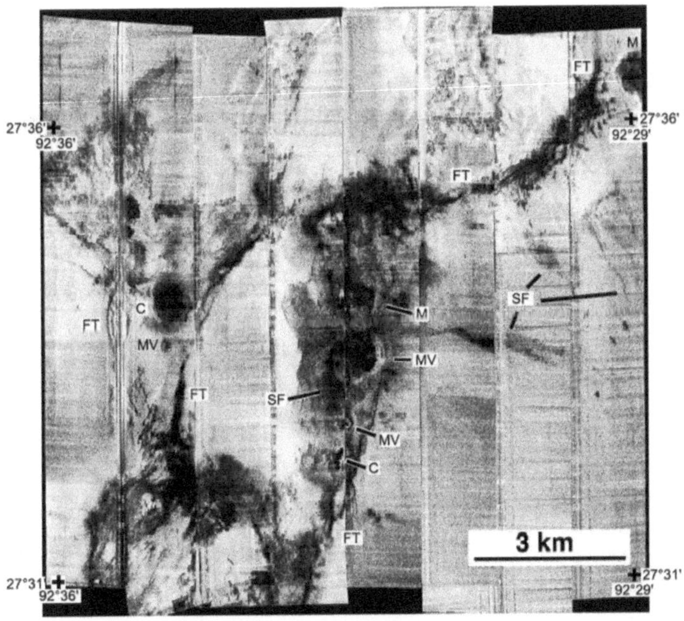

FIGURE 6.9 Side scan sonar mosaic of potential seafloor seep features with the *darker areas* showing high acoustic backscatter, and *lighter areas* show low acoustic backscatter. Features are labeled FT for fault trace, C for crater or pockmark, SF for sediment flow, MV for mud volcano, and M for mound. *From Sager, W.W., MacDonald, R., Hou, R., 2004. Sidescan sonar imagining of hydrocarbon seeps on the Louisiana continental slope. American Association of Petroleum Geologists Bulletin 88, 725–746.*

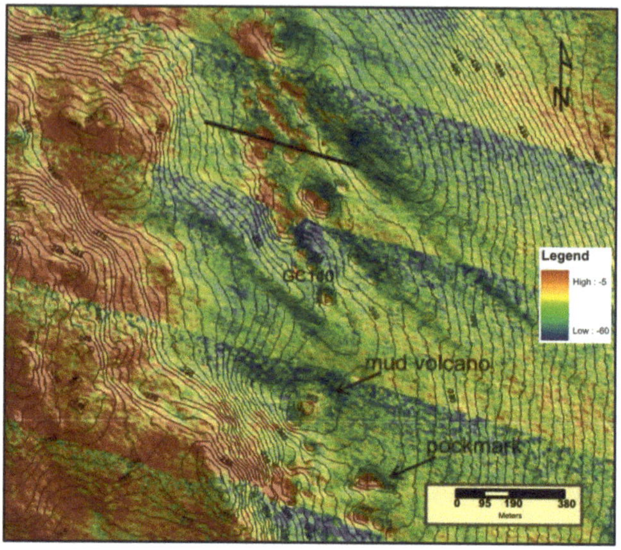

FIGURE 6.10 Contoured multibeam bathymetry with color scaled multibeam backscatter overlay indicating a mud volcano and pockmarks associated with seepage. *From Zhang, Z., McConnell, D.R., 2010. Backscatter characterization of seep-associated seafloor features in the vicinity of Bush Hill, northwest Green Canyon, Gulf of Mexico. In: Proceeding of the Offshore Technology Conference Paper 20662, 13 p.*

When using either side scan sonar or multibeam sonar for seafloor mapping, subbottom profiling is often done in tandem to provide a view into the near-surface sediments to help identify potential seep related features. Subbottom profiling sonar systems can characterize the stratigraphy of sediment or rock below the seafloor, including recognizing the chaotic or amorphous character of wipe-out zones that may indicate the presence of shallow gas charging or hydrates, as illustrated in Fig. 6.11 (Orange et al., 1999; De Beukelaer et al., 2003). Unfortunately, subbottom profiling is confined to a single line along the center of the side can or multibeam swath and may not cover all potential seep features mapped.

There have been some seep studies carried out using autonomous underwater vehicles carrying side scan, multibeam, and subbottom profiling sonars (e.g., Dembicki and Samuel, 2007, 2008). This combination of high-resolution multibeam bathymetry, side scan sonar, and subbottom profiling provides details about the nature of the seafloor features that cannot be obtained by ship mounted instruments. This facilitates the differentiation of seeps from sediment/structural features significantly enhancing the probability that the hydrocarbon seeps, if present, can be recognized and efficiently sampled (Dembicki and Samuel, 2007).

Once potential seafloor seep features are identified by acoustic mapping, it is necessary to link these features to migration pathways from subsurface structures to the surface. This is done by integrating the seafloor maps with conventional deeper imaging seismic, as illustrated with the seismic section in Fig. 6.12. If found, these pathways establish a cause and effect relationship between the potential subsurface accumulations and the potential seafloor seep features.

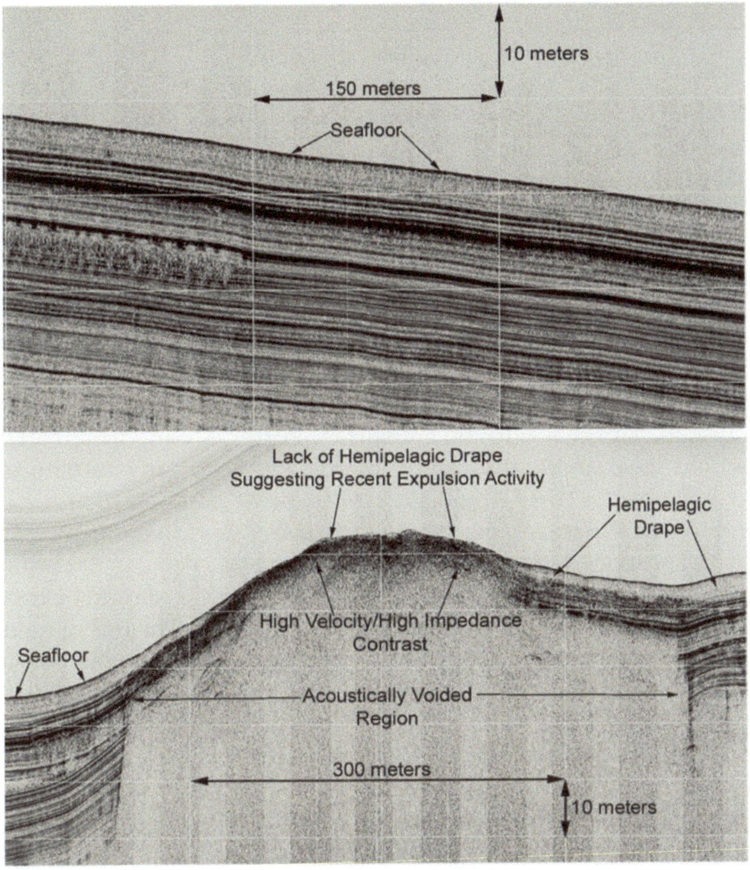

FIGURE 6.11 Subbottom profiler records showing an area with no seepage (top image) and an area with seepage (bottom image). *After Dembicki, Jr., H., Samuel, B.M., 2007. Identification, characterization, and ground-truthing of deep-water thermogenic hydrocarbon macro-seepage utilizing high-resolution AUV geophysical data. In: Proceeding of the Offshore Technology Conference Paper 18556, 11 p.*

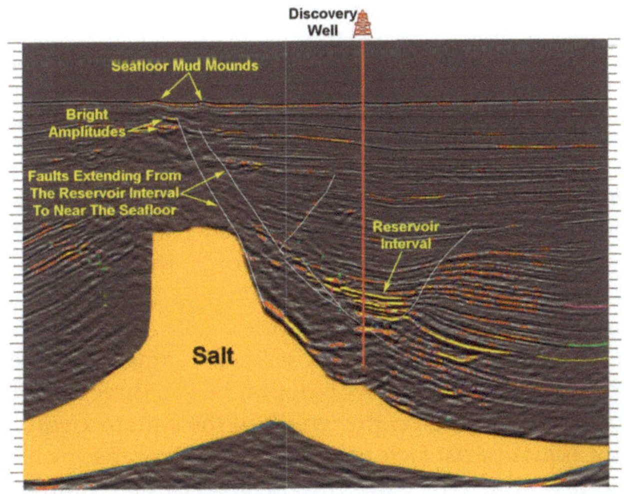

FIGURE 6.12 A 2-D cross-section extracted from a 3-D seismic survey showing faults extending up from a reservoir interval to near the seafloor associated with shallow bright amplitudes and seafloor mud mounding suggest the potential for hydrocarbon seepage. *From Dembicki, Jr., H., Samuel, B.M., 2008. Improving the detection and analysis of seafloor macro-seeps: an example from the Marco Polo field, Gulf of Mexico, USA. In: International Petroleum Technology Conference Proceedings Paper 12124, 18 p.*

SAMPLING POTENTIAL SEAFLOOR SEEP SITES

After potential seafloor seep locations are picked from the seafloor mapping, they need to be sampled to confirm they are indeed seep features. Sampling is typically done with a piston coring device. Seafloor seep features range in size from a few meters to more than 500 m in diameter. Even though a seep feature may be fairly large, the area of active seepage with a high concentration of thermogenic hydrocarbons is often confined to an area less than a few tens of meters in diameter. Therefore, positioning the coring device prior to the drop is critical for successful sampling.

Initial positioning using GPS can place the ship's surface location to within ±5 m above the seafloor feature. In calm seas with no subsurface currents, core drops can be within about 10 m of the specified location depending on depth. However, any motion of the ship may propagate cable sway of the coring device that can be as great as 10 m for every 1000 m of water depth. Add to that the possibility for deflection by subsurface currents, and it is easy to envision missing the seafloor target with a blind core drop.

To mitigate these problems, a ship mounted subbottom profiler should be used to confirm the presence and location of the seep feature prior to the core drop. The use of an ultra-short baseline acoustic positioning system is also recommended. This uses an acoustical beacon attached to the coring device to track its bottom location. Using both these devices, the ship's location can be adjusted to ensure that the seafloor location the core is taken at is within the boundary of the seep feature. If the seep feature is small (less than a few meters in diameter), multiple cores may be taken to compensate for any uncertainty in the positioning (Orange et al., 2008).

The piston coring device should be equipped with at least a 6-m core barrel with a removable liner. It should be triggered to drop about 3 m above the seafloor to ensure appropriate penetration into the sediments. Once the core is taken, it needs be brought to the surface as quickly as possible. This is especially important if hydrate is encountered. Given time, the higher temperatures and lower pressures in the shallower part of the water column may decompose the hydrate on the trip to the surface, and the expanding gas may extrude the core.

Once the piston coring device is back on the ship, the core liner is removed for sampling. The sediments are extruded from the core barrel liner in a clean area away from possible hydrocarbon sources. The recovered core is measured and quickly examined. The sediment's type, appearance, color, and consistency (water content, cohesion) are recorded. The presence/absence of the odor of oil and/or hydrogen sulfide (H_2S) is noted. Features such as oil staining, authigenic carbonates, hydrates, gas bubbling, bioturbation, sedimentary structures, organisms, gas fractures, gas cut, etc., are also documented.

The near-surface sediments in the core may be contaminated with pollution and marine organisms, and thermogenic hydrocarbons may not migrate all the way to the surface and may only be in the lower portion of the core. Alteration of the seeped hydrocarbons may also diminish deeper into the sediment column. This makes the lower portion of the seafloor core the focus of sampling using a sampling plan modified after Abrams (2013). The bottom 10 cm of the core is usually removed to avoid any possible contamination. Because the geochemical signature of the sediment can change with depth, two to three 20-cm sections are selected for geochemical analysis, depending on the length of the core recovered. Each 20-cm section is subdivided into four equal parts: two parts are sealed in cans for headspace gas analysis (see Chapter 3 for the procedure for collecting headspace gas samples), and two parts are preserved (usually flattened between aluminum foil sheets and stored in sealed plastic bags) for high-molecular weight hydrocarbons analysis. These duplicate samples at each depth are used for quality control studies, reproducibility checks, and as a backup. The samples should be frozen at a temperature of −15°C or lower to inhibit microbial activity.

Occasionally, thermogenic hydrocarbons may occupy a zone of porous sediment in the core, or an interval may exhibit staining. Similarly, gas bubbling and fractures may indicate zones of potentially high gas saturation in the core. In these situations, sampling the obvious oil or gas saturation takes precedence, and fixed interval sampling may be abandoned.

If hydrates are encountered, the gas they contain can be sampled using a simple procedure. To prepare, have a large bucket with clean, fresh seawater close to the core processing area with a supply of clean headspace gas cans on hand. When hydrate is observed, quickly transport the hydrate to the bucket of seawater. Completely submerge and fill a headspace gas can with water, and invert the can after making sure there is no atmospheric gas in it. Hold the hydrate chunk under the open mouth of the can until the can is partially filled with gas from the dissociating hydrate. Leave at least 2 in. (5 cm) of water in the can. Carefully place the lid of the can over the opening of the inverted can, and seal by pressing the can and lid against the bottom of the bucket. Repeat the procedure, filling at least two cans with gas. The cans of dissociated hydrate gas and seawater should be frozen and stored upside down for transport with the headspace gas samples.

In the past after geochemical sampling, the remaining core material was usually dumped overboard, but this is changing. In a 4.0-m core, the geochemical sampling only accounts for about 15% of the recovered sediment. Important observations can still be made on the remaining core material by looking for physical evidence of macroseepage such as degraded oil structures, remnants of chemosynthetic organisms/communities, and authigenic carbonate formation (Dembicki and Samuel, 2008). If the core material cannot be preserved for later examination, this is accomplished by wet sieving the remaining core material on the ship through a 1.0-mm mesh and collecting the trapped debris for later analysis.

AGI passive adsorbent samplers described earlier in the section on Direction Indicators of Hydrocarbon Microseepage can also be used with seafloor cores. Sediments from the seafloor core are placed in a jar with the adsorber for a fixed period of time and the analyzed in the same fashion as the microseepage samples (Schrynemeeckers, 2015).

Remember, some seafloor cores also need to be taken in areas away from potential seep features. These essential samples are needed to establish the sediment's hydrocarbon background signal for comparison. Without them, it may be difficult to recognize the seep signal from the background signal.

ANALYZING SEAFLOOR SEDIMENTS FOR THERMOGENIC HYDROCARBONS

Once back in the laboratory, the seafloor core samples are analyzed in multiple ways. The first objective of these analyses is to establish if any seeped thermogenic hydrocarbons are present in the sediments. If thermogenic hydrocarbons are found, the second objective is to use the geochemical characteristics of these thermogenic hydrocarbons to learn as much as possible about the origin of the petroleum in the leaking subsurface accumulation. This includes determining if the seeped petroleum can be correlated to known accumulation in the area and doing a crude oil inversion study to determine the characteristics of the source rock that generated the petroleum and estimating the maturity level of that generation (see Chapter 4).

To accomplish the first object, the core samples are used in a series of screening analyses. The process usually begins with the canned sediment samples being analyzed for their headspace gas content. This analysis is similar to the headspace gas analysis described in Chapter 3. From this procedure, the composition and concentration of the hydrocarbon gases and carbon dioxide are reported. The cans of sediment are kept sealed in the event they may be needed later for isotopic analysis.

Determining whether thermogenic gas is present is based on the concentration and composition. Some typical seafloor core headspace gas data are shown in Fig. 6.13. Thermogenic gas will often overwhelm (>10,000 ppm) the in situ background biogenic gas (<1000 ppm), and the biogenic gas is composed almost entirely of methane, usually >99.95% (Bernard, 1978). In contrast, thermogenic gas typically contains >2.5% C_2+ compounds, except for very mature gases (Bernard, 1978). Biodegradation of the gas, however, may consume much of the C_2+ compounds making the composition less indicative (Dembicki, 2013b). Mixed source gases are commonly encountered at seafloor seep.

The sediment collected for high-molecular weight hydrocarbon analysis first needs to be dried and then solvent extracted. The extract is used for whole extract GC and total scanning fluorescence (TSF) analysis. The whole extract gas chromatograms are examined for characteristics of background organic matter (BOM) as well as thermogenic hydrocarbons. It is highly likely that any thermogenic hydrocarbons present in the seafloor sediments will exhibit some degree of biodegradation from microbial activity at the seafloor and in near

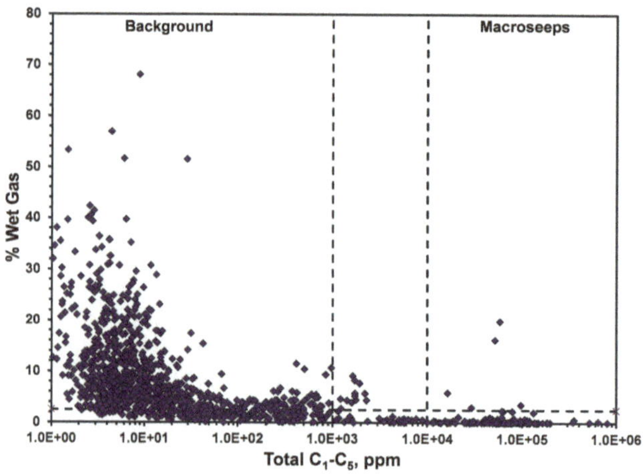

FIGURE 6.13 A typical headspace gas dataset from seafloor cores. Note that macroseepage does not always show the wet gas expected from thermogenic hydrocarbons. *From Dembicki Jr., H., 2013b. Interpreting geochemical data from seafloor hydrocarbons seeps: you can't always get what you want. In: Proceeding of the Offshore Technology Conference Paper 24237 22 p.*

seafloor sediments (refer to Chapter 4 for guidelines used to recognize biodegradation in whole oil gas chromatograms).

The BOM signal is obtained from those cores taken away from suspected seepage specifically to establish the background. BOM typically consists primarily of recent organic matter (ROM). Major contributors to the sediment are marine planktonic and benthic organisms. In addition, contributions from terrestrial higher plant material are frequently observed in marine sediments and transported into the marine environment by both fluvial and aeolian processes. The BOM may also contain eroded, transported, and redeposited (reworked) organic matter from preexisting ancient sediments including source rocks and coals. Though less commonly observed, reworked seeped hydrocarbons may also be included in the BOM. This could be the result of subsea redistribution of seeped hydrocarbons by mass-wasting events or contributions from coastal seeps to nearby benthic sediments. A careful appraisal of the components of the BOM is necessary for proper evaluation of the seafloor cores (Dembicki, 2010).

Assessments of thermogenic versus BOM start with a visual inspection of the whole extract gas chromatograms noting the resolvable peaks, the extent of biodegradation, and the shape of the unresolved complex mixture (UCM) (Logan et al., 2009). ROM generally displays a limited series of peaks in the C_{15}–C_{18} range and a low to moderate UCM. The n-alkane peaks can exhibit an odd carbon predominance in the C_{25}–C_{33} range (high plant input), and some of the peaks are often identifiable as unsaturated hydrocarbons. Thermogenic hydrocarbons, if unaltered, will display a smooth peak envelope skewed toward the low-molecular weight end similar to a crude oil. If biodegraded, the chromatograms will exhibit a reduction of the prominent peaks to varying degrees and the area of the UCM is often large, as discussed in Chapter 4. Quantitative estimates of the amount of material in the UCM and the ratio of thermogenic-to-biogenic n-alkane peaks are also used in the assessment.

TSF is also done using the solvent extract. It looks for aromatic hydrocarbons that fluoresce when exposed to ultraviolet light. Crude oil has a higher concentration and different composition of aromatic hydrocarbons than a sediment's BOM and as a result will give different fluorescence spectra. The sample's extract is irradiated with light from 200 to 500 nm at 10 nm intervals with the fluorescence emission spectra (wavelengths and intensities) recorded at each excitation wavelength (Brooks et al., 1983; Barwise and Hay, 1996). A 3-D excitation/emission/intensity "map" is recorded as well as the maximum intensity of emission and R1, the ratio of emission intensity at 360 nm compared to emission at 320 nm when excitation is 270 nm. These parameters along with the spectra are used to assess the presence/absence of thermogenic hydrocarbons.

If the screening data indicate the presence of thermogenic hydrocarbons, more detailed analysis of the headspace gas and solvent extractable organic matter should be done. For the headspace gas, carbon isotope analysis is usually done if the concentration of the component gases are high enough. Usually methane is the main focus of the isotopic analysis. As previously stated, the methane in biogenic gas is isotopically light, $\delta^{13}C$ −90 to −60‰, while the methane in thermogenic gas is isotopically heavier, $\delta^{13}C$ −60 to −20‰. Some caution is needed in making these interpretations based on methane alone. There is a potential for methanogenesis during the biodegradation of associated seeped oil that can add secondary biogenic methane to the sediments (Dembicki, 2013b). If the concentration of ethane and propane are high enough, their carbon isotope ratios may be used to estimate the maturity of the gas. Again, some caution must be exercised, and biodegradation of the ethane and propane may also alter the carbon isotope ratios of these gases.

For the extract from the high-molecular weight hydrocarbon sample, the detailed analysis consists of separation by liquid chromatography and biomarker analysis of the saturate and aromatic fractions. The biomarker analysis follows the same procedures and the resulting data used in the same way as outlined in the sections on oil correlation and crude oil inversion in Chapter 4. The amount of information that can be gleaned from the biomarker data will depend on the concentration of the seepage and the extent of biodegradation experienced. If the amount of seeped hydrocarbons is low, the BOM in the seafloor sediments can mask the biomarker distribution of the seeped oil (Hood et al., 2002). It is also critical to have the biomarker distributions from the BOM to assess how much interference exists (Dembicki, 2010). The biomarkers from the ROM component in the BOM will contain sterols, sterenes, hopenes, 17β, 21β hopanes, and 17β, 21α hopanes, as well as functional group compounds that can be distinguished from the thermogenic biomarkers by careful examination of the gas chromatography–mass spectrometry (GC–MS) data (Dembicki, 2013a). However, the reworked organic matter whether from source rocks, coals, or nearby seeps may have biomarker compounds similar to those found in migrated thermogenic hydrocarbons and can obscure the true seep signal.

While in many instances the biodegradation experienced by a seeped oil may not be severe enough to alter the biomarkers, biodegradation can be so intense that no recognizable biomarker compounds can be found, as demonstrated in Fig. 6.14 (Dembicki, 2010). This often depends on how much seepage is occurring. When the seepage flux rate is high, the microbial alteration cannot keep up with the seeped hydrocarbons arriving at the surface, and the severity of the biodegradation is reduced. But when the seepage flux rate is low, microbial alteration can be extensive and impact most biomarker compound groups. In general, the hopanes and regular steranes are less resistant to near-surface microbial alteration, while the tricyclic/tetracyclic terpanes, diasteranes, and the monoaromatic and triaromatic steroids are more resistant (Dembicki, 2010), which is similar to the pattern of biodegradation observed in subsurface oils (Wenger and Isaksen, 2002; Peters et al., 2005).

SEA SURFACE SLICKS

When slicks are observed in SAR images or from seismic vessels, it is not always possible to put together a complete seafloor seep study to find evidence to confirm the presence of a working petroleum system. There may not be enough time before the drilling schedule begins, budgetary restricts may prevent including a seafloor seep study in the work plan, or the logistic for a study is difficult to pull together. Yet there is a need to reduce the exploration risk and find some direct evidence of hydrocarbon generation and migration. To provide confirmation of charge, it may be possible to use 3-D seismic data in conjunction with a sea surface slick sampling program. By sampling and analyzing the slicks, the presence of thermogenic hydrocarbons can be confirmed. The hydrocarbon in the slicks may be correlated to know production in the area, or crude oil inversion can provide information about the source rock that has generated it. To verify that the slicks were related to the subsurface, a seafloor

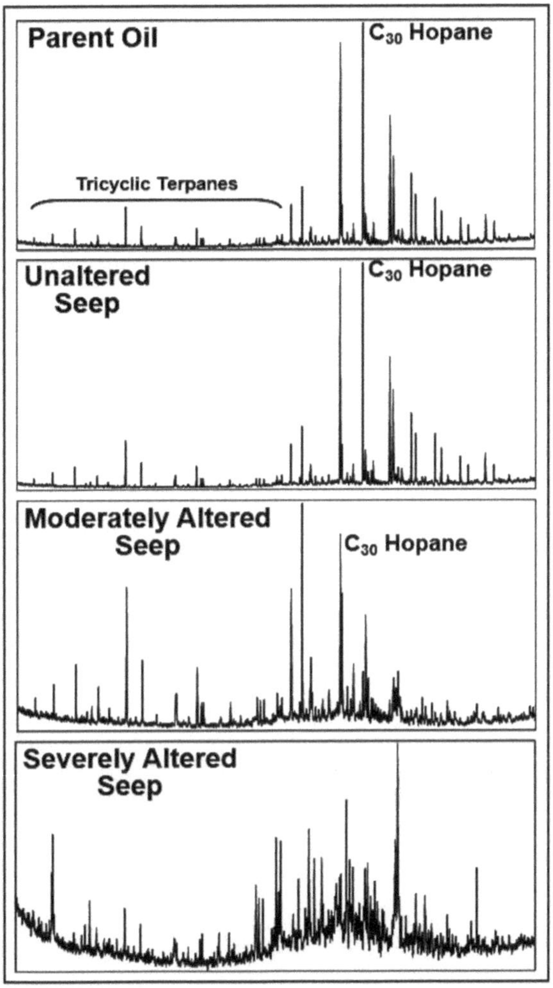

FIGURE 6.14 Biomarker data from a series of seafloor seeps and their parent oil showing increasing alteration by biodegradation. The most severely altered no longer has any recognizable biomarkers compounds. *Modified from Dembicki, Jr., H., 2013b. Interpreting geochemical data from seafloor hydrocarbons seeps: you can't always get what you want. In: Proceeding of the Offshore Technology Conference Paper 24237, 22 p.*

extraction from the 3-D seismic data can be used to identify bathymetric features consistent with seafloor hydrocarbon seepage. The apparent origins of the slicks on the sea surface should be compared to the locations of these seafloor features to see if they are coincident. Finally, the 3-D seismic imaging can be used to demonstrate that there are potential migration pathways from the suspected charged traps to these seafloor features. This combination of data provides a high level of confidence that an active petroleum system is present, and the seismically imaged traps are charged. Dembicki (2014) provides an example of this type of study in the eastern Black Sea.

Oil Slick Sampling and Analysis

Sea surface slicks are very thin layers of oil on the water's surface, as shown in Fig. 6.15. As a result, simply scooping up the water to sample the slick will not usually retrieve enough of the oil for analysis (Abrams and Logan, 2014). To increase the amount of material recovered, two methods have been used that give good results. The first is slick sampling with Nybolt strips. Nybolt is a polyamide bolting cloth used primarily for cell culture work. For slick sampling, strips of Nybolt are attached to the end of a long pole and passed repeatedly through the slick (MacDonald et al., 1993; Dembicki, 2014). The polyamide has an affinity to the oil which is adsorbed on the strips. After exposure to the slick, the strip of Nybolt is removed from the pole and placed in a glass jar with Teflon-lined cap for storage and transport. If possible, samples should be kept frozen until analysis.

Although Nybolt is an effective sampling material, it is difficult to obtain. A readily available alternative is an oil slick sampling system recommended by the United States Coast Guard (Model 5080 General Oceanics Oil Sampling Kit, http://www.generaloceanics.com). It consists of a highly porous polymer net of TFE-fluorocarbon mounted on a disposable ring. It is used in the same fashion as Nybolt strips. The net assembly can be attached to a pole and is repeatedly passed through the oil slick to collect a petroleum sheen sample. The net is then removed and placed in a glass jar with Teflon-lined cap, provided in the kit, for storage and transport. If possible, samples should be kept frozen until analysis. This system has been found to be a very effective means of sampling slicks (Abrams and Logan, 2014).

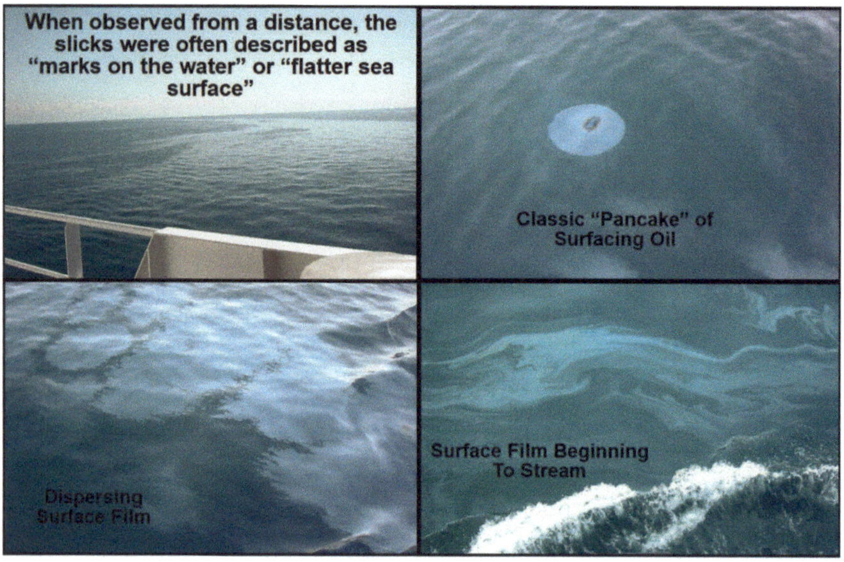

FIGURE 6.15 Ship board observations of sea surface slicks. *From Dembicki, Jr., H., 2014. Confirming the presence of a working petroleum system in the eastern Black Sea Basin, offshore Georgia using SAR imaging, sea surface slick sampling, and geophysical seafloor characterization. American Association of Petroleum Geologists Search and Discovery Article #10610 (2014), 21 p.*

AGI has also developed a Passive Sorbent-Based Slick Sampler (Abrams and Logan, 2014). The samplers are attached to the line on a fishing rod, cast out into the oil slick, and exposed to the slick for a minimum of 2 min. Samplers are returned in special glass containers provided by AGI for analysis by GC–MS. While useful for differentiating natural oil slicks from pollution, the data are difficult to use for oil correlation and oil inversion studies.

Due to rapid loss of volatile hydrocarbons at the sea surface, light hydrocarbon analysis is not an option with sea surface slick samples. The sampling medium, either the Nybolt strips or Teflon net should be solvent extracted to recover the adsorbed oil. The extract material is then analyzed by GC to determine what the distribution of compounds looks like and if there are characteristics of thermogenic hydrocarbons. If the extracts do look thermogenic, they should then be analyzed by GC–MS to look for diagnostic biomarkers.

Remember that the oil will have lost the light end from evaporation and will likely have experienced some biodegradation when evaluating whether they appear to be thermogenic in origin. It is also not unusual during sampling for some biological materials to be captured in addition to the slick material. If adequate sample material is available, separation by liquid chromatography on a short silica gel column into saturate, aromatic, and N-S-O fractions may help reduce these interferences. If the slick has thickened to form "mousse" or tar, separation by liquid chromatography will improve the data by avoiding interferences from many polar compounds and asphaltenes. Slick data from these analyses should be interpreted following the standard oil correlation and crude oil inversion methods for interpretation of the biomarker data discussed in Chapter 4.

Offshore Gas Seep Sampling and Analysis

While searching for and sampling oil slicks in shallow water, occasionally, gas seeps may be encountered. When a gas seep is observed, samples can be collected by completely submerging and filling a headspace gas can with water, then inverting the can over the gas bubble stream. Fill the can partially with the gas, leaving at least 2 in. (5 cm) of water in the can. Carefully place the lid of the can over the opening of the inverted can, press the lid on as tightly as possible, bring the can on to the deck, and finish seal it by tamping the lid firmly in place. The cans of captured gas and seawater should be frozen and stored upside down for transport to the laboratory.

The gas seep sample can be analyzed for the composition of the gas and the stable isotope signature of individual gas component. The gas data from these analyses may be interpreted following the standard natural gas interpretation guidelines discussed in Chapter 4.

Beach Tar Sampling and Analysis

As oil slicks age, loss of volatiles concentrates the high-molecular weight components eventually forming tar blobs or balls. Prevailing winds and currents will sometimes wash the tar onto nearby beaches. Beach tar can be collected by simply picking it up and placing it in either aluminum foil or a glass jar with Teflon-lined cap for storage and transport. If possible, samples should be kept frozen until analysis. Like surface slicks, beach tar may be from pollution from petroleum products spilled into the marine environment or biological organic matter such as algal blooms or fish oil slicks, as well as natural oil seepage. A detailed evaluation of the data is needed to confirm that the beach tar is from seeping thermogenic hydrocarbons. Follow the sample analysis guideline set down for sea surface slick samples outlined above.

Be cautious when using beach tar samples. While the tar could be locally sourced, it could also have drifted a great distance prior to being deposited on the beach. It is best to use beach tar only in situations where active seepage has been reported or observed in the vicinity. The study by Peters et al. (2008) is a good example of how beach tar data can be used in conjunction with seeps and subsurface data to gain insight into a petroleum system.

References

Abrams, M.A., 1992. Geophysical and geochemical evidence for subsurface hydrocarbon leakage in the Bering Sea, Alaska. Marine and Petroleum Geology Bulletin 9, 208–221.

Abrams, M.A., Segall, M.P., Burtell, S.G., 2001. Best practices for detecting, identifying, and characterizing near-surface migration of hydrocarbons within marine sediments. In: Proceedings of the Offshore Technology Conference Paper 13039 14 p.

Abrams, M.A., 2013. Best practices for the collection, analysis, and interpretation of seabed geochemical samples to evaluate subsurface hydrocarbon generation and entrapment. In: Proceeding of the Offshore Technology Conference Paper 24219 21 p.

Abrams, M.A., Logan, G., 2014. Geochemical evaluation of ocean surface slick methods to ground truth satellite seepage anomalies for seepage detection. American Association of Petroleum Geologists Search and Discovery Article #40604 (2010) 18 p.

Barnard, A., Sager, W.W., Snow, J.E., Max, M.D., 2015. Subsea gas emissions from the Barbados Accretionary Complex. Marine and Petroleum Geology 64, 31–42.

Barwise, T., Hay, S., 1996. Predicting oil properties from core fluorescence. In: Schumacher, D., Abrams, M.A. (Eds.), Hydrocarbon Migration and its Near Surface Expression, American Association of Petroleum Geologists Memoir, vol. 66, pp. 363–371.

Bernard, B.D., 1978. Light Hydrocarbons in Marine Sediments (Ph.D. thesis). Texas A&M University, College Station, Texas. 144 p.

Bernard, B.B., Brooks, J.M., Sackett, W.M., 1976. Natural gas seepage in the Gulf of Mexico. Earth and Planetary Science Letters 31, 48–54.

Bjoroy, M., Ferriday, I.L., 2002. Surface geochemistry as an exploration tool: a comparison of results using different analytical techniques. American association petroleum Geologists Hedberg Conference "Near-Surface hydrocarbon migration: mechanisms and seepage rates", April 7–10, 2002, Vancouver BC, Canada, Abstracts. American Association Petroleum Geologists Search and Discovery Article #90006 1 p.

Bolchert, G., Weimer, P., McBride, B.C., 2000. Structural and stratigraphic controls on petroleum seeps, Green Canyon and Ewing Bank, northern Gulf of Mexico: implications for petroleum migration. Gulf Coast Association of Geological Societies Transactions 50, 65–74.

Brooks, J.M., Kennicutt II, M.C., Bernard, L.A., Genoux, G.J., Carey, B.D., 1983. Applications of total scanning fluorescence to exploration geochemistry. In: Proceeding of the Offshore Technology Conference Paper 4624 8 p.

Calhoun, G.C., Hawkins, J.L., 2002. Effects of earth tides on vertical migration. American association petroleum Geologists Hedberg Conference "Near-Surface hydrocarbon migration: mechanisms and seepage rates" April 7–10, 2002, Vancouver BC, Canada, Abstracts. American Association Petroleum Geologists Search and Discovery Article #90006 1 p.

Clarke, R.H., Grant, A.I., Macpherson, M.T., Stevens, D.G., Stevenson, M., 1988. Petroleum exploration with BP's airborne laser fluorosensor. In: Proceedings of the 17th Annual Convention of the Indonesian Petroleum Association, vol. 1, pp. 387–395.

Connolly, D.L., Brouwer, F., Walraven, D., 2008. Detecting fault-related hydrocarbon migration pathways in seismic data: implications for fault-seal, pressure, and charge prediction. Gulf Coast Association of Geological Societies Transactions 58, 191–203.

De Beukelaer, S.M., MacDonald, I.R., Guinnasso Jr., N.L., Murray, J.A., 2003. Distinct side-scan sonar, RADARSAT SAR, and acoustic profiler signatures of gas and oil seeps on the Gulf of Mexico slope. Geo-Marine Letters 23, 177–186.

Dembicki Jr., H., 2010. Recognizing and compensating for interference from the sediment's background organic matter and biodegradation during interpretation of biomarker data from seafloor hydrocarbon seeps: an example from the Marco Polo area seeps, Gulf of Mexico, USA. Marine and Petroleum Geology 27, 1936–1951.

Dembicki Jr., H., 2013a. Analysis and interpretation of biomarkers from seafloor hydrocarbon seeps. In: Aminzadeh, F., Berge, T.B., Connolly, D.L. (Eds.), Hydrocarbon Seepage: From Source to Surface, Society of Exploration Geophysicists Geophysical Developments Series No. 16, Chapter 9, pp. 155–160.

Dembicki Jr., H., 2013b. Interpreting geochemical data from seafloor hydrocarbons seeps: you can't always get what you want. In: Proceeding of the Offshore Technology Conference Paper 24237 22 p.

Dembicki Jr., H., 2014. Confirming the presence of a working petroleum system in the eastern Black Sea Basin, offshore Georgia using SAR imaging, sea surface slick sampling, and geophysical seafloor characterization. American Association of Petroleum Geologists Search and Discovery Article #10610 (2014) 21 p.

Dembicki Jr., H., Samuel, B.M., 2007. Identification, characterization, and ground-truthing of deepwater thermogenic hydrocarbon macro-seepage utilizing high-resolution AUV geophysical data. In: Proceeding of the Offshore Technology Conference Paper 18556 11 p.

Dembicki Jr., H., Samuel, B.M., 2008. Improving the detection and analysis of seafloor macro-seeps: an example from the Marco Polo field, Gulf of Mexico, USA. In: International Petroleum Technology Conference Proceedings Paper 12124 18 p.

Fang, J., 1991. Isotopic evidence for petroleum-derived carbonates in the Gulf of Mexico. Gulf Coast Association of Geological Societies Transactions 41, 276–282.

Fertl, W.H., Chilingar, G.V., 1988. Total organic carbon content determined from well logs. Society of Petroleum Engineers Formation Evaluation 3, 407–419.

Gallagher, A.V., 1984. Iodine: a pathfinder for petroleum deposits. In: Davidson, M.J., Gottlieb, B.M. (Eds.), Unconventional Methods in Exploration for Petroleum and Natural Gas, vol. III. Southern Methodist University, Dallas, Texas, pp. 148–159.

Garcia-Pineda, O., MacDonald, I., Silva, M., Shedd, W., Asl, S.D., Schumaker, B., 2016. Transience and persistence of natural hydrocarbon seepage in Mississippi Canyon, Gulf of Mexico. Deep Sea Research Part II: Topical Studies in Oceanography. http://dx.doi.org/10.1016/j.dsr2.2015.05.011i.

Geyer, R.A., Sweet Jr., W.M., 1973. Natural hydrocarbon seepage in the Gulf of Mexico. Gulf Coast Association of Geological Societies Transactions 23, 158–169.

Harbert, W., Jones, V.T., Izzo, J., Anderson, T.H., 2006. Analysis of light hydrocarbons in soil gases, Lost River region, West Virginia: relation to stratigraphy and geological structures. American Association of Petroleum Geologists Bulletin 90, 715–734.

Hellwig, D., April 2011. Using chemical sampling to decide where to drill. Digital Energy Journal (30), 1–12.

Hitzman, D.C., Schumacher, D., Clavareau, L., 2009. Strategies for surface geochemical surveys in Southeast Asia: best practice designs and recent case studies. In: Proceedings of the 33rd Annual Convention of the Indonesian Petroleum Association, pp. 395–404.

Holysh, S., Tóth, J., 1996. Flow of formation waters: likely cause of poor definition of soil gas anomalies over oil fields in east-central Alberta. In: Schumacher, D., Abrams, M.A. (Eds.), Hydrocarbon Migration and its Near-surface Expression, American Association Petroleum Geologists Memoir, vol. 66, pp. 255–277.

Hood, K.C., Wenger, L.M., Gross, O.P., Harrison, S.C., 2002. Hydrocarbon systems analysis of the northern Gulf of Mexico: delineating of hydrocarbon migration pathways using seeps and seismic imaging. In: Schumacher, D., LeSchack, L.A. (Eds.), Surface Exploration Case Histories: Applications of Geochemistry, Magnetics, and Remote Sensing, American Association of Petroleum Geologists Studies in Geology No. 48, pp. 25–40.

Horvitz, L., 1939. On geochemical prospecting. Geophysics 4, 210–228.

Horvitz, L., 1969. Hydrocarbon prospecting after thirty years. In: Heroy, W.B. (Ed.), Unconventional Methods in Exploration for Petroleum and Natural Gas. Southern Methodist University Press, Dallas, pp. 205–218.

International Tanker Owners Pollution Federation, 2011. Fate of Marine Oil Spills. Technical Information Paper 2. 12 p. http://www.itopf.com/fileadmin/data/Documents/TIPS%20TAPS/TIP2FateofMarineOilSpills.pdf.

Jeffrey, D.A., Zarella, W.M., 1970. Geochemical prospecting at sea. American Association Petroleum Geologists Bulletin 54, 853–854.

Jones, A.T., Logan, G.A., Kennard, J.M., Rollet, N., 2005. Reassessing potential origins of synthetic aperture radar (SAR) slicks from the Timor Sea region of the North West Shelf on the basis of field and ancillary data. Australian Petroleum Production & Exploration Association Journal 45, 311–331.

Jones, V.T., Drozd, R.J., 1983. Predictions of oil or gas potential by near-surface geochemistry. American Association Petroleum Geologists Bulletin 67, 932–952.

Jones, V.T., LeBlanc Jr., R.J., 2004. Moore-Johnson (Morrow) Field, Greeley County Kansas: a successful integration of surface soil gas geochemistry with subsurface geology and geophysics. American Association Petroleum Geologists Search and Discovery Article #20022 (2004) 30 p.

Jones, V.T., Matthews, M.D., Richers, D.M., 2000. Light hydrocarbons for petroleum and gas prospecting. In: Govett, G.J.S. (Ed.), Handbook of Exploration GeochemistryGeochemical Remote Sensing of the Subsurface, vol. 7, pp. 133–212.

Kanaa, T.F.N., Mercier, G., Tonye, E., 2005. Sea surface slicks characterization in SAR images. Oceans 2005-Europe 1, 686–691.

Klusman, R.W., 1993. Soil Gas and Related Methods for Natural Resource Exploration. John Wiley & Sons, Chichester, England. 483 p.

Klusman, R.W., Webster, J.D., 1981. Meteorological noise in crustal gas emission and relevance to geochemical exploration. Journal of Geochemical Exploration 15, 63–76.

Klusman, R.W., Saeed, M.A., 1996. Comparison of light hydrocarbon microseepage mechanisms. In: Schumacher, D., Abrams, M.A. (Eds.), Hydrocarbon Migration and its Near-surface Expression: American Association Petroleum Geologists Memoir, vol. 66, pp. 157–168.

Klusman, R.W., Voorhees, K.J., 1983. A new development in petroleum exploration technology. Colorado School of Mines Magazine 73 (3), 6–10.

Kvenvolden, K.A., Barnard, L.A., 1982. Hydrates of natural gas in continental margins: environmental processes: model investigations of margin environmental and tectonic processes. In: Watkins, J.D., Drake, C.L. (Eds.), American Association of Petroleum Geologists Memoir 34, Studies in Continental Margin Geology, pp. 631–640.

Leaver, J.S., Thomasson, M.R., 2002. Case studies relating soil-iodine geochemistry to subsequent drilling results. In: Schumacher, D., LeSchack, L.A. (Eds.), Surface Exploration Case Histories: Applications of Geochemistry, Magnetics, and Remote Sensing, American Association Petroleum Geologists Studies in Geology No. 48 and SEG Geophysical References Series No. 11, pp. 41–57.

Lennon, M., Babichenko, S., Thomas, N., Mariette, V., Mercier, G., Lisin, A., 2006. Detection and mapping of oil slicks in the sea by combined use of hyperspectral imagery and laser-induced fluorescence. European Association of Remote Sensing Laboratories eProceedings 5, 120–128.

Lennon, M., Thomas, N., Mariette, V., Babichenko, S., Mercier, G., 2005. Oil slick detection and characterization by satellite and airborne sensors: experimental results with SAR, hyperspectral and lidar data. Geoscience and Remote Sensing Symposium Proceedings 1, 25–29.

LeSchack, L.A., Van Alstine, D.R., 2002. High-resolution ground-magnetic (HRGM) and radiometric surveys for hydrocarbon exploration: six case histories in Western Canada. In: Schumacher, D., LeSchack, L.A. (Eds.), Surface Exploration Case Histories: Applications of Geochemistry, Magnetics, and Remote SensingAmerican Association Petroleum Geologists Studies in Geology No. 48 and SEG Geophysical References Series No. 11, pp. 67–156.

Ligtenberg, H., 2005. Detection of fluid migration pathways in seismic data: implications for fault seal analysis. Basin Research 17, 141–153.

Link, W.K., 1952. Significance of oil and gas seeps in world oil exploration. American Association of Petroleum Geologists Bulletin 36, 1505–1540.

Logan, G.A., Abrams, M.A., Dahdah, N., Grosjean, E., 2009. Examining laboratory methods for evaluating migrated high-molecular-weight hydrocarbons in marine sediments as indicators of subsurface hydrocarbon generation and entrapment. Organic Geochemistry 40, 365–375.

MacDonald, I.R., Guinasso Jr., N.L., Ackleson, S.G., Amos, J.F., Duckworth, R., Sassen, R., Brooks, J.M., 1993. Natural oil slicks in the Gulf of Mexico visible from space. Journal of Geophysical Research 98, 16,351–16,364.

MacDonald, I.R., Leifer, I., Sassen, R., Stine, P., Mitchell, R., Guinasso Jr., N., 2002. Transfer of hydrocarbons from natural seeps to the water column and atmosphere. Geofluids 2.

MacDonald, I.R., Reilly Jr., J.F., Best, S.E., Venkataramaiah, R., Sassen, R., Guinasso Jr., N.L., Amos, J., 1996. Remote sensing inventory of active oil seeps and chemosynthetic communities in the northern Gulf of Mexico. In: Schumacher, D., Abrams, M.A. (Eds.), Hydrocarbon Migration and its Near-surface Expression: American Association Petroleum Geologists Memoir, vol. 66, pp. 27–37.

Machel, H.G., 1996. Magnetic contrasts as a result of hydrocarbon seepage and migration. In: Schumacher, D., Abrams, M.A. (Eds.), Hydrocarbon Migration and its Near-surface Expression: American Association Petroleum Geologists Memoir, vol. 66, pp. 99–109.

Machel, H.G., Burton, E.A., 1991. Causes and spatial distribution of anomalous magnetization in hydrocarbon seepage environments. American Association Petroleum Geologists Bulletin 75, 1864–1876.

Matthews, M.D., 1996. Importance of sampling design and density in target recognition. In: Schumacher, D., Abrams, M.A. (Eds.), Hydrocarbon Migration and its Near-surface Expression, American Association of Petroleum Geologists Memoir, vol. 66, pp. 243–253.

Mousseau, R.J., Glezen, W.H., February 11–12, 1980. The Gulf Marine Hydrocarbon Sampling and Analysis System: Proceedings of a Symposium and Workshop on Water Sampling while Underway. National Academy Press, Washington DC, pp. 167–192.

Mousseau, R.J., May 1980. Marine hydrocarbon prospecting geochemistry. In: Proceedings of the Indonesian Petroleum Association Ninth Annual Convention, pp. 367–378.

O'Brien, G.W., Cowley, R., Quaife, P., Morse, M., 2002. Characterizing hydrocarbon migration and fault-seal integrity in Australia's Timor Sea via multiple, integrated remote sensing technologies. In: Schumacher, D., LeSchack, L.A. (Eds.), Surface Exploration Case Histories: Applications of Geochemistry, Magnetics, and Remote Sensing, American Association of Petroleum Geologists Studies in Geology No. 48 and SEG Geophysical References Series No. 11, pp. 393–413.

Oehler, D.Z., Sternberg, B.K., 1984. Seepage-induced anomalies, "false" anomalies, and implications for electrical prospecting. American Association of Petroleum Geologists Bulletin 68, 1121–1145.

Orange, D.L., Angell, M.M., Lapp, D., 1999. Using seafloor mapping (bathymetry and backscatter) and high resolution sub-bottom profiling for both exploration and production: detecting seeps, mapping geohazards, and managing data overload with GIS. In: Proceeding of the Offshore Technology Conference Paper 10870 17 p.

Orange, D.L., Teas, P.A., Decker, J., 2010. Multibeam backscatter – insights into marine geological processes and hydrocarbon seepage. In: Proceeding of the Offshore Technology Conference Paper 20860 22 p.

Orange, D.L., Teas, P.A., Decker, J., Baillie, P., GIlleran, P., Levey, M.D., 2008. The utilisation of SeaSeep Surveys (a defense/hydrography spin-off) to identify and sample hydrocarbon seeps in offshore frontier basins. In: International Petroleum Technology Conference Proceedings Paper 12839 12 p.

Peters, K.E., Hostettler, F.D., Lorenson, T.D., Rosenbauer, R.J., 2008. Families of Miocene Monterey crude oil, seep, and tarball samples, coastal California. AAPG Bulletin 92 p. 1131–1152 11.

Peters, K.E., Walters, C.C., Moldowan, J.M., 2005. The Biomarker Guide, Volume 2: Biomarkers and Isotopes in Petroleum Exploration and Earth History, second ed. Cambridge University Press, Cambridge. 683 p.

Rasheed, M.A., Srinivasa Rao, P.L., Annapurna, B., Zaheer Hasan, S., 2015. Implication of soil gas method for prospecting of hydrocarbon microseepage. International Journal of Petroleum and Petrochemical Engineering 1, 31–41.

Rasheed, M.A., Patil, D.J., Dayal, A.M., 2013. Microbial techniques for hydrocarbon exploration. In: Kutchero, V. (Ed.), Hydrocarbon. InTech. http://dx.doi.org/10.5772/50885. 16 p. http://www.intechopen.com/books/hydrocarbon/microbial-techniques-for-hydrocarbon-exploration.

Rice, G.K., Belt Jr., J.Q., Berg, G.E., 2002. Soil-gas hydrocarbon pattern changes during a west Texas waterflood. In: Schumacher, D., LeSchack, L.A. (Eds.), Surface Exploration Case Histories: Applications of Geochemistry, Magnetics, and Remote Sensing, American Association of Petroleum Geologists Studies in Geology No. 48 and SEG Geophysical References Series No. 11, pp. 157–174.

Roberts, H.H., 1995. High resolution surficial geology of the Louisiana middle-to-upper continental slope. Gulf Coast Association of Geological Societies Transactions 45, 503–508.

Roberts, H.H., 1996. Surface amplitude data: 3D-seismic for interpretation of sea floor geology (Louisiana Slope). Gulf Coast Association of Geological Societies Transactions 46, 353–362.

Roberts, H.H., Coleman, J.M., Hunt Jr., J., Shedd, W.W., 2000. Surface amplitude mapping of 3D seismic for improved interpretation of seafloor geology and biology from remotely sensed data. Gulf Coast Association of Geological Societies Transactions 50, 495–504.

Roberts, H.H., Doyle, E.H., Booth, J.R., Clark, B.J., Kaluza, M.J., Hartsook, A., 1996. 3D-seismic amplitude analysis of the seafloor: an important interpretive method for improved geohazards evaluations. In: Proceeding of the Offshore Technology Conference Paper 7988 10 p.

Roberts, H.H., Hardage, B.A., Shedd, W.W., Hunt, J., 2006. Seafloor reflectivity; an important seismic property for interpreting fluid/gas expulsion geology and the presence of gas hydrate. Leading Edge 25, 620–628.

Sager, W.W., MacDonald, R., Hou, R., 2004. Sidescan sonar imagining of hydrocarbon seeps on the Louisiana continental slope. American Association of Petroleum Geologists Bulletin 88, 725–746.

Sassen, R., Brooks, J.M., MacDonald, I.R., Kennicutt II, M.C., Guinasso Jr., N.L., Requejo, A.G., 1993. Association of oil seeps and chemosynthetic communities with oil discoveries, upper continental slope, Gulf of Mexico. Gulf Coast Association of Geological Societies Transactions 43, 349–355.

Sassen, R., Sweet, S.T., Milkov, A.V., DeFreitas, D.A., Salata, G.G., McDade, E.C., 1999. Geology and geochemistry of gas hydrates, Central Gulf of Mexico continental slope. Gulf Coast Association of Geological Societies Transactions 49, 462–469.

Schrynemeeckers, R., 2015. Improving petroleum system identification in an offshore salt environment: Gulf of Mexico and Red Sea Case Studies. In: Presented at the American Association of Petroleum Geologists Geoscience Technology Workshop, Sixth Annual Deepwater and Shelf Reservoir, Houston, Texas, January 27-28, 2015: American Association of Petroleum Geologists Search and Discovery Article #41607(2015) 46 p.

Schrynemeeckers, R., Silliman, A., 2014. Combining surface geochemical surveys and downhole geochemical logging for mapping liquid and gas hydrocarbons in the Utica Shale. In: Presented at the American Association of Petroleum Geologists Annual Convention and Exhibition, Houston, Texas, April 6-9, 2014: American Association of Petroleum Geologists Search and Discovery Article #80399(2014) 6 p.

Schumacher, D., 1996. Hydrocarbon-induced alteration of soils and sediments. In: Schumacher, D., Abrams, M.A. (Eds.), Hydrocarbon Migration and its Near-surface Expression: American Association Petroleum Geologists Memoir, vol. 66, pp. 71–89.

Sumner, J.S., 1978. Principles of Induced Polarization for Geophysical Exploration. Elsevier Scientific, Amsterdam. 277 pp.

Sweets Jr., W.E., 1974. Marine acoustical seep detection. American Association of Petroleum Geologists Bulletin 58, 1133–1136.

Thrasher, J., Fleet, A.J., Hay, S.J., Hovland, M., Düppenbecker, S., 1996. Understanding geology as the key to using seepage in exploration: spectrum of seepage styles. In: Schumacher, D., Abrams, M.A. (Eds.), Hydrocarbon Migration and its Near-surface Expression, American Association Petroleum Geologists Memoir, vol. 66, pp. 223–241.

Tóth, J., 1996. Thoughts of a hydrogeologist on vertical migration and near-surface geochemical exploration for petroleum. In: Schumacher, D., Abrams, M.A. (Eds.), Hydrocarbon Migration and its Near-surface Expression, American Association Petroleum Geologists Memoir, vol. 66, pp. 279–283.

Tucker, J.D., Hitzman, D.C., 1996. Long-term and seasonal trends in the response of hydrocarbon-utilizing microbes to light hydrocarbon gases in shallow soils. In: Schumacher, D., Abrams, M.A. (Eds.), Hydrocarbon Migration and its Near Surface Expression, American Association Petroleum Geologists Memoir, vol. 66, pp. 353–357.

von der Dick, H.H., Barrett, K.R., Bosman, D.A., 2002. Numerically reconstructed methane-seep signal in soil gases over Devonian gas pools and prospects (northeast British Columbia): surface microseeps and postsurvey discovery. In: Schumacher, D., LeSchack, L.A. (Eds.), Surface Exploration Case Histories: Applications of Geochemistry, Magnetics, and Remote SensingAmerican Association Petroleum Geologists Studies in Geology No. 48 and Society of Exploration Geophysicists Geophysical References Series No. 11, , pp. 193–207.

Voorhees, K.J., Klusman, R.W., 1986. Apparatus and method for geochemical prospecting. United States Patent No. 4573354, 19 p.

Wenger, L.M., Isaksen, G.H., 2002. Control of hydrocarbon seepage intensity on level of biodegradation in sea bottom sediments. Organic Geochemistry 33, 1277–1292.

Williams, A., 1996. Detecting leaking oilfields with ALF, the Airborne Laser Fluorosensor: case histories and latest developments. Geological Society of Malaysia Bulletin 59, 125–129.

Williams, A., Lawrence, G., 2002. The role of satellite seep detection in exploring the South Atlantic's ultradeep water. In: Schumacher, D., LeSchack, L.A. (Eds.), Surface Exploration Case Histories: Applications of Geochemistry, Magnetics, and Remote SensingAmerican Association Petroleum Geologists Studies in Geology No. 48 and SEG Geophysical References Series No. 11, pp. 327–344.

Zhang, Z., McConnell, D.R., 2010. Backscatter characterization of seep-associated seafloor features in the vicinity of Bush Hill, northwest Green Canyon, Gulf of Mexico. In: Proceeding of the Offshore Technology Conference Paper 20662 13 p.

CHAPTER

7

Unconventional Resources

OUTLINE

Introduction	**253**
Coalbed Methane	**254**
Petroleum Geochemistry and Coalbed Methane	*258*
Shale Gas	**259**
Petroleum Geochemistry and Shale Gas	*261*
Additional Support for Shale Gas Plays	*264*
Shale Oil	**265**
Hybrid Systems	**267**
Petroleum Geochemistry and Hybrid Systems	*269*
References	**269**

INTRODUCTION

Since the mid-1990s, unconventional resources have grown to dominate the oil and gas industry in North America, but the actual definition of unconventional resources can be a bit vague. Most petroleum exploration geoscientists consider unconventional plays as hydrocarbon resources that cannot be produced at economic flow rates or that do not produce economic volumes without some form of artificial stimulation, such as hydraulic fracturing, and the use of special recovery processes and technologies, such as horizontal drilling. While this accurately describes these resources, it is basically an engineering definition and does not address the geologic aspects of these plays.

In contrast, the United States Geological Survey (USGS) would prefer to classify hydrocarbon accumulations into two major types, conventional and continuous, based on geology. Conventional accumulations are described as discrete fields localized in structural and stratigraphic traps by the buoyancy of oil or natural gas in water (USGS, 1995). Continuous accumulations are oil or natural gas accumulations that have large spatial dimensions and indistinctly defined boundaries and which exist more or less independently of the water column (USGS, 1995). With respect to the previous engineering definition, much of what the USGS would consider continuous resources has become synonymous with unconventional.

Practical Petroleum Geochemistry for Exploration and Production
http://dx.doi.org/10.1016/B978-0-12-803350-0.00007-6

The best way to envision the fundamental nature of these resources is to combine the two definitions. Unconventional resources are oil or natural gas accumulations that have large spatial dimensions, indistinct boundaries, and are essentially independent of the water column. They cannot be produced at economic flow rates, and/or they do not produce economic volumes unless some form of artificial stimulation and special recovery technologies are employed. Based on this definition, unconventional resources should include CBM, tight gas sands, heavy oil/oil sands, shale gas, tight oil, oil shale/shale oil, and gas hydrates.

In addition to the definition, there is also some vagueness in the terminology used to discuss and describe unconventional resources. A good example is a 2012 article in the American Association of Petroleum Geologists Explorer about the growing number of shale plays (Durham, 2012). It listed 20 "shale plays," but only eight were actual shales. The remainder were plays with sandstone, carbonate, or mixed lithology reservoirs, including a granite wash play. From the article, the author appears to have meant to say plays where horizontal drilling and hydraulic fracturing are needed, but used the term "shale plays" as a shortcut. Ironically, in the very next issue of the Explorer, there was an article pointing out inaccuracies and inconsistencies in some geologic terminology, including that used in unconventional plays (Boak, 2012).

And it is not only definitions and terminology that can be misleading. The petroleum industry has been developing the scientific concepts and methods for conventional petroleum exploration and production for well over 150 years. In contrast, our investigations of the scientific concepts and methods for unconventional petroleum exploration and production have been going on in earnest only since the mid- to late 1990s. As a result, it is important to realize that not all the information found in the early literature on unconventional resources will stand the test of time. It is especially important for this topic area to read critically to separate speculation from fact so as to avoid embracing ideas before they have been validated, or, in some cases, even tested.

The following discussions of unconventional resources will endeavor to stick to tested ideas and concepts to describe what the components of these plays are, how the plays work, and how petroleum geochemistry can assist. These discussions will be concerned with coalbed methane, shale gas, shale oil, and hybrid plays/tight oil. While tight gas sands and heavy oil/oil sands can be classed as unconventional, the development of these mature play types is more engineering oriented, with little input from petroleum geochemistry. And gas hydrates, while representing a significant potential resource, are only in the very early experimental stages of exploitation.

COALBED METHANE

For as long as coal has been mined in the subsurface, the presence of gas in coal seams has been known. This gas created problems in the mines when it would accumulate threatening both asphyxia (hence the need for the canary in the mine) and explosive ignition. Eventually, boreholes were drilled into the coal seams to vent off the gas ahead of the mining operations to reduce the chance of this happening. In time, this gas was being collected and used locally instead of being vented, giving rise to the concept of CBM.

The CBM segment of the oil and gas industry did not begin significant growth until the US federal government provided price control exemptions and tax incentives in the late 1970s to make it more competitive. These incentives also resulted in technological advances in CBM production methods. CBM is still a significant source of natural gas in the United States, as well as Australia, Canada, and China and is being developed in other coal-rich countries.

CBM is actually a misnomer. The gas recovered from coal seams is essentially natural gas. While it does consist mainly of methane, it typically contains other hydrocarbon gases, like ethane and propane and the nonhydrocarbon gases nitrogen and carbon dioxide (Kim, 1973). CBM is also known as coalbed gas (CBG), coalbed natural gas, and coal seam gas.

Coal beds for the most part are self-sourcing reservoirs. The coal generates gas and heavier hydrocarbons in place as it matures in a similar fashion to source rocks (Rice, 1997) and hold these hydrocarbons mainly by adsorption. Because of the high adsorptive capacity of coals, they can also trap biogenic methane early in their history or capture migrating thermogenic hydrocarbons from other sources (Weniger et al., 2012). If the coal is uplifted to a shallow enough depth and meteoric water percolates down to it, microbial activity biodegrading the coal's bitumen can contribute secondary biogenic gas to the CBG (Scott et al., 1994).

With respect to being a reservoir, the porosity of coals is typically less than 10%, making it less than desirable as a conventional reservoir. Instead, the vast majority of the gas is stored adsorbed on the surface of the organic matter in the coal matrix. Only small amounts of gas are stored as free gas in porosity and natural fractures (cleats) or as solution gas in water in the cleats and pores. The adsorption capacity of coal increases with its rank (maturity) (Kim, 1977), as shown in Fig. 7.1, which parallels the generation trend. Under normal pressure conditions, coal beds can store more adsorbed gas than typical sandstones can store in primary pores.

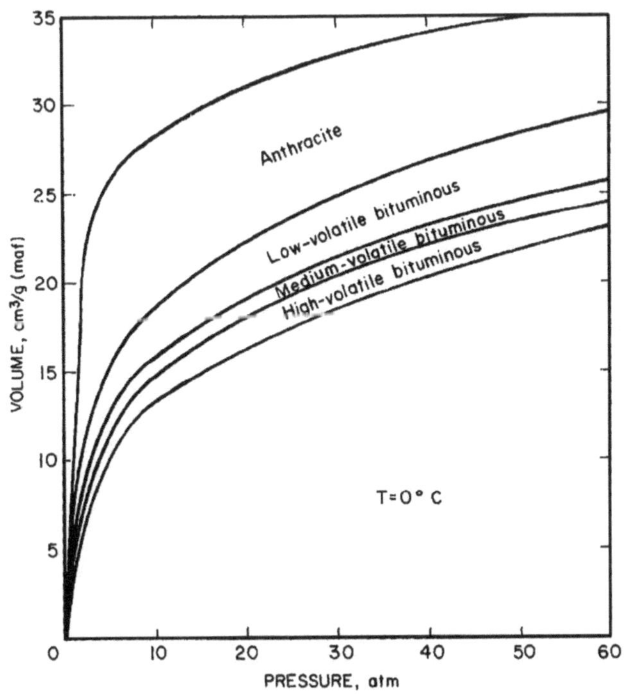

FIGURE 7.1 Variation of amount of gas adsorbed by with rank. *From Kim, A.G., 1977. Estimating Methane Content of Bituminous Coalbeds from Adsorption Data. United States Bureau of Mines Report of Investigations 8245, 26 p.*

The natural fractures in coals known as cleats are important features of the coal for the production of CBM. These fractures usually occur in two sets that are usually at 90 degrees to each other and perpendicular to bedding, as shown in Fig. 7.2. Early cleat development probably occurs at vitrinite reflectance values between approximately 0.3% and 0.5% Ro. However, cleats formed at vitrinite reflectance values around 0.3% Ro may become annealed, while those formed between 0.4% and 0.5% Ro are more likely to persist (Laubach et al., 1998). This suggests cleat formation significant to CBM begins in the sub-bituminous stage.

The geology of the coal basin and coal seam reservoirs needs to be studied in considerable detail with respect to the coal's rank, thickness, lateral extent, and degree of fracturing to determine if it is a candidate for development. It is also necessary to have an estimate of the amount of gas in the coal for the economic assessment. CBG content is most often determined by measuring the volume of gas released from a coal sample recovered from a coring operation and sealed into a desorption canister. This approach allows the gases in the coal sample to desorb under controlled laboratory conditions until a defined low desorption rate cutoff point is reached (Yee et al., 1993). The sample may then be crushed to measure the amount of residual gas remaining within the sample. The lost gas is the gas released from the coal during its collection prior to being sealed into a desorption canister. Lost gas is estimated from the measured desorbed gas volume data by extrapolation.

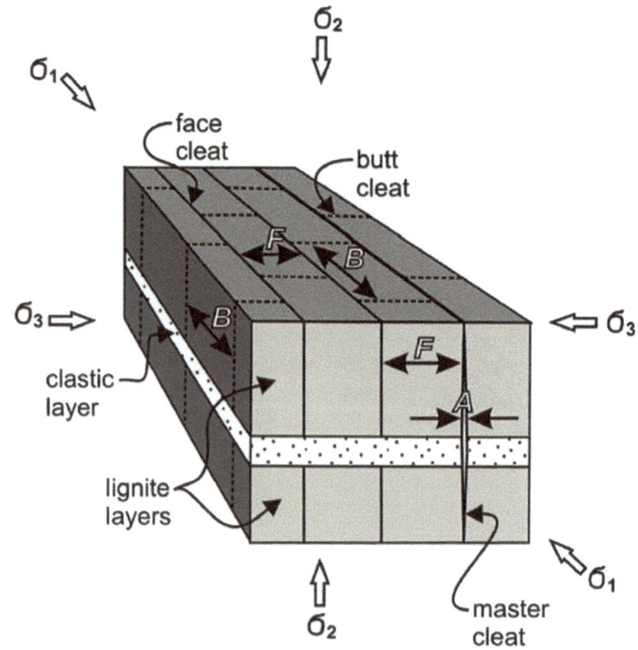

FIGURE 7.2 Model of cleat formation in a lignite. *A*, cleat aperture; *F*, face-cleat spacing; *B*, butt-cleat spacing; σ_1, largest stress; σ_2, intermediate stress; σ_3, smallest stress. *From Widera, M., 2014. What are cleats? Preliminary studies from the Konin lignite mine, Miocene of central Poland. Geologos 20, 3–12.*

The companion data to the estimated CBG content are the coal's adsorption capacity. Adsorption capacity is defined as the volume of gas that can be adsorbed per unit mass of coal at standard pressure and temperature conditions. It is usually expressed in standard cubic feet (SCF) of gas/ton of coal. The capacity to adsorb is dependent mainly on the rank (maturity) and moisture content of coal. Typical values for adsorption capacity are usually between 100 and 800 SCF/ton (Donnez, 2012). Adsorption capacity is determined from adsorption isotherms. An adsorption isotherm describes the variation in gas adsorption as a function of pressure at a constant temperature. It indicates the maximum volume of methane that a coal can store under equilibrium conditions at a given pressure and temperature (Kim, 1977). A coal sample is first purged of any gas in a vacuum. A given quantity of gas is then injected into the container at a specific temperature and pressure and circulated until an adsorption equilibrium is reached (Yee et al., 1993). The amount of adsorbed gas is calculated by difference after measuring the amount of unadsorbed gas. The procedure is repeated at least six times at different pressures to build an adsorption isotherm curve, as shown in Fig. 7.3. Based on the experimentally determined isotherm curve, additional curves at different temperature can be inferred.

When a coal is oversaturated with gas, it will yield gas on initial production, as shown in Fig. 7.4, but most coals are undersaturated. To initiate desorption of gas and production in undersaturated coals, formation pressure must be reduced to the critical desorption pressure, as illustrated in Fig. 7.4. This is done by pumping out and producing the water from the cleats. Dewatering leads to lower pressure and allows gas desorption from the surfaces of the coal's organic matter matrix. Typical water and gas production curves are shown in Fig. 7.5. Actual gas production is usually delayed while the water is being removed from the cleat system. As the water production declines, gas production will increase. To increase the near-borehole permeability and accelerate the dewatering process, hydraulic fracturing may be used. Water removal may continue for several years, and water disposal must be accommodated in the development plan.

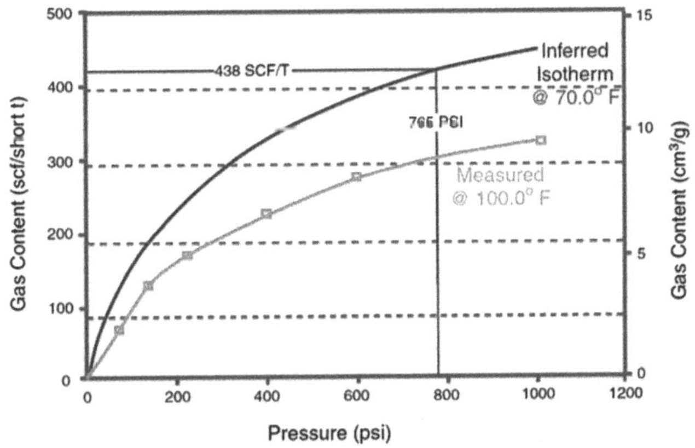

FIGURE 7.3 An example of an adsorption isotherm from experimental measurements and an inferred isotherm extrapolated from the data. *From Dallegge, T.A., Barker, C.E., 2000. Coal-bed methane gas-in-place resource estimates using sorption isotherms and burial history reconstruction: an example from the Ferron Sandstone Member of the Mancos Shale, Utah. In: Kirschbaum, M.A., Roberts, L.N.R., Biewick, L.R.H., (Eds.), Geologic Assessment of Coal in the Colorado Plateau: Arizona, Colorado, New Mexico, and Utah. United States Geological Survey Professional Paper 1625–B, Chapter L, 26 p.*

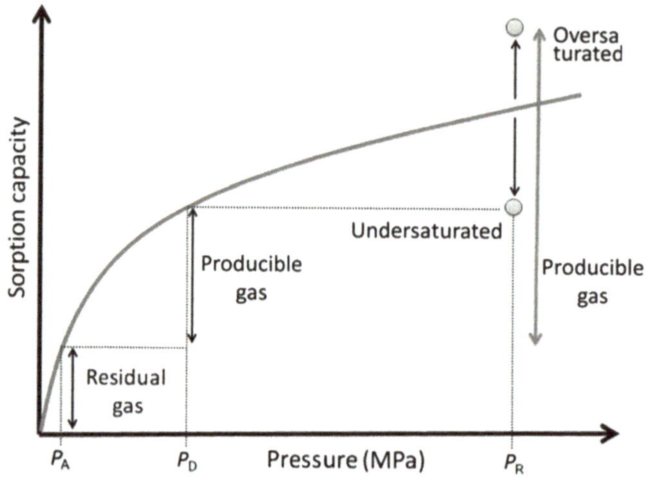

FIGURE 7.4 The relationship of oversaturated and undersaturated coals to a methane sorption. Figure labels include P_R for the initial desorption pressure, P_D for critical desorption pressure, and P_A for the abandonment pressure. The oversaturated coal will produce immediately. The pressure in the undersaturated coal must be reduced to the critical desorption pressure to begin gas production. *From Gensterblum, Y., Merkel, A., Busch, A., Krooss, B.M., Littke, R., 2014. Gas saturation and CO_2 enhancement potential of coalbed methane reservoirs as a function of depth. American Association of Petroleum Geologists Bulletin 98, 395–420.*

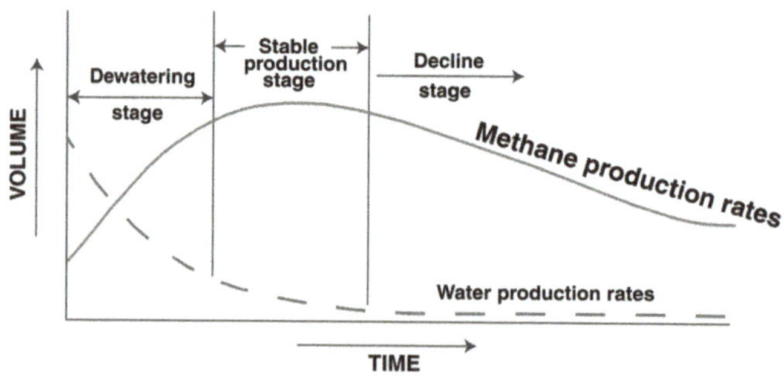

FIGURE 7.5 Typical production curves for a coalbed methane well showing relative volumes of methane and water through time. *Modified from Kuuskraa, V.A., Brandenberg, C.F., 1989. Coalbed methane sparks a new energy industry: Oil and Gas Journal 87 (41), 49–56.*

Petroleum Geochemistry and Coalbed Methane

The main input into CBM projects from petroleum geochemistry is in the maturity assessment of the coal. As stated earlier, the maturity, or rank, of the coal influences the amount of hydrocarbons that have been generated and adsorbed by the coal, the coal's adsorption capacity, and the development of the cleats. This maturity is best measured by vitrinite reflectance. Unlike in source rock analysis where vitrinite reflectance should be used as a trend tool, vitrinite reflectance from an individual coal seam is a more definitive maturity indicator. Coal seam vitrinite reflectance from core is not impacted by caving and reworked vitrinite like source rock samples from cuttings. In most coals, in situ vitrinite is abundant and easily recognized under the microscope. While a maturity trend with depth may not be necessary for coal seam vitrinite reflectance data, measurements on multiple sample from the coal seam at each location is highly recommended. If the coal seam under development has a significant depth change over the area of interest, a depth trend can be built from the individual locations sampled. This trend can then be used to predict the maturity of the coal in areas not yet sampled based on the depth of the seam.

Another input into CBM projects from petroleum geochemistry is the composition and origin of the gas. Analysis of the gas is important for determining the value of the resource. A dry biogenic gas has less value than a wetter thermogenic gas. Gas composition will also indicate how much nonhydrocarbon gases are present. These also impact the value of the resource. Nonhydrocarbon gases dilute the hydrocarbons, diminishing the overall value of the gas. If the concentrations are high enough, steps may need to be taken to remove the nonhydrocarbon gases prior to sending the gas to market.

The composition and isotopic signature of the hydrocarbon gases are important indicators of the origin of the gas. They help to determine if the gas adsorbed on the coal came from early biogenic processes, self-sourced from generation within the coal, captured migration hydrocarbons, or secondary biogenic gas. By combining the maturity information from the coal and isotopic signatures of the gases (see Chapter 4 for guidelines), the origin may be determined. Biogenic gas will have a distinct isotopic signature. The maturity of thermogenic gas should be similar to the maturity of the coal if the gas is self-sourced. If the gas is more mature than the coal, captured migrating gas may be the source. Some caution is urged in making these determinations. Just as in conventional gas reservoirs, coalbed gas may originate from more than one source.

SHALE GAS

Although shale gas has made a big impact on the oil and gas industry only since the late 1990s, the concept has been around for quite some time. In the United States, shale gas was first extracted as a resource by William Hart in Fredonia, NY in 1825 (Lash and Lash, 2014). The well was located on natural gas seepage, was dug by hand with shovel to a depth of only 27 ft deep, and accessed low-pressure natural gas in the fractures in the Dunkirk shale (Lash and Lash, 2014). Early accounts of the enterprise describe the gas being delivered using a pipeline made of hollowed out logs sealed with tar and rags. Additional drilling beginning in 1857 by Preston Barmore targeted fractures in the shale, then recognized as the conduits for the gas migration. Barmore also used gunpowder to induce artificial fractures as a means of stimulating the well and enhancing production (Lash and Lash, 2014). This early development of a shale gas resource was more advanced than many would have anticipated for the time and all before Drake's oil well in 1859.

It was not until the mid-1970s that shale gas resources began to be considered for development again. After the Arab Oil Embargo of 1973, interest was aroused in addition fossil fuel sources, including gas, within the United States, and shale gas was one of the resources considered. In 1976, the Morgantown Energy Research Center (MERC), part of the Energy Research and Development Agency (ERDA) of the Department of Energy (DOE), funded the Eastern Gas Shale Project to define the resource potential and test production methods for the

Upper Devonian shales in the Appalachian, Illinois, and Michigan Basins (DOE, 2007). These sediments were relatively low maturity and, although the project gained important insight into shale gas, it was not commercially successful. Some of the technology developed in this project included directional drilling in shales (the precursor to horizontal drilling), the use of massive hydraulic fracturing in shales, and foam fracturing technology (DOE, 2007; Wang and Krupnick, 2013).

In 1986, the United States Department of Energy partnered with private gas companies to complete the first successful multizoned fractured horizontal well in shale (DOE, 2007). The results were promising and encouraged additional development and investment. In 1991, the DOE and Gas Research Institute (GRI) subsidized Mitchell Energy's first horizontal well drilled in the Barnett Shale, but it was not until 1998 that Mitchell Energy achieved the first economically successful shale gas well with an innovative process called slick-water fracturing (Wang and Krupnick, 2013).

At first, it was thought that shales, specifically source rock shales, could be drilled and completed in a similar manner to what was used in the Barnett and gas would flow. And since the early success in the Barnett, numerous shale gas plays have been tested, mainly in North America. While limited success has been achieved in some of these plays, most have not proven to be viable, and only the Barnett, Haynesville, Eagle Ford, and Marcellus shales have become major commercial successes. The question quickly asked was why these four source rock shales and not all source rock shales. The answer lies in the anatomy of a gas shale.

A gas shale, as shown in the model in Fig. 7.6, needs to be a source rock that is rich enough and mature enough to be generating mainly gas. But the key to a shale gas resource is more than the source rock component. It also needs to be the reservoir rock and a type of stratigraphic trap to contain that gas. Early on, it was thought that shale gas was like CBM, and the gas was stored mainly adsorbed onto kerogen. However, the volumes of gas being produced and the characteristics of the production demonstrated that adsorption could not account for all the gas. Instead, conventional storage in porosity was indicated as the main source of the gas being produced. In addition, the gas shale must itself have some seal-like properties to prevent loss of the gas, as well as being overlain and underlain by effective seal intervals to further contain the gas. The seal aspect of shale gas plays is often overlooked or minimized but is of vital importance (Dembicki and Madren, 2014). While many source rocks can generate gas, only a select few have been able to retain and contain the gas for exploitation as a gas shale.

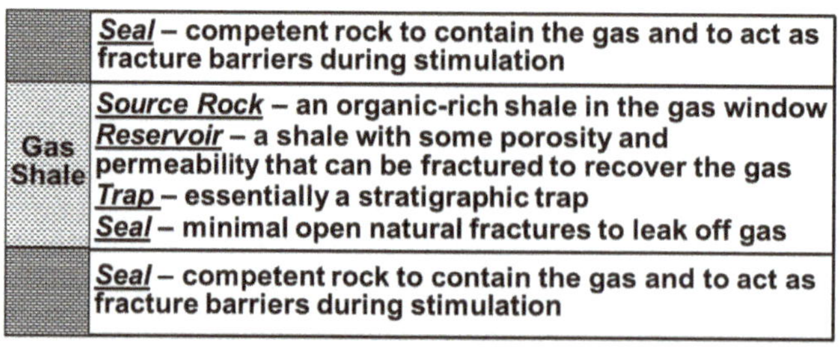

FIGURE 7.6 An idealized model of a gas shale source rock/reservoir petroleum system. *From Dembicki Jr., H., Madren, J.D., 2014. Lessons learned from the Floyd shale play. Journal of Unconventional Oil and Gas Resources 7, 1–10.*

Besides the proper anatomy, a gas shale has to have certain properties to flow gas in commercial quantities. Gas shales have inherently low natural permeabilities which would result in low recovery efficiencies in conventional vertical wells. Despite the inherently low permeabilities in shales, hydrocarbon gas can flow at commercial rates if adequate pressure is present, if the hydrocarbons stay in a single gas phase, and if the surface area over which the natural permeability can deliver the gas is increased.

Overpressure is a good early indicator of potential shale gas success. The presence of overpressure in a gas shale indicates that the seal capacity within and around the shale is more than adequate to retain and contain the generated gas. This pressure also provides the reservoir energy needed to drive the gas flow during production.

Single phase production is also a key element for success. Shale gas plays with associated liquids are much more economically attractive than gas only systems. The ability to recover condensate and natural gas liquids often overshadows the gas production in many of these plays. But being in a single gas phase during production is extremely important for liquid-rich plays. The viscosity of a gas phase is low, allowing the gas to move more freely within the inherent porosity and permeability of the shale. If the shale gas contains any liquids, as long as these liquids remain in a gas solution flow will not be inhibited. However, if the liquids drop out of the gas solution, the separate liquid phase can block or restrict pore throats obstructing or stopping flow. Maintaining the formation pressure to preserve single phase flow is therefore critical.

To increase the effective surface area over which the permeability of a shale can function and thereby improve gas recovery, horizontal wells and hydraulic fracturing are required. Hydraulic fracturing pumps fluids at extremely high pressures out through perforated sections of the wellbore to fracture the surrounding formation and inject sand, or other proppants, into the cracks to hold them open. This fracturing does not significantly increase the permeability by connecting pores in the rock. Effective permeability is increased by opening migration pathways out into the rock away from the borehole that increases the surface area exposure of the matrix permeability for more efficient drainage.

While this description of shale gas system is adequate for a basic geologic understanding, there is still much more to be learned. This is especially true with respect to the engineering aspects of these plays. We have only been working on shale gas in earnest for a few decades. And it needs to be remembered that our understanding of shale gas continues to evolve.

Petroleum Geochemistry and Shale Gas

Because shale gas systems are first and foremost source rocks, defining source rock properties appropriate for these plays is the main assistance petroleum geochemistry can provide. The main concern of gas shales as with most source rocks is to assess whether there is sufficient organic richness, whether the proper kerogen type is present, and if a high enough thermal maturity has been attained. These assessments can be made with the standard source rock evaluation techniques described in Chapter 3 but with some adjustments made for this type of play. As with standard source rock evaluation, it is best to start with maturity.

To determine if a high enough thermal maturity has been attained, vitrinite reflectance measurements are usually used. If vitrinite is lacking, thermal alteration index and/or Rock-Eval T_{max} can be employed with the proper cautions. If the shale is from the Lower Paleozoic, maturity may be roughly determined by using the conodont alteration index. After

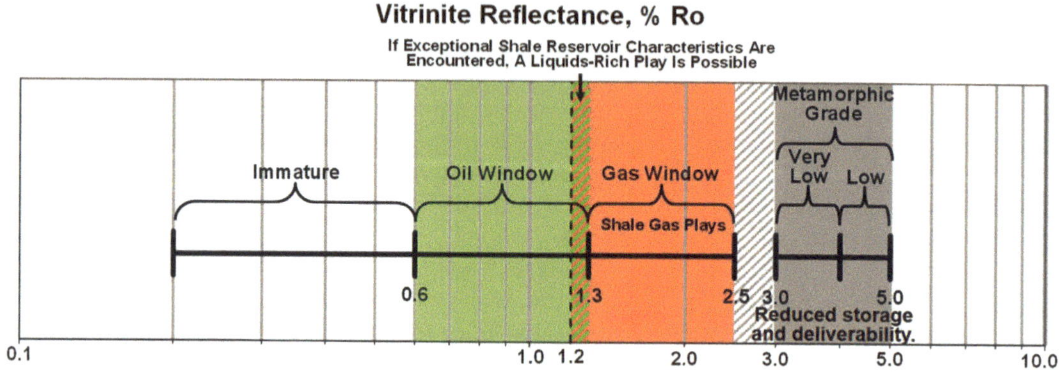

FIGURE 7.7 Maturity window for shale gas plays. *From Dembicki Jr., H., 2014. Challenges to Black Oil Production from Shales. American Association of Petroleum Geologists Search and Discovery. Article # 80355, 22 p.*

reviewing the shale gas literature for how vitrinite reflectance is used as a maturity indicator, it was found that there is a tendency to only make measurements in the interval of interest and not to build maturity trends over an adequate depth interval. This practice can result in misleading and inaccurate maturity assessments due to caved vitrinite, reworked vitrinite, and/or kerogen particles misidentified as vitrinite (Dembicki, 2013). A maturity trend should always be used, and it is best to place the thermal maturity data in the context of the sediment's burial history for proper interpretation (Dembicki, 2009).

The target maturity range for shale gas systems is from about 1.3–2.5% Ro vitrinite reflectance or equivalent (Dembicki, 2014), as shown in Fig. 7.7. Most often, maturities below about 1.3% Ro have had insufficient gas generated, and any liquids present may not be in gas solution. Occasionally, a liquid-rich play may be possible between about 1.2 and 1.3% Ro but only if exception reservoir properties are encountered. Above 2.5% Ro, the sediments begin their transition into metamorphic rock, resulting in reduced storage and deliverability.

To determine if there is sufficient organic richness, total organic carbon (TOC) is usually employed. Because the gas shales under investigation are usually at an advanced maturity, less than half the original TOC likely remains in the sediment (Dembicki, 2009), and a different approach is needed for interpretation. The assumption is made that the appropriate kerogen type was originally present in the sediments, and the TOC indicates the amount of remnant organic matter. The minimum remnant TOC for charging a gas shale based on field observations appears to be 2.0% (Dembicki, 2014), as shown in Fig. 7.8. There is also an upper limit for the TOC that needs to be considered. Shale gas works as a free gas resource where the bulk of the gas is stored in the porosity. This is in contrast to CBG where the gas is stored adsorbed on the organic matrix. Coals begin at about 40% TOC, as shown in Fig. 7.8, and the majority of the shale gas systems investigated have average TOCs of about 10% or less. Based on this, it is assumed that there is a transition from free gas to adsorbed gas storage that occurs somewhere between 10% and 40% TOC. At what point this transition begins is unclear. As more shale gas systems are investigated, some additional clarity concerning this transition should be discovered.

Rock-Eval pyrolysis is also used in shale gas evaluations but not entirely in the traditional manner. T_{max} is still used as a maturity indicator, as mentioned earlier; however, the S2 and HI

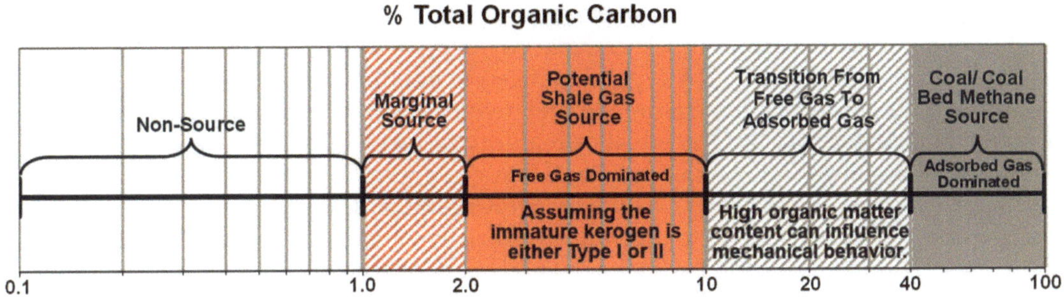

FIGURE 7.8 Total organic carbon window for shale gas plays. *From Dembicki Jr., H., 2014. Challenges to Black Oil Production from Shales. American Association of Petroleum Geologists Search and Discovery. Article # 80355, 22 p.*

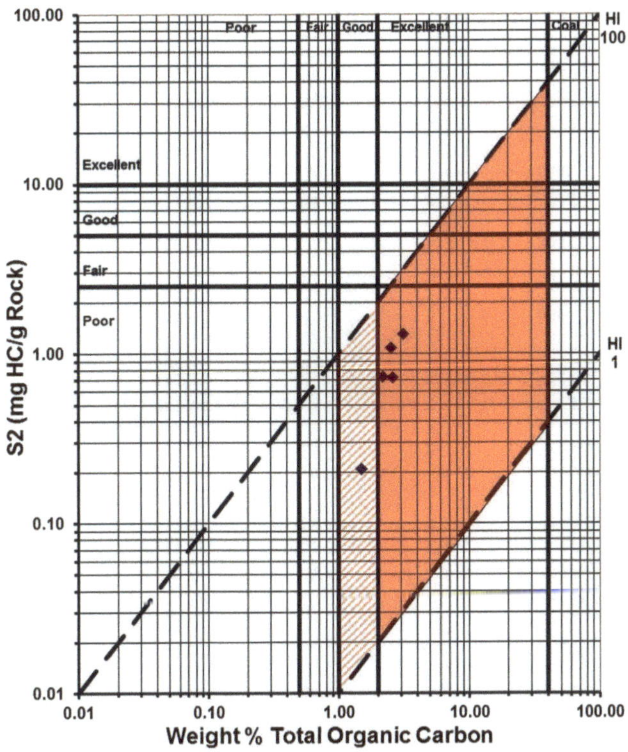

FIGURE 7.9 A cross plot of % total organic carbon (TOC) and Rock-Eval for use in gas shale evaluation. The TOC indicates adequate richness, while the S2/HI values indicate maturity. *After Dembicki Jr., H., Madren, J.D., 2014. Lessons learned from the Floyd shale play. Journal of Unconventional Oil and Gas Resources 7, 1–10.*

are also used as maturity indicators more than indicators of richness and kerogen type as for conventional source rocks. Based on field observations, the hydrogen index, HI, is less than 100 when source rocks are mature enough to be considered gas shales. A convenient method for displaying both the TOC and the Rock-Eval S2/HI data for shale gas is the modified cross plot of TOC and S2 shown in Fig. 7.9.

To determine whether the proper kerogen type is present for a shale gas assessment is not as straightforward as in most source rock evaluations. As discussed in Chapter 2, Type I or Type II kerogen will yield more gas than a Type III kerogen in a source rock of equivalent richness. As a result, Type I and Type II kerogens are the preferred kerogen type for gas shales, as well as for conventional source rocks. But the advance maturity of potential gas shales would make any sediment that originally contained Type I or II kerogen appear to contain Type III gas-prone or even Type IV inert, as shown in Fig. 7.10. It is therefore necessary to search for less mature locations to sample the potential gas shale to determine the original kerogen type that was present. Standard analysis and interpretation by Rock-Eval and pyrolysis–gas chromatography, as described in Chapter 3, are recommended. Be aware that lateral changes in kerogen type may occur in the stratigraphic interval under study, so some caution is warranted when using this approach.

Additional Support for Shale Gas Plays

In addition to the geochemical aspects of the source rock component of a gas shale, the reservoir properties must be considered. The traditional characteristics of porosity and

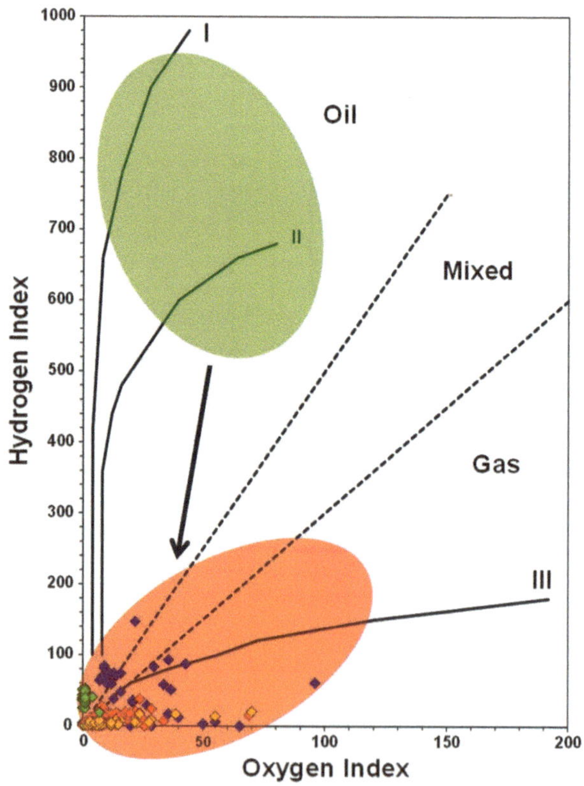

FIGURE 7.10 Shale gas sediments that were originally oil prone appear gas prone due to their maturity. *From Dembicki, Jr., H., 2013. Shale Gas Geochemistry Mythbusting. American Association of Petroleum Geologists Search and Discovery. Article #80294, 23 p.*

permeability are of prime importance. Because of the inherently low porosity and permeability encountered in shales, conventional methods for measuring porosity and permeability used in sandstone are not practical or accurate. The GRI method provides an alternate means of indirectly measuring these properties. This method utilizes a crushed sediment sample where the amount and rate that gas under pressure can be forced into and recovered from the rock's pore network is measured. This information can then be used to estimate a relative porosity and permeability (Guidry et al., 1995). Because the results are relative porosity and permeability, caution is needed when using the data to make resource assessments.

The mineralogy of the rock matrix, its depositional fabric, and any sensitivity to drilling and completion fluids are equally important. The mineralogy can be determined by X-ray diffraction (XRD) analysis. Keep in mind that XRD data with respect to clays are often semi-quantitative at best (Poppe et al., 2001). Depositional fabric can be observed in thin section of the shale or by scanning electron microscopy techniques such as QEMSCAN. Fluid sensitivity testing measures the reaction of the rock, especially the clay component, to exposure to drilling and completion fluids. Testing methods, such as capillary suction time testing and Brinell hardness testing (Carman and Lant, 2010), can track the changes in rock properties before and after exposure to these fluids.

And the mechanical behavior of the rock must also be understood. While many authors have related mechanical behavior of shales to bulk clay content (e.g., Jarvie et al., 2007; Wang and Gale, 2009), this approach is problematic. It does not take into consideration differences in mechanic behavior of different clay types (Ikari et al., 2009) as well as the influence of depositional fabric on the ability to fracture the rock. To better address the geomechanical properties of shales, rock mechanics results from triaxial compressional testing are needed to determine if the sediments are ductile or not and to indicate how the rock will behave during stimulation. Many look to Poisson's ratio and Young's modulus as indicators of hydraulic fracture behavior. While these parameters are important for understanding natural fractures in constant pore pressures environments and provide some insight into how rocks react during hydraulic fracturing, the minimum effective stress and tensile strength are also critical to predicting how rocks behave during hydraulic fracturing (Molenda et al., 2013) and need to be considered.

SHALE OIL

Shale oil is a term that has many meanings in the oil and gas industry. It can refer to the oil produced from retorting oil shales, such as the Eocene Green River Shale in Colorado, Wyoming, and Utah or the kukersite in the Middle Ordovician shales in Estonia (Kann et al., 2013). While there is great potential in exploiting oil shales, this topic will be left for future discussion as the technologies develop.

Shale oil may also signify oil recovered from highly fractured shales, such as in the Cretaceous section of the Raton Basin in New Mexico (Mallory, 1977; Woodward, 1984), the Monterey formation in the California coastal basins (Regan, 1953; Truex, 1972), and the shales in the Frontier Formation in the Salt Creek Field (Mills, 1923; Thorn et al., 1931). These shales are usually highly siliceous (brittle) and have experienced one or more periods of tectonic deformation to account for the high degree of fracturing. Some of these fracture reservoirs, such as the Monterey, are likely self-sourced. These shales are source rocks mature enough to generate and migrate oil. The filling of the fractures with oil is merely a very short distance migration. Other fracture shale reservoirs, like those in the Salt Creek Field, may not be source rocks or are not mature enough to generate oil. They contain oil migrated into them from other source rocks. In either situation, the commercial viability of these reservoirs will depend on the volume of fracture porosity available and the production characteristics of the oil in place. Horizontal wells into these fractured shale reservoirs may well be the key to attaining adequate production in these plays.

Another use of the term shale oil is in reference to liquids associated with shale gas production. These liquids, as discussed earlier, are condensed out of the shale's gas production at the surface, rather than being a separate liquid phase that is being produced out of the shale. But because they are condensate liquids from the gas, they are not truly produced like an oil and will not be part of this discussion.

Shale oil can also refer to oil production from a hybrid unconventional play. These mixed lithology plays can be very productive and will be discussed in the next section.

And the final use of the expression shale oil is for oil extracted from shales in liquid phase using techniques similar to those employed in shale gas production, i.e., horizontal wells and hydraulic fracturing. To better understand how this might occur, it is necessary to describe the parameters that define this potential resource.

First of all, the shales in these plays must be at maturities less than about 1.2–1.3% Ro vitrinite reflectance, or it will be in the gas window and a potential shale gas resource. This places the shale in the oil window and suggests the liquid phase available would most likely be a black oil. The flow behavior of a black oil in a conventional reservoir rock is much different than a gas, and the production of a black oil from a tight conventional reservoir is much more difficult than producing gas from a tight conventional reservoir. Considering that shales are much tighter than even tight conventional reservoirs, the question we need to ask is can we get black oil to flow from a shale in the oil window at sustained commercial rates. To answer that question, we need to determine if black oil can escape from shales, under what conditions does this occur, and are commercial flow rate possible/probable under these conditions.

We know that expulsion and migration of black oil from shales is possible. If it was not, there would be no conventional oil accumulations. As discussed in Chapter 2, black oil flow from a generating source rock is initiated at ~0.7–0.9% Ro vitrinite reflectance after the source rock reaches a minimum hydrocarbon saturation and overcomes the adsorption of the oil on the kerogen and mineral matrix. But oil accumulations form by oil flowing (migrating) out of source rocks over geologic time with flow rates that are very low relative to production. So what are the main obstacles to shale oil achieving commercial flow rates?

The rate of production of recoverable oil is determined by a number of factors including the permeability of the rocks, the viscosity of the oil, and the strength of natural drives. When the reservoir rocks are "tight," such as shales, oil generally cannot flow easily in real time, but when the rocks are more permeable, such as sandstones, oil can flow more freely. As shown in Fig. 7.11, shales are actually more like seals than they are like reservoir rocks. Flow of oil is often helped by natural pressures/forces within and surrounding the reservoir including pressure exerted by natural gas that may be dissolved in the oil, natural gas present above the oil (a gas cap in a two phase reservoir), and/or water below the oil (not realistic in shales), as

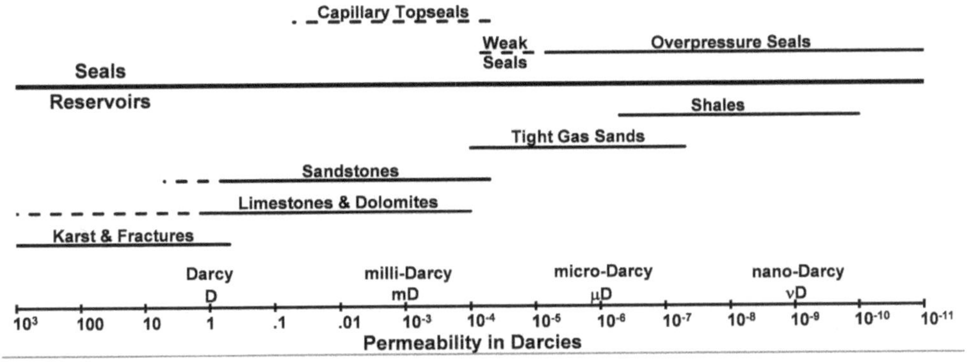

FIGURE 7.11 A comparison of conventional versus unconventional reservoir permeabilities. Note that shale reservoirs are more similar to seals than they are to conventional sand and carbonate reservoirs. *Modified after Lainda, N., 2012. Basic formation evaluation of shale gas play. In: Presentation for the American Association of Petroleum Geologist, Universitas Gadjah Mada Student Chapter, 32 p. https://ugmsc.files.wordpress.com/2010/08/presentation-to-ugm.pdf.*

well as gravity. As discussed in Chapter 4, black oil is in single phase with the pressure greater than the bubble point. Black oil's viscosity typically ranges from about 2 to 100 cp, and the initial GOR is usually less than 2000 scf/STB reflecting the fact that it is undersaturated with gas. Even with horizontal wells and hydraulic fracturing, these characteristics are not conducive to sustained commercial flow rates from shale. Instead, wells drilled for shale oil (in the oil window) may have some initial production, but most often, rates fall off rapidly after a few days or weeks. And if the formation pressure drops below the bubble point, the gas coming out of solution will result in a decrease in GOR of the liquid phase and an increase in its viscosity, making it even more difficult for the liquid to move out of the pore spaces.

The shale's permeability, the oil's viscosity, and the lack of reservoir energy all work against sustained commercial flow rates for black oil from a shale in the oil window. While these shale oil plays are unlikely due to these circumstances, there may be some specific set of circumstances that make it work. However, the best chance for liquids in true shale plays are shale gas plays rich in condensate and natural gas liquids where the maturities are in the range of about 1.2–1.3 to about 1.5% Ro vitrinite reflectance.

HYBRID SYSTEMS

True shale plays get their production directly out of shales that act as both the source rock and the reservoir. In contrast, hybrid systems consist of source rocks adjacent to and interbedded with silty sands, fine-grained sandstones, or porous carbonate with these more porous rocks acting as the reservoir for the oil and gas generated by the source rocks. Hybrid plays are best characterized as tight oil plays. They differ from conventional oil and gas reservoirs in that their permeabilities are generally lower, reservoir thicknesses are usually smaller, and horizontal well and hydraulic fracturing are needed to produce sustained commercial flow rates. In many of these hybrid plays, the presence of the oil and gas in the reservoir interval is known early on in the exploration history of the basin. But it is only since the development of horizontal drilling and reservoir stimulation technologies that these resources could be tapped.

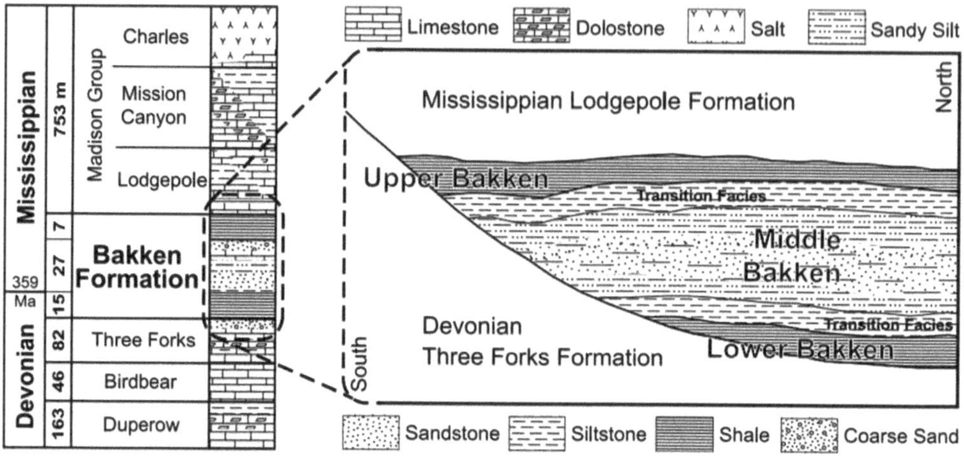

FIGURE 7.12 Stratigraphic column and lithofacies of the Bakken formation in the basin. *From Kuhn, P.P., di Primio, R., Hill, R., Lawrence, J.R., Horsfield, B., October 2012. Three-dimensional modeling study of the low-permeability petroleum system of the Bakken formation. American Association of Petroleum Geologists Bulletin 96 (10), 1867–1897.*

A prime example of a simple hybrid system is the Bakken Formation. While some production in the Bakken is from open fractured systems associated with the anticlinal structures in the Williston Basin (Dow, 1974), the majority of the recent production comes from a hybrid play. The Bakken Formation consists of three members, as shown in Fig. 7.12. The upper and lower members consist of organic-rich shales (up to 20% TOC) containing oil-prone kerogen, while the middle member varies in lithology from dolomite to sand to silty sand to shaley lime (Dow, 1974). Although the porosity and permeability are low in the middle member as compared to a conventional reservoir, there is adequate hydrocarbon storage in these sediments for substantial amounts of the oil and gas generated by the upper and lower members. Deliverability from the middle member is provided by horizontal drilling within this zone and hydraulic fracturing to add to the tectonic fractures.

Other hybrid systems are more complex and consist of a series of source rock intervals interbedded with reservoir intervals. An example of this type is the Niobrara in the Denver–Julesburg Basin. It consists of series of chalk reservoir intervals interbedded with organic-rich marly shale source rocks (Sonnenberg, 2011), as shown in Fig. 7.13. Just as in the Bakken, hydrocarbon generation in the adjacent source rock intervals charge the reservoir zones. And again, the horizontal wells target these reservoir facies. Depending on the thickness and rock properties of the interbedded source rocks, hydraulic fracturing in one of the reservoir intervals may spread vertically and also access reservoir intervals above and/or below the target zone.

There is great potential in searching for hybrid systems in mature basin. Many of these systems have already been encountered during the initial drilling of prolific petroleum basins around the world. Because of their low production potential at that time, they were either ignored or considered bypassed pay. By going back through old drilling records and wireline log data, these plays can be recognized and pursued. And hybrid system can be productive in both the gas and oil window, so they are not limited with respect to maturity. Because of the opportunities they provide, hybrid system should be considered for inclusion in most exploration portfolios.

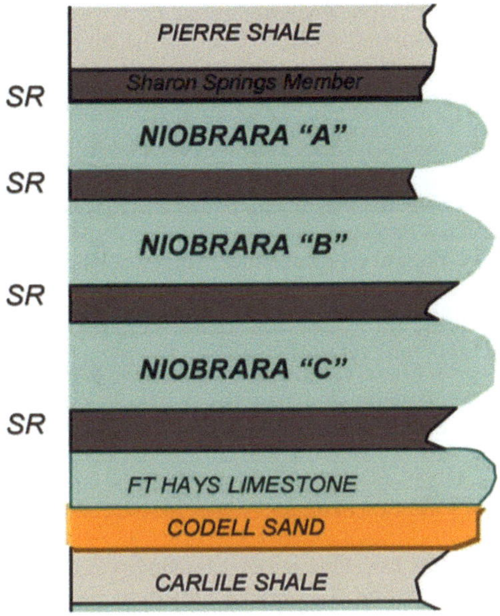

FIGURE 7.13 Stratigraphic column of the Niobrara Formation in the Denver–Julesburg Basin. Note the marly shale source rock intervals labeled SR interbedded with the chalk reservoir units labeled Niobrara "A," "B," and "C." *Modified from Sonnenberg, S.A., 2011. The Niobrara Petroleum System, a Major Tight Resource Play in the Rocky Mountain Region. American Association of Petroleum Geologists Search and Discovery. Article #10355, 31 p.*

Petroleum Geochemistry and Hybrid Systems

Hybrid systems are essentially a conventional source rock–reservoir pair where the migration distance is very short. As such, petroleum geochemistry would be used in the same manner as it would be in a conventional exploration play. Organic-rich oil-prone source rocks are the target, and the key issues are still the maturity, source richness, and kerogen type. Maturity needs to be established preferably with vitrinite reflectance, and it should be based on a maturity trend with depth and not just isolate data from the interval of interest. Source richness evaluations would use the standard TOC and Rock-Eval data interpretation schemes. And kerogen type should be established using a combination of Rock-Eval and pyrolysis–gas chromatography data. If the samples available from the play are in or near the gas window, efforts should be made to obtain less mature representative sample to evaluate kerogen type. If any information is needed about the recovered oil and/or gas and their relationship to other petroleum discovered in the vicinity, the standard oil and gas analytical protocols and interpretations can be applied.

References

Boak, J., 2012. Common wording vs. historical terminology. American Association of Petroleum Geologists Explorer 33 (8), 42–43.

Carman, P.S., Lant, K.S., 2010. Making the case for shale clay stabilization. In: Society of Petroleum Engineers Conference Paper 139030, 8 p..

Dallegge, T.A., Barker, C.E., 2000. Coal-bed methane gas-in-place resource estimates using sorption isotherms and burial history reconstruction: an example from the Ferron Sandstone Member of the Mancos Shale, Utah. In: Kirschbaum, M.A., Roberts, L.N.R., Biewick, L.R.H. (Eds.), Geologic Assessment of Coal in the Colorado Plateau: Arizona, Colorado, New Mexico, and Utah United States Geological Survey Professional Paper 1625–B, Chapter L, 26 p.

Dembicki Jr., H., Madren, J.D., 2014. Lessons learned from the Floyd shale play. Journal of Unconventional Oil and Gas Resources 7, 1–10.

Dembicki Jr., H., 2009. Three common source rock evaluation errors made by geologists during prospect or play appraisals. American Association of Petroleum Geologists Bulletin 93, 341–356.

Dembicki Jr., H., 2013. Shale Gas Geochemistry Mythbusting. American Association of Petroleum Geologists Search and Discovery. Article #80294, 23 p.

Dembicki Jr., H., 2014. Challenges to Black Oil Production from Shales. American Association of Petroleum Geologists Search and Discovery. Article # 80355, 22 p.

Donnez, P., 2012. Essentials of Reservoir Engineering, vol. 2. Editions Technip, Paris. 512 p.

Dow, W.G., 1974. Application of oil-correlation and source-rock data to exploration in Williston Basin. American Association of Petroleum Geologists Bulletin 58, 1253–1262.

Durham, L.S., 2012. Shale list grows. American Association of Petroleum Geologists Explorer 33 (7) p. 12, 14, and 16.

Gensterblum, Y., Merkel, A., Busch, A., Krooss, B.M., Littke, R., 2014. Gas saturation and CO_2 enhancement potential of coalbed methane reservoirs as a function of depth. American Association of Petroleum Geologists Bulletin 98, 395–420.

Guidry, F.K., Luffel, D.L., Curtis, J.B., 1995. Development of Laboratory and Petrophysical Techniques for Evaluating Shale Reservoirs (Final Report) GRI-95/0496. Gas Research Institute. 304 p.

Ikari, M.J., Saffer, D.M., Marone, C., 2009. Frictional and hydrological properties of clay-rich fault gouge. Journal of Geophysical Research 114. B05409, 18 p. http://dx.doi.org/10.1029/2008JB006089.

Jarvie, D., Hill, R.J., Ruble, T.E., Pollastro, R.M., 2007. Unconventional shale-gas systems: the Mississippian Barnett Shale of north-central Texas as one model for thermogenic shale-gas assessment. American Association of Petroleum Geologists Bulletin 91, 475–499.

Kann, J., Raukas, A., Siirde, A., 2013. About the gasification of kukersite oil shale. Oil Shale 30, 283–293.

Kim, A.G., 1973. The Composition of Coalbed Gas. United States Bureau of Mines Report of Investigations 7762, 12 p.

Kim, A.G., 1977. Estimating Methane Content of Bituminous Coalbeds from Adsorption Data. United States Bureau of Mines Report of Investigations 8245, 26 p.

Kuhn, P.P., di Primio, R., Hill, R., Lawrence, J.R., Horsfield, B., October 2012. Three-dimensional modeling study of the low-permeability petroleum system of the Bakken formation. American Association of Petroleum Geologists Bulletin 96 (10), 1867–1897.

Kuuskraa, V.A., Brandenberg, C.F., 1989. Coalbed methane sparks a new energy industry. Oil and Gas Journal 87 (41), 49–56.

Lainda, N., 2012. Basic formation evaluation of shale gas play. In: Presentation for the American Association of Petroleum Geologist, Universitas Gadjah Mada Student Chapter, 32 p. https://ugmsc.files.wordpress.com/2010/08/presentation-to-ugm.pdf.

Lash, G.G., Lash, E.P., 2014. Early History of the Natural Gas Industry, Fredonia. American Association of Petroleum Geologists Search and Discovery, New York. Article #70000, 44 p.

Laubach, S.E., Marrett, R.A., Olson, J.E., Scott, A.R., 1998. Characteristics and origins of coal cleat: A review. International Journal of Coal Geology 35, 175–207.

Mallory, W.W., 1977. Oil and Gas from Fractured Shale Reservoirs in Colorado and Northwest New Mexico, vol. 1. Rocky Mountain Association of Geologists Special Publication. 38 p.

Mills, R.V.A., 1923. Natural gas as a factor in oil migration and accumulation in the vicinity of Faults. American Association of Petroleum Geologists Bulletin 7, 14–24.

Molenda, M., Stöckhert, F., Brenne, S., Alber, M., 2013. Comparison of hydraulic and conventional tensile strength tests. In: Bunger, A.P., McLennan, J., Jeffrey, R. (Eds.), Effective and Sustainable Hydraulic Fracturing. InTech, Rijeka, Croatia, pp. 981–992. http://dx.doi.org/10.5772/56300.

Poppe, L.J., Paskevich, V.F., Hathaway, J.C., Blackwood, D.S., 2001. A Laboratory Manual for X-ray Powder Diffraction. United States Geological Survey Open-File, Report 01–041, 88 p.

Regan Jr., L.J., 1953. Fractured shale reservoirs of California. American Association of Petroleum Geologists Bulletin 37, 201–216.

Rice, D.D., 1997. Composition and origins of coalbed gas. In: Law, B.E., Rice, D.D. (Eds.), Hydrocarbons from Coal. American Association of Petroleum Geologists Studies in Geology 38, pp. 159–184.

Scott, A.R., Kaiser, W.R., Ayers Jr., W.B., 1994. Thermogenic and secondary biogenic gases, San Juan Basin, Colorado and New Mexico – implications for coalbed gas producibility. American Association of Petroleum Geologists Bulletin 78, 1186–1209.

Sonnenberg, S.A., 2011. The Niobrara Petroleum System, a Major Tight Resource Play in the Rocky Mountain Region. American Association of Petroleum Geologists Search and Discovery. Article #10355, 31 p.

Thorn, W.T. Jr., Spieker, E.M., Stabler, H., 1931. The Significance of Geologic Conditions in Naval Petroleum Reserve N. 3, Wyoming. United States Geological Survey Professional Paper 163, 80 p.

Truex, J.N., 1972. Fractured shale and basement reservoir, long Beach Unit, California. American Association of Petroleum Geologists Bulletin 56, 1931–1938.

United States Department of Energy (DOE), 2007. DOE's Unconventional Gas Research Programs 1976–1995: An Archive of Important Results. United States Department of Energy. 149 p.

United States Geological Survey (USGS) National Oil and Gas Resource Assessment Team, 1995. 1995 National Assessment of United States Oil and Gas Resources: United States Geological Survey Circular 1118. 20 p.

Wang, F.P., Gale, J.F.W., 2009. Screening criteria for shale-gas systems. Gulf Coast Association of Geological Societies Transactions 59, 779–793.

Wang, Z., Krupnick, A., 2013. A Retrospective Review of Shale Gas Development in the United States: What Led to the Boom? Resources for the Future. Discussion Paper 13-12, 42 p. http://www.rff.org/files/sharepoint/WorkImages/Download/RFF-DP-13-12.pdf.

Weniger, P., Francu, J., Krooss, B.M., Buzek, F., Hemza, P., Littke, R., 2012. Geochemical and stable carbon isotopic composition of coal-related gases from the SW Upper Silesian Coal Basin, Czech Republic. Organic Geochemistry 53, 153–165.

Widera, M., 2014. What are cleats? Preliminary studies from the Konin lignite mine, Miocene of central Poland. Geologos 20, 3–12.

Woodward, L.A., 1984. Potential for significant oil and gas fracture reservoirs in Cretaceous rocks of Raton Basin, New Mexico. American Association of Petroleum Geologists Bulletin 68, 628–636.

Yee, D., Seidle, J.P., Hanson, W.B., 1993. Gas sorption on coal and measurement of gas content. In: Law, B.E., Rice, D.D. (Eds.), Hydrocarbons from Coal, American Association of Petroleum Geologists Studies in Geology, vol. 38, pp. 159–184.

CHAPTER

8

Basin Modeling

OUTLINE

Introduction	273
Burial History	275
Thermal History	281
Modeling Maturation, Hydrocarbon Generation, and Expulsion	286
Maturation Modeling	286
Hydrocarbon Generation Modeling	288
Expulsion Modeling	291
Modeling Migration	292
Ray-Path Modeling	292
Darcy Flow Modeling	293
Invasion Percolation Modeling	294
Predicting Preservation	294
1-D Model Results	295
Burial History Diagrams	296
Depth Profiles	298
Time Profiles	298
Model Validation	299
Sensitivity Analysis	300
Volumetric Estimations	302
The Role of Basin Modeling in Unconventional Plays	305
References	305

INTRODUCTION

Basin modeling, sometimes referred to as petroleum systems modeling, uses the depth, age, and lithologic description of a stratigraphic sequence, cross section, or an entire basin in conjunction with information about the thermal history of the basin setting to simulate its geologic history and predict petroleum generation, expulsion, migration, and accumulation. The primary goals of basin modeling are to (1) determine if, when, where, how much, and what type of hydrocarbons have been generated and expelled by a source rock; (2) to be able to compare the timing of generation and expulsion with the timing of trap development; (3) to be able to trace potential migration pathway from source areas to trap areas; and (4) to be able to estimate the amount of hydrocarbons that are filling a trap.

The ability to achieve these goals depends on the type of model used and the amount of data available. One dimensional (1-D) basin models use a point on the earth's surface and the stratigraphic sequence below it representing a well, pseudowell, or seismic shot point as the

http://dx.doi.org/10.1016/B978-0-12-803350-0.00008-8

basis for the simulation. The main questions addressed by 1-D models are if and when hydrocarbon generation and expulsion may have occurred. Two dimensional (2-D) models use a cross section (vertical slice) of stratigraphy to build the simulation. In addition to predictions about if and when generation and expulsion may have occurred, potential vertical migration pathways may be predicted. Another form of 2-D modeling using a horizon surface(s) (map view) can also be used. This approach relies on a series of 1-D models to provide information about hydrocarbon generation and expulsion with the horizon surfaces used to predict migration pathways on those surfaces. While useful, it is not a true 2-D simulation. A three dimensional (3-D) basin model uses a cube or other basin volume for the simulation. With a 3-D model, potential migration pathways in three dimensions can be visualized and volumes of hydrocarbons that may have migrated and been trapped can be estimated.

The complexity of the models and the amount of data required (or assumptions made) increases with the number of dimensions almost exponentially. While 1-D models can be defined and executed using only a simple stratigraphic sequence and some basic knowledge (or assumptions) about the source rock and temperature history, 2-D and 3-D models require more information about rock and fluid properties in addition to more complex descriptions of the geology and heat flow with these added dimensions. And while 1-D models may be built and run by nonspecialist exploration geologists with some training, the running of 2-D and 3-D models is usually delegated to subject matter specialists.

Contrary to popular opinion, basin models do not actually prove anything. They can only suggest things about a petroleum system with varying degrees of certainty. As a result, there are five fundamental concepts that need to be kept in mind for effective and appropriate use of basin modeling. *Concept 1*: Always remember that basin models are simulations, they are recreations of what could have occurred, and they never represent what actually happened. At best, they are close approximations of what happened. *Concept 2*: Modeling is an iterative process and is never finished. New data can always be added, and a model can always be improved. However, the model may reach a point where additional data will not be available or new data may not significantly change the results. *Concept 3*: There are no unique solutions in basin modeling. When matching measured data with predicted values, remember you are matching the end products of the model to the end points of the natural processes. Many paths can lead to the same destination. *Concept 4*: Basin models should be thought provoking. They allow geoscientists to ask "What if…" questions and to be creative and examine a variety of alternative geologic scenarios. For petroleum explorationists, this is an opportunity to experiment to gain a better understanding of the factors controlling the model's results. And, *Concept 5*: Basin models do not actually provide answers but rather reduce the questions we might have and limit uncertainty. Use modeling to eliminate possible scenarios and not to just support your ideas. If these basic concepts can be kept in mind, interpretations from basin models can be useful and help to guide an exploration program.

Statistician George Box pointed out that "all models are wrong" (Box, 1976) and later added "some models are useful" (Box and Draper, 1987). And Link (1954) observed machines (i.e., computers and software) cannot replace the human element in geologic reasoning, but they might be able to help. So it should be the goal of basin modelers to work carefully and use good geologic sense to make basin models that are useful and help reduce exploration risk.

With these ideas in mind, the following discussions will briefly cover the fundamental concepts that go into developing a basin model. The emphasis will be on 1-D models as a tool that most exploration geologists will be able to use in their daily work. All these concepts are also applicable to 2-D and 3-D models, and information about some of the aspects specific to 2-D and 3-D basin modeling will also be included when appropriate. For more in-depth discussions of the theories and mechanics of 2-D and 3-D basin modeling, the reader is referred to Hantschel and Kauerauf (2009).

BURIAL HISTORY

A burial history simulates the sedimentation events represented in a stratigraphic column. This stratigraphic column can be based on actual well data, inferred from seismic data or postulated from outcrop data. The burial history will consist of a depth–time plot that will represent the geologic events that define the stratigraphic sequence. The curves in the burial history represent either the tops and/or bottoms of stratigraphic intervals (formations) or unconformity surfaces. The sedimentation events portrayed by each segments of a burial history curve represent either deposition, erosion, or nondeposition (depositional hiatus).

To demonstrate how burial history curves are constructed, consider the hypothetical stratigraphic column shown in Fig. 8.1. This stratigraphic column consists of five depositional events, with each event involving the deposition of a thickness of sediment over a period of time. There are also two unconformities present. The first unconformity, at 2000 ft, is a depositional hiatus with a time gap from 2 to 3 MYBP when no sediment deposition occurred. The second unconformity, at 6000 ft, has a time gap from 8 to 12 MYBP. It is an erosional unconformity where from 12 to 10 MYBP, 2000 ft of sediment was deposited followed by 2000 ft of erosion from 10 to 8 MYBP.

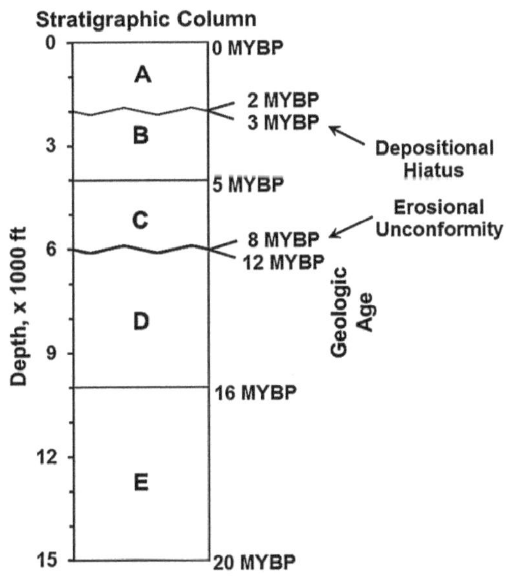

FIGURE 8.1 A hypothetical stratigraphic column containing both a depositional hiatus and an erosional unconformity to be used in the construction of a burial history.

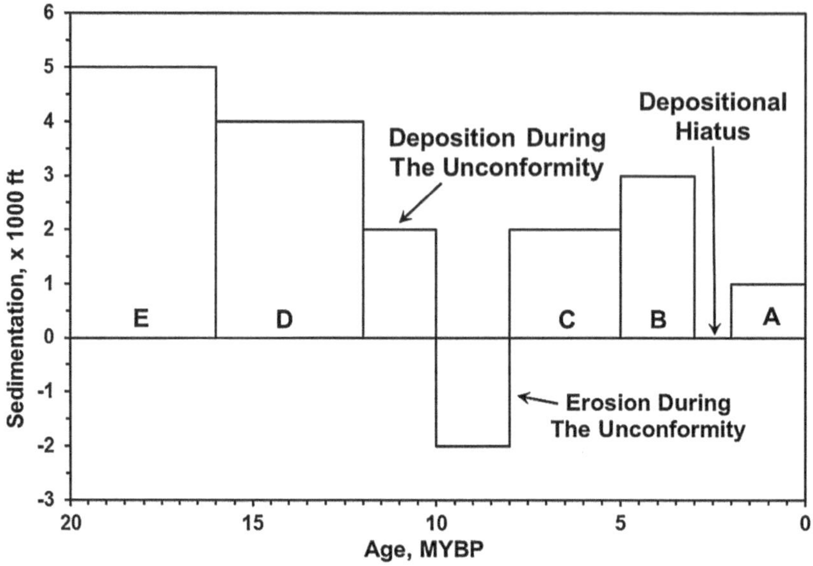

FIGURE 8.2 A sedimentation history for the stratigraphic column shown in Fig. 8.1 to be used in the construction of a burial history.

To represent this stratigraphic column in a burial history, it is sometimes useful to break it down into a sedimentation history as shown in Fig. 8.2. Each depositional event is plotted as an amount of sediment lain down over a certain time period. Erosional events are negative sedimentation events and depositional hiatus events plot along the zero line. To build a burial history diagram, simply start on the left side of the sedimentation history plot and move right one step at a time.

The process begins by depositing interval E, starting at 0 depth and 20 MYBP on the burial history diagram in Fig. 8.3. Interval E deposits 5000 ft of sediment in 4 million years with the burial history curve segment ending a depth of 5000 ft at 16 MYBP. Next comes deposition of interval D, 4000 ft of sediment in 4 million years. The second segment of the burial history curve ends at a depth 9000 ft at 12 MYBP. Between intervals D and C is the erosional unconformity. Deposition of 2000 ft of sediment occurs between 12 MYBP and 10 MYBP, so that the third segment of the burial history curve ends at a depth of 11,000 ft at 10 MYBP. This is followed by the erosional phase of the unconformity where 2000 ft of sediment is removed between 10 MYBP and 8 MYBP. This brings the end of the fourth segment of the burial history curve to a depth of 9000 ft at 8 MYBP. This is followed by the deposition of intervals C and B bringing the burial history curve to 11,000 ft at 5 MYBP and 14,000 ft at 3 MYBP, respectively. After the deposition of interval B, a period on nondeposition occurs between 3 and 2 MYBP. This segment of the burial history curve remains at 14,000 ft for this time period. And finally, interval A deposits 1000 ft of sediment between 2 MYBP and the present to complete the burial history curve at a depth of 15,000 ft at 0 MYBP. To draw the other burial history curves in Fig. 8.3, continue this process moving to the right one depositional interval for each successive curve.

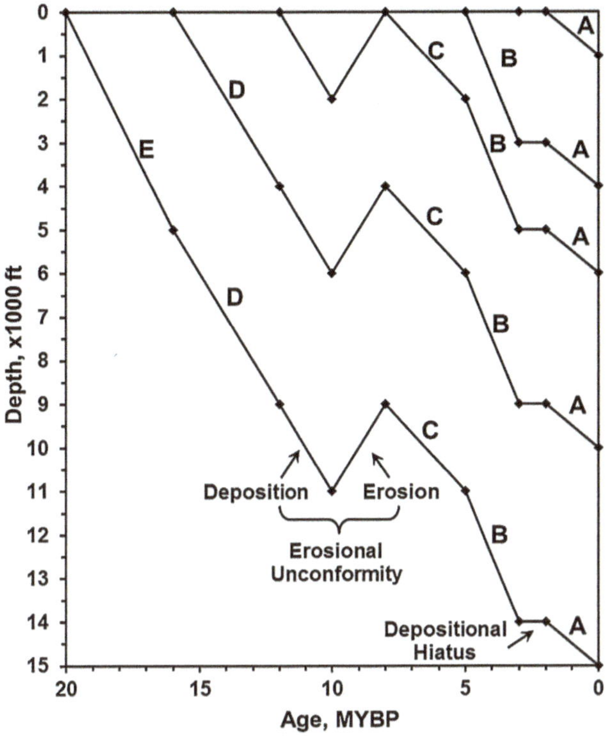

FIGURE 8.3 The burial history based on the stratigraphic column in Fig. 8.1 and the sedimentation history in Fig. 8.2.

While the resulting burial history plot is very useful as a snapshot of the sedimentation events that are represented in the stratigraphic column, the depiction of these events is not accurate. As the sediments are being deposited, they have an initial high porosity, and the grains are loosely packed. With successive burial, the weight of the overburden being deposited progressively forces out some of the pore fluid collapsing the pore space and reducing the thickness of the sediment, as shown in Fig. 8.4 and is usually referred to as compaction due to porosity reduction, or mechanical compaction. For more accurate burial histories, it is necessary to do compaction corrections to the sediment thicknesses as the overburden is added to the sediment column.

To correct for compaction, researchers have developed empirical relationships to help predict changes in porosity, such the exponential model of Sclater and Christie (1980), the reciprocal model of Falvey and Middleton (1981), and the argillaceous sediment model of Butler and Baldwin (1985), shown in Fig. 8.5. These relationships recognize that different lithologies will have different initial porosities and different rates of compaction with increasing burial, as shown in Fig. 8.6. It is also important to remember that the depositional units used in burial histories are seldom made up of one lithology. While it would be more accurate to subdivide these units into subunits containing only one lithology, it is not practical. A stratigraphic column would likely consist of thousands of depositional units

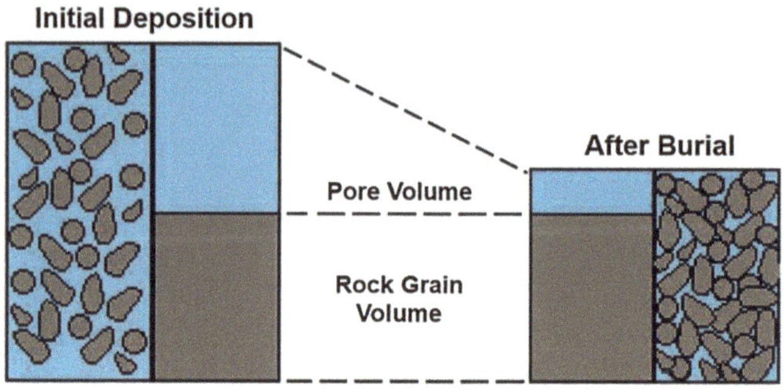

FIGURE 8.4 Compaction due to porosity reduction *After Crews, S.G., 2001, Personal Communication.*

Exponential Model of Sclater and Christie (1980)

$$P = P_0 \exp(-Kz)$$

Reciprocal Model of Falvey and Middleton (1981)

$$1/P = 1/P_0 + Kz$$

Argillaceous Sediment of Baldwin and Bulter (1985)

$$z = 6.02\, S^{6.35}$$

Where P is the porosity, P_0 is the initial porosity, K is the lithology's compaction factor, z is depth, and S is solidity, the inverse of porosity (1-P)

FIGURE 8.5 Some empirical relationships describing porosity reduction with depth linked to sediment compaction.

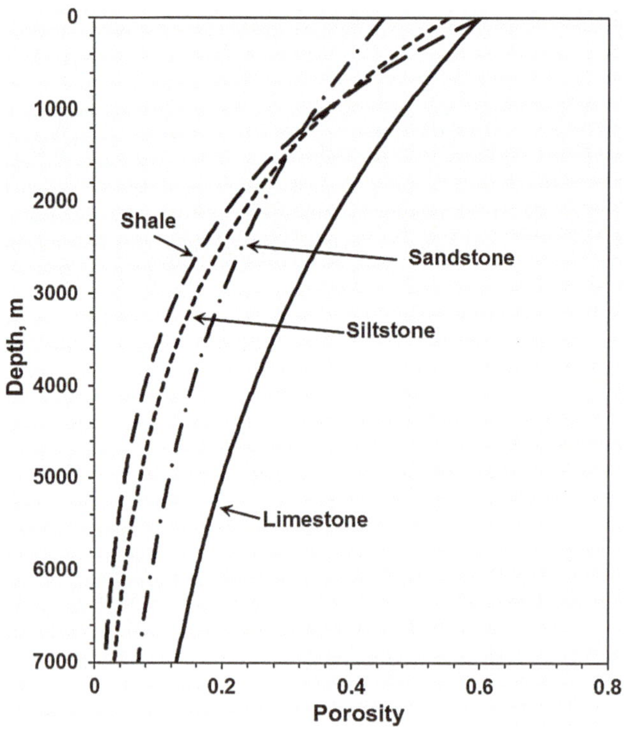

FIGURE 8.6 Depth versus porosity plot for a series of lithologies based on the Sclater and Christie (1980) exponential model using empirically determined initial porosities and compaction factors.

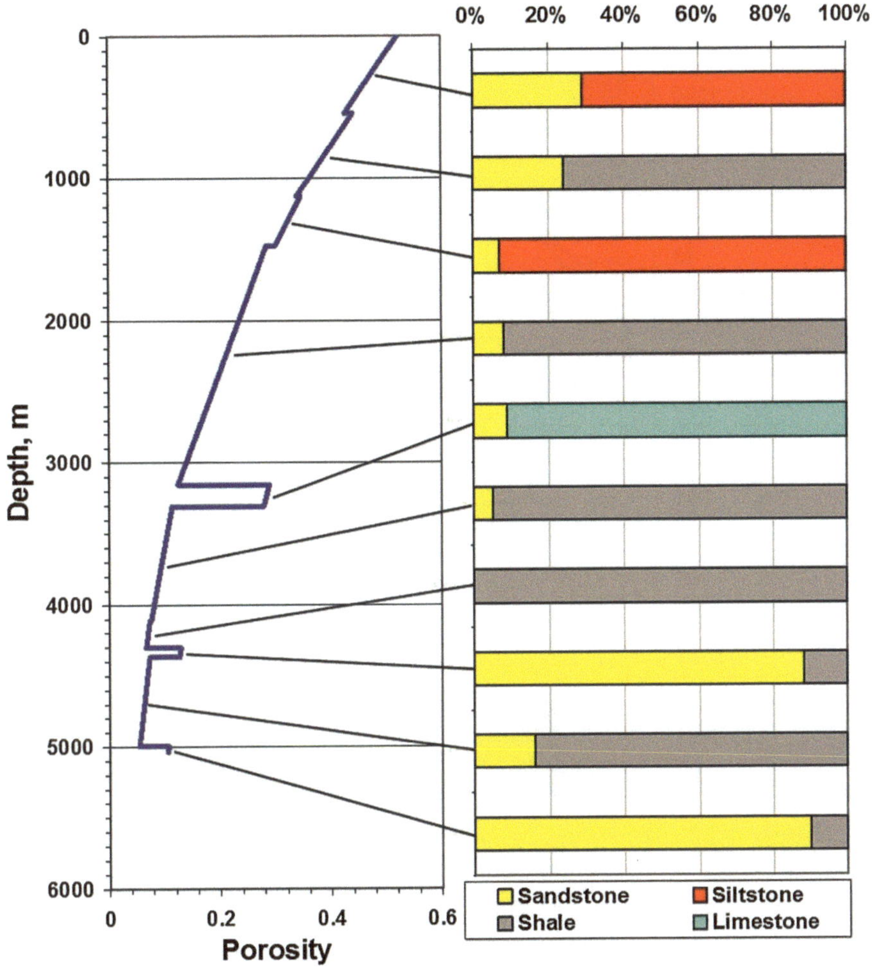

FIGURE 8.7 Depth trend for porosity changes with mixed lithologies.

instead of tens of units, and absolute age assignments to these smaller sediment packages would be impossible to make. Instead, porosity reduction for the entire unit is calculated based on the lithologic makeup of the stratigraphic interval. And these mixed lithologies are treated as a homogeneous mixture of the rock types present and not as discrete beds of each rock type. An example of porosity changes in a stratigraphic sequence with mixed lithologies is shown in Fig. 8.7.

For fluids to leave the sediment during compaction, the sediment must possess adequate permeability. To properly correct for compaction, the permeability of the sediments must be factored into the process. Most compaction models use a porosity–permeability relationship, usually the Kozeny–Carman equation (Ungerer et al., 1990) for the sediment and apply Darcy's law to predict a fluid flow rate. An example of a set of porosity–permeability relationships used in compaction correction is shown in Fig. 8.8.

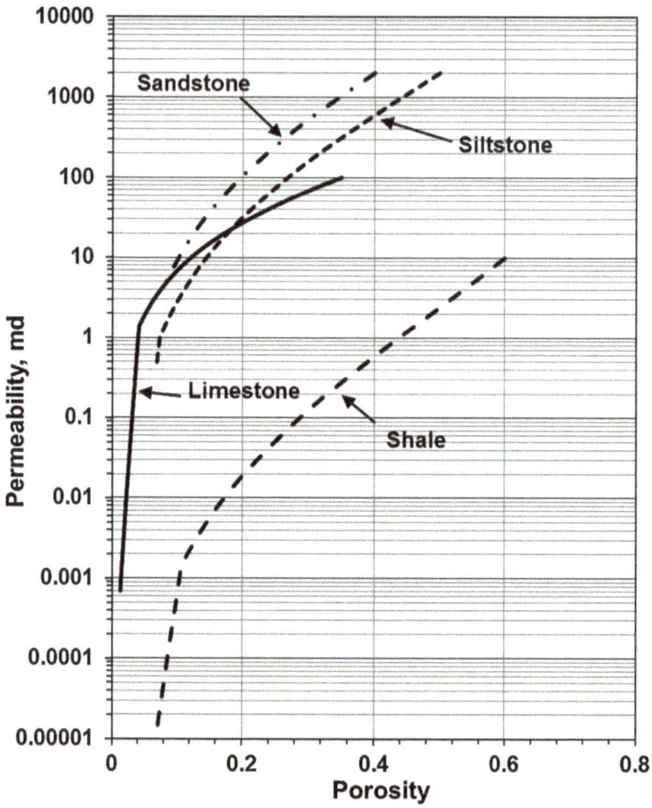

FIGURE 8.8 Kozeny–Carman relationship of porosity with permeability.

While mechanical compaction is the main cause of porosity loss and volume reduction in sediments, other factors may also influence the process. During rapid burial, low-permeability sediments (e.g., shales), may not be able to loose fluid at high enough rates resulting in excess fluid pressure (overpressure) and higher porosities than expected. Cementation can result in a more rigid grain framework halting compaction. And clay diagenesis and authigenic mineral growth can fill pores, while pressure solution (stylolitization) can eliminate pore space and reduce rock volume. It is difficult to predict and simulate these processes in basin modeling, and they are frequently ignored.

The actual process of compaction correction is a fairly complex mathematical procedure, especially for mixed lithologies and is usually handled by basin modeling software. The need for correcting for compaction can be clearly seen in the example in Fig. 8.9. Looking at both the corrected and uncorrected burial history curves, they begin and end at the same points. However, the corrected curve is always deeper than the uncorrected curve. Because temperature increases with depth, the temperature experienced by the compaction corrected curve is always greater than that experienced by the uncorrected curve. And this will have a major impact on the modeling of maturation and hydrocarbon generation as discussed later.

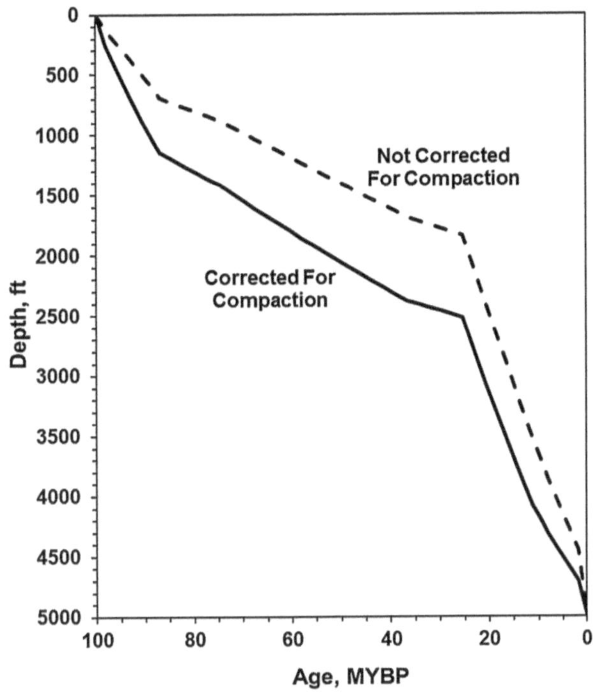

FIGURE 8.9 Burial history curve for a horizon with and without compaction correction.

THERMAL HISTORY

The burial history puts a stratigraphic sequence into a time–depth context. However, as stated in Chapter 2, organic matter maturation and hydrocarbon generation are kinetic processes controlled by time and temperature. To model maturation and generation, it will be necessary to convert the depth component in the burial history to temperature to arrive at a thermal history.

The thermal history is a simulation of the heat flow and temperatures experienced by sediments in a stratigraphic column during their burial history. It is usually expressed as the time–temperature histories of geologic events in a stratigraphic sequence. The thermal history is controlled by the surface temperature, heat flow, and thermal properties of the sediments, as well as influences from igneous bodies and/or circulating fluids.

A simplistic approach for approximating the thermal history that was used early in the development of basin models was to employ the surface temperature and bottom hole temperature (BHT) to calculate a geothermal gradient. This defined a linear relationship of temperature with depth where geothermal gradient = (BHT—surface temperature)/depth, usually expressed in either °F/100 ft or °C/km. The BHT is measured during wireline logging runs and reported in the log headers. The BHT needs to be adjusted for the chilling of the borehole by circulating drilling fluids using a correction such as the Horner Plot method (Horner, 1951; Fertl and Wichmann, 1977) as shown in Fig. 8.10.

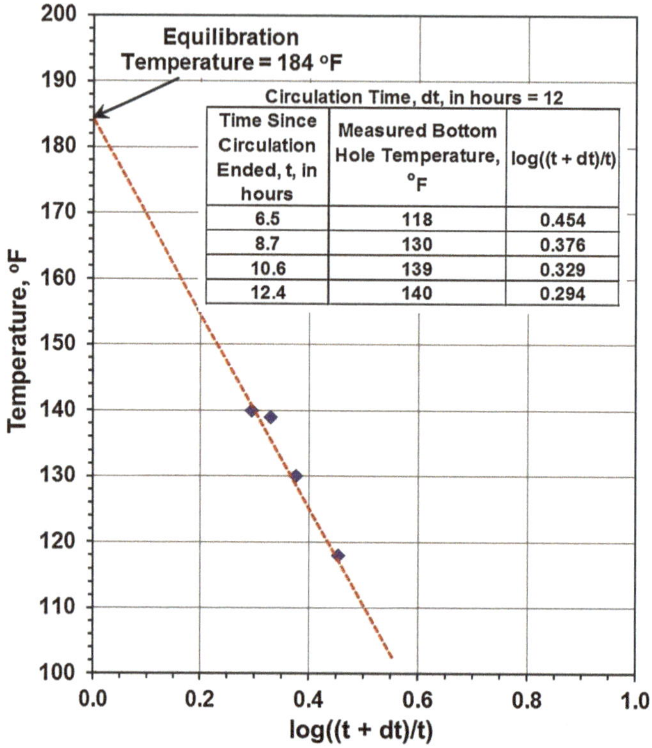

FIGURE 8.10 An example of a Horner plot correction of bottom hole temperatures.

Setting the surface temperature is not as straightforward as it might seem. Onshore, the mean annual surface temperature is often suggested for the surface temperature (Gretener, 1981). However, solar heating at the surface, climatic conditions, and the thermal properties of the surface sediments may make the mean annual surface temperature an inaccurate estimator for the surface temperature in some settings. Guidance may be gained from near surface groundwater and cave air temperatures to help constrain the surface temperature. For the onshore, experience has shown that a surface temperature of 10°C (50°F) is usually a good starting point.

In offshore settings, the surface temperature is the temperature at the sediment–water interface. This temperature will vary with water depth and latitude (Pickard, 1963). Below about 500 m, where most exploration activity is currently focused, typical deep ocean temperatures will vary from -1° to 10°C, with the colder temperatures usually occurring at depths greater than 4000 m or in the higher latitudes (Beardsmore and Cull, 2001). An estimate of the sediment–water interface temperature can be calculated from depth and latitude using an equation proposed by Beardsmore and Cull (2001). However, a sediment surface temperature of 4°C (39°F) is usually a good starting approximation for offshore deep water basin models.

Geothermal gradients assume the thermal properties of the sediments in the stratigraphic sequence are constant with depth. However, when compared to measured high-resolution temperature profiles in wells, geothermal gradients do not accurately estimate subsurface temperatures. This is because the thermal properties of sediments change with depth and need to be accounted for to more accurately simulate the thermal history. As a result, it is more appropriate to look at the thermal history of sediments from the perspective of heat flow.

To begin this discussion of heat flow, it is useful to review the concept of temperature. Temperature is usually thought of as the degree of hotness or coldness of a body or environment. However, it is more accurately a measure of the average heat of a mass or environment. Temperature can also be viewed as the property of a body or a region of space (environment) that determines whether or not there will be a net flow of heat either into it or out of it from a neighboring body or region and in which direction (if any) the heat will flow. If there is no heat flow, that body or region is said to be in thermal equilibrium and remains at the same temperature. If there is a flow of heat, the direction of the flow is from a body or region of higher temperature to a body or region of lower temperature.

For a stratigraphic sequence, the movement of heat is from within the earth to the surface, where it is dissipated into the atmosphere, surface water, and eventually space by radiation. Under steady state conditions, if we consider the geothermal gradient as the change in temperature (dT) over a depth interval (dz), or $d\mathrm{T}/d\mathrm{Z}$, then the heat flow, Q, is equal to k ($d\mathrm{T}/dz$), where k is the thermal conductivity. The thermal conductivity is a measure of the ability of a substance to conduct heat. High thermal conductivity indicates a good conductor, while a low thermal conductivity indicates an insulator. Under steady state conditions, the conductive heat flow is considered constant from the bottom to the top of a sediment interval, and the local geothermal gradient is inversely proportional to the local thermal conductivity.

Basin development is, however, a dynamic set of processes that are seldom constant through time. As such, most stratigraphic sequences are deposited under transient heat flow conditions with the potential for brief period of steady state conditions. Under transient conditions, the conductive heat flow is not constant from the bottom to the top of the sediment interval. If sedimentation is rapid, the large volume of cool sediments being deposited requires part of the heat flow coming up from the basement to be used to heat these newly deposited sediments, and the surface heat flow is lower than the bottom heat flow. If erosion is rapid, sediments exposed at the surface contain excess heat that must be dissipated to return to a steady state resulting in the surface heat flow being higher than the bottom heat flow. These effects do not result in sudden changes in the thermal profile. Rather, thermal inertia will smooth out these changes over time. As a result, additional terms must be added to the heat flow equation to incorporate these conditions.

For transient heat flow, it is necessary to consider the heat capacity, or specific heat, of a substance to define its thermal inertia. Heat capacity is a measure of how heat flow affects the temperature of a system and is usually expressed as the amount of heat required to raise the temperature of a mass by a given number of degrees. High heat capacity indicates temperature will change more slowly with additional heating (more heat can be adsorbed), while low heat capacity indicates that temperature will change more quickly with additional heating (less heat can be adsorbed). The thermal inertia, I, represents the ability of a material to conduct and store heat and is a measure of the responsiveness of the material to variations in temperature. It is defined as $\mathrm{I} = (\mathrm{k}\,\rho\,\mathrm{c})^{1/2}$, where k is the thermal conductivity, ρ is the material's density, and c is the heat capacity.

The thermal conductivity and heat capacity are also dynamic quantities. When discussing the thermal conductivity and heat capacity of sediments, it is necessary to consider both the matrix and bulk properties. Matrix thermal conductivity and heat capacity refers to property of the sediment grains and is dependent on the mineralogy of the grains. In contrast, the bulk thermal conductivity and heat capacity refers to the property of the entire sediment including both the sediment grains and the interstitial fluid. As shown in Fig. 8.11, the matrix thermal conductivity and heat capacity for a specific lithology do not change with depth, but the bulk thermal conductivity and heat capacity will change with the porosity and fluid content. For an in-depth discussion of these thermal properties of rocks and their fluids, consult Robertson (1988).

The calculation of thermal histories using heat flow and the thermal properties of sediments is computationally intensive. Fortunately, basin modeling software will use the lithologic definitions of the stratigraphy along with the porosity predictions from the burial history as input into the thermal history calculations of thermal conductivity and heat capacity. These data are then used in conjunction with surface temperature and estimates of the basal heat flow to calculate the thermal history. Guidance for input values for the heat flow for an area can be obtained from any one of the numerous online database available.

While most of the heat flow is the result of heat conducted up from the mantle, there can also be a significant contribution from radioactive decay within crust from both basement

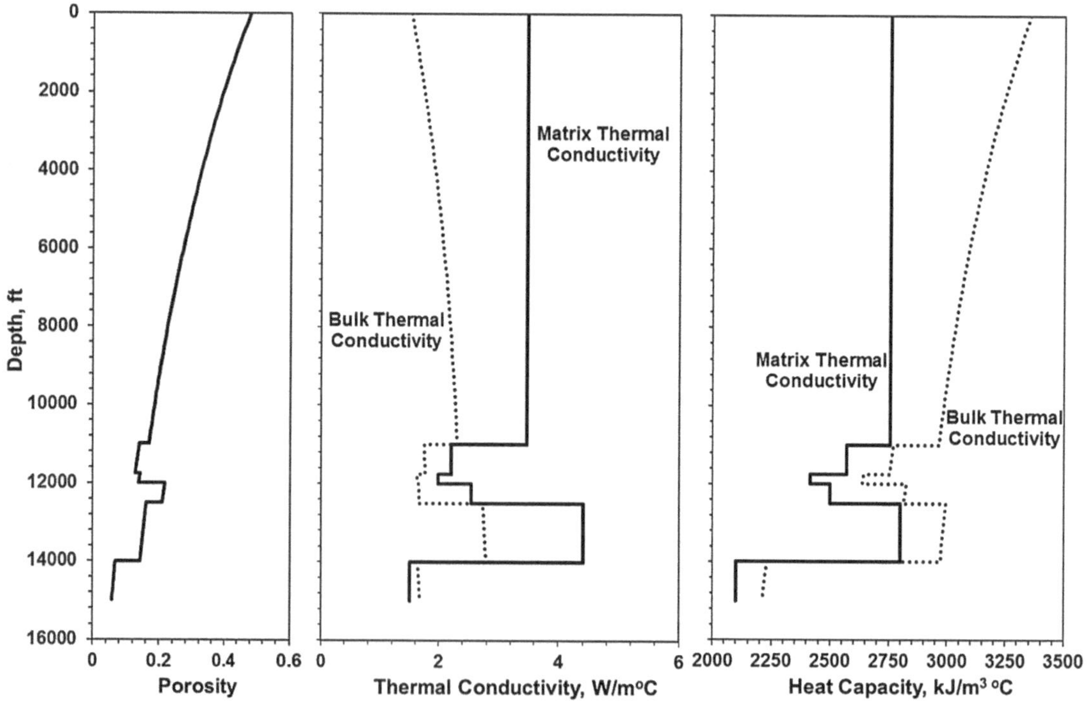

FIGURE 8.11 Comparison of porosity changes with the matrix and bulk thermal conductivities and heat capacities of a stratigraphic sequence. Note that while the matrix properties of the sediment are constant with depth, the bulk properties will change significantly as the fluid content changes.

and sediments. Basement rocks composed of granite and rhyolite have an average radiogenic heat production of 2.5 μW/m^3, while basalt and gabbro have an average radiogenic heat production of about 0.3 μW/m^3 (Pollack, 1982). Radiogenic heat contributions from sediment depend on the uranium, thorium, and potassium contents and can be estimated from gamma ray log response (Bucker and Rybach, 1996).

Short-term heat sources such as igneous intrusives (dikes and sills) and circulating fluids (e.g., hydrothermal fluids) can also influence the thermal history. Igneous intrusions are for the most part localized short-duration events. A high heat pulse is introduced into the sediment column, and it dissipates quickly with respect to geologic time. They can result in volatilization of sediment pore fluids and diagenetic effects up to and including metamorphism (Esposito and Whitney, 1995).

Diapiric salt column can also have a significant impact on heat flow. Evaporitic minerals have exceptionally high thermal conductivities as compared to other sediments. This high thermal conductivity can draw heat away from surrounding sediments by providing a low resistance conduit for heat flow (Beardsmore and Cull, 2001).

And it should be remembered that the surface temperature and heat flow do not have to remain constant through time. Surface temperature may change with changing water depth or with changing latitude. Heat flow may increase, decrease, ramp up, spike, or any combination of events. Heat flow can also change with geologic events, e.g., igneous activity. Predicting surface temperature and heat flow changes with time is often difficult and imprecise, and some experimentation may be required to arrive at a workable scenario.

A classic example of heat flow changing with time is the rifting model put forward by McKenzie (1978). At the start of rifting, a heat spike occurs, as shown in Fig. 8.12, due to the

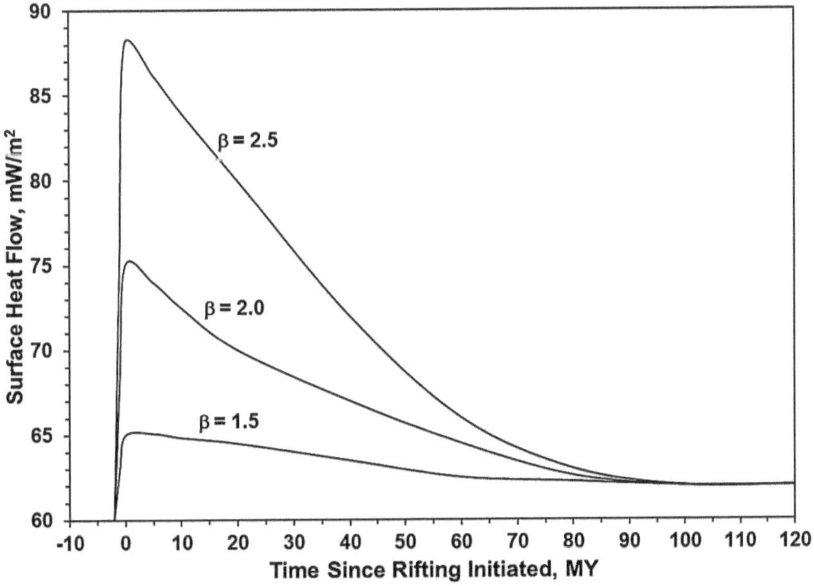

FIGURE 8.12 Influence of the Beta factor on the heat flow history during a rifting event.

upwelling of the asthenosphere and accompanying crustal thinning. This heat flow spike is described by the Beta factor, which represents the amount of stretching that the crust underwent prior to faulting, breakup, and subsidence. The greater the amount of stretching, the greater the amount of crustal thinning and higher the heat flow. The model estimates the rate of decay of the heat flow spike as it returns to lower heat flows. Often, it is necessary to vary the Beta factor to arrive at a reasonable heat flow history for the rifting event.

MODELING MATURATION, HYDROCARBON GENERATION, AND EXPULSION

After the burial history and thermal history modeling puts a stratigraphic sequence into a depth–time–temperature context, it is possible to use the time–temperature data for the sequence to simulate the maturity, hydrocarbon generation, and hydrocarbon expulsion histories of the source rocks. Maturity modeling can predict a stratigraphic horizon's current maturity and reconstruct its maturation history. The maturity is usually expressed as estimated vitrinite reflectance in % Ro equivalence. Hydrocarbon generation modeling calculates how much, what type, and when the generation of oil and gas may have occurred in the source rocks. It is linked to the source rock's richness and the kerogen types it contains. And expulsion modeling uses the hydrocarbon generation model results and estimated porosity/permeability to predict when and how much of the generated hydrocarbon can move from a source rock toward a carrier bed. While the modeling of all three of these processes are interrelated, they are usually handled separately within the basin modeling software. The following discussions outline the main methods used in these processes and any major concerns associated with them.

Maturation Modeling

In the early days of basin modeling in the 1970s, the scientific development of models to predict maturity was primarily in the hands of petroleum geochemists. Some of the early models consisted of simple time–temperature relationship based on the Arrhenius equation, such as the one proposed by Connan (1974). They were often basin specific and validated with empirical data. These types of relationships were sometimes converted to graphic solutions, like the one shown in Fig. 8.13, and provided maturity predictions tools based on effective heating time and maximum temperature experienced, similar to the coalification model of Karweil (1955). At the same time, the major petroleum companies were also developing mainframe computer–based proprietary modeling software using simple burial and thermal histories and incorporating these early kinetic models.

A major step forward in basin modeling came in 1980 with the publication of Waples (1980) interpretation of the Lopatin (1971) method and the arrival of the personal computer. Lopatin had developed an approach to predicting coal rank (maturity) using its time–temperature history. This time–temperature index, or TTI, of Lopatin was adapted by Waples to predict maturity in source rocks, as summarized in Fig. 8.14. Waples employed rudimentary burial and temperature histories that could be constructed with a pencil, straightedge, graph paper, and a calculator using stratigraphic columns and geothermal gradients. He also simplified the Lopatin calculation and provided a conversion of TTI

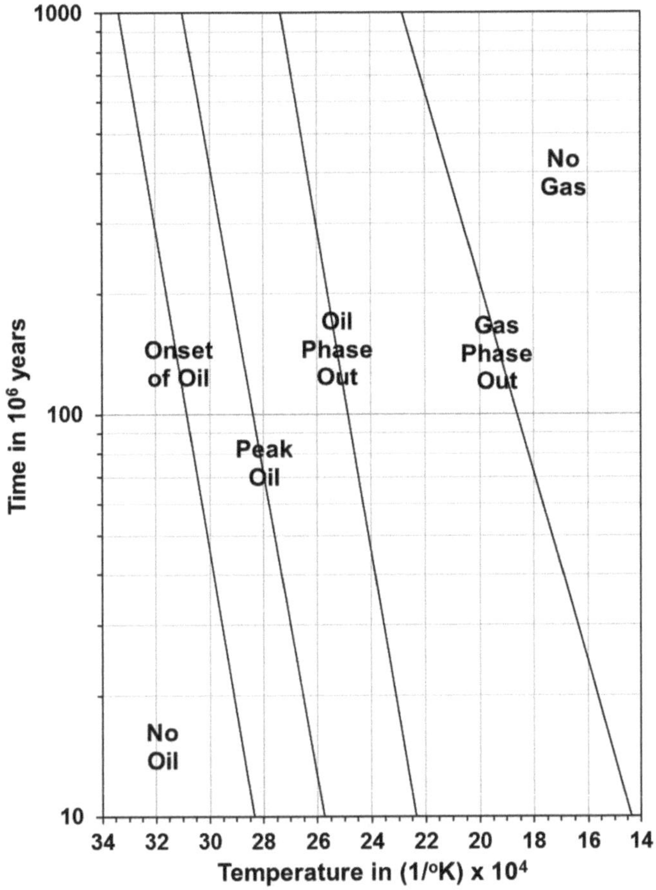

FIGURE 8.13 Graphical time–temperature maturation model based on a simple kinetical model for hydrocarbon generation. The time is effective heating time of the sediment, and the temperature is the maximum temperature experienced.

$$TTI = \Sigma (\Delta T_n)(r^n)$$

where T_n is the time spent in each 10°C temperature interval and r^n is the temperature factor for the interval. Time spent during any decrease in temperature (eg. during an uplift) is not included in the calculation.

TTI	Interpretation	% Ro
15	Onset of oil generation	0.65
75	Peak oil generation	1.00
160	End oil generation	1.30
~500	40° oil preservation deadline	1.75
~1,000	50° oil preservation deadline	2.00
~1,500	Wet gas preservation deadline	2.20
>65,000	Dry gas preservation deadline	4.80

Temperature Intervals in °C	Temperature Factor
30-40	2^{-7}
40-50	2^{-6}
50-60	2^{-5}
60-70	2^{-4}
70-80	2^{-3}
80-90	2^{-2}
90-100	2^{-1}
100-110	2^{0}
110-120	2^{1}
etc.	etc.

FIGURE 8.14 Summary of the TTI calculation method. *After Waples, D.W., 1980, Time and temperature in petroleum formation: application of Lopatin's method to petroleum exploration. American Association of Petroleum Geologists Bulletin 64 (6), 916–926.*

to equivalent vitrinite reflectance. The significance of this method was not necessarily the science it represented but that it put the ability to model maturity into the hands of geologists. And with personal computers becoming available at about the same time, it was only a short time before the basin modeling had migrated from the mainframes to the desktop computers when the first commercial 1-D basin modeling software packages became available. This software was intended for the geologists to do the modeling, not specialists. Petroleum geochemists were still involved with basin modeling but now only in the capacity of developing the concepts and methods.

Waples (1980) TTI method had some significant shortcomings. First, time spent during any decrease in temperature (e.g., during an uplift) is not included in the calculation. During uplift sediment temperature may decrease, but the maturation process does not necessarily stop. Instead the rate of increase may only slow. Another problem was in the temperature steps. In many instances, the 10°C temperature windows used may not capture significant features of the thermal history. And finally, the inability to correct for compaction or use thermal histories based on heat flow also introduced errors.

After the Waples (1980) paper, basin and maturation modeling made some rapid advancements. A prime example is the work by McKenzie (1981). He used a simple modification of the Waples (1980) method to calculate TTI values based on an integration of the time–temperature history. This integration is more geologically accurate accounting for periods of erosion and depositional hiatus as well as depositional events and eliminated the need for the 10°C temperature windows. McKenzie's model also included methods to construct burial histories accounting for compaction and thermal histories using heat flow.

While the integrated TTI approach was an improvement, it still relied on some form of "calibration" to convert the simulated time–temperature histories of sediments into equivalent vitrinite reflectance, and it did not address the actual chemical evolution of the vitrinite itself. This would require the development of kinetic models for vitrinite reflectance prediction. As discussed in Chapter 2, the development of kinetic models for hydrocarbon generation was started in the late 1960s. But it was not until the late 1980s and early 1990s that a number of kinetic schemes for prediction of vitrinite reflectance were proposed including Burnham and Sweeney (1989), Larter (1989), Sweeney and Burnham (1990), and Suzuki et al. (1993). Among them, the EASY%Ro model of Sweeney and Burnham (1990) has gained the most widespread acceptance. The kinetic parameters for this model, shown in Fig. 8.15, are a condensed version of an earlier VITRIMAT model (Burnham and Sweeney, 1989) and require a less complex set of calculations. EASY%Ro has been shown to be a robust model for predicting vitrinite reflectance in a variety of basin setting and can handle special circumstances such as the influences of igneous intrusion and hydrothermal fluids.

Hydrocarbon Generation Modeling

The modeling of hydrocarbon generation uses the burial and thermal histories of a sediment column to simulate the oil and gas generating reactions in source rocks. As shown in Fig. 8.16, kerogen thermally decomposes to produce oil, gas, and a carbon residue (char), and the oil can further thermally decompose to form more gas and carbon residue, with all these reactions being governed by first-order Arrhenius kinetics. Details of this model and the kinetic process can be found in Chapter 2 in the section on Maturation and Hydrocarbon Generation.

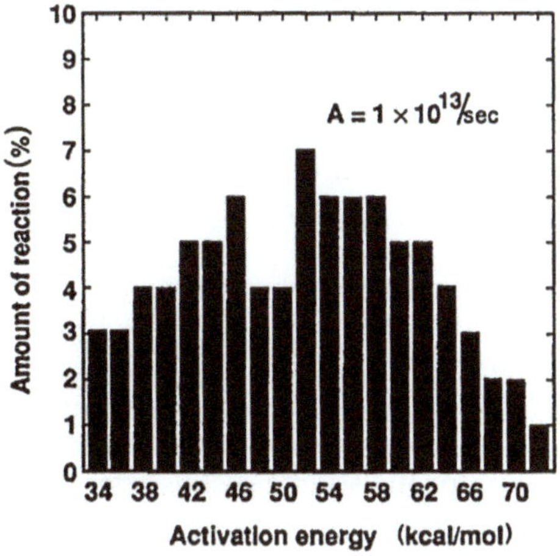

FIGURE 8.15 Kinetic parameters for the EASY%Ro vitrinite reflectance model *From Sweeney J.J., Burnham A.K., 1990. Evaluation of a simple model of vitrinite reflectance based on chemical kinetics. American Association of Petroleum Geologists Bulletin 74, 1559–1570.*

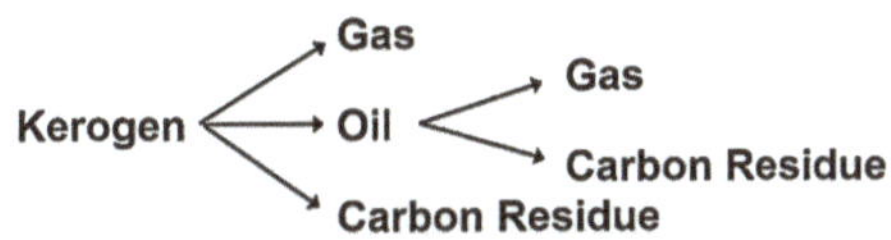

FIGURE 8.16 The simple (3 component) model for kerogen generation of oil, gas, and a carbon residue.

Early work to define the hydrocarbon generation kinetic parameters, such as Tissot (1969) and Tissot and Espitalie (1975) laid the groundwork for the understanding of the hydrocarbon generation process and how it could be simulated. And the development of computational methods for efficiently solving initial value differential equations, such as Gear (1971) and Balarin (1977), made it possible to use the first-order Arrhenius kinetics to model hydrocarbon generation.

Numerous investigations have contributed kinetic parameters for modeling hydrocarbon generation for the major chemical kerogen types over the years. These include Tissot et al. (1987), Braun et al. (1991), Behar et al. (1992), Tegelaar and Noble (1994), Pepper and Corvi (1995), and Behar et al. (1997). These kinetic parameters are for the simple hydrocarbon generation model, shown in Fig. 8.16 and are often available in commercial modeling software.

Because source rocks contain a mixture of kerogen types, most modeling softwares accommodate kerogen mixtures in their hydrocarbon generation models. Most software also allows the input of kinetic parameters determined specifically for the source rock being modeled. These measured, or custom, kinetics are derived from a series of heating experiments, usually using modified Rock-Eval analyses as described by Tissot et al. (1987) to produce hydrocarbon generation results from at least three different heating rates. The source rock samples used must be immature to properly assess the complete generative history of the kerogen. The results from these heating experiments are subjected to mathematical analysis using methods described by Ungerer and Pelet (1987), Braun and Burnham (1990) and others to converge on a set activation energies and frequency factors specific for the kerogen. Custom kinetics for many individual source rocks can be also found in the literature. A few examples are the Monterey formation (Jarvie and Lundell, 2001), Green River (Reynolds and Burnham, 1995), Phosphoria (Reynolds and Burnham, 1995; Lewan and Ruble, 2002), La Luna (Sweeney et al., 1995), and Kimmeridge Clay/Draupne formation (Reynolds and Burnham, 1995; Vandenbroucke et al., 1999).

There is some disagreement over whether these custom kinetics are useful in basin modeling. When doing custom kinetics, a few discrete samples from the source rocks are selected from an immature part of the basin. While the kinetic parameters derived may be representative of those samples, it is unknown how representative they are for that source rock in the generative kitchen. They could actually be misleading. It may be that an approximate solution based on a kerogen mixture for the source rock is adequate for constraining the hydrocarbon generation portion of the model (Curiale and Friberg, 2012). Considering all the other uncertainties that may be associated with the input data to a basin model, using a kerogen mixture instead of custom kinetics seems appropriate.

There are also more complex models for hydrocarbon generation where the gas component is separated into methane (C_1) and wet gas (C_2–C_4), and the oil component is separated into light (C_5–C_{14}) and heavy (C_{15}+) fractions, as shown in Fig. 8.17. These so-called five-component models require special compositional kinetic parameters be derived for individual kerogens using a complex series of analyses, such as those described by Behar et al. (1997), Vandenbroucke et al. (1999), and Dieckmann et al. (2000). These complex models are not typically used in routine basin modeling applications but are employed in special circumstances where the source rock is well defined, and more detailed information is sought from the modeling.

The results from hydrocarbon generation models will depend on the amount and type of organic matter in the source rock. Model inputs should be based on measurements from immature samples or estimated for the immature sediment based on analogs. The source rock input parameters usually consist of the total organic carbon (TOC) as a measure of the total organic matter, the Rock-Eval S2 or hydrogen index (HI) as a measure of the amount of

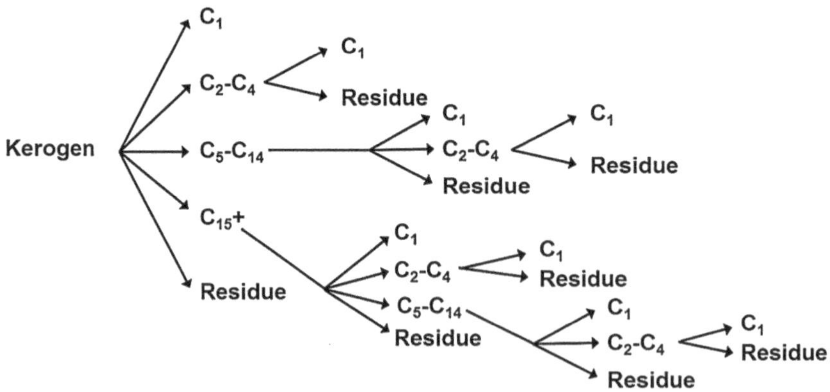

FIGURE 8.17 The five-component model for kerogen generation of methane (C_1), wet gas (C_2–C_4), light oil (C_5–C_{14}), heavy oil (C_{15}+), and carbon residue.

hydrocarbon that can be generated with respect to the TOC, and the kerogen type(s) to define the appropriate kinetic parameters to be used.

Initially, hydrocarbon generation was modeled as a closed system. Closed system models assume the source rock is "closed," and none of the generated hydrocarbons are removed. Because the oil never leaves the source rock, it will eventually be converted in place into gas. This is not a realistic model for a natural system where the generated hydrocarbons have the potential to leave the source rock to form a hydrocarbon accumulation.

Eventually, open system models were developed that would allow for some of the generated oil and gas to leave the source rock according to the expulsion criteria specified. By allowing oil to leave the source rock and migrate off, more oil will be preserved, and the oil converted to gas in the source rock is limited to the residual or nonmigrated oil.

Expulsion Modeling

From the discussion in Chapter 2, expulsion from a source rock occurs when hydrocarbons move out into the pore spaces and form a contiguous oil-wet migration pathway along which hydrocarbons can leave the source rock. Any hydrocarbons generated above the amount needed to maintain the minimum hydrocarbon saturation that formed the pathway are available for expulsion. Expulsion can be aided when pore fluids, both water and petroleum, become overpressured due to compaction, tectonic stress, thermal expansion of water, and hydrocarbon generation.

If and when expulsion occurs is influenced by several factors, such as the amount and type of organic matter in the source rock, the type of sediment, and the sedimentation rate. The more organic matter present, the more hydrocarbons that can be generated allowing the minimum saturation to be reached sooner. Certain kerogen types generate more hydrocarbons than others, and different kerogen types generate hydrocarbons at different points in their time–temperature history, which also influences when the minimum saturation can be reached. The lithology controls the porosity and permeability evolution of the sediment which in turn controls the pore volume needed to be filled by the hydrocarbons. Sedimentation rate can influence the thermal history, overpressure development, and rate of hydrocarbon generation.

Some early basin models simply linked expulsion to maturity indicators, such as vitrinite reflectance or transformation ratio. At a given maturity level, often at a vitrinite reflectance of about 0.7% Ro, the expulsion was thought to begin regardless of the richness of the source rock or the kerogen type. These approaches to expulsion prediction were inadequate and often misleading.

As expulsion became better understood, models linked to the expulsion mechanism were developed. The main expulsion model used in 1-D basin modeling software is the porosity saturation model. It uses an estimate of the hydrocarbon saturation of the pore spaces in a source rock (Ungerer et al., 1988b) based on hydrocarbon generation and porosity reduction/compaction model results. Once the saturation exceeds a threshold value, usually 20–25%, any additional hydrocarbon generated is expelled.

A more rigorous approach to expulsion uses capillary entry pressure and permeability in Darcy type flow to calculate if and how much fluid might be expelled from a sediment (Hantschel and Kauerauf, 2009). This approach requires accurate permeability predictions and some knowledge of the fluid's viscosity. While there are versions of this expulsion modeling approach suited for 1-D applications (Nakayama, 1987), it is most often employed in 2-D and 3-D basin models.

MODELING MIGRATION

Migration modeling predicts the pathways by which hydrocarbons move from the source rock to traps and from one trap to another to form accumulations. From the discussion in Chapter 2 in the section Petroleum Migration, oil migration occurs along restricted pathways or conduits formed after the oil has reached a high enough saturation in the reservoir rock for its buoyancy to overcome the capillary pressure in the pore throats. In a reservoir rock, oil moves vertically until it reaches the top of the reservoir interval, the seal, and then moves updip via narrow, restricted pathways until it reaches the trap and begins to accumulate. Multiple conduits are likely to form, and because these conduits are limited in diameter, the amount of oil lost during migration is limited to the irreducible oil saturation in the conduit.

1-D basin models do not address migration. The best a 1-D model can do is estimate if, when, and how much petroleum has been expelled from the source rock and is available for migration. Migration is a 2-D or 3-D problem. 2-D modeling can demonstrate potential migration pathways either on a surface (in map view) for lateral movement in a specific carrier bed or in cross section for limited observations of both horizontal and vertical migration in a single plane. 3-D modeling of migration can demonstrate potential migration pathways within a volume and represents the only true form of migration modeling.

The three main approaches to simulating petroleum migration from source rock to trap are ray-path modeling, Darcy flow, and invasion percolation. While this discussion of basin modeling is focused on the application of 1-D modeling techniques, the following brief descriptions of these three migration modeling methods are intended to provide a basic understanding.

Ray-Path Modeling

Ray-path, or flow path, migration modeling is based on the principles that hydrocarbon migration is driven by buoyancy, and the dip of the migration surface controls the pathway. The basic concepts of the model are summarized in Fig. 8.18. In its simplest form, ray-path modeling can be thought of as using reserve dip vectors on the migration surface. The modeling is performed on a specific depth surface at a specific point in time. The migration surfaces are based on current and paleostructure maps of the top of a carrier bed/base of a seal. Hydrocarbons generated and expelled from an underlying source rock move vertically into the carrier bed. Points where oil and gas may have reached the carrier bed are designated, usually over generative areas in the source rock where expulsion should have taken place. The model assumes oil and gas will move updip through the carrier bed along that surface due to the buoyancy of the hydrocarbons in the water-wet medium. Flow paths are then calculated taking into consideration the hydrodynamic conditions and the orientation of the potentiometric surface (Hindle, 1997). Any oil reaching traps can accumulate and, if traps are filled, may spill updip to adjacent traps. If column height and seal capacity are factored into the model, vertical leakage may occur from the top of traps or along fault plains allowing the hydrocarbon to migrate and accumulate in a shallower carrier bed (Burley and Scotchman, 1999)

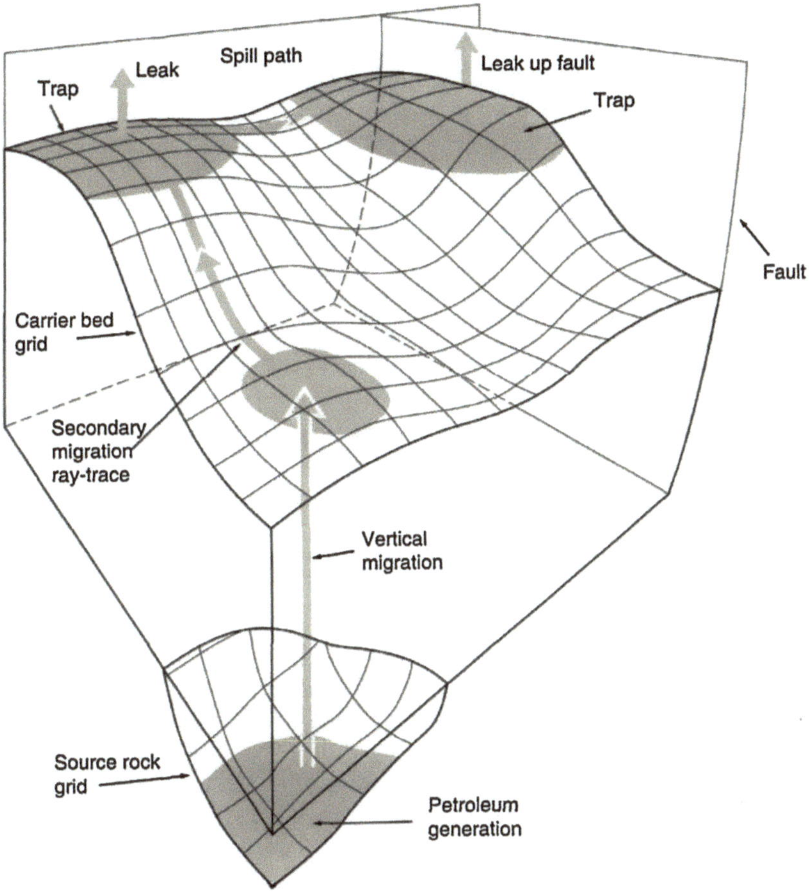

FIGURE 8.18 The basic elements of ray-path migration modeling. *From Burley, S.D., Scotchman, I.C., 1999. Basin modelling applications in reducing risk and maximizing reserves. In: Fleet, A.J., Boldly, S.A.R. (Eds.), Petroleum Geology of Northwest Europe: Proceedings of the 5th Conference. The Geological Society, London, pp. 1309–1311.*

or leak to the surface to form a seep (Hindle, 1997). While rather simple and computationally fast, ray-path modeling can reveal potential patterns of flow path focusing and divergence, areas where accumulations may form, possible migration shadows, and drainage area boundaries.

Darcy Flow Modeling

Darcy flow describes bulk laminar flow of a fluid through a porous medium along a pressure gradient. The amount of flow between two points is directly related to the difference in pressure between the points and the permeability of flow pathways in the rock. Movement is from higher pressure toward lower pressure, and oil and gas migrate as separate phases. The Darcy equation described fluid flow that is analogous to the movement of water in an aquifer (Meinzer, 1936). It also describes the flow in an oil or gas reservoir during production (Dake, 1978) and hydrocarbon leakage through mudrock sequences acting as a seal (Pegaz-Fiornet et al., 2012). While Darcy flow is often used to model migration of hydrocarbons from source rock to trap to form an accumulation in many 2-D and 3-D simulators, it does not describe the style of flow envision for this process (England et al., 1987). The low flow rates along narrow tortuous pathways, some dead-ending at permeability barriers, that potentially flow episodically do not fit the description for Darcy flow.

Invasion Percolation Modeling

Invasion percolation describes the process of one fluid displacing another in a porous medium under the action of capillary forces (Wilkinson and Willemsen, 1983). In the case of hydrocarbon migration, the porous medium is the reservoir rock, and the two fluids are oil and formation water. Percolation is most often thought of as a fluid moving down through a porous medium, but for hydrocarbon migration, because of the buoyancy of petroleum in the water saturated reservoir rock, movement is upward. With buoyancy and capillary pressures being the main driving forces for invasion percolation, the effects of viscosity and permeability are minimized (Carruthers and Ringrose, 1998). Using invasion percolation, it is also possible to model two phase flow, as well as to accommodate multiple charge points operating concurrently.

In the subsurface, the process of invasion percolation is initiated as oil begins to enter the base of the carrier bed. As oil accumulates, it will eventually reach a critical saturation where the buoyancy of the oil overcomes the capillary pressure in the pore throats, and oil movement begins (Carruthers, 2003). It is in this way that the oil moves from pore to pore along the migration pathway. This is consistent with the experimental evidence for secondary migration described by Dembicki and Anderson (1989), Catalan et al. (1992), and Thomas and Clouse (1995).

Currently, many 2-D and 3-D basin modeling software packages include invasion percolation in their tool kits, and it is the preferred method for modeling migration.

PREDICTING PRESERVATION

Basin modeling does not only simulate the generation and migration of hydrocarbons. It can also predict the preservation of hydrocarbons once they fill a reservoir. The two main processes affecting reservoired oil that can be predicted from basin modeling are biodegradation and thermal cracking. Both processes, as discussed in Chapter 4, are temperature dependent and use the temperature history of the reservoir interval to make the predictions. While providing useful insight, these predictions are only qualitative and meant to indicate the potential or lack of potential for thermal cracking or biodegradation to occur.

One of the primary controls on biodegradation is temperature. When a reservoir is at temperatures less than 45°C, the risk for biodegradation is high (Larter et al., 2006). As the reservoir temperature increases, the biodegradation risk decreases to a moderate risk between 45°C and 65°C and down to a low risk between 65°C and 80°C (Larter et al., 2006). At temperatures greater than 80°C, the microbes in the reservoir are considered to be "pasteurized" and little or no risk exists, even if the reservoir subsequently cools (Head et al., 2003). While formation water salinity, nutrient supply, and surface area of the oil–water contact all play important roles in determining whether or not biodegradation actually takes place (Larter et al., 2006), these factors are beyond the scope of basin modeling to predict at this time. As a result, it needs to be remembered that the prediction is for the risk of biodegradation and is not a direct indicator that biodegradation has actually occurred.

With respect to predicting the cracking of oil to gas, the general guidelines for thermal cracking from Hunt (1996) are typically employed. Most crude oils at reservoir temperatures of less than 150°C should be stable. As temperatures increase to greater than 150–160°C, the liquid hydrocarbons begin to be converted to gas, and at higher temperatures (190–200°C) wet gas components (C_2–C_4) may also convert to methane, with complete conversion to methane by about 250°C. These are general guidelines and may vary due to the composition of the crude oil.

Another approach to predicting the thermal stability of oil would be to use one of the kinetic models of oil cracking to gas, such as Ungerer et al. (1988a) or Vandenbroucke et al. (1999). The time–temperature history for the reservoir interval would be the input to run this simulation resulting in a more quantitative prediction. Kinetic parameters could also be derived for specific crude oils, similar to how kinetic parameters are derived for kerogens. In general, this kinetic approach would be preferred; however, it is not widely used in basin modeling software.

1-D MODEL RESULTS

1-D basin models are simple but effective ways to gain insight into the hydrocarbon generation and expulsion process through simulation. As stated earlier, these models can be built and run by nonspecialist exploration geologists with some training. They are also the easiest models to build and constrain, require the least amount of data, and are based on readily available information. And, 1-D models are easily modified to test alternate geologic scenarios.

Typical input for a 1-D basin model consists of a description of the stratigraphic sequence, details of the events represented by any unconformities, surface temperature, geothermal gradient and/or heat flow, and the characteristics of the source rock(s) to be modeled. Each stratigraphic unit in the sequence will require the absolute age and current depth of the top and bottom of the unit, as well as a lithologic description of the sediments it contains. Source rock characteristics needed are TOC, S2 or HI, description of the kerogen types with percentages, and thickness. The TOC, S2, and HI should represent the initial values for the source rock and not necessarily the current values measured. In addition, any measured data, such as vitrinite reflectance and/or corrected subsurface temperatures, should be included for comparison to predicted values. It is also possible in most modeling software to select which decompaction algorithms, kerogen kinetics, vitrinite reflectance model, constraints of the thermal history calculations, etc., that will be used during the modeling run.

The output from a modeling run is a vast amount of data. In addition to all the age, depth, and temperature data for each of the stratigraphic horizons in the model, maturity information, and the amounts of hydrocarbon generation and expelled, properties like porosity, permeability, pressure, bulk thermal conductivity, bulk heat capacity, and many other calculated values are stored and available for inspection. All this information is typically available in either tabular or graphic form. Many basin modeling software packages allow the user to design custom tables of model results according to needs. These tables can be exported for analysis by other software. For graphic displays, burial history diagrams can be generated with a large number of data overlays and color mapping options, and any calculated value can usually be plotted versus depth, time, or any other calculated value.

The main goal of 1-D basin modeling is to determine if, when, relatively how much, and what type of hydrocarbons may have been generated and expelled by a potential source rock, as well as understand the geologic events that occurred. This can usually be accomplished using burial history diagrams (depth–time plots), depth profiles of predicted parameters, and time profiles of predicted parameters.

Burial History Diagrams

Burial history diagrams are depth–time plots that represent the geologic events in a stratigraphic sequence. The lines on the diagram represent surfaces that denote the tops and/or bottoms of stratigraphic units (formations) or unconformity surfaces. These surfaces have been corrected for sediment compaction. When viewing a burial history diagram, such as the one shown in Fig. 8.19, depth increases down the diagram, while age increases to the left.

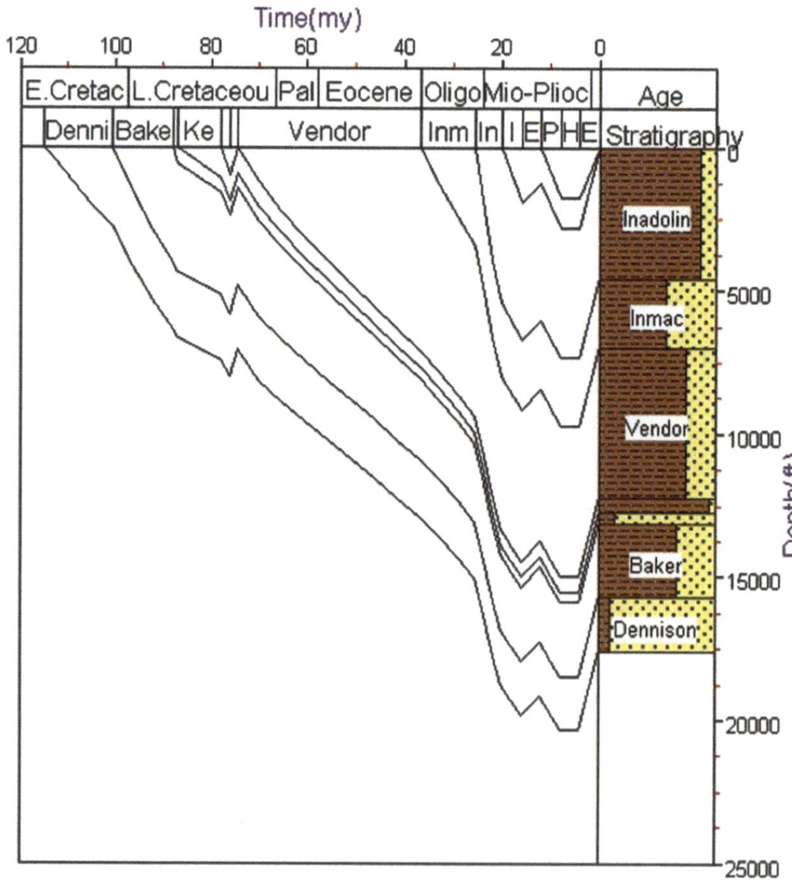

FIGURE 8.19 An example burial history diagram from a basin modeling run.

Stratigraphic surfaces go from zero depth on the left to their present depth of burial on the right axis. Line segments pointing down (increasing depth) indicate deposition, while line segments pointing up (decreasing depth) indicate erosion. Horizontal line segments indicate intervals of no deposition or erosion.

On the burial history diagrams, basin modeling results can be overlain either as a color map or contour lines. Most frequent results used as overlays are predictions of maturity/vitrinite reflectance, temperature, and hydrocarbon generation/expulsion. As shown in Fig. 8.20, predicted vitrinite reflectance data are often contoured over the burial history providing a record of maturation change with depth through time. Temperature data are also useful when view in this format. Color mapping of attributes can also be used. In Fig. 8.20, the amount of oil expulsion, confined to the source rock interval, is color mapped to provide insight into what geologic events may be the drivers for this process. Gas expulsion and transformation ratio are also useful when viewed in this format.

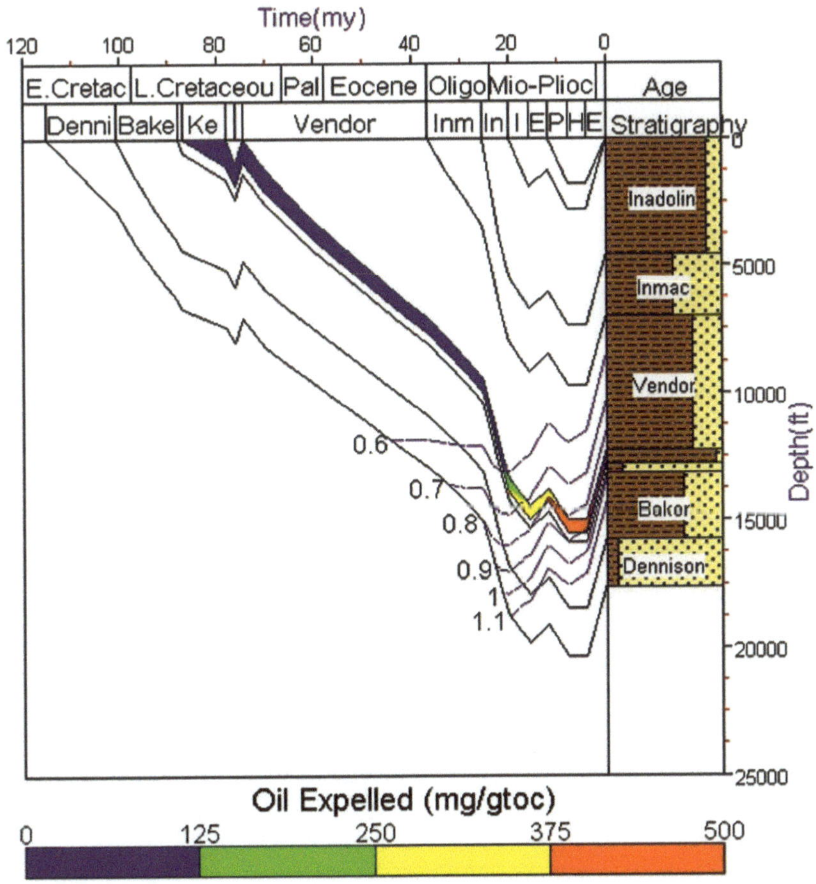

FIGURE 8.20 An example burial history diagram from a basin modeling run showing vitrinite reflectance as contoured values overlain on the curves and the color mapping of oil expelled from the source rock interval.

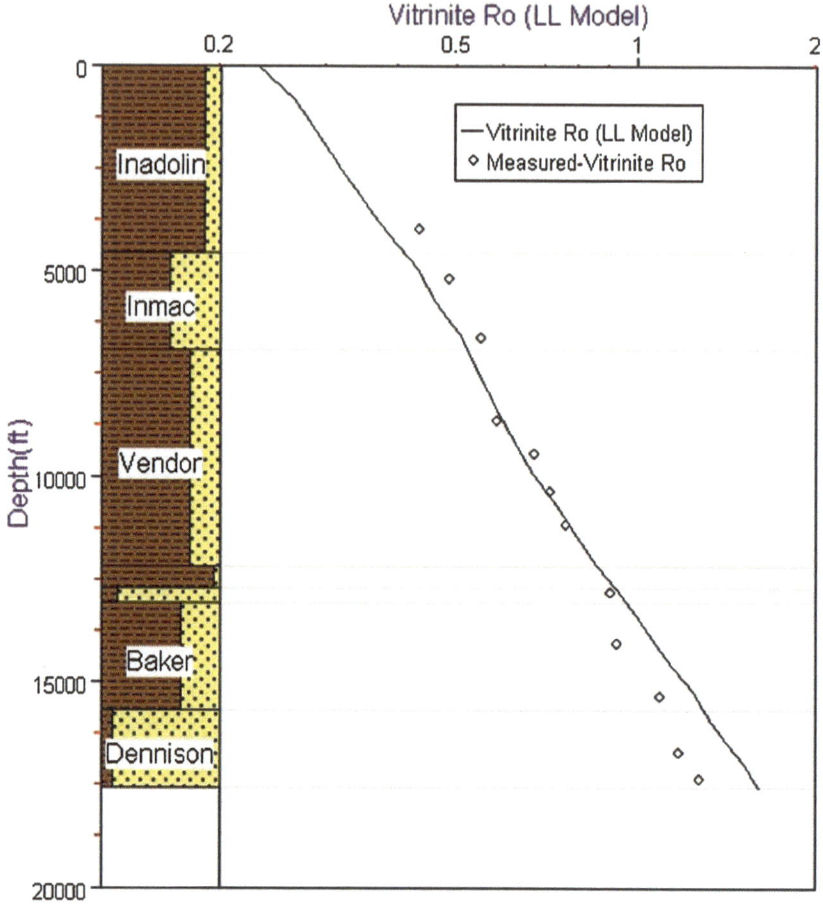

FIGURE 8.21 An example of a depth plot at present day for predicted and measured vitrinite reflectance.

Depth Profiles

Depth profiles plot depth on the vertical axis, with depth increasing from top to bottom, and predicted attributes on the horizontal axis. These plots can be made with predicted attributes for present day or at any time in the past. In addition, measured values of the attributes can also be added to depth profiles made for present day. The example in Fig. 8.21 shows the predicted and measured vitrinite reflectance at present. These plots can be used in conjunction with depth plots of measured and predicted temperature data for model validation, as discussed later. Other calculated values such predicted pressure or porosity could be plotted in a similar fashion.

Time Profiles

Time profiles are plotted for a single horizon/depth with time along the horizontal axis, with present day on the right and age increasing to the left and the calculated attribute(s) on

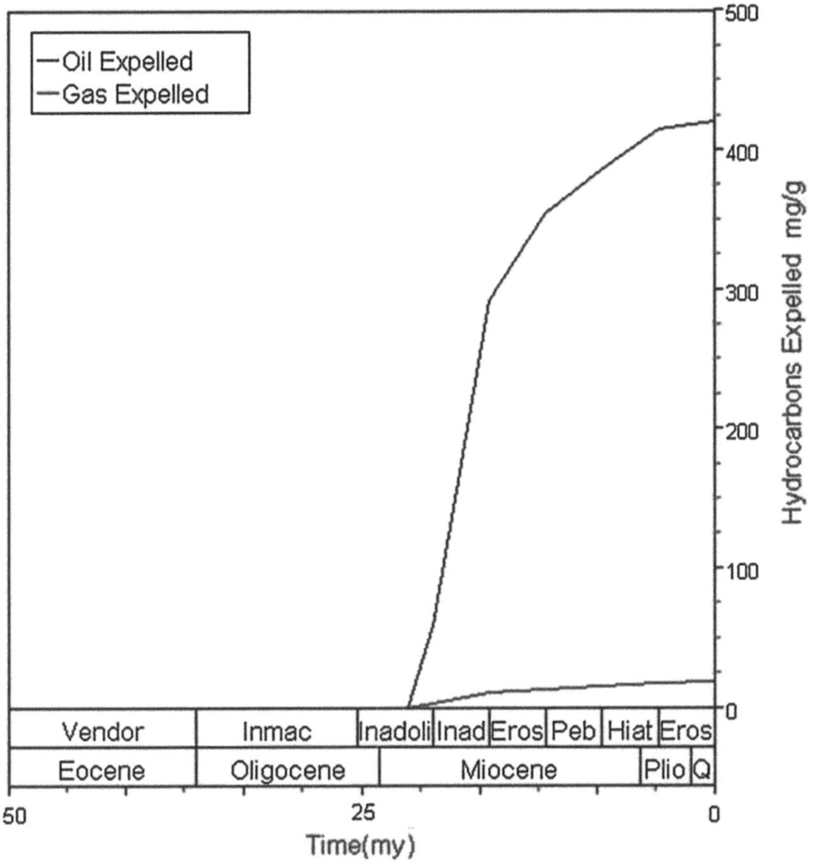

FIGURE 8.22 An example of a time plot at present day for predicted oil and gas expelled. Note that the curve presents cumulative expulsion for the hydrocarbons.

the vertical axis. This provides a means of tracing processes for a given horizon through time. These plots are used to answer questions primarily about the timing of maturity and hydrocarbon expulsion needed to evaluate a potential source rock. The example in Fig. 8.22 shows oil and gas expulsion curves for the source rock highlighted on the burial history in Fig. 8.20. The plot shows the time when expulsion is initiated, and the curves represent cumulative expulsion for the hydrocarbons. Other calculated values such as predicted vitrinite reflectance, temperature, or pressure, could be plotted in a similar fashion.

MODEL VALIDATION

To help assure that a basin model is reasonable for the geologic conditions it represents, predicted results from the model are compared to measured data from the subsurface. This is often referred to incorrectly as calibration. Calibration is defined as fixing or correcting the graduations of a measuring device, usually with known standards. Because basin models are not measuring devices, but rather simulations, models cannot be calibrated. Instead, the term validation, to prove to be well grounded on principles or evidence, is more appropriate.

Validation uses measured well data to "ground truth" the model results. For example, a given thermal history may result in the matching of measured with predicted values. However, that thermal history may not provide a unique solution. Other thermal histories may result in similar matches between measured and predicted values. So the thermal history is valid for the model as compared to the measured values but not the only possible solution. Validations are usually done with present day temperatures and/or vitrinite reflectance data.

Using present day temperatures are only good for validating present day heat flow. It is sometimes also difficult to get accurate temperature measurements due to chilling of the borehole by drilling fluids and inadequate means of correcting these temperatures (see the discussion in the earlier section on Thermal History). If only present day temperatures are used, there is greater uncertainty concerning the validity of the entire thermal history.

Validation with vitrinite reflectance is preferred because vitrinite reflectance represents the cumulative thermal history and not just the present day conditions. While it is difficult to discern details of the thermal history from the vitrinite reflectance profile alone, agreement between the measured and predicted values suggests that the thermal history scenario used in the model is at least one of the possible thermal histories that would give these results.

The validation process is also a means for adjusting the input parameters for the thermal history. In the example in Fig. 8.23, the initial thermal history scenario was off for both the present day BHT and measured vitrinite reflectance data. Through several iterations, adjustments were made to both the surface temperature and heat flow to arrive at a thermal history that could be validated with the measured data. This approach is frequently used in the early stages of basin model development.

SENSITIVITY ANALYSIS

Often, there is substantial uncertainty about basin model inputs. This usually occurs when the model is based on a stratigraphic column derived from seismic data and/or when actual source rock data are unavailable to use as input. In these situations, assumptions and guesses about the input data are used to fill in the blanks for the basin model. Even when real data are available for a basin model's input, there may be some uncertainty about how representative that data may actually be. Or, there may be a need to test alternate hypotheses with the basin model to account for possible changing conditions in the basin or prospect area. Sensitivity analysis is one way to address these issues.

Sensitivity analysis is a way to measure susceptibility to changes in the model's input and to determine how sensitive the results are to these variations. Often, sensitivity analysis is used to test parameters that contain the most uncertainty. Typical parameters that are tested include heat flow, the amount of sediment lost at an unconformity and source rock characteristics such as TOC, S2 or HI, and kerogen type. However, any of the model's input can be investigated. This testing is often done by making model runs using the maximum, minimum, most likely values for the input parameter and comparing results. At other times, one or two input parameters are selected to address specific questions.

Some examples of investigating sensitivities are shown in Fig. 8.24 for heat flow and Fig. 8.25 for kerogen type and TOC. In the first example, the modeler used a constant heat flow through

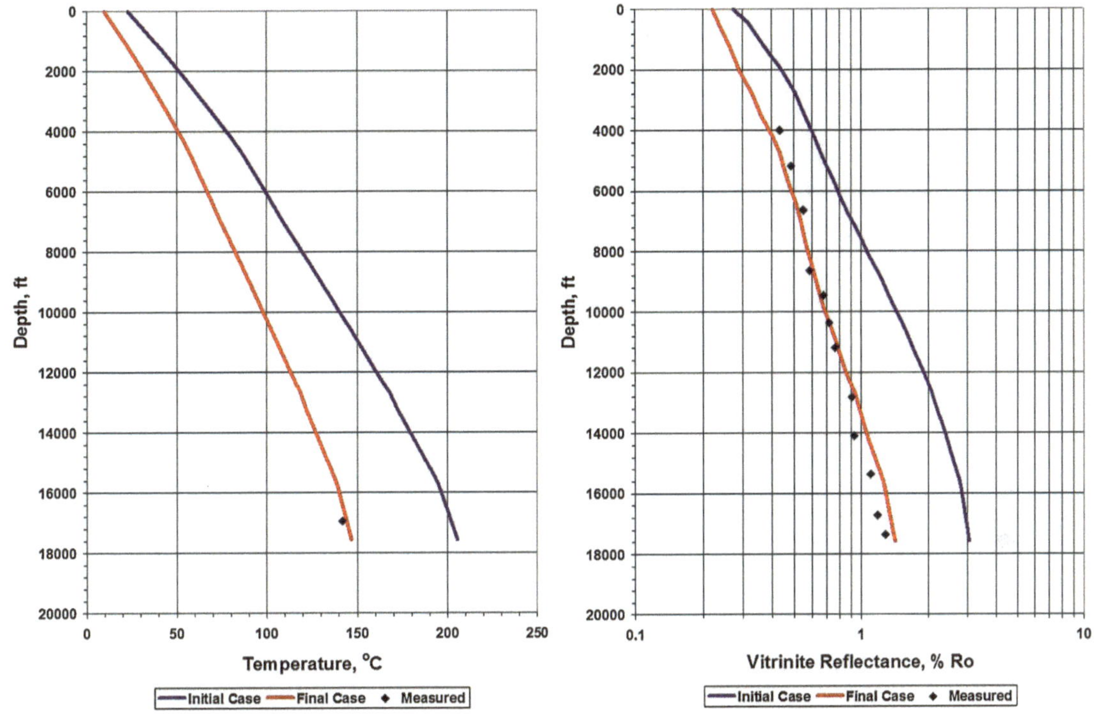

FIGURE 8.23 An example of model validation using measured and predicted subsurface temperature and vitrinite reflectance. The initial case used a surface temperature 23°C and heat flow 60 mW/m². The final case used a surface temperature 10°C and heat flow 45 mW/m² to get a reasonable match between measured and predicted values.

time of 45 mW/m² to obtain a reasonable fit of the predicted vitrinite reflectance to measured data. To test the impact of the heat flow on the vitrinite reflectance prediction, additional model runs were made using heat flows of 40 mW/m² and 50 mW/m². The results shows that 45 mW/m² still gives the best fit to the measured vitrinite data with the predictions using heat flows of 40 mW/m² and 50 mW/m² bracketing the measured data. But there are some departures in the predicted vitrinite reflectance with the measured data in the shallow and deep sections of the stratigraphic column at 45 mW/m² that come close to the predicted results from the other heat flows. This may be indicating that a constant heat flow scenario may not be the best simulation for this model and a change in heat flow with time may be more appropriate.

In the second example in Fig. 8.25, the modeler initially chose source rock parameters of 4% TOC and a kerogen made up 80% Type II and 20% Type III. Recognizing that these source rock characteristics may be optimistic, additional modeling runs were made with 1%, 2%, and 3% TOC and a kerogen mixture of 60% Type II and 40% Type III. The results show that not only does the decreases in TOC and in oil-prone kerogen content impact the amount of oil expelled, they also have an influence on when the expulsion is initiated. Using this series of models, minimum success cases for TOC and kerogen type can be made to assist in the risking of the prospect.

While testing individual or small groups of input parameters can address specific questions, a more efficient approach would be to test the uncertainty of all the input by using

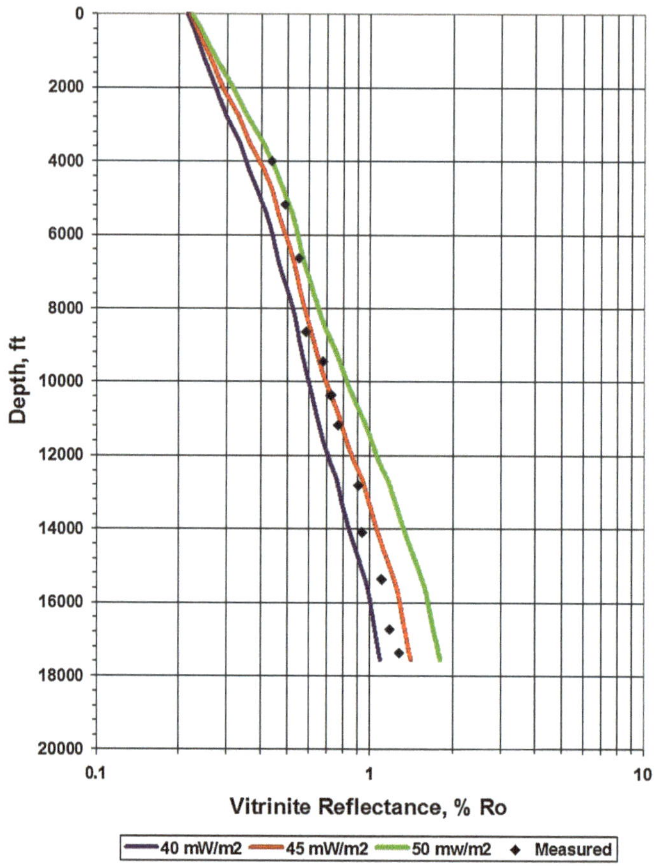

FIGURE 8.24 An example of a sensitivity analysis for heat flow.

distributions of input values for parameters and processing the model with a Monte Carlo, or similar, simulation. This allows the identification of key inputs that have the highest sensitivity and elimination of those that contribute little variance to the results. This type of sensitivity analysis is beginning to become available in basin modeling software packages and will likely see more widespread use in the future. For more details about this method for identifying and quantifying uncertainty in basin models, the reader is directed to Thomsen (1998) and Hicks et al. (2012) for a more in-depth discussion.

VOLUMETRIC ESTIMATIONS

Volumetric estimations assess the amount of oil and gas a source rock may have generated, expelled, and migrated toward one or more traps in a basin. These estimations are made to help explain observed accumulations, to determine if additional accumulations are possible, or to indicate if one or more individual traps can be adequately filled by a particular source

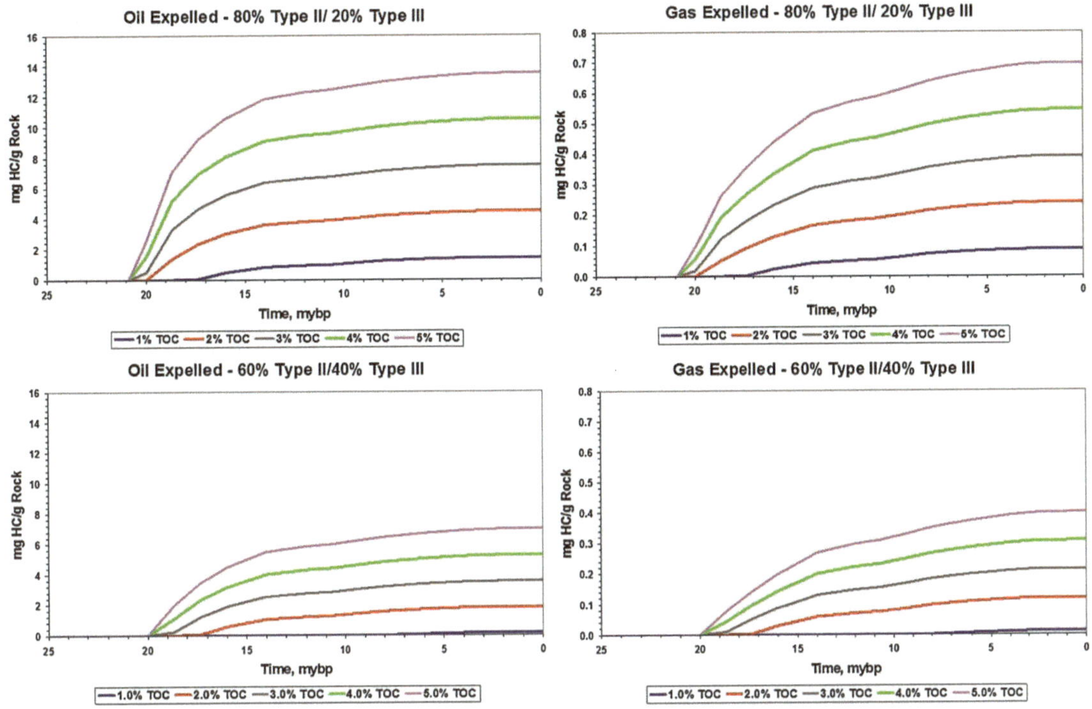

FIGURE 8.25 An example of a sensitivity analysis for total organic carbon (TOC) and kerogen type.

rock. On a basin wide scale, the entire generative "kitchen" for the source rock might be considered, but for an individual trap, only that part of the generative "kitchen" that drains into the trap should be included. Although more accurate predictions of the volumes of oil and gas generated, expelled, and migrated can be done with a 3-D model, 1-D and 2-D model results when carefully applied may be used to adequately describe these processes.

The input required includes a description of the source rock consisting of initial % TOC, initial S2 or HI, the kerogen type(s) present, and the rock type. All of these values are also part of the input to a 1-D basin model. In addition, the average effective source rock thickness and the prospect drainage (fetch) area are needed. The estimate of the API gravity of the oil generated is required to determine its density. From the 1-D model results, the cumulative oil and the cumulative gas expelled are used. An example volumetric calculation is shown in Fig. 8.26. Note the need for attention to the units of measure used in each step. Other volumetric calculations are also available, such as Bishop et al. (1983) and Schmoker (1994); however, these do not require the output of basin models.

Volumetric estimates usually result in very large, often unrealistic, amounts of oil and gas available to fill one or more traps. Remember, these estimates are for the amounts of hydrocarbons expelled. By some estimates, up to 90% of the expelled hydrocarbons are likely lost during migration. In addition, it may only be possible to recover 20–30% of the hydrocarbons from the reservoir. Therefore, these volumetric estimates should be used with extreme caution and with a realistic view of what the numbers actually convey.

Input:

Basin Model Input: **TOC = 2.0%; Initial HI = 500; Type II kerogen; Rock type = 100% shale with a density of**

Basin Model Output: **Cumulative oil expulsion = 440.0 mg oil/g TOC; Cumulative gas expulsion = 10.0 mg gas/g TOC**

Other Calculation Input: **Source Rock Density = 2.76 g/cc; Average effective source rock thickness = 180 ft (54.9 m); Prospect drainage area = 4.24 sq. miles (10.98 sq. km); Assume 35° API Gravity oil (density = 0.85 g/cc)**

The calculations:

1. **Calculate mg HC/g Rock for Oil and Gas**
   440 mg Oil/g TOC/(100/2.0) = 8.8 mg Oil/ g Rock
   10 mg Gas/g TOC/(100/2.0) = 0.2 mg Gas/g Rock
2. **Calculate cc HC/cc Rock for Oil and Gas**
   ((8.8 mg Oil/g Rock/1000)/0.85 g/cc) x 2.76 g/cc = 0.0285741 cc Oil/cc Rock
   ((8.8 mg Gas/g Rock/1000)/0.0054 g/cc) x 2.76 g/cc = 0.1022222 cc Gas/cc Rock
3. **Calculate Source Rock Volume**
   ((4.24 sq miles x 2.788x10^7 sq ft/sq mi) x 180 ft) x 2.832x10^4 cc/cu ft = 6.0259x10^{14} cc
4. **Calculate Barrels of Oil Available**
   (6.0259x10^{14} cc Source Rock x 0.0285741 cc Oil/cc Source Rock) x 6.2905x10^{-6} barrels/cc = 1.0831x10^8 barrels of oil
5. **Calculate Cubic Feet of Gas Available**
   (6.0259x10^{14} cc Source Rock x 0.1022222 cc Gas/cc Source Rock) x 3.531x10^{-5} cubic feet/cc = 2.175x10^9 cu ft of gas

FIGURE 8.26 An example volumetric estimation calculation using output for a 1-D basin model. *HI*, hydrogen index; *TOC*, Total organic carbon.

The causes of some of these large results can be attributed to the basin model input, assumptions made about the migration pathway, and overestimation of source rock effective thickness and kitchen/drainage area. Very often, basin modeling input is optimistic. Initial TOC and S2/HI values may be too high or the kerogen type may not be correct and/or dilution by Type III and/or Type IV kerogen may not have been accounted for. In addition, lateral and/or vertical variations in the TOC, S2/HI, and kerogen type may have occurred.

Effective source rock thickness used may be too high. The entire thickness of the source rock interval may not be capable of generating sufficient oil to migrate out and to fill a trap. As a result, only the effective source rock thickness should be considered (Dembicki and Pirkle, 1985). Think of effective source rock thickness like net sand versus gross sand thickness in reservoir facies. If the total thickness of the source rock interval is approximately 100 m, the effective source rock thickness may be a few tens of meters at best. Generative kitchen/migration drainage areas may be too large. It is tempting to include all the fetch area where the source rock is mature enough to generate hydrocarbons. However, the fetch area should be confined to where expulsion is likely (Sluijk and Parker, 1988). And the migration efficiency used for the expelled hydrocarbons may be too high providing a more optimistic view of the amount of petroleum reaching the trap. These uncertainties in the inputs to the volumetric calculation need to be addressed to better constrain the estimates and keep them in the realm of reality.

Applying these estimates should be done with some restraint. They are not a means of estimating the absolute trap-filling capacity of a source rock but are best thought of as relative indicators. One way to look at these results is that very large numbers for the oil and gas available for migration indicates robust source rock generation and expulsion. This initial appraisal suggests conditions for charging of the trap are favorable. If the volumetric estimates are low numbers, this would suggest there may be some concerns about charging and a more in-depth look at the source rock and/or maturity is needed. These estimates may also be useful for relative comparisons of two or more prospects for ranking purposes. Again, it is not recommended to use these volumetric estimations as absolute indicators of the amount of petroleum that may be in a given trap. There is simply too much uncertainty in these estimations for that application.

THE ROLE OF BASIN MODELING IN UNCONVENTIONAL PLAYS

Basin modeling should not be overlooked as a tool for assessing unconventional resources. As discussed in Chapter 7, unconventional plays are maturity driven. In the lower maturity plays, natural gas liquids and condensate are desirable products. But below a vitrinite reflectance of 1.3% Ro, PVT behavior of the hydrocarbons present will influence producibility (GOR, liquid dropout potential). In higher maturity plays, phase behavior is less of a problem. However, above a vitrinite reflectance of 2.5% Ro, loss of porosity and permeability from early metamorphism will likely limit gas available for production. Basin modeling can be used as a means of predicting maturity where data are lacking or filling in gaps in data-poor areas.

In addition, these models can be used to estimate temperature and pressure conditions in target intervals that can assist in predicting the phase behavior of reservoir fluids. And time–temperature histories from the models can be useful for predicting potential diagenetic mineral transformation, such as the opal-CT to quartz transition (Dralus et al., 2011), quartz cementation (Lander and Walderhaug, 1999), and the smectite to illite transformation (Velde and Vasseur, 1992), that can effect porosity, permeability, geomechanical properties, and fluid sensitivity.

Basin modeling is also important in understanding the impact of the burial history on unconventional resources. Many of the successful shale gas plays, including the Barnett, Eagle Ford, Haynesville, and Marcellus, have experienced significant uplift after going into the gas window and reaching maximum burial depth. Uplift likely assists in the development of overpressure in the stored gas and aids in production.

References

Balarin, M., 1977. Improved approximations of the exponential integral in tempering kinetics. Journal of Thermal Analysis and Calorimetry 12, 169–177.

Butler, B., Baldwin, C.O., 1985. Compaction curves. American Association of Petroleum Geologists Bulletin 69 (4), 622–626.

Beardsmore, G.R., Cull, J.P., 2001. Crustal Heat Flow: A Guide to Measurement and Modelling. Cambridge University Press, Cambridge. 324 p.

Behar, F., Kressmann, S., Rudkiewicz, J.L., Vandenbroucke, M., 1992. Experimental simulation in a confined system and kinetic modeling of kerogen and oil cracking. Organic Geochemistry 19, 173–189.

Behar, F., Vandenbrouke, M., Tang, Y., Marquis, F., Espitalie, J., 1997. Thermal cracking of kerogen in open and closed systems: determination of kinetic parameters and stoichiometric coefficients for oil and gas generation. Organic Geochemistry 26, 321–339.

Bishop, R.S., Gehman Jr., H.M., Young, A., 1983. Concepts for estimating hydrocarbon accumulation and dispersion. American Association of Petroleum Geologists Bulletin 67, 337–348.

Box, G.E.P., 1976. Science and statistics. Journal of the American Statistical Association 71, 791–799.

Box, G.E.P., Draper, N.R., 1987. Empirical Model-building and Response Surfaces. John Wiley & Sons, Hoboken, New Jersey. 688 p.

Braun, R.L., Burnham, A.K., 1990. KINETICS: A Computer Program to Analyze Chemical Reaction Data: Lawrence Livermore National Laboratory Report UCID-21588, Rev. 1. 11 p.

Braun, R.L., Burnham, A.K., Reynolds, J.G., Clarkson, J.E., 1991. Pyrolysis kinetics for lacustrine and marine source rocks by programmed micropyrolysis. Energy and Fuels 5, 192–204.

Bucker, C., Rybach, L., 1996. A simple method to determine heat production from gamma ray logs. Marine and Petroleum Geology 13, 373–375.

Burley, S.D., Scotchman, I.C., 1999. Basin modelling applications in reducing risk and maximizing reserves. In: Fleet, A.J., Boldly, S.A.R. (Eds.), Petroleum Geology of Northwest Europe: Proceedings of the 5th Conference. The Geological Society, London, pp. 1309–1311.

Burnham, A.K., Sweeney, J.J., 1989. A chemical kinetic model of vitrinite maturation and reflectance. Geochimica et Cosmochimica Acta 53, 2649–2657.

Carruthers, D.J., 2003. Modeling of secondary petroleum migration using invasion percolation techniques. In: Duppenbecker, S., Marzi, R. (Eds.), Multidimensional Basin Modeling, American Association of Petroleum Geologists Datapages Discovery Series No. 7, pp. 21–37.

Carruthers, D., Ringrose, P., 1998. Secondary oil migration: oil-rock contact volumes, flow behavior and rates. In: Parnell, J. (Ed.), Dating and Duration of Fluid Flow and Fluid-rock Interaction. Geological Society Special Publications, No. 144, London, pp. 205–220.

Catalan, L., Xiaowen, F., Chatzis, I., Dullien, F.A., 1992. An experimental study of secondary oil migration: American Association of Petroleum. Geologists Bulletin 76, 638–650.

Connan, J., 1974. Time-temperature relation in oil genesis. American Association of Petroleum Geologists Bulletin 2516–2521.

Crews, S.G., 2001. Personal Communication.

Curiale, J., Friberg, L., 2012. Measured kinetics – are they useful Basin model inputs? Oral presentation at the American Association of Petroleum Geologists Annual Convention and Exhibition, April 22–25, 2012, Long Beach, California, abstract. American Association of Petroleum Geologists Search and Discovery 1 Article #90142.

Dake, L.P., 1978. Fundamentals of Reservoir Engineering. Elsevier, Amsterdam. 437 p.

Dembicki Jr., H., Anderson, M.J., 1989. Secondary migration of oil: experiments supporting efficient movement of separate, buoyant oil phase along limited conduits. American Association of Petroleum Geologists Bulletin 73, 1018–1021.

Dembicki Jr., H., Pirkle, F.L., 1985. Regional source rock mapping using a source potential rating index. American Association of Petroleum Geologists Bulletin 69, 567–581.

Dieckmann, V., Horsfield, B., Schenk, H.J., 2000. Heating rate dependency of petroleum-forming reactions: implications for compositional kinetic predictions. Organic Geochemistry 31, 1333–1348.

Dralus, D., Peters, K.E., Lewan, M.D., Schenk, O., Herron, M., Tsuchida, K., 2011. Kinetics of the opal-CT to quartz phase transition control diagenetic traps in Siliceous Shale source Rock from the San Joaquin Basin and Hokkaido. American Association of Petroleum Geologists Search and Discovery 25 Article #40771.

England, W.A., Mackenzie, A.S., Mann, D.M., Quigley, T.M., 1987. The movement and entrapment of petroleum fluids in the subsurface. Journal of the Geological Society, London 144, 327–347.

Esposito, K.J., Whitney, G., 1995. Thermal effects of thin igneous intrusions on diagenetic reactions in a tertiary basin of southwestern Washington. United States Geological Survey Bulletin 2085-C 48.

Fertl, W.H., Wichmann, P.A., 1977. How to determine static BHT from well log data. World Oil 184, 105–106.

Falvey, D.A., Middleton, M.F., 1981. Passive continental margins: evidence for a prebreakup deep crustal metamorphic subsidence mechanism. In: Oceanologica Acta, Proceedings of 26th IUGG SP, pp. 103–114.

Gear, C.W., 1971. Numerical Initial Value Problems in Ordinary Differential Equations. Prentice-Hall, Englewood Cliffs, New Jersey. 253 p.

Gretener, P.E., 1981. Geothermics, Using Temperature in Hydrocarbon Exploration. American Association of Petroleum Geologists Education Course Note Series, vol. 17, 156 p.

Hantschel, T., Kauerauf, A.I., 2009. Fundamentals of Basin and Petroleum Systems Modeling. Springer, Heidelberg, Germany. 476 p.

Head, I.M., Jones, D.M., Larter, S.R., 2003. Biological activity in the deep subsurface and the origin of heavy oil. Nature 426, 344–352.

Hicks Jr., P.J., Fraticelli, C.M., Shosa, J.D., Hardy, M.J., Townsley, M.B., 2012. Identifying and quantifying significant uncertainties in basin modeling. In: Peters, K.E., Curry, D.J., Kacewicz, M. (Eds.), Basin Modeling: New Horizons in Research and Applications, American Association of Petroleum Geologists Hedberg Series, No. 4, pp. 207–219.

Hindle, A., 1997. Petroleum migration pathways and charge concentration: a three dimensional model. American Association of Petroleum Geologists Bulletin 81, 1451–1481.

Horner, D.R., 1951. Pressure build-up in wells. In: Proceedings of the Third World Petroleum Congress, Lieden, The Netherlands, Sec. II, pp. 503–521.

Hunt, J.M., 1996. Petroleum Geochemistry and Geology, second ed. W.H. Freeman, New York. 743 pp.

Jarvie, D.M., Lundell, L.L., 2001. Amount, type, and kinetics of thermal transformation of organic matter in the Miocene Monterey Formation. In: Isaacs, C.M., Rullkotter, J. (Eds.), The Monterey Formation: From Rocks to Molecules. Columbia University Press, New York, pp. 268–295.

Karweil, J., 1955. Die Metamorphose der Kohlen vom Standpunkt der physikalischen Chemie. Zeitschrift der Deutschen Geologischen Gesellschaft 107, 132–139.

Lander, R.H., Walderhaug, O., 1999. Predicting porosity through simulating sandstone compaction and quartz cementation. American Association of Petroleum Geologists Bulletin 83, 433–449.

Larter, S.R., 1989. Chemical models of vitrinite reflectance evolution. Geologische Rundschau 78, 349–359.

Larter, S., Huang, H., Adams, J., Bennett, B., Jokanola, O., Oldenburg, T., Jones, M., Head, I., Riediger, C., Fowler, M., 2006. The controls on the composition of biodegraded oils in the deep subsurface: Part II-Geological controls on subsurface biodegradation fluxes and constraints on reservoir-fluid property prediction. American Association of Petroleum Geologists Bulletin 90, 921–938.

Lewan, M.D., Ruble, T.E., 2002. Comparison of petroleum generation kinetics by isothermal hydrous and nonisothermal open-system pyrolysis. Organic Geochemistry 33, 1457–1475.

Link, W.K., 1954. Robot geology. American Association of Petroleum Geologists Bulletin 38, 2411.

Lopatin, N.V., 1971. Temperature and geologic time as factors in coalification (in Russian): Akademii Nauk SSSR Izvestiya. Ser Geologicheskaya 3, 95–106.

McKenzie, D.P., 1978. Some remarks on the development of sedimentary basins. Earth Planetary Science Letters 40, 25–32.

McKenzie, D.P., 1981. The variation of temperature with time and hydrocarbon maturation in sedimentary basins formed by extension. Earth and Planetary Science Letters 55, 87–98.

Meinzer, D.K., 1936. Movements of ground water. American Association of Petroleum Geologists Bulletin 20, 701–725.

Nakayama, K., 1987. Hydrocarbon-expulsion model and its application to Niigata area, Japan. American Association of Petroleum Geologists Bulletin 71, 810–821.

Pepper, A.S., Corvi, P.J., 1995. Simple kinetic models of petroleum formation: Part I— oil and gas generation from kerogen. Marine and Petroleum Geology 12, 291–319.

Pegaz-Fiornet, S., Carpentier, B., Michel, A., Wolf, S., 2012. Comparison between the different approaches of secondary and tertiary hydrocarbon migration modeling in basin simulators. In: Peters, K.E., Curry, D.J., Kacewicz, M. (Eds.), Basin Modeling: New Horizons in Research and Applications, American Association of Petroleum Geologists Hedberg Series, no. 4, pp. 221–236.

Pickard, G.L., 1963. Descriptive Physical Oceanography. Pergamon Press, New York. 200 p.

Pollack, H.N., 1982. The heat flow from the continents. Annual Review of Earth and Planetary Sciences 10, 459–481.

Reynolds, J.G., Burnham, A.K., 1995. Comparison of kinetic analysis of source rocks and kerogen concentrates. Organic Geochemistry 23, 11–19.

Robertson, E.C., 1988. Thermal Properties of Rocks. United States Geological Survey Open-File Report 88–441. 110 p.

Schmoker, J.W., 1994. Volumetric calculation of hydrocarbons generated. In: Magoon, L.B., Dow, W.G. (Eds.), The Petroleum System – from Source to Trap, 60. American Association of Petroleum Geologists Memoir, pp. 323–326.

Sclater, J.G., Christie, P.A.F., 1980. Continental stretching: an explanation of post Mid-Cretaceous subsidence of the Central North Sea. Journal of Geophysical Research 85, 3711–3739.

Sluijk, D., Parker, J.R., 1988. Comparison of predrilling predictions with postdrilling outcomes, using Shell's prospect appraisal system. In: Association of Petroleum Geologists Studies in Geology. Oil and Gas Assessment, vol. 21. Methods and Applications American, pp. 55–58.

Suzuki, N., Matsubayashi, H., Waples, D.W., 1993. A simpler kinetic model of vitrinite reflectance. American Association of Petroleum Geologists Bulletin 77, 1502–1508.

Sweeney, J., Braun, R.L., Burnham, A.K., Talukdar, S., Vallejos, C., 1995. Chemical kinetic model of hydrocarbon generation, expulsion, and destruction applied to the Maracaibo Basin, Venezuela. American Association of Petroleum Geologists Bulletin 79, 1515–1532.

Sweeney, J.J., Burnham, A.K., 1990. Evaluation of a simple model of vitrinite reflectance based on chemical kinetics. American Association of Petroleum Geologists Bulletin 74, 1559–1570.

Tegelaar, E.W., Noble, R.A., 1994. Kinetics of hydrocarbon generation as a function of the molecular structure of kerogen as revealed by pyrolysis-gas chromatography. Organic Geochemistry 22, 543–574.

Thomas, M.M., Clouse, J.A., 1995. Scaled physical model of secondary oil migration. American Association of Petroleum Geologists Bulletin 79, 19–29.

Thomsen, R.O., 1998. Aspects of applied basin modelling: sensitivity analysis and scientific risk. In: Duppenbecker, S.J., Ilffe, J.E. (Eds.), Basin Modelling: Practice and Progress. The Geological Society Special Publications, No. 141, London, pp. 209–221.

Tissot, B., 1969. Premières données sur les mécanismes et la cinétique de la formation du pétrole dans les bassins sédimentaires. Simulation d'un schema réactionnel sur ordinateur. Revue Institut Francais Petrole 24, 470–501.

Tissot, B.P., Espitalie, J., 1975. L'evolution thermique de la matiere organique des sediments: applications d'une simulation mathematique. Revue de l'Institut Franccais du Petrole 30, 743–777.

Tissot, B.P., Pelet, R., Ungerer, P.H., 1987. Thermal history of sedimentary basins, maturation indices, and kinetics of oil and gas generation. American Association of Petroleum Geologists Bulletin 71, 1445–1466.

Ungerer, P., Behar, F., Villalba, M., Heum, O.R., Audibert, A., 1988a. Kinetic modelling of oil cracking. Organic Geochemistry 13, 857–868.

Ungerer, P., Burrus, J., Doligez, B., Chenet, P.Y., Bessis, F., 1990. Basin evaluation by integrated two-dimensional modelling of heat transfer, fluid flow, hydrocarbon generation and migration. American Association of Petroleum Geologists Bulletin 74, 309–335.

Ungerer, P., Espitalie, J., Behar, F., Eggen, S., 1988b. Modelisation mathemetique des interactions entre craquage thermique et migration lors de la formation du petrole et du gaz. Comptes Rendus de l'Académie des Sciences, Series II 927–934.

Ungerer, P., Pelet, R., 1987. Extrapolation of oil and gas formation kinetics from laboratory experiments to sedimentary basins. Nature 327, 52–54.

Vandenbroucke, M., Behar, F., Rudkiewicz, J.L., 1999. Kinetic modelling of petroleum formation and cracking: implications from the high pressure/high temperature Elgin Field (UK, North Sea). Organic Geochemistry 30, 1105–1125.

Velde, B., Vasseur, G., 1992. Estimation of the diagenetic smectite to illite transformation in time-temperature space. American Mineralogist 77, 967–976.

Waples, D.W., 1980. Time and temperature in petroleum formation: application of Lopatin's method to petroleum exploration. American Association of Petroleum Geologists Bulletin 64 (6), 916–926.

Wilkinson, D., Willemsen, J.F., 1983. Invasion percolation: a new form of percolation theory. Journal of Physics A: Mathematical and General 16, 3365–3376.

CHAPTER

9

Petroleum System Concepts and Tools

OUTLINE

Introduction	309	Plays and Prospects	317
Elements and Processes	310	A Working Petroleum System	318
Temporal Aspects	312	Risking	320
Spatial Aspects	315	References	322

INTRODUCTION

When geologists discuss petroleum systems, they are usually referring to the concepts presented in the paper by Magoon and Dow (1994a). While this was an important paper for formalizing the ideas behind petroleum systems, these concepts were being applied to exploration at least since the early 1970s, as evidenced by publications such as Dow (1974). Often under names like hydrocarbon systems and oil systems, the philosophy behind petroleum systems has been used informally by geologists as the foundation for most exploration for decades.

A petroleum system as defined by Magoon and Dow (1994a) consists of a pod of active source rock and all its genetically related oil and gas accumulations. It includes all the geologic elements and processes that are necessary if an oil and/or gas accumulation is to exist. According to the paper, these elements are a source rock, a reservoir rock, a seal on the reservoir, and overburden to bury the system. Magoon and Dow (1994a) only consider two processes, trap formation and a combination of generation–migration–accumulation of hydrocarbons. For a petroleum system to exist, these elements and processes must also occur in time and space such that everything is working in concert to form a petroleum accumulation.

Magoon and Dow (1994a) classified petroleum systems into known, hypothetical, and speculative based on their level of certainty. A known petroleum system requires a positive oil-source rock or gas-source rock correlation to be made. This would be more typical of mature exploration areas. A hypothetical petroleum system only needs geochemical evidence for a source rock and perhaps an accumulation but no petroleum-source correlation. This

Practical Petroleum Geochemistry for Exploration and Production
http://dx.doi.org/10.1016/B978-0-12-803350-0.00009-X

could describe a new play in a mature basin or a developing play in an underexplored basin. And their speculative petroleum system only requires geological or geophysical evidence that the elements may exist, which is more typical of frontier basin.

In the early days of the unconventional boom of the late 1990s and early 2000s, many petroleum geologists thought that unconventional resources did not fit the criteria for petroleum systems. However, when you examine coal bed methane, shale gas, or hybrid systems, it is easy to see that they possess all the elements of a petroleum system, and the same processes are needed to form both conventional and unconventional accumulations. It is simply a matter of thinking about the elements and processes in a different context and realizing some of the petroleum systems elements and processes may not be as critical in some of the unconventional resource plays. For example, when the source rock and the reservoir are the same rock, migration still needs to occur, but the distances may only be from the kerogen to the pore space.

As fundamental as the petroleum system concept is, it is simply a starting point for more detailed analysis of the elements and processes needed to form a petroleum accumulation and a means of linking them into a coherent framework to identify the key risks as well as strengths in plays and prospects. And when used appropriately, it can keep exploration programs out of avoidable trouble, as well as contribute to the program's overall success. For examples of how the petroleum system concept has been applied successfully to exploration, the reader is referred to the case studies in Parts V and VI of Magoon and Dow (1994b).

To conclude this survey of the exploration and production applications of petroleum geochemistry, the following discussion focuses on the geochemical aspects of the petroleum system and explores how they might be expanded upon to provide more detailed information and insight into exploration problems. The discussion will take a more pragmatic approach to some of the fundamental concepts and show how they might be changed. It will also suggest some tools that utilize geochemical data to further define the petroleum system and make it more useful in exploration. And it reviews some of the concepts that must be considered when using petroleum geochemistry in the risking of exploration projects.

ELEMENTS AND PROCESSES

Magoon and Dow (1994a) recognized source rocks, reservoir rocks, seals, and overburden as elements in a petroleum system. The source rock is defined as a sediment that contains adequate organic matter prone to hydrocarbon generation to form a petroleum accumulation. The reservoir rock is a lithologic unit with adequate porosity and permeability in which moveable hydrocarbons can accumulate. Seals provide barriers on reservoir rocks through which hydrocarbons cannot move effectively. And overburden is the sediment that buries the system and provides the means by which generation, expulsion, and migration can occur.

In exploration, one of the first steps in defining a prospect or play is to identify the physical attributes needed to form petroleum accumulations. While the classic elements of Magoon and Dow (1994a) are essential, there is good reason to expand this list. A more complete list of physical entities needed to form a petroleum accumulation would include the addition of a migration pathway and a trap. The migration pathway is defined as the avenues within the rocks that allow hydrocarbons to move from source rock to reservoir rock and to eventually fill a trap. And the trap is the structural and/or stratigraphic configuration that focuses hydrocarbons into an accumulation. The inclusion of these two additional elements would complete the "checklist" of physical attributes exploration geoscientists would need to look for to begin the exploration process and, once found, allow them to focus on the processes involved.

Magoon and Dow (1994a) only consider two processes, trap formation and a combination of generation–migration–accumulation of hydrocarbons. Trap formation is defined as the structural and/or sedimentological events that form a containment for hydrocarbons and is an essential process for the formation of petroleum accumulation. Trap formation should not be confused with the physical entity that is the trap. As a process, trap formation has a temporal aspect that is important in determining if an accumulation eventually forms or not, while the trap itself is merely the physical container for the accumulation.

Similarly, generation, migration, and accumulation also are all essential processes, and it is more helpful to look at the component parts as separate processes rather than a composite. Generation is concerned with the transformation of kerogen under the influence of time and temperature to form gas, oil, and a carbon-rich inert residue. Expulsion is the movement of hydrocarbons out of the source rock and is related to the amount of hydrocarbon generated and porosity and permeability of the source rock. Migration is the movement of hydrocarbons along a carrier system, the migration path, after it leaves the source rock to travel to a reservoir and hopefully eventually filling a trap. It can also include the movement of petroleum from one trap to another. And accumulation is the creation of a volume of hydrocarbon by migration into a trap faster than the trap leaks.

While generation, expulsion, migration, and accumulation are linked, they have different controls and should be considered separately. For example, hydrocarbon generation may occur, but the amount generated may not be enough to initiate expulsion. Expelled hydrocarbons may not find their way to a migration pathway. And migration can be dispersive, rather than focused, resulting in a lack of accumulation. It simply makes more sense to consider each part separately rather than to simply look at a combined end result.

In addition to breaking out the component parts of the generation–migration–accumulation process, preservation should also be considered. Magoon and Dow (1994a) only considered preservation time in their original petroleum system concept. It was defined as the time after accumulation is complete until the present during which hydrocarbons can be lost, altered, or destroyed. However, preservation should be considered as an assessment of whether the hydrocarbons that accumulate in a trap will remain unaltered by actions such as remigration, thermal cracking, biodegradation, and/or water washing (Blanc and Connan, 1994). And the potential alteration or loss does not start when the accumulation is completed but rather when the first hydrocarbons reach the trap.

These expanded lists of elements and processes are summarized in Fig. 9.1. These suggested changes comprise a more comprehensive assessments of the component parts of a petroleum system and are perhaps an evolutionary step in applying petroleum systems concepts in exploration.

Elements:	Processes:
• Source Rock	• Trap Formation
• Reservoir Rock	• Generation
• Seal	• Expulsion
• Overburden	• Migration
• Migration Pathway	• Accumulation
• Trap	• Preservation

FIGURE 9.1 Expanded lists of the elements and processes that make up a petroleum system.

TEMPORAL ASPECTS

As stated earlier, for a petroleum system to exist, the elements and processes must occur in the proper temporal sequence for a hydrocarbon accumulation to be possible. For example, to capture the maximum amount of hydrocarbon, expulsion/migration should occur after trap formation. To assess the timing of events, a graphical display called the critical events chart (also called the petroleum systems event chart) is used. An events chart shows the temporal relation of the essential elements and processes of a petroleum system and also includes the preservation time and the critical moment for the system (Magoon and Dow, 1994a).

A typical critical events chart is shown in Fig. 9.2. The time lines in the chart are used to compare the times that the processes occurred with the times that the elements formed. While there is usually only one active source rock for each petroleum system, there may be many reservoir rocks and each requires its own seal. The timing of trap formation should come from the structural analysis of geological depictions, such as cross sections and structure maps. The best assessments of the timing of generation, migration, and accumulation may come from basin modeling results. And the critical moment is the time that best depicts the most significant aspect of the generation, migration, and accumulation of hydrocarbons in a petroleum system. This is usually the initiation of expulsion/migration and is often estimated from basin model results.

While critical events chart are extremely useful for displaying and understanding the temporal components of petroleum systems, there are some potential improvements that can be made to enhance their value. One of these improvements would simply be to use the expanded list of elements and processes (see Fig. 9.1). Their use would provide more detail in the critical events chart for interpretation. Another is to merge the burial history diagrams with critical events charts (e.g., Abdel-Fattah et al., 2015), as shown in Fig. 9.3. The depth–time curves in the burial history diagrams represent the geologic events in the stratigraphic sequence found in both the petroleum system and the basin model. When combined with the critical events chart, the elements and processes in the petroleum system can be placed in the context of the geological history of the area.

A third improvement would be to use more detailed information to describe the processes to give better assessments of the petroleum system as well as contribute to the understanding of the risk associated with the play/prospect. An example of providing more detail is given in Fig. 9.4. It shows three ways to display information about the timing of expulsion in the critical events chart. The top time line shows the traditional methods of blocking out the time

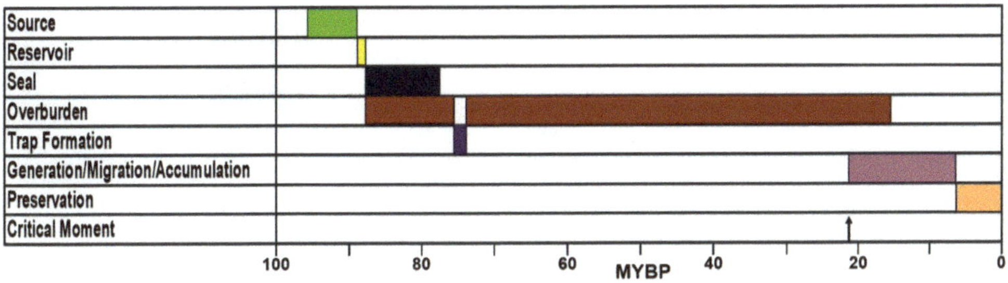

FIGURE 9.2 A critical events chart after the method of Magoon and Dow (1994a).

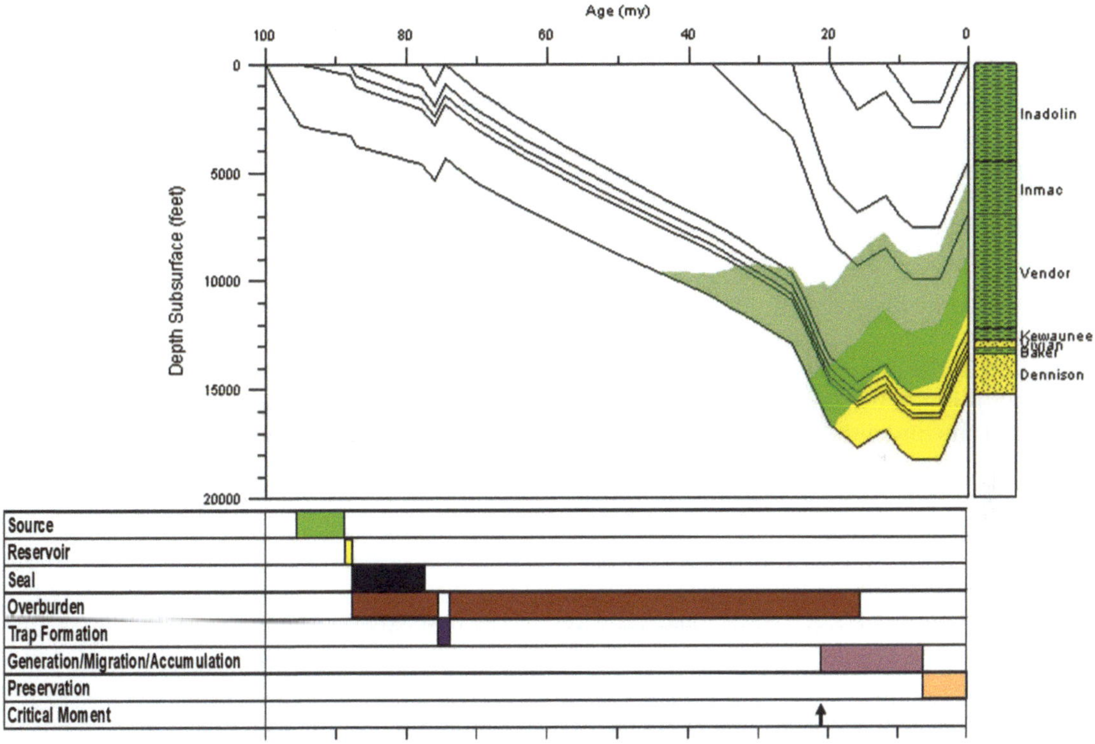

FIGURE 9.3 A critical events chart merged with a burial history diagram.

interval during which the event occurs. This provides information about beginning and ending times for the event but does not convey any information about the details of what happened when. Occasionally, the process is shown as a curve or a ramp to depict the progression of the process from beginning to end (e.g., Schlakker et al., 2012) or as changing colors to represent the transitions from oil to wet gas to dry gas (e.g., Pollastro, 2003). In contrast, the middle time line shows cumulative expulsion curves. These curves not only indicate when the expulsion occurred but also gives information about the amount of hydrocarbon expelled through time. A different perspective is displayed in the bottom time line. It shows the timing

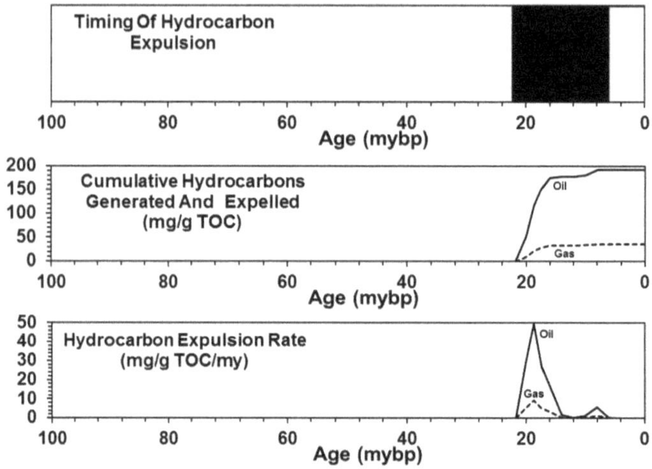

FIGURE 9.4 Alternate methods of displaying data in the critical events chart. The top shows the traditional methods of blocking out the time interval the event occurs. The middle shows a cumulative expulsion curve that not only indicates when the expulsion occurred but gives the amount expelled through time. And the bottom shows the timing of expulsion expressed as a rate. *TOC*, Total organic carbon.

of expulsion expressed as a rate and clearly indicates that expulsion occurred as two pulses, a major event from 22 to 14 mybp and a minor event from 10 to 6 mybp. The information conveyed in the lower two time lines, derived from basin modeling results, provides a better understanding of the expulsion process in this petroleum system than does the simple blocking out of the time interval. These time lines would also be a useful means of comparing two or more source rock to assess their contributions to a petroleum system.

Another example of adding more information to a critical events chart panel is the preservation risk time line, shown in Fig. 9.5. Instead of simply showing the time interval when preservation of the reservoired hydrocarbons is at risk, the reservoir temperature predicted from a basin model is plotted over time. The general guidelines for biodegradation from Larter et al. (2006) and thermal cracking from Hunt (1996) are applied. The vertical dotted line signals the time when the reservoir began to fill, and the area to the right of that line indicates the time of alteration risk for the hydrocarbons. In the example, the temperature curve suggests good preservation of the hydrocarbons. This approach provides useful information about the potential preservation of the hydrocarbons that has real value especially when applied to individual prospects.

To make the critical events chart more useful and provide an improved understanding of the temporal aspects of the petroleum system, geoscientists should consider expanding the list of elements and processes considered to provide more in-depth information about them. Some of the additions to the critical events charts not discussed already could include cementation/porosity enhancement, fracturing, seal integrity/failure, structural disturbance/remigration, and hydrodynamic flushing/water-washing/displacement. The critical events chart should be a flexible tool and adapted to the circumstances observed in the petroleum system. Much of the information needed to expand the critical events chart may already be available from whatever basin modeling that has been done on the petroleum system. And all this information can be readily applied to risking at the play and prospect levels.

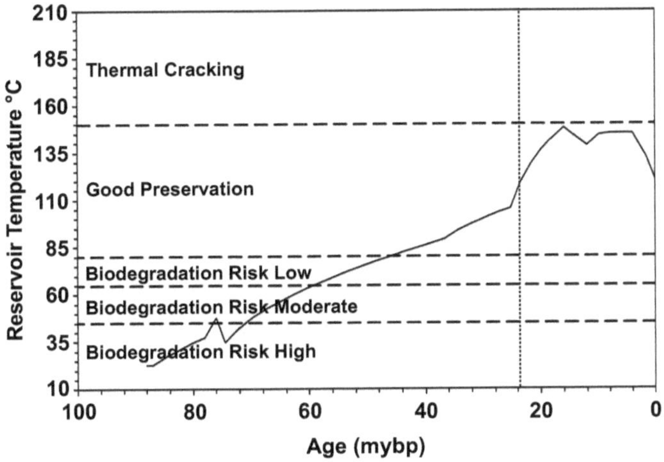

FIGURE 9.5 A preservation panel for a critical events chart based on reservoir temperature derived from basin modeling. The vertical scale was expanded to facilitate viewing. *The general temperature guidelines for biodegradation were taken from Larter, S., Huang, H., Adams, J., Bennett, B., Jokanola, O., Oldenburg, T., Jones, M., Head, I., Riediger, C., Fowler, M., 2006. The controls on the composition of biodegraded oils in the deep subsurface: Part II-Geological controls on subsurface biodegradation fluxes and constraints on reservoir-fluid property prediction. American Association of Petroleum Geologists Bulletin 90, 921–938, while the temperature guidelines for thermal cracking came from Hunt, J.M., 1996. Petroleum Geochemistry and Geology, second ed. W.H. Freeman, New York, 743 pp.*

SPATIAL ASPECTS

Within the construct of the petroleum system concept of Magoon and Dow (1994a), spatial aspects to be considered include both stratigraphic and geographic perspectives. The stratigraphic relationships are defined primarily with a stratigraphic column and further illustrated in cross sections and burial histories, while the geographic relationships are elucidated with maps. Looking at the present-day aspects of a petroleum system may not reveal all the details required to accurately assess its potential or fully explain the hydrocarbon accumulations that may have been previously discovered. Instead, examining the petroleum system at some time(s) in the past may better reveal the interplay and evolution of elements and processes that lead to the formation of hydrocarbon accumulations.

The time in the past recommended by Magoon and Dow (1994a) for inspecting the spatial components of the petroleum system is the critical moment. If the critical moment is being defined as the onset of hydrocarbon expulsion, and migration, cross sections and structure maps at that time will help define and assess the effectiveness of the migration pathways from source to trap. In addition to structure maps on top of the reservoir interval, structure maps on top of the source rock with an overlay of predicted expulsion can help show the limits of the actively generating and expelling source rock that is essential for defining the charging area for the migration. This combination of maps showing the structure and area of expulsion can then be used for ray path migration modeling as described in Chapter 8.

While the critical moment is a seminal point in time for the petroleum system, there may also be times between the critical moment and the present that should also be considered.

Any event that might impact the generation, expulsion, or migration of the hydrocarbons or any structural events that may reconfigure the trap and/or migration pathways should also be considered for paleostructure maps, paleocross sections, and burial history diagrams. This series of time steps are important in defining the evolution of the petroleum system as well as illustrating the filling history of potential traps.

In addition to the standard mapping of the petroleum system, there are other approaches that can be used to give a more regional view of some of the geochemical issues. Dembicki and Pirkle (1985) proposed a method to combine source rock thickness, organic richness, and thermal maturity into a single mappable parameter to indicate areas of potential hydrocarbon generation. Average percent organic carbon or Rock-Eval S2 values are multiplied by the effective source rock thickness to give a richness factor. The richness factor is then scaled using maturity to give source potential ratings for oil and/or gas generation. The resulting ratings provide semiquantitative measures of source potential that can be mapped to provide a regional assessment. By incorporating basin modeling, maps could be produced to show the evolution of source potential through geologic time. And, with some modification to the maturity scaling factors, the source potential rating index can be easily applied in unconventional play screenings.

Demaison and Huizinga (1991) proposed a similar mapping method using their source potential index. It combined the Rock-Eval S1 and S2 values and multiplied the sum by the effective source rock thickness and density to provide a relative ranking of hydrocarbon generation potential. It was designed as a means of comparing two or more source rocks in a basin as part of their genetic classification of petroleum systems.

A more generic type of petroleum system mapping is derived from play fairway analysis. A play fairway is the area where a particular play type is expected to occur, and play fairway analysis is used to identify and rank areas within the fairway by examining the spatial distribution of the elements of the play (White, 1988; Allen and Allen, 2005). A minimum of three elements are typically used to define a play: a source rock, potential reservoir, and a seal, or caprock. Additional elements may be added to the map set depending on the geologic circumstances in the basin or play. The goal of this approach is to provide an assessment of the elements of the play resulting in a more objective final product that can be easily understood.

The method is usually referred to as common risk segment (CRS) mapping (Allen and Allen, 2005). All the elements being considered are mapped at the same scale using a simple color code for the confidence level or the probability that the element has of being effective in that area. These maps are often called traffic light, or stop light, maps because they use green, yellow, and red to depict high confidence, moderate confidence, and low confidence, respectively. If probabilities are used, the green, yellow, and red indicate high probability/low risk, moderate probability/moderate risk, and low probability/high risk, respectively. Actual probability ranges can also be assigned to these map colors.

Once the individual element maps are complete, they are overlain to produce a resultant map (White, 1988). The example shown in Fig. 9.6 uses a CRS approach for the source rock component of the petroleum system. Organic richness, kerogen type, and maturity are the input elements. Map interpretation uses a Venn diagram approach to evaluate the area on a point by point basis. For any point in the map area, if there is red on any of the input maps, the overall rating is red. If there is green on all the input maps, the overall rating is green. If there is yellow on all the input maps, the overall rating is yellow. For any combination of

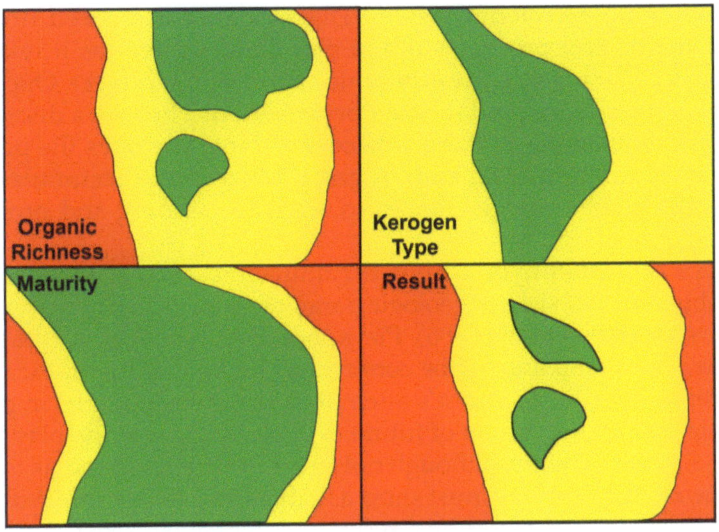

FIGURE 9.6 A simple example of common risk segment mapping for the source rock component.

greens and yellows on the input maps, the overall rating is yellow. On the result map, the green region(s) represents the lowest risk exploration area(s) where efforts should be concentrated, while the red region(s) is the high-risk area(s) and should be avoided. The yellow region(s) may or may not be prospective and will likely require additional study to make that determination.

Some examples of how this mapping style is applied to exploration can be found in Grant et al. (1996) and Chen et al. (2002). Although a very useful tool for visualizing exploration risk, it is not used extensively but should be considered as part of any petroleum systems analysis.

PLAYS AND PROSPECTS

While the petroleum system is an extremely useful concept, it can be too broad for the evaluation of small parts of sedimentary basins of interest for petroleum exploration. Petroleum systems also do not take into consideration any of the economic concerns of exploration and consider only whether oil and gas accumulation exist and why (Magoon and Dow, 1994a). Exploration, however, is a commercial endeavor rather than an academic exercise. The goal is to find large enough accumulations of oil and gas that can be produced economically and make a profit. As a result, exploration programs may still use petroleum systems concepts, but their focus is on prospects and plays. It is necessary, therefore, to define what is specifically meant by a prospect and play and see how petroleum systems can help the exploration geoscientists assess them.

Prospects are present-day structural or stratigraphic features that can be mapped and may contain commercial quantities of petroleum. They represent potential traps. To justify the drilling of a prospect, evidence must be gathered that suggests it could be the product of a functioning petroleum system. This evidence is derived from the examination of all the elements and processes of the petroleum system as it relates to that particular trap but must also include the economic aspect of the project. Once a prospect has been drilled, it no longer exists, and it either becomes a dry hole or a discovery. Success at the prospect level is not simply making a discovery. Not only does the petroleum system have to form an accumulation of oil and/or gas, the size of the accumulation determined by delineation wells must be large enough to offset the finding and production costs to make a profit.

Plays are groups of related hydrocarbon accumulations and/or prospects that are characterized by combinations of similar geologic parameters such as charge type, reservoir-seal couplet, and trap style (Baker et al., 1986). Plays should have a clear geographic distribution that can be defined on a map and/or be confined to limited stratigraphic intervals (White, 1993). The play represents the geologist's attempt to recognize patterns in petroleum occurrence that can help predict the results of future exploration. It is at the play level where economic factors are first considered. Although not as rigorously reviewed as for a prospect, the economic characteristics of the play must show a high potential for profitability (Doust, 2010). Success at the play level is measured by a series of successive discoveries that confirms the viability of the play concept.

A WORKING PETROLEUM SYSTEM

As shown earlier, Magoon and Dow (1994a) classified petroleum systems into known, hypothetical, and speculative based on their level of certainty. While these terms accurately describe these classes of petroleum systems as they are defined, personal experience has shown that they are not necessarily acceptable terms for use in commercial exploration work. As an alternative, calling a petroleum system demonstrated instead of known, potential instead of hypothetical, and conceptual instead of speculative have been found to be more suitable when dealing with exploration management.

In an exploration context, the concepts of petroleum systems need to be applied in a more practical manner. Petroleum systems cannot simply form oil and gas accumulations but need to form oil and gas accumulations large enough to be commercially exploited and give a return on investment. Often, this is referred to as a working petroleum system. A working petroleum system is a demonstrated petroleum system that can be shown to have formed an economically producible petroleum accumulation.

According to Magoon and Dow (1994a), the only acceptable proof for a known, or demonstrated, petroleum system is a petroleum-source rock correlation. However, source rocks are not often encountered during the early phases of exploration drilling, e.g., the Gulf of Mexico. In fact, source rocks may never be directly identified for a petroleum system if there are no commercial targets in the stratigraphic vicinity of the source rock. Wells are drilled to find producible hydrocarbons and not source rocks, so the ability to make petroleum to source rock correlation may be severely limited. But the presence of economically producible petroleum does confirm the presence of a working petroleum system, even if all the details of that system cannot be stated with certainty. So in the context of a working petroleum system, the definition of a demonstrated petroleum system should be expanded to include the presence of economically producible petroleum, as well as a petroleum-source rock correlation.

In an exploration program, proof of a working petroleum system is sought to minimize the perceived risk. As stated earlier, the only proof of working petroleum system is a petroleum-source rock correlation or producible petroleum. Anything short of this is merely evidence of a working petroleum system and not proof. But over the years in published papers, conference presentations, and exploration project reviews, many geoscientists have invoked shows, seeps, and hydrocarbon-filled fluid inclusion as proof of a working petroleum system. Should these lines of evidence be considered proof of a working petroleum system?

As discussed in Chapter 5, shows are any noncommercial quantities of oil or gas encountered while drilling and can range from large volumes of hydrocarbons below the commercial threshold for the area down to molecular level traces of thermogenic hydrocarbons in sediments. Using the Schowalter and Hess (1982) classification for shows, only the continuous phase oil or gas show where observable oil staining or saturation is encountered should be considered evidence for a working petroleum system. If this type of show represented a substantial volume and was still considered noncommercial only because of environmental conditions, remote location, or other economic factors, it may qualify as proof of a working petroleum system. The overall volume, trap size, and the potential rate of production would have to be taken into consideration to make this determination.

Migration of hydrocarbons out of a source rock is basically a dispersive process. For this reason, it is unlikely that a seep can form from migration of hydrocarbons directly out of a source rock. As discussed in Chapter 6, it is thought that an accumulation must form that eventually leaks for hydrocarbons to reach the surface. As a result, the presence of seeps indicates that a subsurface oil/gas accumulation is very likely but cannot tell you how big the accumulation is, where it is in the subsurface, or what its source rock might be. Because of this, seeps may strongly suggest an active petroleum system is present but does not necessarily constitute proof of a working petroleum system. Seeps are therefore classed as evidence of a working petroleum system and risked accordingly. But if the seeped oil or gas is analyzed geochemically and can be correlated a source rock or a previously discovered subsurface petroleum accumulation in the area, it can then be used as proof of a working petroleum system.

Hydrocarbon-bearing fluid inclusions outside of reservoirs are often portrayed as being the result of trapping migrating hydrocarbons during the diagenetic formation of quartz or carbonate cements (Buruss, 1981). The implications of this are that hydrocarbons have been generated in sufficient quantities to be able to migrate and perhaps form an accumulation. But this is not always the case. Sedimentary organic matter is pervasive in the subsurface and will generate hydrocarbons once sufficient thermal maturity has been reached. The hydrocarbons in a fluid inclusion could have easily come from a few centimeters or meters away and does not have to be from migrating hydrocarbons on their way to form an accumulation. Even if the hydrocarbon-filled fluid inclusions are formed from migrating hydrocarbons, it is impossible to judge if this migration has given rise to a commercial accumulation. Because of the uncertainty concerning the origin of the hydrocarbons, hydrocarbon-bearing fluid inclusions cannot be used as proof of a working petroleum system and might not even be evidence that one may exist.

RISKING

An exploration program will have a fixed amount of capital it can use to pursue a set of opportunities in the exploration portfolio. To determine where this capital is best spent, it is necessary to make some form of quantitative evaluations to objectively compare these opportunities. There is an abundance of published methods for quantitative prospect and play evaluation including Sluijk and Nederlof (1984), Chen and Fang (1993), White (1993), Otis and Schneidermann (1997), Snow et al. (1997), Rose (2001) and Chen and Osadetz (2006), to name just a few. Some like Mackenzie and Quigley (1988) focus primarily on the geochemical aspects of prospect appraisal. While all of these methods could be used for prospect and play appraisal, the reality is that individual companies have their own approaches to this problem, either using proprietary methods or commercially available products. And most have adopted some form of risk analysis (for an in-depth discussion of risk analysis in exploration, the reader is referred to Rose, 2001).

Risk can be defined a state of uncertainty where some possible outcomes have an undesired effect or significant loss. What is actually being sort in risk analysis is some measure of this uncertainty to arrive at a level of confidence. This is usually reached by employing a probabilistic approach applied to a petroleum systems analysis of the prospect or play as part of the assessment. Using a probabilistic approach, risking a prospect or play gives the chance that it will be successful, often referred to as the geologic chance of success, or Pg. In addition to Pg, economic factors and above grounds risks (e.g., political stability, market issues, etc.) will also factor into the final decision. However, for the purposes of this discussion, only geologic factors and Pg are considered.

To calculate an overall Pg, the independent chance of success of each of the components of the assessment is multiplied by each other. Typically, four to five components are used to assess geologic risk (e.g., Baker et al., 1986; White, 1993; Otis and Schneidermann, 1997). To accommodate all the elements and processes of a petroleum system, some of these factors are combined.

Pg represents two very different concepts when applied to prospects and plays. For an individual prospect, Pg indicates the chance that all the elements and processes of a petroleum system are present and operating in the proper temporal and spatial relationships. In contrast, risking a play gives the chance that at least one prospect somewhere within the play area will have all the elements and processes of a petroleum system present and operating in the proper temporal and spatial relationships. The success of a single prospect thereby proves the play concept. For the play, the risk is effectively distributed over the entire play area and, as a result, the "burden of proof" for a play is considerably lower than it is for a prospect.

One obstacle to effective risking of prospects and plays is that exploration geoscientists are often overly optimistic about their project. They tend to overlook shortcoming in some aspects of the petroleum system to get their prospect on the drilling schedule. This competition for a share of the exploration budget is counter to the premise that risking helps to objectively rate the opportunities in the exploration portfolio to indicate which of the opportunities will be pursued. As a result, explorationists and their management need to be diligent about being realistic when assigning risk.

With respect to petroleum geochemistry, most risking schemes will focus on source rocks, maturity, expulsion/migration, and preservation in their deliberations. Risking scenarios usually combine the aspects of the source rock with maturity into a single composited item. The evaluation of expulsion/migration is concerned with both the pathway from source to trap and the timing of expulsion/migration as compared to the timing of trap formation. And preservation is often grouped with containment/trap integrity issues. It is easy to see the value of a petroleum systems analysis and tools such as critical events charts and CRS maps in making assessments of these factors.

Just as in source rock evaluation as described in Chapter 3, to assess source rock quality and richness, it is necessary to look at organic content with respect to reactive organic matter, not just total organic matter present. Kerogen type and hydrogen content of the kerogen must also be included to estimate how much and what type(s) of hydrocarbon may be generated. And to insure an accurate appraisal is made, any data used in these assessments must be compensated, or adjusted, for the thermal maturity.

When considering maturity in risking, it is necessary to remember what maturity indicates about hydrocarbon generation and expulsion. Maturity indicates if generation is possible and what type(s) of hydrocarbon could be formed. But it cannot directly indicate when and how much hydrocarbon has been generated. And maturity does not indicate when migration occurred. Generation and expulsion are not coincidental, and expulsion is related to the amount of hydrocarbon generated and the porosity and permeability of the source rock. To estimate when expulsion may have occurred and how much hydrocarbon might have been available for migration, judicious use of basin modeling is needed.

When considering the volume of source rock available, remember the entire thickness of the source rock interval may not be capable of generating sufficient oil to migrate out and fill a trap. Only the effective source rock thickness should be considered. For example, the total thickness of the source rock interval might be 150 m. However, the effective source rock thickness might only be tens of meters. This effective thickness may not be continuous but rather occurs in more than one discrete bed. Think of effective source rock thickness versus total source rock thickness in a similar manner to net sand thickness versus gross sand thickness in a reservoir. And the effective source rock area also needs to be considered. While the area of generation for the source rock within the prospect's fetch area may be large, the area where hydrocarbon is being expelled may be more limited. As mentioned in the discussion of volumetric estimations in Chapter 8, if not properly constrained the effective source rock thickness and effective fetch, or drainage, area can be easily overestimated and lead to unrealistic assessments of the amount of hydrocarbons available to fill the trap.

As part of the risking process, it is usually necessary to set the minimum recoverable volume that will be contained in the prospect for the exploration well to be considered a success. It is very tempting to look at this minimum volume as the amount of hydrocarbon that needs to be generated by the source rock for a success case. In fact, it is far from it. Consider that a minimum of about 20% of the generated hydrocarbons likely remains in the source rock and that some researchers have estimated up to 90% of the expelled hydrocarbons can be lost during migration. Also consider that only 20–30% of the hydrocarbons may be recoverable from the reservoir. So whatever the minimum recoverable volume of oil is, there needs to be a substantially larger amount generated to result in this minimum.

The final thing to consider is what should be done post drilling. Once a well is drilled or a play is tested, there should be a review of the risking and petroleum systems analysis. This should occur whether the project was successful or not and utilize all the new data produced. Unfortunately, these postmortems are not very popular exercises. In the success case, there is often the attitude that "we got it right," and there is no need for any analysis. When failure occurs, no one wants to prolong the agony of being wrong and would rather distance themselves from the unsuccessful project. This is unfortunate because there is a substantial amount that can be learned in these exercises. In the success case, the review will point out those aspects of the petroleum system that were accurately represented in the original analysis and risking. It will also show what may have been wrong, where improvements could be made in future projects, and how to build on the success. In the failure case, the review should be directed toward finding the cause or causes of the project's failure. Often times, those elements of the petroleum system that had the highest risk (most uncertainty) predrilling are not the cause of the project failure. By identifying the cause of failure, similar errors may be avoided in the future. It may be that the play concept is sound but just not appropriate where it was applied. Do not overlook this as a learning opportunity. The well results are already paid for, use them.

References

Abdel-Fattah, M., Gameel, M., Awad, S., Ismaila, A., 2015. Seismic interpretation of the Aptian Alamein Dolomite in the Razzak oil field, western Desert, Egypt. Arabian Journal of Geosciences 8, 4669–4684.

Allen, P.A., Allen, J.R., 2005. Basin analysis: Principles and Application to Petroleum Play Assessment, second ed. Wiley-Blackwell, Hoboken, New Jersey. 560 p.

Baker, R.A., Gehman, H.M., James, W.R., White, D.A., 1986. Geologic field number and size assessment of oil and gas plays. In: Rice, D.D. (Ed.), Oil and Gas Assessments – Methods and Applications, American Association of Petroleum Geologists Studies in Geology, vol. 21, pp. 25–31.

Blanc, P., Connan, J., 1994. Preservation, degradation, and destruction of trapped oil. In: Magoon, L.B., Dow, W.G. (Eds.), The Petroleum System-from Source to Trap. American Association of Petroleum Geologists Memoir, vol. 60, pp. 237–247.

Burruss, R.C., 1981. Hydrocarbon fluid inclusions in studies of sedimentary diagenesis. In: Hollister, L.S., Crawford, M.L. (Eds.), Fluid Inclusions: Applications to Petrology, vol. 6. Mineralogical Association of Canada Short Course Notes, pp. 138–156.

Chen, H.C., Fang, J.H., 1993. A new method for prospect appraisal. American Association of Petroleum Geologists Bulletin 77, 9–18.

Chen, Z., Embry, F.A., Osadetz, K.G., Hannigan, P.K., 2002. Hydrocarbon favourability mapping using fuzzy integration: western Sverdrup Basin. Bulletin of Canadian Petroleum Geology 50, 492–506.

Chen, Z., Osadetz, K.G., 2006. Geological risk mapping and prospect evaluation using multivariate and Bayesian statistical methods, western Sverdrup Basin of Canada. American Association of Petroleum Geologists Bulletin 90, 859–872.

Demaison, G., Huizinga, B.J., 1991. Genetic classification of petroleum systems. American Association of Petroleum Geologists Bulletin 75, 1626–1643.

Dembicki Jr., H., Pirkle, F.L., 1985. Regional source rock mapping using a source potential rating index. American Association of Petroleum Geologists Bulletin 69, 567–581.

Doust, H., 2010. The exploration play: what do we mean by it? American Association of Petroleum Geologists Bulletin 94, 1657–1672.

Dow, W.G., 1974. Application of oil correlation and source rock data to exploration in Williston basin. American Association of Petroleum Geologists Bulletin 58 (7), 1253–1262.

Grant, S., Milton, N., Thompson, M., 1996. Play fairway analysis and risk mapping: an example using the Middle Jurassic Brent Group in the northern North Sea. In: Dore, A.G., Sinding-Larsen, R. (Eds.), Quantification and Prediction of Petroleum Resources. vol. 6. Norwegian Petroleum Society Special Publication, pp. 167–181.

Hunt, J.M., 1996. Petroleum Geochemistry and Geology, second ed. W.H. Freeman, New York. 743 pp.

Larter, S., Huang, H., Adams, J., Bennett, B., Jokanola, O., Oldenburg, T., Jones, M., Head, I., Riediger, C., Fowler, M., 2006. The controls on the composition of biodegraded oils in the deep subsurface: Part II-Geological controls on subsurface biodegradation fluxes and constraints on reservoir-fluid property prediction. American Association of Petroleum Geologists Bulletin 90, 921–938.

Mackenzie, A.S., Quigley, T.M., 1988. Principles of geochemical prospect appraisal. American Association of Petroleum Geologists Bulletin 72, 399–415.

Magoon, L.B., Dow, W.G., 1994a. The petroleum system. In: Magoon, L.B., Dow, W.G. (Eds.), The Petroleum System-from Source to Trap. American Association of Petroleum Geologists Memoir, vol. 60, pp. 3–24.

Magoon, L.B., Dow, W.G., 1994b. The petroleum system-from source to trap. In: Magoon, L.B., Dow, W.G. (Eds.), American Association of Petroleum Geologists Memoir, vol. 60 644 p.

Otis, R.M., Schneidermann, N., 1997. A process for evaluating exploration prospects. American Association of Petroleum Geologists Bulletin 81, 1087–1109.

Pollastro, R.M., 2003. Total petroleum systems of the Paleozoic and Jurassic, Greater Ghawar Uplift and adjoining provinces of Central Saudi Arabia and Northern Arabian-Persian Gulf. U. S. Geological Survey Bulletin 2202-H 100 p.

Rose, P., 2001. Risk analysis and management of petroleum exploration ventures. American Association of Petroleum Geologists Methods in Exploration No. 12 164 p.

Schlakker, A., Csizmeg, J., Pogácsás, G., Horti, A., October 23–26, 2012. Burial, Thermal and Maturation History in the Northern Viking Graben (North Sea). American Association of Petroleum Geologists Search and Discovery Article #50545. Adapted from Poster Presentation at American Association of Petroleum Geologists International Convention and Exhibition, Milan, Italy, p. 201.

Schowalter, T.T., Hess, P.D., 1982. Interpretation of subsurface hydrocarbon shows. American Association of Petroleum Geologists Bulletin 66, 1302–1327.

Sluijk, D., Nederlof, M.H., 1984. Worldwide geological experience as a systematic basis for prospect appraisal. In: Demaison, G. (Ed.), Petroleum Geochemistry and Basin Evaluation, American Association of Petroleum Geologists Memoir, vol. 35, pp. 15–26.

Snow, J.H., Dore, A.G., Dorn-Lopez, D.W., 1997. Risk analysis and full-cycle probabilistic modeling of prospect: a prototype system developed for the Norwegian shelf. In: Dore, A.G., Sinding-Larsen, R. (Eds.), Quantitative Prediction and Evaluation of Petroleum Resources, vol. 6. Norwegian Petroleum Society Special Publication, pp. 135–166.

White, D.A., 1988. Oil and gas play maps in exploration and assessment. American Association of Petroleum Geologists Bulletin 72, 944–949.

White, D.A., 1993. Geologic risking guide for prospects and plays. American Association of Petroleum Geologists Bulletin 77, 2048–2061.

Index

'*Note*: Page numbers followed by "f" indicate figures and "t" indicate tables.'

A

Allochthonous organic matter, 20
API gravity, 136
Aromatic hydrocarbons
  aromatic and naphthenoaromatic structures, 11–12, 12f
  benzene structure, 10, 11f
Asphaltenes, 12–14, 13f–14f, 200–202, 202f

B

Background organic matter (BOM) signal, 243
Bakken Formation, 268, 268f
Basin modeling
  burial history
    burial history diagram, 276, 277f
    compaction correction, 280, 281f
    depth–time plot, 275
    depth *vs.* porosity plot, 277–279, 278f
    hypothetical stratigraphic column, 275, 275f
    Kozeny–Carman relationship, 279, 280f
    mechanical compaction, 280
    mixed lithologies, 277–279, 279f
    porosity reduction, 277, 278f
    porosity reduction, sediment compaction depth, 277–279, 278f
    sedimentation history, 276, 276f
  concepts, 274
  depth–time–temperature context, 286
  1-D model results
    burial history diagrams, 296–297, 296f–297f
    depth profiles, 298, 298f
    goal, 296
    properties, 295
    software packages, 295
    stratigraphic sequence, 295
    time profiles, 298–299, 299f
  expulsion modeling, 291–292
  goals, 273
  hydrocarbon generation modeling. *See* Hydrocarbon generation modeling
  maturation modeling
    graphical time–temperature maturation model, 286, 287f
    kinetic models, 288, 289f
    TTI calculation method, 286–288, 287f
  migration modeling
    Darcy flow modeling, 293–294
    1-D basin models, 292
    invasion percolation modeling, 294
    ray-path modeling, 292–293, 293f
    reservoir rock, 292
  model validation, 299–300, 301f
  one dimensional (1-D) basin models, 273–274
  predicting preservation, 294–295
  sensitivity analysis, 300–302, 302f–303f
  thermal history
    beta factor, 285–286, 285f
    bottom hole temperature (BHT), 281
    diapiric salt column, 285
    evaporitic minerals, 285
    geothermal gradients, 282–283
    heat capacity, 284
    Horner plot correction, 281, 282f
    matrix and bulk thermal conductivities, 284, 284f
    matrix thermal conductivity, 284
    offshore settings, 282
    radioactive decay, 284–285
    short-term heat sources, 285
    solar heating, 282
    stratigraphic sequence, 283
    surface temperature, 281
    thermal equilibrium, 283
    time–temperature histories, 281
    transient heat flow, 283
  three dimensional (3-D) basin model, 273–274
  two dimensional (2-D) models, 273–274
  unconventional plays, 305
  volumetric estimations, 302–305, 304f
Biodegradation
  biogenic gas, 144–146
  biogenic methane gas, 144–146
  intensity, 144, 145f
  oil gas chromatograms, 144, 146f
  oil–water contact, 143
  paleo-pasteurization, 143
  properties, 143–144, 144f
  reservoir sediments, 142–143
  saturated hydrocarbons, 144
  water-soluble nutrients, 142–143

Biomarker ratios, 156–158, 157f, 168f
advantages, 169–170
C_{22}/C_{21} tricyclic terpane ratio, 168
C_{29} steranes ratios, 168
disadvantages, 170
four cross-plots, 167–168, 169f
lacustrine *vs.* marine indicators, 169
maturity indicators, 167–168
source rock characteristics, 167
Biomarkers, 2, 151, 151f–152f
analysis, 152–154, 153f–154f, 244
data, 244, 245f
Bisnorhopane, 161–162, 161f
Bitumen, 5–6, 25, 80–81

C

Chemical fossils, 2
Coalbed methane (CBM)
adsorption capacity, 257
adsorption isotherm curve, 257, 257f
gas adsorbtion, 255, 255f
lignite formation, 256, 256f
oversaturated and undersaturated coals, 257, 258f
petroleum geochemistry, 258–259
production curves, 257, 258f
self-sourcing reservoirs, 255
US federal government, 254–255
Covalent bonds, 6–7, 7f
Crude oil and natural gas data
biodegradation. *See* Biodegradation
biogenic *vs.* thermogenic
biodegraded oil, 176
biogenic gas, 174
exploration program, 176
methane $\delta^{13}C$ *vs.* C_2^+ hydrocarbon gases, 174–175, 174f
microbial decomposition, 175–176
oil-bearing reservoir, 176
thermal decomposition, 173–174
bulk properties
American Petroleum Institute's (API) gravity, 136
cloud point, 137
flash point, 138
gas specific gravity, 137
geochemical analysis, 136
heating value, 139
nickel (Ni), 138
pour point, 137
S-A-R-A, 138
solution gas–oil ratio, 138–139
solution oil–gas ratio, 138–139
vanadium (V) contents, 138
viscosity, 137
weight percent sulfur, 138
carbon isotope, 173
contamination, 149–150
crude oil inversion
biomarker ratios. *See* Biomarker ratios
C_{29} regular sterane compounds, 164, 165f
diamondoids, 165–167, 166f–167f
diasteranes, 164
individual biomarker indicators, 160–163, 161f–162f
marine organic matter, 164
relative abundance, 163
reservoired crude oils, 160
steranes, 163–164
ternary plot, 163, 163f
thiophenes, 164–165, 166f
deasphalting, 147–148
devolatilization, 148
gas-to-gas and gas-to-source rock correlations, 175f, 177f–182f, 178–183
gas-washing, 147, 148f
maturity assessments, 183
natural gases, composition, 173
nonhydrocarbon gases, 173
oil correlation and oil inversion studies
biomarker compound distributions, 172
biomarker data, 170
drilling muds, 171
hydrocarbon generation and expulsion, 171–172
maturation process, 172
multiple correlation/inversion parameters, 170–171
physical sampling, 172
thermal alteration, 171
oil-to-oil and oil-to-source rock correlations
biomarker analysis, 152–154, 153f–154f
biomarker ratios, 156–158, 157f
biomarkers, 151, 151f–152f
carbon isotope ratios, 158, 159f
distillation methods, 150–151
geochemical compounds, 150
mass chromatograms, 155, 155f
multiple parameter approach, 154–155
multivariate statistical techniques, 159–160
radar plot, 158, 158f
sterane data, 156, 157f
structural analysis, 150
tricyclic and tetracyclic terpanes, normalized distributions, 156, 156f
oil-to-source rock/gas-to-source rock correlation, 136

phase behavior
black oils reservoirs, 140
bubble point curve, 140
dew point curve, 140
dry gas reservoirs, 140–141
gas condensate reservoirs, 140
PVT analysis, 141
temperature–pressure diagram, 139–140, 139f
volatile oil reservoirs, 140
wet gas reservoirs, 140
process, 141
stable isotopes, 173
thermal alteration, 141–142, 142f
thermochemical sulfate reduction, 149
thermogenic natural gas, 176–178
water-washing, 147

D
Darcy flow modeling, 293–294
Decarboxylation, definition, 14–15
Diamondoids, 165–167, 166f–167f

F
Fischer–Tropsch reactions, 19, 20f

G
Gammacerane, 160–161, 161f
Gas chromatography (GC) analysis
contamination recognization, 87–88, 88f
CPI calculation, 84, 87f
liquid column chromatography, 82
n-paraffin distributions and CPI, 84, 87f
retention time, 83
saturate fraction, 83, 83f
source quality interpretations, 83–84, 85f
source rock extract gas chromatograms, 84, 86f
vapor pressure, 82–83
Geochemistry, definition, 4–5

H
Haworth mud gas parameters, 192–194, 194f–195f
Headspace gas analysis
canned cuttings, 88, 89f
headspace gas parameters *vs.* depth, 88–89, 90f
quart size metal paint, 88–89
source richness interpretations, 89–90
thermal maturity interpretations, 90–91
High-molecular weight hydrocarbon analysis, 242–243
High-temperature gas chromatography, 199, 199f
Horner plot correction, 281, 282f
Hydrocarbon generation modeling
closed system models, 291
discrete samples, 290
first-order Arrhenius kinetics, 289
five-component model, 290, 290f
kerogen generation, 288, 289f
modified Rock-Eval analyses, 289–290
total organic carbon (TOC), 290–291
Hydrocarbon generation process, 2
Hydrocarbon microseepage
direct indicators
adsorbers, 221–222
soil gas, 220–221
indirect indicators
gamma-ray spectrometry, 223
induced polarization (IP), 223–224
iodine, 223
magnetic contrasts, 224
microbial, 222–223
resistivity, 223–224
Hydrocarbons
five- and six-membered ring cyclic saturated hydrocarbons, 10, 11f
isomers, 8, 8f
saturated hydrocarbons, 7–8
side chains, 8, 9f
steranes and hopanes, 10, 11f
"stick" notation, 9
straight and branched-chained saturated hydrocarbon, 9–10, 10f
structural notations, 9, 9f
triple bond electrons, 7–8, 8f
unsaturated hydrocarbons, 7–8
Hydrogen index (HI), 68
Hydrolysis, 14–15

I
Invasion percolation modeling, 294
Isomers, 8, 8f
Isoprenoidal hydrocarbon botryococcane, 162

K
Kerogen, 6
formation
bitumen, 25
classification, 26
diagenetic sulfurization, 28
hydrocarbon-generating kerogen, 28
hydrogen-rich organic matter, 25–26
organic matter *vs.* sediment and preservation conditions, 25–26, 26f
solvent-soluble/solvent-insoluble sedimentary organic materials, 25, 25f
type III kerogen structure, 27F, 27–28
type II kerogen structure, 27F, 27–28
type I kerogen structure, 27F, 27–28
type IV kerogen, 28–29
isolation, 95–96

Kimmeridgian source rock, 120, 121f
Kinetic isotope effect, 15
Kozeny–Carman relationship, 279, 280f

L

Lower Paleozoic maturity indicators, 110–111

M

Macroseepage, 218–220
  offshore, 230–231
  onshore, 229–230
Microseepage
  background concentrations, 225
  ethane concentration map, 225, 226f
  ethane/ethane ratio, 225, 227f
  exploration program/petroleum systems analysis, 228–229
  faults and fracture zones, 225–228
  geochemical anomalies, 225, 229f
  grid sampling, 224–225
  leaked hydrocarbons, 225–228
  porous and permeable sediments, 225–228
  sampling denisity, 224
  signal-to-noise ratio, 224
  transect surface geochemistry data, 225, 228f
Mitchell Energy, 260
Mud gas analysis, 191–192, 192f
  data interpretation
    background gas, 194
    Haworth mud gas parameters, 192–194, 194f–195f
    Pixler plot, 193
    real-time evaluation, 195
    recycled gas, 194–195

N

Naphthenes/cycloalkanes, 10, 11f
Niobrara Formation, 268, 269f
Nonhydrocarbon gases, 173
  carbon dioxide, 322
  helium, 52–53
  hydrogen, 53, 54f
  hydrogen sulfide, 51-52, 52F
  nitrogen, 51
Nonmetallic elements, 6
N-S-O compounds, 12, 13f

O

Offshore macroseepage, 230–231
Oil and gas deposits, 6
Oil shale, 6
Oleanane, 162, 162f
Onshore macroseepage, 229–230

P

Petroleum accumulations
  allochthonous organic matter, 20
  coarser grained sediment, 23, 24f
  eolian process, 20
  first-order Arrhenius kinetics, 56
  Fischer–Tropsch reactions, 19, 20f
  kerogen formation. *See* Kerogen formation
  marine and terrestrial primary production, 21, 22f
  maturation and hydrocarbon generation
    biogenic methane, 319
    chemical kerogen types, 321
    conversion reaction, 42F, 320
    elemental composition, 318
    first-order Arrhenius reaction kinetics, 42F, 40F, 320
    gas-prone source rocks, 321
    "Gaussian-like" distribution, 321
    iterative statistical procedure, 320
    kerogen type *vs.* temperature, 44F, 321
    kinetic parameters, 43F, 321
    maturation indicators, 319
    oil and gas generation, 40F, 319
    oil composition, 321
    oil-prone source rock, 321
    oil window concept, 40F, 320
    saturated hydrocarbon gas chromatograms, 38F, 319
    thermal maturation, 34F, 318
    type II kerogen structure, 36F, 318
    van Krevelen diagram plots, 37F, 318
    wet gas content and methane carbon isotope ratios, 37F, 318–319
  migration
    associated gas generation, 45-46
    buoyancy force, 47
    carrier system, 45
    contiguous oil-wet migration pathway, 45, 46F
    expulsion efficiency, 46-47
    factors, 45
    oil migration, 48, 47, 48F
    porosity and permeability, 46
    remigration, 48, 49F
    reservoired hydrocarbons, 48-50
    reservoir/trap filling process, 48, 49F
    solvent extraction organic matter, 47, 47F
    types, 45-46
  nonhydrocarbon gases, 56. *See also* Nonhydrocarbon gases
  nutrient-rich ocean bottom waters, 21
  oil-prone source rocks, 53-55, 55F
  organic matter-rich sediments, 20

oxic/anoxic boundary, 55, 22–23
primary biological productivity, 21
recycled/reworked organic matter, 20
reservoir/trap system, 56
sedimentary basins, 19
sedimentation rate, 21–22, 24f
sediment–water interface, 22
source rock deposition
deltaic settings, 32F, 316–317
depositional settings, 30F, 315
lacustrine deposition, rift basins, 317
negative and positive water balance, 31F, 316
open marine anoxic settings, 317–318
oxygen-minimum layers, 315–316
quality and distribution, 33F, 317
silled basins settings, 316
transport mechanisms, 20, 21f
Viking shale, grain size fractions, 23, 23f
zooplankton, 22
Petroleum, definition, 4
Petroleum system
definition, 5, 5f, 309
drilling schedule, 320
effective source rock thickness *vs.* total source rock thickness, 321
elements and processes, 310–311, 312f
exploration programs, 310
expulsion/migration evaluation, 321
failure analysis, 322
hypothetical petroleum system, 309–310
plays and prospects, 317–318
risk analysis, 320
spatial aspects, 315–317, 317f
speculative petroleum system, 309–310
temporal aspects
biodegradation, 314
critical events chart, 312–314, 313f–315f
cumulative expulsion curves, 312–314
temporal sequence, 312
working petroleum system, 318–319
Polycrystalline diamond compact (PDC) drill bit, 126
Pyrobitumen, 6
Pyrolysis–gas chromatography (PGC)
immature source rocks, 92–93, 92f
kerogen-type assessments, 91–92
preanalysis solvent extraction, 94–95
progressive changes, 93, 94f–95f
qualitative and quantitative analysis, 91
Rock-Eval pseudo-van Krevelen diagram, 93–94
source rock analysis, 91, 91f
type III gas-prone kerogen, 92–93
type IIS kerogens, 93
Pyrolysis method, 2

R

Random drilling, exploration efficiency, 3, 3f
Ray-path modeling, 292–293, 293f
Reduction, definition, 14–15
Reservoir geochemistry
asphaltenes, 200–202, 202f
enhanced oil recovery, 210–211, 211f
geochemical methods, 190
high-molecular weight waxes, 199–200, 199f
pay zone detection
continuous phase oil/gas, 190–191
drilling petrophysical tools, 191
fluorescence and cut, 195–196
gas chromatography, 197–198, 198f
isolated droplets of oil/gas, 190–191
isotubes, 196–197
mud gas analysis. *See* Mud gas analysis
reservoir rock, 190
Rock-Eval pyrolysis, 197, 197f
solvent extraction, 197–198, 198f
thermal extraction–gas chromatography, 198–199
periodic sampling, 208–210, 209f
petroleum production and field development, 189
production allocation, 207–208, 207f
production problems, 208–210, 209f
reservoir continuity
definition, 202
gas samples, 206
methane *vs.* ethane, 203, 203f
mud gas composition, 202
mud gas Pixler ratios, 203, 204f
oil samples, 203–206, 205f–206f
reservoir geochemistry, 212–213
reservoir souring, 212
Rock-Eval pyrolysis, 197, 197f
H/C and O/C atomic ratio, 75, 75f
HI *vs.* T_{max} cross-plot, 78–79, 78f
hydrocarbon-generating capacity, 75–77
hydrogen index (HI), 68
maturation series, 72, 72f
mixtures and mineral matrix effects, 80
organic drilling mud additives, 71–72
parameters *vs.* depth, 68, 69f
pseudo-van Krevelen plot, 75, 76f
semiquantitative source richness interpretation, 68–70, 70f
source evaluation programs, 70–71, 70f
source quality interpretation, 74, 74f
temperature profile, 67–68
thermal decomposition, 67
thermal maturity interpretations, 72–74, 73f–74f
total analysis time, 68
total organic carbon (TOC)-S2 cross-plot, 70–71, 71f, 79, 79f

Rock-Eval pyrolysis (*Continued*)
type I and type II kerogens with type III kerogen, 75–77, 76f
type I, type II, and type III kerogens with type IV kerogen, 77, 77f
volatile organic material, 68

S

Saturated hydrocarbons, 7–8, 144
Saturates-Aromatics-Resins-Asphaltenes (S-A-R-A) analysis, 81
Sedimentary basins, 19
Sedimentary organic matter, 5–6
reactions, 14–15, 15f
Shale gas
hydraulic fracturing, 261
mechanical behavior, 265
Mitchell Energy, 260
natural gas seepage, 259
petroleum geochemistry
maturity window, 262, 262f
sediments, 264, 264f
total organic carbon (TOC) and Rock-Eval S2/HI data, 262–263, 263f
total organic carbon window, 262, 263f
vitrinite reflectance measurement, 261–262
porosity and permeability, 264–265
recovery efficiencies, 261
single phase production, 261
source rock/reservoir petroleum system, 260, 260f
United States Department of Energy, 260
X-ray diffraction (XRD) analysis, 265
Shale oil, 265–267, 267f
Source rock evaluation
bitumen, 80–81
bottomhole temperature and pressure measurements, 126
characteristics, 62
comprehensive petroleum system analysis, 128
conodont alteration index, 117–118, 118f
density/resistivity cross-plot quick-look method, 120, 122f
drilling mud composition, 126
elemental analysis, 96–98, 97f–98f
flow chart, 127–128, 127f
gas chromatography (GC), 81, 82f. *See also* Gas chromatography (GC) analysis
geochemical analyses, 127
graptolite, scolecodont, and chitinozoan reflectance, 110–111, 111f
headspace gas analysis. *See* Headspace gas analysis
kerogen fluorescence
caved sediments and contamination, 116
parameter, 115–116, 117f
qualitative assessment/quantitative spectral analysis, 115
quantitative spectral fluorescence, 117
UV fluorescence spectra, 115–116, 116f
vitrinite reflectance range, 115, 116f
kerogen isolation, 95–96
Kimmeridgian source rock, 120, 121f
laboratory-based methods, 63
liptinite/sporinite/exinite reflectance, 109, 110f
maturation analysis, 109
maturity interpretations
burial history, 107–108, 108f
hydrocarbon generation, 107, 107f, 109
mean reflectance values, 106
oil and gas generation, 106, 107f
multiple parameter approach, 62–63
N-S-O/resin fraction, 81
outcrop samples, 125–126
polycrystalline diamond compact (PDC), 126
pyrolysis–gas chromatography (PGC). *See* Pyrolysis–gas chromatography (PGC)
quick look ΔIog R method, 120, 123f
Rock-Eval pyrolysis, 67f. *See also* Rock-Eval pyrolysis
sampling program
core and sidewall cores, 65
drill cuttings, 63
drilling mud sample, 65
gas chromatography (GC) analysis, 64
headspace gas/canned cuttings, 64–65
mud additive sample, 65
unwashed cuttings, 64
washed cuttings, 64
saturate fraction, 81
Saturates-Aromatics-Resins-Asphaltenes (S-A-R-A) analysis, 81
sediments lithology, 126
solid bitumen reflectance, 110
source quality, 62, 128
source richness, 62
quality interpretations, 81–82, 121–122, 123f–124f
Soxhlet extraction apparatus, 80–81, 80f
thermal alteration index (TAI), 113–115, 114f
thermal maturity, 62, 128
interpretations, 122–125, 125f
total organic carbon (TOC), 66–67, 66f
true effective source rocks, 62
visual kerogen typing
amorphous kerogen, 112–113
image analysis–based system, 113
organic petrography, 111–112
transmitted and reflected light classification schemes, 111–112, 113f
transmitted light microscopy, 111–112, 112f

vitrinite reflectance
anisotropy, 106
cavings and reworked vitrinite, 100–101, 101f–102f
data interpretation, 99–100, 100f
with depth, 99
geothermal gradient/heat flow, 103–104, 104f
igneous intrusives, 104, 105f
maturity indicator, 98
measurement of, 98–99, 99f
organic-rich source rocks, 98
pre-Devonian sediments, 100
reverse faults, 103, 103f
suppression and misidentification, 104–106, 105f
unconformities/normal faults, 101–103, 102f
wireline log interpretations
formation density, 119
gamma ray, 119
logging tools, 118–119
resistivity, 120
sonic transit time, 119–120
wireline log data, 118–119
Stable isotopes, 15–16, 16f
Surface geochemistry
direct/indirect near-surface expressions, 218
hydrocarbon microseepage. *See* Hydrocarbon microseepage
hydrocarbon seepage, 218
macroseepage, 218–220
offshore, 230–231
onshore, 229–230
microseepage, 218. *See also* Microseepage
petroleum systems and fluid flow regimes, 217
sampling potential seafloor seep sites, 240–241
seafloor seep sites
acoustic mapping, 235
bathymetry and near-surface amplitudes, 235–236, 236f
Coal Oil Point Seeps, 231–232, 231f
3-D seismic survey, 238, 239f
multibeam bathymetry and backscatter data, 237, 238f
seafloor features, 234–235
sea surface slicks, 232–234, 233f
side scan sonar mosaic, 237, 237f
subbottom profiling sonar systems, 238, 239f
water column "sniffers,", 234
sea surface slicks
beach tar sampling and analysis, 247–248
3-D seismic imaging, 244–245
offshore gas seep sampling and analysis, 247
oil slick sampling and analysis, 246–247, 246f
seepage styles, 218, 218f
thermogenic hydrocarbons, seafloor sediments
background organic matter (BOM) signal, 243
biomarker analysis, 244
biomarker data, 244, 245f
gas chromatography–mass spectrometry (GC–MS) data, 244
geochemical characteristics, 241–242
headspace gas dataset, 242, 242f
high-molecular weight hydrocarbon analysis, 242–243
isotopic analysis, 243–244
total scanning fluorescence (TSF) analysis, 243

T

Thermal alteration index (TAI), 96, 113–115, 114f
Tight oil reservoir, 4
Total scanning fluorescence (TSF) analysis, 243
Type III kerogen structure, 27F, 27–28
Type II kerogen structure, 27F, 27–28
Type I kerogen structure, 27F, 27–28
Type IV kerogen, 28–29

U

Unconventional resources
coalbed methane (CBM). *See* Coalbed methane (CBM)
definition, 254
hybrid systems
Bakken Formation, 268, 268f
horizontal drilling and reservoir stimulation technologies, 267
Niobrara Formation, 268, 269f
petroleum geochemistry, 269
shale gas. *See* Shale gas
shale oil, 265–267, 267f
United States Department of Energy, 260
United States Geological Survey (USGS), 253
Unsaturated hydrocarbons, 7–8

V

Viking shale, grain size fractions, 23, 23f

Kolibri 360 GmbH
Ferdinand-Jühlke-Straße 7
99095 Erfurt, Deutschland
produktsicherheit@kolibri360.de